Molecular Biology
of Chromosome Function

Kenneth W. Adolph
Editor

Molecular Biology of Chromosome Function

With 110 Illustrations and 3 Color Plates

Springer-Verlag New York Berlin Heidelberg
London Paris Tokyo Hong Kong

Kenneth W. Adolph
Department of Biochemistry, University of Minnesota Medical School
Minneapolis, MN 55455, USA

Library of Congress Cataloging-in-Publication Data
Molecular biology of chromosome function/Kenneth W. Adolph, editor.
    p.   cm.
    Includes index.
    ISBN-13:978-1-4612-8192-4   e-ISBN-13:978-1-4612-3652-8
    DOI: 10.1007/978-1-4612-3652-8

    1. Chromosomes. 2. Molecular biology.   I. Adolph, Kenneth W.,
    1944–
    QH600.M63   1989
    574.87′322—dc20   89-6367
Printed on acid-free paper.

© 1989 by Springer-Verlag New York Inc.
Softcover reprint of the hardcover 1st edition 1989

Typeset by TCSystems, Inc., Shippensburg, Pennsylvania.

9 8 7 6 5 4 3 2 1

ISBN-13:978-1-4612-8192-4

General Introduction

Chromosomes have structure, determined by the interactions of proteins with DNA, and chromosomes have functions, in particular, replication of DNA and transcription of messenger RNA. Chromosome structure and function are not separate topics, since chromosome organization profoundly influences the activity of the genome in replication and transcription. This is especially clear for higher cells, including human cells, in which chromatin fibers are created by the binding of histone proteins to the DNA, and folding of the fibers produces mitotic chromosomes and interphase nuclei. The intricate organization of DNA in higher cells is now recognized as being closely involved with genome activity. Many fundamental results have originated from studies of bacterial and viral systems, which have been systems of choice because of their less complex life cycles. The processes of replication and transcription show differences between the higher and simpler systems (e.g., different enzymes and protein factors are involved). But the parallels are as striking as the differences in detail. Even for bacteria and viruses, a full understanding of these processes will require integrating the results of molecular biology with those of structural biology and cell biology.

Three important subjects are covered in this volume: DNA replication and recombination, gene transcription, and chromosome organization. The sections dealing with replication and transcription examine recent results obtained by applying the techniques of molecular biology and biochemistry. Eukaryotic, prokaryotic, and viral systems are discussed. The emphasis in the third section, which concerns chromosome organization, is on molecular cell biology. The information that is presented was derived from techniques of structural biology and biophysics, including computer graphics and X-ray crystallography, as well as biochemistry and cell biology. The book begins with chapters describing fundamental topics on DNA replication and recombination: DNA precursor synthesis, replication fork propagation, and the role of recA protein. Viral replication (SV40, HIV) is also covered. Chapters then follow that review basic aspects of the process of gene transcription: repressor proteins, RNA polymerases, transcription termination, pre-mRNA splicing, ribonucleo-proteins. The topics that conclude the book concern chromosome structure: the 3-D arrangement of human mitotic chromosomes, DNA binding proteins, histones, and protamines.

The book therefore provides a broad overview of centrally important subjects regarding the functions and structure of chromosomes. It is hoped that the work will prove to be of value for a variety of interested readers.

Contents

Contributors

Kenneth W. Adolph, Department of Biochemistry, University of Minnesota Medical School, Minneapolis, Minnesota, USA

Rod Balhorn, Biomedical Sciences Division, Lawrence Livermore National Laboratory, Livermore, California, USA

Stanley F. Barnett, Department of Molecular Biology, Vanderbilt University, Nashville, Tennessee, USA

Ekkehard K.F. Bautz, Institute of Molecular Genetics, University of Heidelberg, Heidelberg, FRG

Richard R. Burgess, McArdle Laboratory for Cancer Research, University of Wisconsin-Madison, Madison, Wisconsin, USA

Michael M. Cox, Department of Biochemistry, College of Agricultural and Life Sciences, University of Wisconsin-Madison, Madison, Wisconsin, USA

Russell J. DiGate, Program in Molecular Biology, Sloan-Kettering Institute, Memorial Sloan-Kettering Cancer Center, New York, New York, USA

Alicia J. Dombroski, Department of Biochemistry, University of Rochester Medical Center, Rochester, New York, USA

John J. Furth, Department of Pathology and Laboratory Medicine, University of Pennsylvania School of Medicine, Philadelphia, Pennsylvania, USA

Michael Grunstein, Molecular Biology Institute and Department of Biology, University of California, Los Angeles, California, USA

Min Han, Molecular Biology Institute and Department of Biology, University of California, Los Angeles, California, USA

Norman B. Hecht, Department of Biology, Tufts University, Medford, Massachusetts, USA

Paul Kayne, Molecular Biology Institute and Department of Biology, University of California, Los Angeles, California, USA

Kathleen M. Keating, Department of Molecular Biophysics and Biochemistry, Yale University School of Medicine, New Haven, Connecticut, USA

Ung-Jin Kim, Molecular Biology Institute and Department of Biology, University of California, Los Angeles, California, USA

William H. Konigsberg, Department of Molecular Biophysics and Biochemistry, Yale University School of Medicine, New Haven, Connecticut, USA

Wallace M. LeStourgeon, Department of Molecular Biology, Vanderbilt University, Nashville, Tennessee, USA

Arnold Jay Levine, Department of Biology, Princeton University, Princeton, New Jersey, USA

James L. Manley, Department of Biological Sciences, Columbia University, New York, New York, USA

Kenneth J. Marians, Program in Molecular Biology, Sloan-Kettering Institute, Memorial Sloan-Kettering Cancer Center, New York, New York, USA

Christopher K. Mathews, Department of Biochemistry and Biophysics, Oregon State University, Corvallis, Oregon, USA

Kathleen S. Matthews, Department of Biochemistry, Rice University, Houston, Texas, USA

Charles S. McHenry, Department of Biochemistry, Biophysics and Genetics, and University of Colorado Health Sciences Center, Denver, Colorado, USA

Alexander McPherson, Department of Biochemistry, University of California at Riverside, Riverside, California, USA

Jonathan C.S. Noble, Department of Biological Sciences, Columbia University, New York, New York, USA

Stephanie J. Northington, Department of Molecular Biology, Vanderbilt University, Nashville, Tennessee, USA

Gabriele Petersen, Institute of Molecular Genetics, University of Heidelberg, Heidelberg, FRG

Terry Platt, Department of Biochemistry, University of Rochester Medical Center, Rochester, New York, USA

Tillman Schuster, Molecular Biology Institute and Department of Biology, University of California, Los Angeles, California, USA

Yousif Shamoo, Department of Molecular Biophysics and Biochemistry, Yale University School of Medicine, New Haven, Connecticut, USA

Nancy E. Thompson, McArdle Laboratory for Cancer Research, University of Wisconsin-Madison, Madison, Wisconsin, USA

Kyle L. Wick, Department of Biochemistry, Rice University, Houston, Texas, USA

Kenneth R. Williams, Department of Molecular Biophysics and Biochemistry, Yale University School of Medicine, New Haven, Connecticut, USA

Section I DNA Replication and Recombination

The two basic functions of the genomes of all living systems are replication and transcription. Important topics concerning the replication of DNA and RNA genomes are discussed in this section, while the transcription of messenger RNA from a DNA template is covered in the section that follows. For most organisms, including humans, the interaction of replication proteins with DNA is at the heart of the replication process. But the process is complicated by the organization of the DNA double helix into chromosomes which, in cells of higher organisms, consist of the DNA tightly bound to the histones and nonhistone proteins. The DNA-protein chromatin fibers are, in turn, arranged to form the characteristic structures of chromosomes. Understanding DNA replication in prokaryotes such as the bacterium *Escherichia coli* is complicated by the circular nature of the DNA molecule and by the fact that it is folded into a "nucleoid" in the intact cell. Viral chromosomes, even though tiny compared to those of bacteria and higher cells, also show distinctive features in their replication. This is true for the DNA-containing viruses such as the simian virus 40 (SV40) and for viruses with RNA genomes such as HIV, the human immunodeficiency virus.

Investigations of DNA replication are at an advanced level. The biochemistry of replication for a variety of systems is known in considerable detail. Not only has the enzymology of the DNA polymerases been extensively characterized, but precise roles have been revealed for a number of other important proteins associated with replication. The identification of many of the components involved in duplicating the DNA helix has demonstrated that the process is far from simple. The activity of DNA polymerase itself is but one event in a complex chain of events. The protein-DNA and protein-protein interactions at the replication fork must be considered, as must the roles of additional molecules that regulate DNA replication. These include the enzymes of DNA precursor synthesis.

Besides replication, DNA is involved in additional significant processes. Homologous genetic recombination, discussed in this section, produces DNA rearrangements in the genetic transformation of prokaryotes, in DNA repair, and in other situations. The replication of certain viruses can be considered as model systems that illuminate general aspects of replication; SV40, for instance, contains a minichromosome with histones bound to the DNA to create nucleosomes. However, with retroviruses such as HIV, a genome of RNA is replicated and a DNA duplex is secondarily produced for integration into human chromosomes. Variety is clearly a key word in describing viral replication, as the chapters in this section illustrate.

Chapter 1
Enzymes of DNA Precursor Synthesis and the Control of DNA Replication

Christopher K. Mathews

Although the enzymes of DNA precursor biosynthesis were originally described in the 1950s and 1960s, it is still timely to discuss these proteins with respect to their involvement in regulating DNA replication. This chapter will focus on current research that emphasizes the genetics and cell biology of these enzymes: How do the enzymes interact in cells, with each other, and with replication proteins? What controls reaction fluxes through individual enzymes in vivo? How do these fluxes vary through the cell cycle, and what factors control the variations? What are the effective concentrations of deoxyribonucleoside triphosphates (dNTPs*) at replication sites? How do variations in dNTP pools affect the rate and accuracy of DNA replication? What are the metabolic and genetic consequences of inhibiting a particular enzyme with an antimetabolite? Metabolic inhibitors provide useful probes for understanding compartmentation and control. In turn, this understanding could lead to the more effective use of antimetabolites as anticancer, antimicrobial, or antiviral agents.

Although direct roles of dNTPs in regulating DNA replication have not been established, the synthesis of DNA precursors is closely coordinated with DNA replication in most organisms studied. Such coordination might be expected a priori, simply from the facts that DNA replication is localized both in time and in space within a cell and that replication uses precursors that have few if any additional metabolic roles. This chapter will focus on mechanisms involved in this coordination and on the extent to which dNTPs are involved in controlling the rate and fidelity of DNA replication.

---

* Abbreviations: rNMP, rNDP, rNTP, ribonucleoside mono-, di-, and tri-phosphate, respectively; dNMP, dNDP, dNTP, deoxyribonucleoside mono-, di-, and triphosphate, respectively.

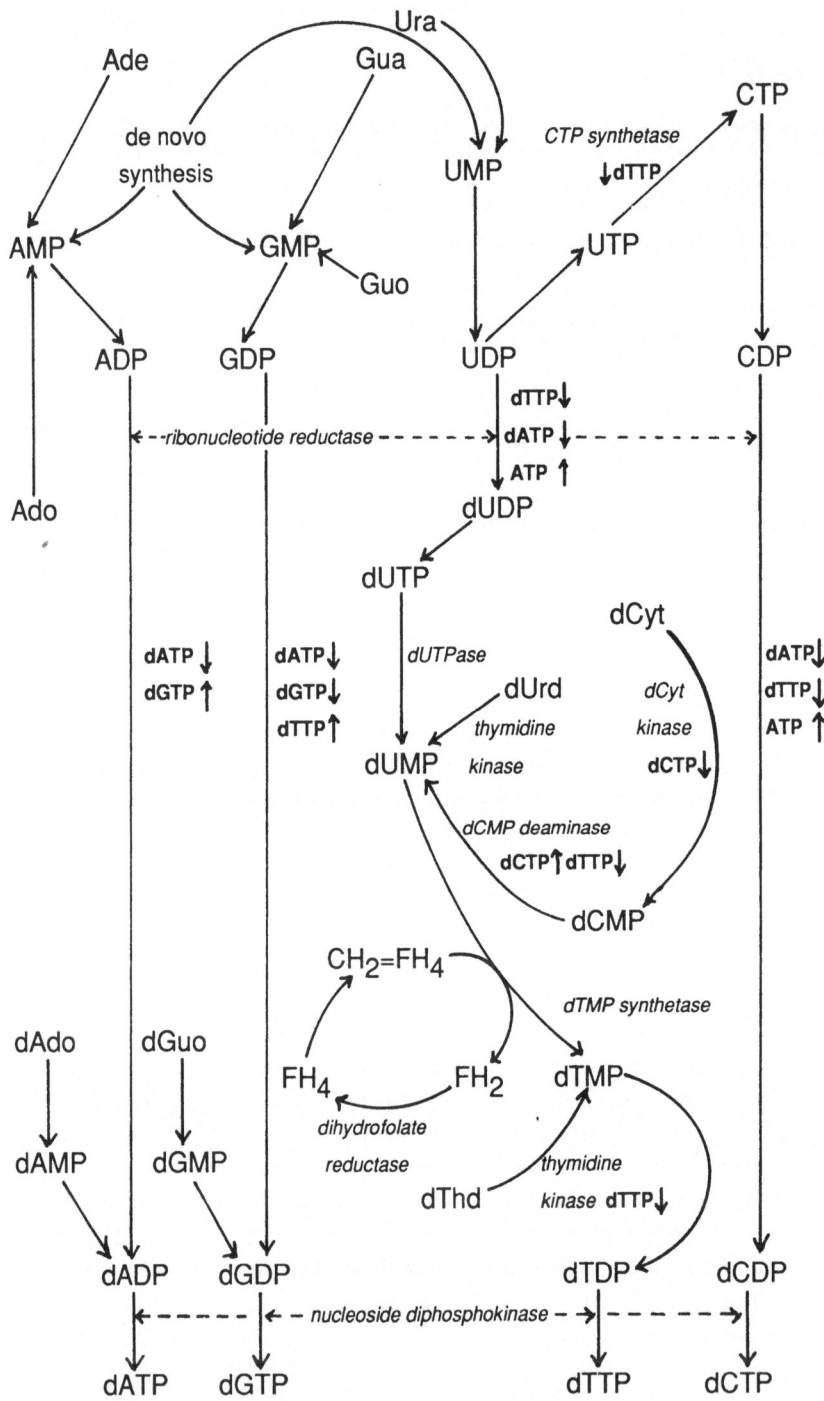

Figure 1.1. Pathways of deoxyribonucleoside triphosphate biosynthesis in mammalian cells. Major enzymes and allosteric effectors are identified. Arrows indicate whether each allosteric effect is positive or negative.

Pathways of dNTP Biosynthesis and Regulation of the Enzymes

Figure 1.1 summarizes the enzymatic pathways of dNTP biosynthesis in mammalian cells and identifies effectors for reactions known to be controlled in allosteric fashion. Some variations are found in microbial systems and in virus-infected cells, and a few of these variations will be noted. The enzymatic reactions in these pathways are well described in most biochemistry textbooks. A particularly useful review is provided in Chapter 1 of Kornberg's excellent book [1].

The pathways in Figure 1 can be represented more simply as follows:

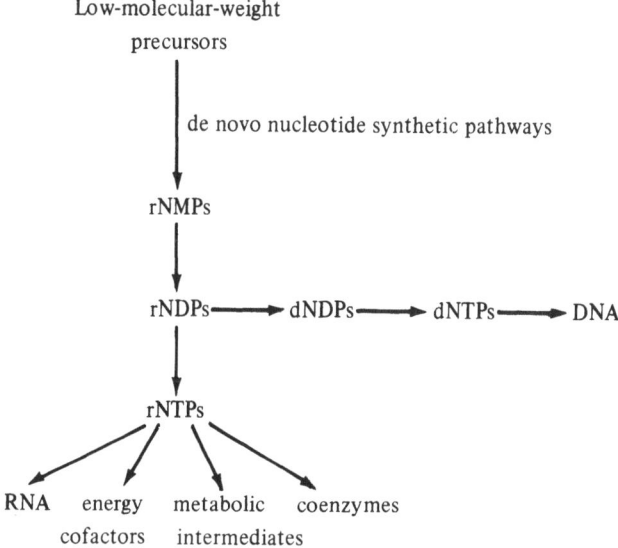

In most cells RNA is 5 to 10 times more abundant than DNA. Because of this and the numerous metabolic roles played by ribonucleotides, only a small percentage of the carbon flux through the de novo purine and pyrimidine synthetic pathways is diverted toward deoxyribonucleotides. Great interest has been paid to ribonucleotide reductase, because it catalyzes the first committed step in DNA replication. Additional interest derives from the fact that the reaction—reduction of a nucleotide ribose moiety to deoxyribose—is mechanistically unusual. Early attention focused also on the enzyme that generates the other distinctive functional group in DNA—i.e., thymidylate synthase—which forms the methyl group of thymine nucleotides. Both ribonucleotide reductase and thymidylate synthase displayed low activity when assayed in extracts during early investigations, and it was suggested that either or both enzymes are rate limiting for DNA replication.

Although our understanding of the control of DNA replication is far from complete, developments in the 1970s and 1980s have indicated that the major regulated step is initiation from replicon origins, with replicative chain elongation being a secondary target. Therefore, the simple idea that thymidylate synthase or ribonucleotide reductase could be rate limiting for DNA replication has been abandoned. However, ribonucleotide reductase does play an important regulatory role of a more complex nature, through the fact that this one enzyme system participates in the synthesis of all four DNA precursors. Through interactions of nucleoside triphosphates with two distinct allosteric sites on the enzyme molecule, the relative abilities of the enzyme to reduce ADP, CDP, UDP, and GDP are modulated so that the cell produces dNTPs in the correct relative proportions needed for DNA replication [reviewed in 2]. The biological importance of these mechanisms was underscored in a series of important papers from Martin's laboratory [cf. 3]. These investigations showed that mutations altering the control sites of mouse ribonucleotide reductase led to dNTP pool imbalances and genetic abnormalities, including elevated spontaneous mutation rates.

Two other mechanisms control ribonucleotide reductase. First, the enzyme contains a tyrosine free radical that plays an essential role in catalysis. Reichard's laboratory has characterized an enzyme system in *E. coli* that generates the radical and that may control the proportion of enzyme in the active and inactive forms [reviewed in 4], although the nature of the control has not yet been elucidated. Second, the level of the enzyme protein is closely coordinated with the proliferative state of the cell. In synchronized mammalian cells the amount of enzyme protein increases at least 10-fold as cells enter S phase. The availability of cDNA probes has led to a demonstration that posttranscriptional mechanisms are involved in this regulation [5].

By contrast, thymidylate synthase is not considered to be an important regulatory target. No allosteric effectors are known, and it seems likely that flux through this reaction is determined by intracellular concentrations of substrates and products. However, two other enzymes of dNTP biosynthesis are, like ribonucleotide reductase, regulatory targets that help to maintain appropriate dNTP pool sizes. Deoxycytidylate deaminase is a branch point between dCTP and dTTP biosynthesis. Mutations that affect the control of this enzyme by dCTP and dTTP (positive and negative, respectively) alter the balance of dCTP and dTTP pools and similarly effect spontaneous mutation rates [6,7]. The other enzyme, CTP synthetase, is involved in ribonucleotide biosynthesis. However, mutations affecting the feedback inhibition of this enzyme by dTTP have a complex phenotype that includes both elevated cytidine and deoxycytidine nucleotide pools and increased spontaneous mutation rates [8,9].

In addition to the above three regulated enzymes, some of the nucleotide salvage enzymes, particularly thymidine kinase and deoxycytidine kinase, are subject to allosteric control in most organisms (see Fig. 1.1). However, rather than a detailed presentation of these regulatory mechanisms, our major goal is to understand how the products of these pathways, the deoxyribonucleotides, participate in regulating or coordinating the synthesis of DNA.

dNTPs as Regulators of DNA Replication Rates

Are dNTP Levels Rate Limiting?

The most direct way for DNA replication rates to be controlled by DNA precursors would be simple substrate-level control of DNA polymerase activity. The fact that dNTP pool sizes increase as cycling cells approach S phase is consistent with the existence of substrate level control, although it is still impossible to distinguish between causes and effects of cell-cycle-related dNTP pool size variations.

In general, dNTP pool determinations from many types of cells suggest that average intracellular dNTP concentrations are relatively low—in the 10 to 100 μM range—when pool sizes are divided by average intracellular volumes. While in some cases these values fall below apparent saturating concentrations for replicative DNA polymerases, interpretation of the data is complicated by several factors. These include the difficulty of estimating the true intracellular volume into which a nucleotide is dispersed, the likelihood of dNTP pool compartmentation, and the difficulty of accurately estimating kinetic constants for DNA replication complexes from in vitro experiments with purified DNA polymerases. In mammalian cells, for example, earlier work on the substrate concentration dependence of permeabilized cell systems suggested that dNTP levels in the range of 50 to 100 μM were needed to saturate the replication machinery [reviewed in 10]. Since estimated dNTP concentrations in mammalian cells fall below these levels, it seemed possible that replication rates are determined by dNTP concentrations. However, more careful studies on concentration dependence, both of replicative polymerases (α and δ) and of permeabilized cells [11], have yielded lower values than most of those previously reported. K_M values fall in the 1 to 5 μM range, with saturation achieved by 10 to 20 μM of each nucleotide. In addition, data from our laboratory [12] indicate that most of the dNTP content of mammalian cells is present at approximately equal concentrations in the nucleus and in cytoplasm (except for small, metabolically distinct pools found in mitochondria [13]). Thus, it is likely that mammalian replication sites are supplied with dNTPs at saturating concentrations for replicative DNA polymerases. On the other hand, further attention should probably be paid to dGTP levels. In nearly every cell analyzed, dGTP is the least abundant of the four dNTPs, with pool sizes one to two orders of magnitude lower than those of the most abundant dNTPs. Our estimates of intranuclear dGTP levels in S phase Chinese hamster ovary cells are on the order of 10 μM, which may fall below the level needed to saturate the replicative apparatus; a similar possibility is suggested by data of Dresler et al. [11]. It would be worthwhile to manipulate cells to expand dGTP pools and determine the effects on replication rates or length of the S phase.

In prokaryotic systems the situation seems a bit more complex. The number of replication sites is limited, with dNTP turnover necessarily being very high at each site. Both in *Escherichia coli* and in T4 phage-infected *E. coli*, replicative chain growth occurs at about 850 nucleotides per second. At the same time, the replication apparatus in prokaryotes has lower affinity for dNTPs than in eukaryotes, with about 250 μM of each dNTP needed to saturate replication forks in T4 phage-infected *E. coli* [14]. While this suggests that the replicative machinery is not saturated, experimental evidence suggests otherwise [15]. As shown in Figure 1.2, a reversible inhibition of DNA replication, brought about by inactivating a temperature-labile replication protein, caused deoxyribonucleotides to accumulate. However, the rate of replication did not increase as a result of this accumulation, suggesting that the replication apparatus is normally saturated with dNTPs. These apparently contradictory conclusions can be resolved if concentration gradients of DNA precursors exist at replication sites.

Multienzyme Complexes for DNA Precursor Biosynthesis

How might dNTP concentration gradients be maintained, particularly in prokaryotic cells, where organelles are not available to establish such gradients? In T4 phage-infected *E. coli*, the enzymes of deoxyribonucleotide biosynthesis interact to form a large multienzyme complex, of molecular weight about 1.5×10^6 [16]. At lease 10 activities remain associated through four fractionation steps, with an enrichment of activity of several hundredfold. Figure 1.3 shows the coelution of several activities from a gel filtration column, including the activity of a three-step reaction sequence. Figure 1.4 identifies the 10 activities we have detected, including the three-step sequence just mentioned (dCTP→dCMP→dUMP→dTMP). Other reaction sequences have been found to be kinetically coupled as well, and the kinetic data suggest that the complex maintains substantial concentration gradients, simply by restricting diffusion of intermediates away from catalytic sites.

Could such a kinetic mechanism maintain dNTP concentration gradients at replication sites? It could, if the "dNTP synthetase" complex were physically linked to the replication apparatus. Thus, distal DNA

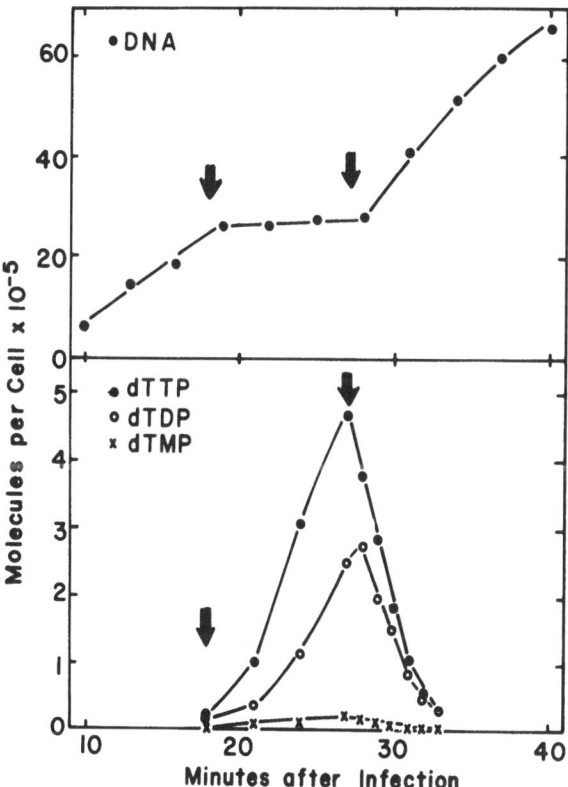

Figure 1.2. Effects of reversible inhibition of DNA synthesis on thymine nucleotide pools (lower panel) and DNA synthesis (upper panel). Data are from
reference 15. *E. coli* was infected at 30°C with T4, under conditions where all
thymine nucleotides are derived from exogenous thymidine. The phage bore a *ts*
mutation in gene 45, essential to replication. At 18 minutes, DNA replication was
arrested by shifting the culture from 30°C to 42°C. The culture was sampled for
thymine nucleotide pool determinations before and after restoration of DNA
synthesis by a shift back to 30°C.

precursors, such as ribonucleotides, would be "channeled" directly into
DNA. This model suggests that replication is fed by small, rapidly turned
over pools of precursors that do not readily mix with the bulk pools.
Considerable data, both in prokaryotes and in eukaryotes, have been
interpreted in terms of this "functional compartmentation" of DNA
precursors. However, in prokaryotic systems there is as yet no evidence
for physical intracellular interactions between dNTP synthetic enzymes
and proteins of the replication machinery. Moreover, because the enzymes of dNTP biosynthesis have much lower V_{max} values than the
replicative DNA polymerase, there would have to be several copies of the
complex per replication fork; a "megacomplex" would be required. On
the other hand, extensive studies by Greenberg and colleagues, of

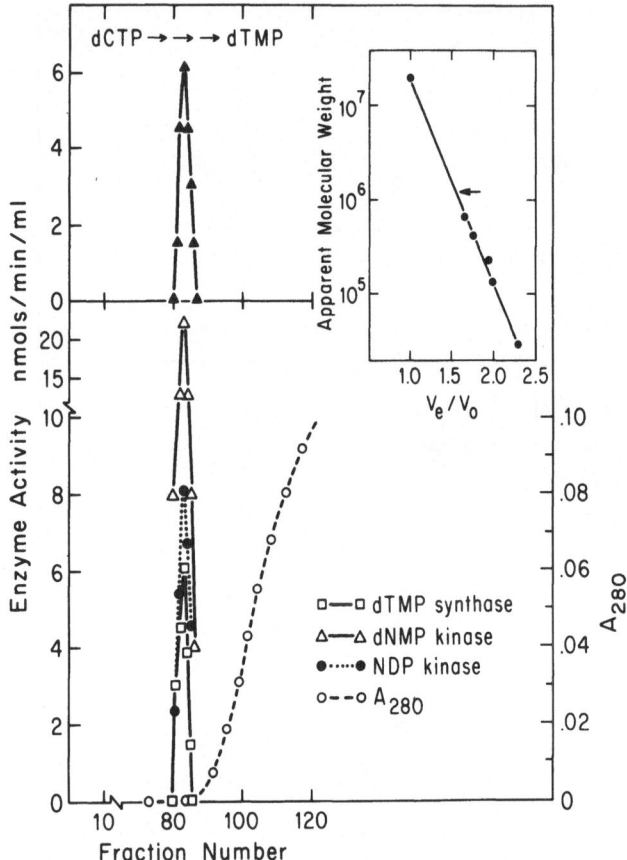

Figure 1.3. Elution profile of T4 phage dNTP synthetase complex from a gel filtration column. Data are from reference 16. Partially purified material was applied to a 3.5 × 82 cm Sephacryl S400 column, and 4.6-ml fractions were collected. Inset: Calibration of the column with high M_r standards; arrow indicates M_r for peak fraction 83. The upper panel shows activity of the conversion of dCTP to dTMP by a three-step sequence (dCTP→dCMP→dUMP→dTMP).

reaction fluxes in vivo and of interactions between genes controlling DNA replication and deoxyribonucleotide synthesis, point to a close coordination between these processes, which they have interpreted as direct coupling [17–19]. However, in the absence of direct physical evidence for such coupling, the question of DNA precursor channeling in prokaryotic systems remains open.

In mammalian cells several observations argue against channeling of DNA precursors to replication sites [reviewed in 20]. First is the likelihood, mentioned above, that intracellular dNTP concentrations exceed those needed to saturate replicative polymerases. Taken together

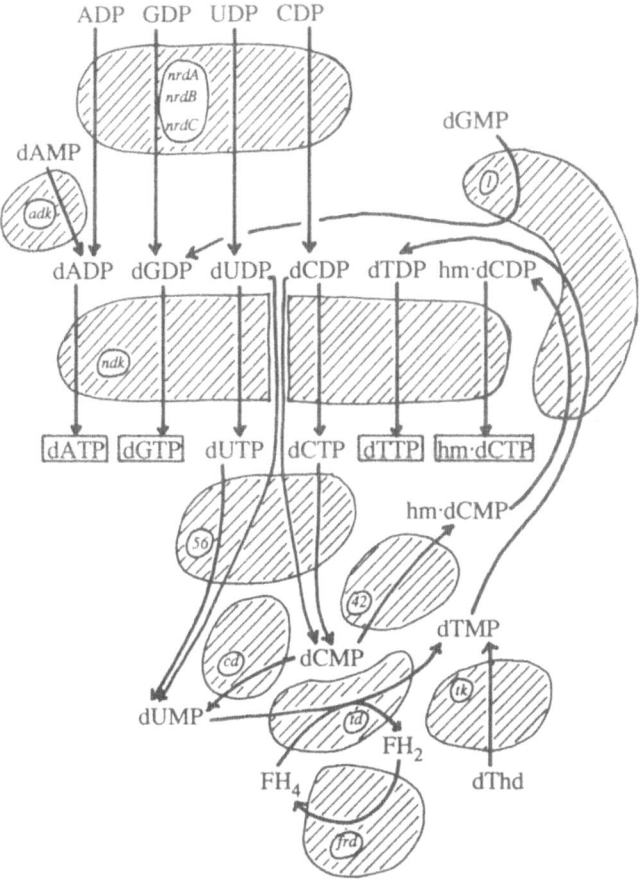

Figure 1.4. The T4 dNTP synthetase complex. Each dNTP product is boxed. The structural gene for each enzyme is shown. *E. coli* genes: *ndk,* nucleoside diphosphokinase; *adk,* adenylate kinase. T4 genes: *nrdA, B,* and *C,* ribonucleotide reductase system; 56, dCTPase-dUTPase; *cd,* dCMP deaminase; 42, dCMP hydroxymethylase; *td,* thymidylate synthase; *frd,* dihydrofolate reductase; 1, deoxyribonucleoside monophosphate kinase; *tk,* thymidine kinase. hm-dCMP, hm-dCDP, and hm-dCTP are 5-hydroxymethyldeoxycytidine mono-, di-, and triphosphate, respectively.

with the fact that eukaryotic replication rates are lower than those in prokaryotes by at least an order of magnitude [1], this suggests that channeling need not be invoked to explain how dNTPs are delivered to replication sites. Second, some forms of dNTP compartmentation can be explained without invoking channeling. For example, Chinese hamster ovary cells in G1 phase contain substantial dNTP pools—pools that are inaccessible to replication forks and that turn over slowly. This type of compartmentation is intercellular, not intracellular [21].

Third and most direct are recent findings on the intracellular locations of dNTP biosynthetic enzymes. Several laboratories have found that enzymes including ribonucleotide reductase, thymidylate synthase, and dUTPase are localized to the cytoplasm of mammalian cells [12,22–24]. Since replication occurs in the nucleus, these data seem to rule out any direct association between these enzymes and the replicative DNA polymerase complexes. At the same time, studies on the rate of labeling of dNTP pools by exogenous nucleosides [21] suggest that dNTPs are synthesized in the cytoplasm of infected cells, with the deoxyribonucleotides moving into the nucleus without establishment of a concentration gradient. Recent work of Wawra [25], involving microinjection of exogenous radiolabeled DNA precursors, leads to a similar conclusion.

dNTP Pool Compartmentation

As noted above, the existence of DNA precursor channeling connotes a form of functional compartmentation. In prokaryotes we can visualize small, highly active dNTP pools formed and utilized at replication sites, with the bulk of the intracellular dNTP existing as a much larger and less active pool. In T4 phage-infected cells this model is supported by the fact that permeabilized cells readily incorporate deoxyribonucleoside monophosphates into DNA, even though triphosphates never accumulate to detectable levels in these cells [26]. In uninfected *E. coli,* two observations are particularly striking. First, pulse labeling studies showed that thymidine labels DNA at maximal rates well before the dTTP pool has reached its maximal radioactivity [27], suggesting that the label must flow into at least two distinct pools, only one of which is used for replication. Second, inactivation of a thermolabile form of ribonucleotide reductase by temperature upshift of a mutant cell culture immediately halts DNA synthesis, even though substantial dNTPs remain [28], suggesting that most of the dNTP content of a cell is not immediately accessible to replication forks.

Several observations point to compartmentation of dNTP pools in mammalian cells. As noted above, the dNTP contents of mitochondria form a physically distinct pool. However, other evidence is more indirect, as shown by the following examples. First, DNA repair is less sensitive to inhibition by hydroxyurea than is DNA replication, suggesting that the two processes draw from different pools of precursors [29,30], although K_M values for repair are some 10-fold lower than those for replication [11]. Second, hydroxyurea inhibition of DNA replication cannot be reversed by adding deoxyribonucleosides, suggesting that dNTPs formed by salvage synthesis are not readily accessible to replication forks and, hence, represent a replication-inactive pool [29,31]. Third, in cells inhibited by mycophenolic acid, an inhibitor of guanine nucleotide synthesis, the block to replication can be reversed by guanine but not deoxyguanosine [32], suggesting that the two precursors enter different pools, with the guanine-derived pool preferentially used for replication. Fourth, in 3T6 mouse fibroblasts, radiolabeled deoxycytidine was found to label DNA at a rate severalfold lower than the rate of DNA synthesis, suggesting that much of the labeled precursor enters a replication-inactive pool [33].

Results from the above four approaches can be interpreted differently. First, as noted earlier, the compartmentation may be intercellular, with the replication-inactive pools being formed in non–S phase cells [21]. Second, kinetic analysis of dNTP pool labeling by exogenous nucleosides in nuclei and in whole cells indicates that the cytoplasmic and nuclear pools constitute a single metabolic compartment [21]. Third, under appropriate conditions a hydroxyurea block to replication can be readily bypassed by exogenous deoxyribonucleosides [34,35]. Finally, that dCTP that is not used for replication in 3T6 cells has another, recently identified fate—incorporation into "liponucleotides" [36], the deoxycytidine analogs of CDP-choline and CDP-ethanolamine. None of these observations rule out the existence in mammalian cells of special, replication-active dNTP pools. However, since the data supporting this model can all be interpreted differently, the model currently has little support.

Are There Modulating Proteins That Bind dNTPs?

While the data discussed earlier argue against substrate-level control of DNA replication in eukaryotic cells, there remains the possibility of a more subtle regulatory mechanism that responds to changes in intracellular dNTP concentrations. It is well known that certain deoxyribonucleosides are effective inhibitors of replication. The mechanism involves allosteric regulation of ribonucleotide reductase. For example, thymidine increases the dTTP pool, which causes allosteric inhibition of the reduction of CDP, thereby inhibiting replication by shrinking the dCTP pool [37]. Grindey and colleagues [38] observed relationships between changes in pool size of the critical dNTP and replication rates. In thymidine-inhibited cells, for example, relatively small decreases in dCTP pools were correlated with much larger inhibitions of DNA synthesis. Comparable data had accumulated throughout the 1970s, from investigations of dNTP pool sizes in cells treated with inhibitors such as 5-fluorodeoxyuridine or methotrexate; inhibition of DNA replication seemed more severe than expected from the decreases in dNTP pools.

Of course, these results could be attributed to compartmentation, with selective depletion of dNTP pools either in the nucleus or in S phase cells. Steinberg et al. [39] proposed an alternative explanation: substrate inhibitions of DNA polymerase α, with the inhibitory dNTPs bound at a site distinct from the catalytic site. Steinberg et al. presented evidence for a dNTP-binding protein among the subunits of DNA polymerase α, and they proposed that it represents a modulating protein, which senses the levels of dNTPs and controls catalytic activity independently of the binding of substrates at the catalytic site. The original observations, made in 1979, have not been pursued by the several laboratories working now with highly purified preparations of DNA polymerase α. However, Wierowski et al. [40] reported that DNA polymerase α activity is stimulated at high concentrations (4 mM) of various nucleotides, notably ATP, but also including CTP, dATP, and dCTP. The effect of dCTP was of special interest, because inhibition of DNA polymerase α by aphidicolin is reported to be competitive specifically with dCTP [41]. To rationalize this apparent anomaly, Nicander and Reichard [42] proposed that dCTP regulates DNA polymerase α activity by binding to an allosteric site and that aphidicolin inhibits by binding to this site, not the catalytic site.

In the meantime, Tan et al. [43] have reported that the activation of DNA polymerase α by ATP was an artifact of improper pH adjustment of the ATP solutions at the high concentrations used by Wierowski et al. [40]. While pH control could be a factor, this does not explain why several nucleotides, including GTP and dGTP, gave no activation when added at 4 mM. Clearly, there is a need for rigorous nucleotide-binding studies of purified DNA polymerase α and any associated polypeptides.

dNTP Pool Turnover as a Possible Regulatory Mechanism

In a recent series of papers, Reichard and his co-workers have provided strong evidence that pyrimidine deoxyribonucleotide pool sizes are controlled by substrate cycles—cycles of nucleotide breakdown and resynthesis [44–47]. The earliest evidence came from experiments that followed the fates of radiolabeled nucleosides or deoxyribonucleosides in cultured mammalian cells. In exponentially growing cells, more than one quarter of dCDP that was formed through the action of ribonucleotide reductase was excreted to the culture medium as deoxycytidine or deoxyuridine. In cells treated with hydroxyurea to inhibit ribonucleotide reductase, much of the labeled deoxyribonucleoside from the medium was taken back into cells and reused for synthesis of dTTP [48]. Most cells contain nucleotidases and deoxyribonucleoside kinases, the enzymes that would most likely participate in substrate cycles, as schematized in Figure 1.5. Since nucleotide kinases have equilibrium constants close to unity, Reichard [4] proposed that the key regulated reactions in these cycles are those that interconvert deoxyribonucleosides with deoxyribonucleotides.

The elegant experiments from Reichard's laboratory clearly demonstrate the quantitative extent of nucleotide breakdown and resynthesis under different conditions. However, the precise regulatory mechanisms remain unknown, partly because the nucleotidases that function in these

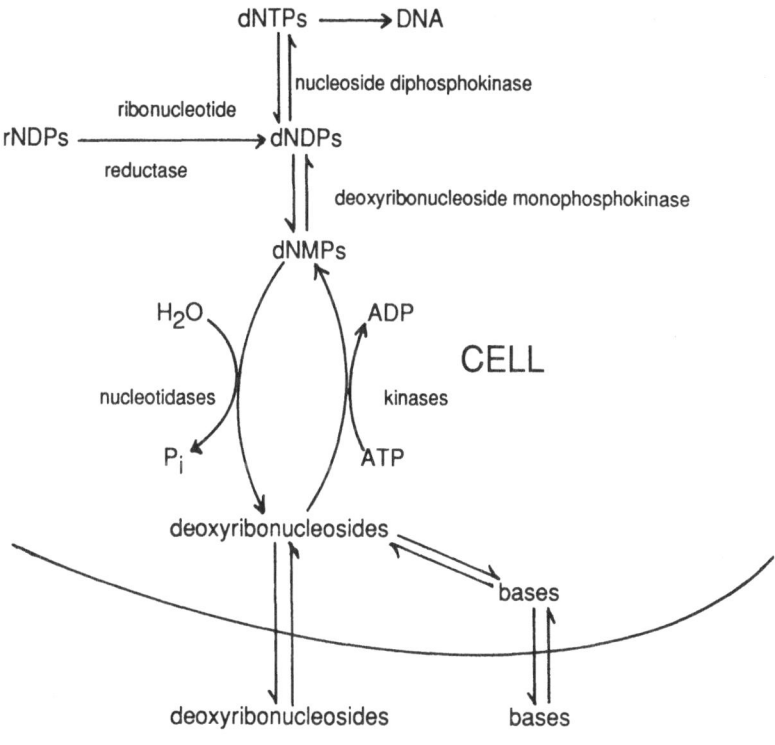

Figure 1.5. A model for regulation of dNTP pool sizes by substrate cycles, as proposed by Reichard [4]. In addition to the phosphatase and kinase reactions depicted, these cycles may involve nucleoside deaminases and phosphorylases.

cycles have not been characterized with respect to regulation and partly because the phenotypes of mutants altered in these enzymes have not been described. Clearly, the maintenance of dNTP pools within defined limits is important to the well-being of a cell, because of the manifold consequences of nucleotide pool imbalances (see next section). Moreover, the importance of nucleotide catabolic pathways has become belatedly acknowledged through awareness of the dramatic consequences of human mutations that inactivate enzymes in these pathways—specifically, the inherited immune system disorders resulting from deficiencies in adenosine/deoxyadenosine deaminase, or purine nucleoside phosphorylase [49].

   An unusual nucleotidase has recently been described in *E. coli* by Seto et al. [50]. This enzyme specifically cleaves dGTP to deoxyguanosine plus tripolyphosphate. Mutants in a gene called *optA* apparently overproduce this enzyme, and these mutants restrict the growth of certain T7 phage strains, evidently because the infected cells cannot retain pools of dGTP [51]. At present there is no evidence that the dGTPase plays a role in regulating dNTP levels or rates of replication. However, Seto et al. have suggested a novel role for the enzyme in maintaining replication fidelity. Since AT→CG transversions are thought to result from incorporation of dGTP in the unusual *syn* conformation opposite a template adenine, Seto et al. suggested that dGTPase may reduce the frequency of this mispairing by specifically cleaving the *syn* rotamer of dGTP. Further analysis of this or other roles for the dGTPase is awaiting isolation and characterization of mutants with reduced activity of the enzyme.

dNTPs as Regulators of DNA Replication Fidelity

The previous section alluded to an important aspect of deoxyribonucleotide metabolism, namely, the importance of maintaining dNTP pools within rather narrow limits. The consequences of dNTP pool imbalance include increased mutagenesis, recombinogenesis, chromosomal abnormalities, induction of latent viruses, and cell death [reviewed in 52,53]. From a mechanistic standpoint, most attention has been paid to mutagenesis induced by dNTP pool imbalances. Thus, it is appropriate to consider roles of deoxyribonucleotides as regulators of the accuracy, as well as the rate, of DNA replication.

As first enunciated by Fersht [54], there are at least two dNTP concentration-dependent steps in replication: (1) the initial incorporation step, in which a non-Watson-Crick base pair can form if a "wrong" nucleotide is present at sufficiently high concentrations to compete with the "right" nucleotide; and (2) the "next nucleotide effect," in which a high concentration of the next nucleotide promotes its incorporation, thereby blocking the 3'-exonucleolytic proofreading of a previously misincorporated nucleotide. Both of these effects have been amply documented through studies of errors induced in replication systems reconstituted from purified prokaryotic proteins.

Understanding of mutagenesis induced by dNTP pool imbalances in intact cells has come more slowly. Meuth and colleagues [8,55] have analyzed a series of mutant genes encoding adenine phosphoribosyltransferase in CHO cells. Mutations that altered restriction patterns of the *aprt* gene were mapped and characterized by Southern blot analysis using an *aprt* gene probe. For many mutations, this procedure yielded sufficient information for cloning and sequencing the mutant DNA region [55,56]. Nucleotide sequences of mutant regions suggested that mutations arose both through incorporation errors and next-nucleotide effects. The latter result was of particular importance because it strongly implied that mammalian DNA replication is subject to 3'-exonucleolytic editing, at a time before any proofreading exonucleases had been characterized in mammalian DNA replication.

Our laboratory has taken a somewhat different approach to pool imbalance-induced mutagenesis [57]. We measure the rates at which defined mutations in the phage T4 *rIIA* or *rIIB* cistrons revert to wild type when certain imbalances are imposed. For example, we can induce a large imbalance by infecting *E. coli* with mutant phage defective in the synthesis of deoxycytidylate deaminase, a condition that greatly increases the 5-hydroxymethyl-dCTP pool and shrinks the dTTP pool. This imbalance specifically stimulates reversion pathways that proceed along an AT→GC transition pathway. For most mutants analyzed, the increase in spontaneous reversion frequency is of the order of 10-fold. However, one mutation, *rIIA*UV215, displays a 1,000-fold increase when present in a dCMP-deaminase-negative background. This implies that pool imbalance mutagenesis is strongly influenced by nucleotide sequences flanking the mutant site. The nature of that sequence context is under investigation.

Finally, we have used this system to ask whether the dNTP synthetase multienzyme complex contributes toward replication fidelity, possibly by maintaining optimal dNTP levels in the microenvironment at replication sites. T4 mutants defective in ribonucleotide reductase induction do not form a functional complex [16]. However, deoxyribonucleotides can be provided at adequate rates in infection of *E. coli* strains that overproduce the bacterial ribonucleotide reductase. Under these conditions, we find that *rII* mutations revert at four- to fivefold higher rates than seen in normal infections [58]. These preliminary results suggest that the action of the dNTP synthetase complex helps to maintain the high fidelity of normal DNA replication, although the precise role of the complex remains to be determined.

Summary

DNA precursor pools are maintained within defined limits by a combination of mechanisms, including allosteric control of key enzymes, regulation of enzyme synthesis, interconversion of active and inactive forms of ribonucleotide reductase, substrate cycles, and in some systems the functioning of multienzyme complexes for dNTP synthesis. Faulty operation of some of these mechanisms has severe genetic consequences, including increases in spontaneous mutagenesis and chromosome abnormalities, suggesting that the synthesis of DNA precursors somehow controls the accuracy of DNA replication. Control of replication rates by DNA precursor levels is probably somewhat indirect. In prokaryotes it appears that multienzyme complexes help to deliver precursors to replication sites, when simple diffusion of precursors would not be able to maintain substrate saturation of replicative DNA polymerases. In mammalian cells dNTPs are synthesized in the cytoplasm and probably pass freely into the nucleus. Because DNA polymerases α and δ have very low K_M values for dNTPs, it seems likely that the eukaryotic replication apparatus is normally saturated with dNTPs. However, some control may be exerted through low pool sizes of dGTP, or through possible interaction of dNTPs with regulatory sites on DNA polymerase α.

Acknowledgments. Research in our laboratory is supported by NIH research grants AI-15145 and GM-37508.

References

1. Kornberg, A. (1989) *DNA Replication, Third Edition;* W.H. Freeman and Co., San Francisco (in press).
2. Eriksson, S., and Sjöberg, B.-M. (1988) CRC Crit. Rev. Biochem. (in press).
3. Eriksson, S., Gudas, L.J., Ullman, B., Clift, S.M., and Martin, D.W. Jr. (1981) J. Biol. Chem. *256,* 10184–10189.
4. Reichard, P. (1988) Annu. Rev. Biochem. *57,* 349–374.
5. McClarty, G.A., Chan, A.K., Engström, Y., Wright, J.A., and Thelander, L. (1987) Biochemistry *26,* 8004–8011.
6. Weinberg, G., Ullman, B., and Martin, D.W. Jr. (1981) Proc. Natl. Acad. Sci. USA *78,* 2447–2451.
7. Weinberg, G.L., Ullman, B., Wright, C.M., and Martin, D.W. Jr. (1985) Somat. Cell Mol. Genet. *11,* 413–419.
8. Trudel, M., Van Genechten, T., and Meuth, M. (1984) J. Biol. Chem. *259,* 2355–2359.
9. Aronow, B., Watts, T., Lasseter, J., Washtien, W., and Ullman, B. (1984) J. Biol. Chem. *259,* 9035–9043.
10. Mathews, C.K. (1985) In Genetic Consequences of Nucleotide Pool Imbalance, De Serres, F.J., ed. Plenum Press, New York, pp. 47–66.
11. Dresler, S.L., Frattini, M.G., and Robinson-Hill, R. (1988) Biochemistry *27,* 7247–7254.
12. Leeds, J.M., Slabaugh, M.B., and Mathews, C.K. (1985) Mol. Cell. Biol. *5,* 3443–3450.
13. Bestwick, R.K., Moffett, G.L., and Mathews, C.K. (1982) J. Biol. Chem. *257,* 9300–9304.
14. Mathews, C.K., and Sinha, N.K. (1982) Proc. Natl. Acad. Sci. USA *79,* 302–306.
15. Mathews, C.K. (1976) Arch. Biochem. Biophys. *172,* 178–187.
16. Moen, L.K., Howell, M.L., Lasser, G.W., and Mathews, C.K. (1988) J. Mol. Recog. *1,* 48–57.
17. Flanegan, J.B., and Greenberg, G.R. (1977) J. Biol. Chem. *252,* 3019–3027.
18. Chiu, C.-S., Cook, K.S., and Greenberg, G.R. (1982) J. Biol. Chem. *257,* 15087–15097.

19. Wirak, D.O., Cook, K.S., and Greenberg, G.R. (1988) J. Biol. Chem. *263*, 6193–6201.
20. Mathews, C.K., and Slabaugh, M.B. (1986) Exp. Cell Res. *162*, 285–295.
21. Leeds, C.K., and Mathews, C.K. (1987) Mol. Cell. Biol. *7*, 532–534.
22. Engström, Y., Rozell, H.-A., Stemme, S., and Thelander, L. (1984) EMBO J. *3*, 863–867.
23. Kucera, R., and Paulus, H. (1986) Exp. Cell Res. *167*, 417–423.
24. Vilpo, J.A., and Autio-Harmainen, H. (1983) Scand. J. Clin. Lab. Invest. *43*, 583–587.
25. Wawra, E. (1988) J. Biol. Chem. *263*, 9908–9912.
26. Reddy, C.K., and Mathews, C.K. (1978) J. Biol. Chem. *253*, 3461–3467.
27. Pato, M.L. (1979) J. Bacteriol. *140*, 518–524.
28. Manwaring, J.D., and Fuchs, J.A. (1979) J. Bacteriol. *138*, 245–249.
29. Snyder, R.D. (1984) Mutat. Res. *131*, 163–172.
30. Collins, A., and Oates, D.J. (1987) Eur. J. Biochem. *169*, 299–305.
31. Scott, F.E., and Forsdyke, D.R. (1980) Biochem. J. *190*, 721–730.
32. Nguyen, B.T., and Sadee, W. (1986) Biochem. J. *234*, 263–269.
33. Nicander, B., and Reichard, P. (1982) Proc. Natl. Acad. Sci. USA *80*, 1347–1351.
34. Lagergren, J., and Reichard, P. (1987) Biochem. J. *36*, 2985–2991.
35. Slabaugh, M.B., Howell, M.L., Wang, Y., and Mathews, C.K. (in preparation).
36. Spyrou, G., and Reichard, P. (1987) J. Biol. Chem. *262*, 16425–16432.
37. Bjursell, G., and Reichard, P. (1973) J. Biol. Chem. *248*, 3904–3909.
38. Grindey, G.B., Winkler, M., Otten, M., and Steinberg, J.A. (1980) Mol. Pharmacol. *17*, 256–261.
39. Steinberg, J.A., Otten, M., and Grindey, G.B. (1979) Cancer Res. *39*, 4330–4335.
40. Wierowski, J.V., Lawton, K.G., Hockensmith, J.W., and Bambara, R.A. (1983) J. Biol. Chem. *258*, 6250–6254.
41. Huberman, J.A. (1981) Cell *23*, 647–648.
42. Nicander, B., and Reichard, P. (1981) Biochem. Biophys. Res. Commun. *210*, 148–155.
43. Tan, C.-K., So, M.J., Downey, K.M., and So, A.G. (1987) Nucleic Acids Res. *14*, 2269–2278.
44. Nicander, B., and Reichard, P. (1985) J. Biol. Chem. *260*, 5376–5381.
45. Nicander, B., and Reichard, P. (1985) J. Biol. Chem. *260*, 9216–9222.
46. Bianchi, V., Pontis, E., and Reichard, P. (1986) Proc. Natl. Acad. Sci. USA *83*, 986–990.
47. Bianchi, V., Pontis, E., and Reichard, P. (1987) Mol. Cell. Biol. *7*, 4218–4224.
48. Bianchi, V., Pontis, E., and Reichard, P. (1986) J. Biol. Chem. *261*, 16037–16042.
49. Martin, D.W. Jr., and Gelfand, E.W. (1981) Annu. Rev. Biochem. *50*, 845–877.
50. Seto, D., Bhatnagar, S.K., and Bessman, M.J. (1988) J. Biol. Chem. *263*, 1494–1499.
51. Myers, J.A., Beauchamp, B.B., and Richardson, C.C. (1987) J. Biol. Chem. *262*, 5288–5292.
52. Meuth, M. (1984) Mutat. Res. *126*, 107–112.

53. DeSerres, F.J., ed. (1985) Genetic Consequences of Nucleotide Pool Imbalance. Plenum Press, New York.
54. Fersht, A.R. (1979) Proc. Natl. Acad. Sci. USA 76, 4946–4950.
55. Goncalves, O., Drobetsky, E., and Meuth, M. (1984) Mol. Cell. Biol. 4, 1792–1799.
56. Phear, G., Nalbantoglu, J., and Meuth, M. (1987) Proc. Natl. Acad. Sci. USA 84, 4450–4454.
57. Sargent, R.G., and Mathews, C.K. (1987) J. Biol. Chem. 262, 5546–5553.
58. Sargent, R.G., Mun, B.-J., and Mathews, C.K. (1989) Mol. Gen. Genet. in press.

Chapter 2
Replication Fork Propagation in *Escherichia coli*

Russell J. DiGate and Kenneth J. Marians

DNA replication is a complex process involving many enzymatic activities. The efficient and rapid nature of this process requires the coordination and interaction of many proteins, including the organization of these separate entities into multifunctional complexes or machines [1–8]. Much of our current knowledge pertaining to replication fork assembly and movement has come from the study of this process in the bacteria *Escherichia coli* and its bacteriophages.

Many of the proteins involved in DNA replication in *E. coli* have been purified and characterized [for a review see 1]. The study of DNA replication in prokaryotes has reached a high level of sophistication. Many current efforts are designed to elucidate the protein-protein and protein-DNA interactions that govern the DNA replication process. The ultimate goal, however, of completely reconstituting accurate and efficient DNA replication in vitro remains. A schematic representation of a replication fork is presented in Figure 2.1.

The current model of DNA replication has incorporated information from over 30 years of research and taken into consideration the following observations: (1) The synthesis of DNA cannot initiate de novo; DNA polymerases can only elongate a preexisting polynucleotide chain. (2) Because DNA is composed of two antiparallel strands, synthesis of one of the strands (the lagging strand) is discontinuous, resulting in small discrete DNA chains termed Okazaki fragments. (3) Replicative DNA polymerases are generally inefficient at strand displacement synthesis; therefore, another activity must be present to unwind the parental duplex DNA ahead of the advancing replication fork. (4) Since the replicative DNA polymerase in *E. coli*, the DNA polymerase III holoenzyme (Pol III HE), is present in extremely low amounts (10 to 20 molecules per cell [9]), it is likely that there is a mechanism that facilitates the location and binding to a primer terminus of the polypeptides involved in polymerization. In addition, since DNA replication is a rapid process (replication fork movement in *E. coli* has been estimated at 1,000 nucleotides (nt) per second at 37°C) it has been proposed that both leading- and lagging-strand

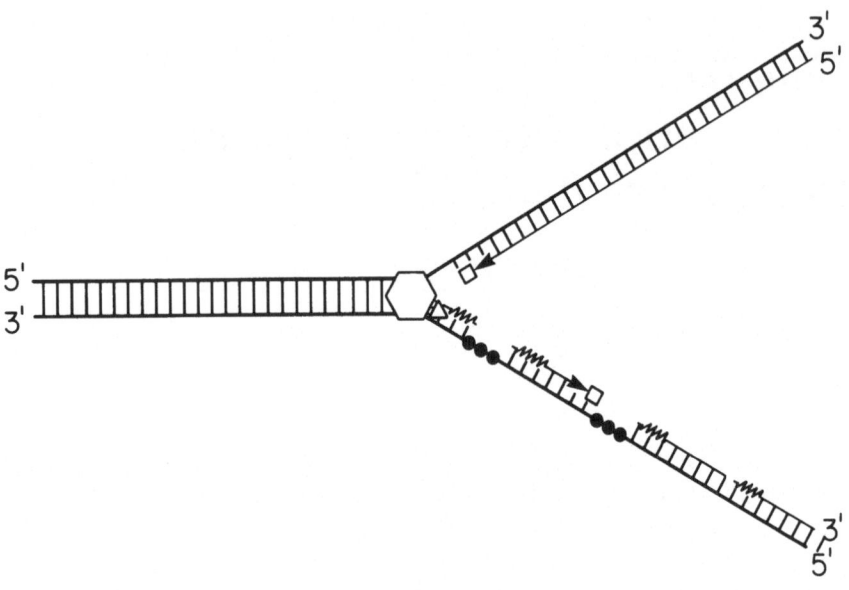

Figure 2.1. Schematic representation of a DNA replication fork. DNA replication at a propagating replication fork occurs continuously on the leading strand (upper branch) and discontinuously on the lagging strand (lower branch). A helicase (○) unwinds the parental duplex DNA ahead of the moving replication fork. The synthesis of Okazaki fragments on the lagging strand initiates with an oligonucleotide primer (ᨇ) synthesized by a DNA primase (△). The DNA polymerase (□) is responsible for DNA synthesis upon both the leading and lagging strands. Single-stranded DNA binding protein (●) removes any inhibitory secondary structure present on the single-stranded parental DNA. This drawing is not to scale.

DNA synthesis occur simultaneously and are in some way coupled (i.e., the replication machinery on one strand communicates information about its progress to the cognate machinery on the other strand, influencing the action of the latter complex) [3–6]. This chapter will discuss the current concepts of replication fork movement in *E. coli,* emphasizing the roles and interactions of the *E. coli* replication proteins.

Proteins Required for Replication Fork Movement

At least five distinct types of enzymatic activity appear to be necessary for efficient replication fork movement in *E. coli*. Studies of the DNA replication requirements of the *E. coli* bacteriophages T_4, T_7, and ϕX174 have revealed a general scheme of the activities required at the replication fork.

DNA synthesis at the replication fork is carried out by a DNA polymerase. Studies with bacteriophages T_7 and T_4 have revealed that replicative DNA polymerases generally consist of a polymerase subunit (the gene 5 protein [10] and the gene 43 protein [11] of T_7 and T_4, respectively) and an auxiliary subunit(s) that confers processivity to the enzyme (*E. coli* encoded thioredoxin in T_7 [12–14]; the gene 44/62 and gene 45 DNA polymerase [3,15] accessory proteins in T_4). The processivity of a DNA polymerase refers to the number of nucleotides added per interaction with a primer terminus. A highly processive DNA polymerase is capable of adding thousands of nucleotides per binding event.

DNA polymerase III is the replicative DNA polymerase of *E. coli* [16]. This enzyme is composed of a three-subunit core enzyme [17] and possibly as many as seven auxiliary subunits [18,19] to form a holoenzyme (DNA Pol III HE). The 140-kD α subunit of the core enzyme, the *dnaE* gene product, contains the actual DNA polymerase activity [20]. The 27-kD ϵ subunit, the *dnaQ* gene product, possesses a $3' \rightarrow 5'$ proofreading exonuclease activity [21] that is greatly stimulated by the presence of the DNA polymerase III α subunit [22]. Neither the function of, nor the gene encoding, the 12-kD θ subunit of the core enzyme has been determined.

At least four other subunits—τ [23,24], β [18,25,104,105], γ [18,19], and δ [18,19], with corresponding molecular weights of 71,48,52, and 35 kD, respectively—are associated with the DNA Pol III HE. The τ subunit and γ subunit are products of the *dnaZX* gene [27,29]. The γ subunit is encoded by the N-terminal three fourths of the *dnaZX* gene. It has been postulated that the γ subunit is derived from the τ subunit by proteolysis [30,31]. The β subunit, the *dnaN* gene product, has been purified to homogeneity and is required for the ATP (or dATP)-dependent initiation of DNA synthesis on primed single-stranded DNA templates [32] by the DNA Pol III HE. The δ subunit has been isolated in association with the holoenzyme [18,19] and in a separate complex that also contains the γ subunit of the DNA Pol III HE [18,26,33]. Recent studies from Kornberg's laboratory [33] suggest that the DNA Pol III HE probably contains three additional subunits—δ' (33kD), χ (15kD), and ψ (12 kD).

Two other activities required for replication fork propagation are a DNA unwinding activity, or helicase, and a DNA primase. Helicases are DNA-dependent nucleoside triphosphatases that possess the unique property of being able to translocate unidirectionally along a DNA strand while displacing DNA fragments hybridized to that strand. DNA primases are specialized RNA polymerases capable of synthesizing short oligonucleotide primers that serve to initiate DNA synthesis [1].

A common feature in prokaryotic DNA replication is the association of the primase and helicase. The gene 4 protein of T_7 is a primase/helicase [34,35]. Highly purified preparations of the gene 4 protein contain polypeptides of 66 and 57 kD [36,37]. The gene 4 region of the T_7 genome contains a single reading frame with a 63-kD coding capacity. The smaller peptide observed in gene 4 preparations appears to be the result of the initiation of protein synthesis at an internal AUG start condon [38]. The smaller gene 4 polypeptide has been molecularly cloned and shown to possess helicase activity, but no primase activity [39]. The larger polypeptide appears to contain both primase and helicase activity [40]. It has been proposed that the active primase/helicase consists of an oligomer of the two forms of the enzyme [40]. Similarly, the T_4 primase/helicase consists of a complex between the T_4 gene 41 helicase [41–44] and gene 61 primase [45–49].

In *E. coli*, the helicase/primase complex appears to be variable, depending on the particular system examined. The replication of plasmids in vitro containing the *E. coli* bacterial origin of DNA replication (*oriC*) [50] requires the *dnaB*-encoded helicase [51] (DNA B protein) in association with the *dnaG*-encoded primase [52] (DNA G protein). However, the priming complex (the primosome) used during the replication of ϕX174 SS(c) → RF DNA appears to be different. The ϕX174 primosome requires replication factor Y (protein n') [53–55]; primosomal proteins n, n'', i; the *dnaC* gene product (DNA C protein), as well as the DNA B and DNA G proteins, for assembly [56–58]. Interestingly, replication factor Y, similar to the DNA B protein, is a helicase [59,60]; however, factor Y translocates in the opposite direction of the DNA B protein.

An additional activity predicted to be necessary for efficient fork propagation is provided by single-stranded DNA binding proteins (SSBs) [61]. The bacteriophage T_4 gene 32 protein [62], T_7 gene 2.5 protein [63], and *E. coli* SSB [64] are examples of this class of proteins. DNA synthesis by DNA polymerase III is quite sensitive to inhibition by secondary structure within the DNA template. At least one function of the 18 kD *E. coli* SSB appears to be the binding of single-stranded DNA, resulting in the removal of this possible inhibitory secondary structure within the template. In some systems, SSBs from heterologous sources cannot fully substitute for their homologous counterparts, suggesting that there may be specific interactions between cognate DNA polymerases and their SSBs as well as between the SSB and the single-stranded DNA [65,66].

The *E. coli* chromosome is a circular 4.4×10^6 bp DNA molecule [67]. The replication of circular DNA imposes certain topological problems. As this replication fork proceeds during replication, positive supercoils accumulate in the unreplicated parental duplex. If these positive supercoils are allowed to remain, DNA replication fork propagation would cease; therefore, DNA topoisomerases, enzymes that alter the linking number of DNA, are required for replication fork propagation in *E. coli*. It has been shown that *E. coli* DNA gyrase is essential for DNA synthesis in vivo [68–72]. DNA gyrase is composed of a tetramer of two subunits, the *gyrA* gene product (105 kD) [73,74] and the *gyrB* (95 kD) gene product [74,75]. DNA gyrase is capable of catalyzing the ATP-dependent supercoiling of relaxed circular (form I') DNA [76], the direct conversion of positive supercoils to negative ones, the catenation of circular duplex DNA [77,88], and the decatenation of singly [77,79] and multiply interlinked DNA dimers [80]. Because of these properties, DNA gyrase is most likely involved in many phases of replication. The role of DNA gyrase in replication fork propagation is presumably to keep the parental duplex ahead of the replication fork negatively supercoiled (underwound), thereby facilitating rapid fork movement and inhibiting the formation of positive supercoils in the unreplicated region of the template. In addition, DNA gyrase is required for the initiation of DNA synthesis [81]. The decatenation properties of DNA gyrase are most likely required for the separation of the fully replicated daughter chromosomes [69,72]. It has also been proposed that *E. coli* topoisomerase I [82] and topoisomerase III [83] may also have a role in chromosomal segregation.

Establishment of the Replication Fork

The establishment of a replication fork can be separated into two distinct phases. In the first phase, the origin region is recognized and "activated." This process involves a mechanism that creates single-stranded regions within the origin region. The second phase of replication fork assembly involves the establishment of the leading- and lagging-strand DNA synthesis apparati. Although these two phases are necessary for the formation of the replication fork, the pathways by which they are achieved vary depending on the DNA to be replicated.

The initiation of Col EI or pBR322 DNA replication requires transcription by RNA polymerase [84]. A transcript initiated from a strong promoter within the origin region (subsequently processed by RNase H [85]) serves as the primer for leading-strand DNA. Transcription and/or DNA replication through the origin region displaces a sequence on the lagging-strand DNA template that catalyzes the formation of a primosome [86]. This unique DNA sequence, termed a primosome assembly site (PAS), was initially recognized during the study of ϕX174 SS(c) → RF DNA synthesis [55]. The PAS, when coated with SSB, is recognized by the site-specific, DNA-dependent ATPase, factor Y (protein n') [54,56]. Factor Y, in conjunction with primosomal proteins n, n″, and i and the DNA B and DNA C proteins, catalyzes the assembly of the preprimosome [reviewed in 1]. The addition of the DNA G primase results in the production of the primosome, a mobile priming apparatus. The primosome is capable of translocating in a 5′→3′ direction on the ϕX174 SS(c) DNA template and occasionally synthesizes a short oligoribonucleotide primer [58]. Once the primosome is assembled, replication fork propagation can occur.

Plasmids containing the *E. coli* origin of replication, *oriC*, use a different mechanism to activate the origin. The initiation of DNA synthesis of *oriC*-containing plasmids and many single-copy plasmids requires the *dnaA* gene product (DNA A protein) [87]. The DNA A protein, in the form of a multimeric complex, binds to specific DNA sequences within *oriC* (DNA A boxes) and induces the formation of single-strand regions within the origin [88]. Subsequently, the DNA B helicase (mediated through the action of the DNA C protein) and the DNA G primase are recruited to the origin by their interaction with the DNA A protein. A priming complex, which consists of only the DNA B and DNA G proteins, then assembles on the lagging-strand DNA template [89,90].

The mechanism of the initiation of leading-strand DNA synthesis at *oriC* is unclear. The replication of *oriC*-containing plasmids in vitro does not require the participation of RNA polymerase [91]. However, RNA polymerase has been shown to be essential for DNA replication in vivo [92]. Thus, it is not clear what role RNA polymerase plays during the initiation of DNA synthesis at *oriC*. If the initiation of DNA synthesis is similar to that employed by Col EI and pBR322 plasmids, then RNA polymerase may synthesize a transcript(s) that could serve as the primer for leading-strand DNA synthesis. In support of this mechanism, several strong promoters exist in the vicinity of *oriC*, and RNA-DNA transitions have been mapped within the origin region [93–96]. Alternatively, it is also possible that RNA transcription through *oriC* serves a structural purpose by displacing a factor(s) that is inhibitory to DNA A protein binding. A likely candidate for such a protein would be the *E. coli* protein HU. The initiation of leading-strand DNA synthesis at *oriC* need not require the use of a RNA transcript as a leading-strand primer. In the initiation of bidirectional replication at *oriC*, the first lagging-strand primer could serve to initiate leading-strand DNA synthesis upon that strand.

At this time, the components that make up the primosome are unclear. The DNA G primase and the DNA B helicase are clearly required; however, it is not known whether the other primosomal proteins (factor Y, n″, n, and i) are present. In the presence of a PAS, the primosomal proteins enhance the loading of the DNA B helicase onto the template [97] (discussed in the following section), but so may the DNA A protein [89,90]. Interestingly, although the replication of *oriC*-containing plasmids in vitro is DNA A protein–dependent and does not require factor Y, n″, n, and i proteins, there are PAS-like sequences located near *oriC* [98], suggesting a possible role for these proteins in the replication of the *E. coli* chromosome.

Replication Fork Propagation: Leading-Strand DNA Synthesis

Once the replication fork has been assembled at the replication origin, replication fork propagation can occur. The synthesis of DNA at the replication fork is inherently asymmetric, since leading-strand DNA synthesis can occur continuously, requiring only one interaction with a primer, whereas lagging-strand DNA synthesis requires the DNA polymerase to continually recycle to multiple primers along the lagging-strand DNA template (see Fig. 2.1). The low abundance of the DNA Pol III HE in the cell requires that the leading-strand enzyme be capable of adding many nucleotides per binding event (i.e., be very processive). The DNA Pol III HE has been shown to be a very processive enzyme, capable of polymerizing at least 5,000 nt [99,100] and as many as 150,000 nt (when coupled with a helicase during rolling-circle DNA synthesis [97]) per interaction with primer. The processivity of the DNA Pol III HE appears to be influenced by its auxiliary subunits. DNA polymerase III core enzyme alone has a processivity of 10 to 15 nt per binding event; however, in the presence of the auxiliary subunits, the processivity of the polymerase dramatically increases. Various subassemblies of DNA Pol III HE have been purified and their processivities determined [99,100]. DNA polymerase III', which consists of the DNA polymerase III core enzyme and the auxiliary τ subunit, has a processivity of 30 to 40 nt. The addition of spermidine, however, increases the processivity to approximately 60 nt. DNA polymerase III*, which contains all of the auxiliary subunits except β, has a processivity of 50 nt. This can be increased to 200 nt in the presence of SSB.

The replication of double-stranded DNA requires a combined action of a DNA polymerase and a DNA helicase. The primosome is capable of functioning as a replication fork helicase. Mok and Marians [101] have shown that large linear multimer DNA products formed during ϕX174 SS(c) → RF DNA replication are a result of rolling-circle DNA synthesis using the primosome as a DNA helicase. The formation of these products was dependent on both factor Y and the other primosomal proteins; in addition, DNA templates that did not contain a PAS could not efficiently support rolling-circle DNA synthesis. The translocation of the primosome in the 5'→3' direction is presumably mediated by the DNA B helicase. The strand displacement activity of the primosome has also recently been measured directly [Lee and Marians, unpublished data]. The DNA B protein has been shown to be a helicase capable of binding to single-stranded DNA and translocating in a 5'→3' direction along the substrate. Helicase activity requires a 5' single-stranded tail on the displaced strand; however, DNA B protein binding is severely inhibited by the presence of SSB [51]. One possible function of the primosomal proteins may be to load the DNA B helicase onto the lagging-strand template in the presence of SSB.

Factor Y is also a DNA helicase; however, translocation mediated by factor Y occurs in the 3'→5' direction [59,60]. The assembly of two helicases of opposite directionality within the same complex presents an apparent paradox. Studies from Kornberg's laboratory have demonstrated that the primosome moves 5'→3' along the DNA in the antielongation direction [58]. The directionality of primosome movement coupled with its priming activity places the primase/helicase on the lagging strand during DNA replication. If this is the case, what is the function of factor Y in the primosome? One possibility is that other than being required for primosome formation, factor Y has no function in the primosome. Another possibility is that factor Y could be transferred to the leading-strand DNA template of the replication fork and in some way serve to couple leading- and lagging-strand DNA synthesis [59]. Or, perhaps the primosome itself is capable of moving in either the 5'→3' or 3'→5' direction, depending on which primosomal helicase is used. Alternatively, it is possible that the helicase activity of factor Y may be the result of an uncoupled reaction. The DNA translocation properties of factor Y may be what are required at the replication fork, not its displacement properties.

A particularly useful substrate for the study of replication fork movement in vitro has been a "tailed form II" DNA molecule [97,102]. This molecule consists of a nicked circular molecule with a heterologous 5' tail (Fig. 2.2). This DNA substrate presumably mimics a replication fork. The DNA polymerase binds to the 3'-OH terminus present within the molecule, and the replication helicase is loaded onto the template via the 5' tail. Mok and Marians [97] have used this substrate to study replication fork movement in vitro with purified replication proteins. The 5' heterologous tail of the substrate contained a φX174 primosome assembly site; therefore, both the φX174 preprimosome and the DNA B helicase alone could be examined as replication fork helicases. When DNA synthesis

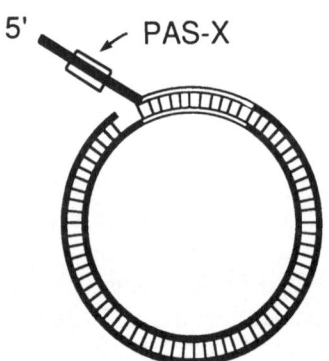

Figure 2.2. Schematic diagram of a tailed form II DNA molecule. Replication fork movement can be examined using a tailed form II DNA substrate. DNA polymerase can bind the 3'-OH terminus present at the gap. The heterologous 5' tail facilitates the binding of a helicase that translocates 5'→3' along the tail and displaces the duplex ahead of the DNA polymerase. The molecule shown above contains PAS-X, the φX174 primosome assembly site. When unwinding is coupled with DNA polymerization, rolling-circle DNA synthesis results in the formation of form II molecules with long single-stranded (when no primase is added) or double-stranded (when DNA primase is present) tails.

was coupled with helicase action, long single-stranded tails were generated by rolling-circle DNA synthesis. The single-stranded tails generated were at least 150,000 nt long. DNA replication was resistant to challenge by excess poly(dt) (which should compete for the preprimosomal proteins and DNA polymerase) and antibody to the β subunit of the DNA Pol III HE (anti β-IgG has been shown to effectively block reinitiation by the DNA Pol III HE), indicating that both the helicase and the DNA polymerase were extremely processive.

When the DNA B helicase was compared to the preprimosome, no difference was observed; i.e., both were capable of supporting processive replication fork movement. Interestingly, when the preprimosome was used as the replication fork helicase, eightfold more DNA was synthesized in the reaction than when the DNA B protein was used as a helicase. This result suggested that either the DNA B helicase translocated at a slower rate than the preprimosome or that it was not recruited to the fork as effectively as the preprimosome. To test the first possibility, the rates of replication fork movement with the preprimosomal helicase and DNA B helicase were compared. The preprimosome and the DNA B protein were both able to support replication fork movement at rates up to 730 nt/sec at 30°C, suggesting that the difference in DNA synthesis promoted by the two helicases was because of the efficiency of helicase loading onto the template. Furthermore, SSB was not required to sustain the rapid rate of DNA synthesis when the DNA B protein was used as the replication fork helicase (SSB was required in the case of the preprimosome, because its formation requires that the PAS be coated with SSB). This suggests that the polymerase and helicase are tightly coupled during the reaction, since polymerization by the DNA Pol III HE is severely inhibited by secondary structure in single-stranded DNA in the absence of SSB [103]. The *E. coli* SSB may be required on the lagging-strand DNA template, where substantial regions of single-stranded DNA may be present.

Replication Fork Movement: Lagging-Strand DNA Synthesis

Lagging-strand DNA synthesis requires properties that are quite different from those of leading-strand DNA synthesis. As discussed previously, leading-strand DNA synthesis can proceed continuously and thus requires a highly processive DNA polymerase. Lagging-strand DNA synthesis, however, occurs discontinuously via the synthesis of short 2,000-nt-long Okazaki fragments [1]; therefore, lagging-strand DNA synthesis is comparatively nonprocessive. How can the DNA Pol III HE be processive on the leading strand and nonprocessive on the lagging strand? Considerable insight into this apparent discrepancy has been ascertained through the study of the effects of the auxiliary holoenzyme subunits on the DNA polymerase III core enzyme. Not only do the auxiliary subunits appear to influence the processivity of the holoenzyme, but they also appear to influence primer recognition by DNA polymerase III. The β, γ, and τ subunits have been purified separately [25–27,104,105]. The γ and δ subunits can be isolated as a complex (the $\gamma \cdot \delta$ complex) containing at least two γ subunits, one (or two) δ subunits, and possibly one each of two other polypeptides, χ and ψ [18,33]. It has been shown that the $\gamma \cdot \delta$ complex and β subunit are capable of forming a preinitiation complex with the 3'-OH terminus of a primer [33,106].

The β subunit alone has a low affinity for a primer terminus [108], so it appears that the $\gamma \cdot \delta$ complex serves to "load" or recruit the β subunit onto the 3'-OH terminus of the primer [33,106,107]. Once formed, this preinitiation complex dramatically stimulates the nucleation of the holoenzyme onto the 3'-OH terminus of a primer and greatly stimulates the recycling time between subsequent initiation events [106].

The advent of recombinant DNA technology and better purification techniques have resulted in the cloning of many of the DNA Pol III HE subunits and the isolation of many of the subassemblies and complexes required for DNA Pol III HE activity. The availability of purified individual subunits and complex assemblies has allowed the properties of the reconstituted holoenzyme to be evaluated.

Maki and Kornberg [24] have demonstrated that the properties of the DNA Pol III HE can vary depending on the subunit composition of the reconstituted enzyme. When the reconstituted enzyme contained core enzyme, β, and the $\gamma \cdot \delta$ complex, rapid and efficient DNA synthesis occurred on a primed circular template; however, this assembly of the polymerase appeared to be hindered by secondary structure present within the template. In contrast, a reconstituted holoenzyme assembly containing this core enzyme, β, and τ was unencumbered by secondary structure or even the presence of additional primers on the template. The ability to reconstitute holoenzymes with different properties from different subunits led Maki and Kornberg [24] to propose that the holoenzyme was composed of two separate entities. One polymerase species, consisting of core enzyme, β, and τ, may be responsible for processive leading-strand DNA synthesis. The other species, consisting of the core enzyme, β, and the $\gamma \cdot \delta$ complex, may be responsible for less processive lagging-strand DNA synthesis. The concept for an asymmetric DNA Pol III HE was originally proposed by Johanson and McHenry [109] to explain the effect of ATPγS on initiation complex formation with the DNA Pol III HE. In the presence of ATPγS, McHenry found that only half as many holoenzyme molecules were capable of forming an initiation complex with a primed DNA template as were capable of forming such a complex in the presence of ATP. Maki and Kornberg [24], however, did not analyze the effect of this inhibitor on the formation of initiation complexes found in the presence of their various reconstituted DNA polymerases.

The Coordination of Leading- and Lagging-Strand DNA Synthesis: Coupling

The coordination of leading- and lagging-strand DNA synthesis in the cell may be extremely important. The low abundance of DNA polymerase in the cell, the need for DNA polymerase to initiate multiple rounds of DNA synthesis on the lagging-strand DNA template, and the high rate of fork movement in vivo have led to the proposal that leading-strand and lagging-strand DNA synthesis may be coupled.

The study of bacteriophage T$_4$ DNA replication has shed considerable light on the coordination of leading- and lagging-strand DNA synthesis. During the replication of T$_4$ DNA, the size distribution of the lagging-strand Okazaki fragments remains relatively constant (centered about 1 kb) [4]. Okazaki fragments are initiated by RNA primers synthesized by the T$_4$ gene 41/61 helicase/primase complex [45,48]. This complex recognizes a specific sequence on the template strand and synthesizes a specific RNA primer (pppApCpNpNpN) that serves to initiate DNA synthesis. Interestingly, an analysis of T$_4$ DNA reveals that potential priming sites occur every 50 to 60 nt along the T$_4$ DNA template [6]. Since Okazaki fragments are 1 to 2 kb in length, there must be a mechanism for selecting primer sites for the initiation of synthesis. It is highly unlikely that the primers are synthesized every 50 to 60 nt along the DNA, and only some are utilized, since it has been shown that the majority of RNA primers that are manufactured are in fact used to initiate DNA synthesis [110].

Another interesting observation is that the size distribution of Okazaki fragments is unaffected by dilution of the T$_4$ DNA polymerase [4]. This suggested that the lagging-strand DNA polymerase was tightly coupled to the replication fork and did not recycle on and off the lagging-strand DNA template. Alberts and co-workers [4] reasoned that if the leading- and lagging-strand DNA polymerases were coupled to one another by protein-protein interactions and thus always remained associated with the replication fork, the size of the lagging-strand DNA products (in the presence of excess auxiliary proteins) should be unaffected by the dilution of the DNA polymerase. If, on the other hand, the lagging-strand DNA polymerase dissociated from the template, the lagging-strand DNA products should become larger. This follows if the selection of primer sites is intimately connected to the actions of the lagging-strand DNA polymerase, i.e., if primer synthesis occured only when the lagging-strand DNA polymerase signaled that it was ready to bind a new 3'-OH terminus. This coordination would seem essential since the primers are too small to remain bound to the DNA without some form of stabilization. Thus, with the concentration of the lagging-strand DNA polymerase very low, if it dissociated from the fork, the distance between the 5' end of the last Okazaki fragment and the 3' end of the newest primer would get very large. This is because the leading-strand side of the fork should continue forward, unwinding the parental duplex. These observations have led Alberts to propose the "trombone" model (Fig. 2.3).

In contrast to the simple model of the replication fork illustrated in Figure 2.1, the lagging-strand DNA polymerase remains at the replication fork during DNA replication. Rather than the polymerase synthesizing DNA away from the fork, the DNA polymerase remains associated or coupled to the replication fork. The lagging-strand DNA template is looped out and through the lagging-strand DNA polymerase, forming the "slide" of the trombone. In this way, the active sites of both DNA polymerases face in the same direction. As the lagging-strand DNA

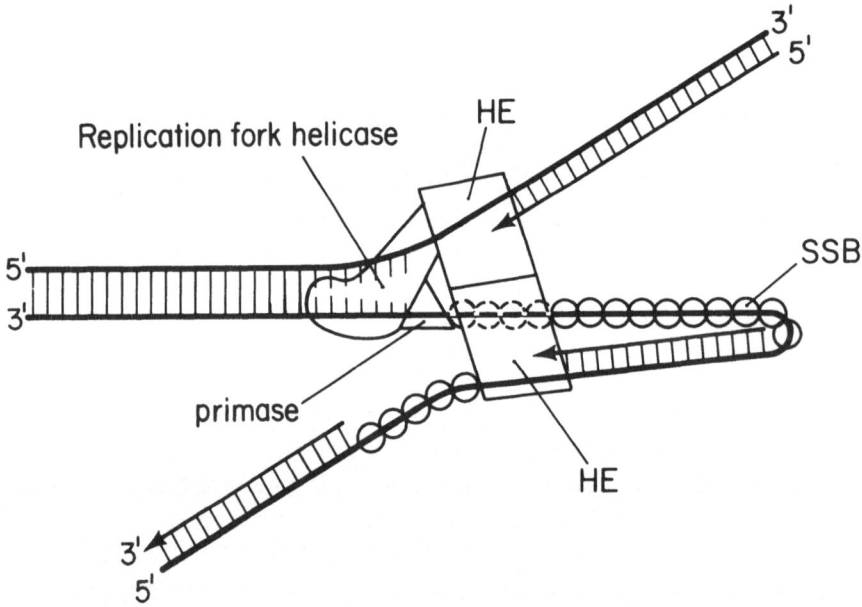

Figure 2.3. The trombone model of the replication fork. The trombone model of Alberts and co-workers [4] proposed that DNA was synthesized by a dimeric holoenzyme. The lagging strand is looped out and through the lagging-strand DNA polymerase, creating a "slide" containing both single- and double-stranded DNA. The helicase/primase complex unwinds the parental duplex and synthesizes the oligonucleotide primers that initiate Okazaki fragment synthesis. This drawing is not to scale. HE, DNA polymerase holoenzyme; SSB, single-stranded DNA-binding protein.

polymerase synthesizes an Okazaki fragment, the loop grows larger until the polymerase meets the 5′ terminus of the previously synthesized Okazaki fragment. At this point, the loop is released, a new primer is synthesized, and the loop is reset at the new primer.

This model makes a number of predictions. One prediction is the presence of a loop. This loop should consist of the lagging-strand DNA template and contain both double- and single-stranded regions in equal proportion. The trombone model also proposes that a new RNA primer is synthesized as the previous Okazaki fragment is terminated. A mechanism such as this would lead to a very homogeneous species of lagging-strand products, since the length of a nascent fragment would be dictated by the previously synthesized Okazaki fragment. This hypothesis has been tested experimentally and appears to require some modification. If the size of a newly synthesized Okazaki fragment is dictated by the previously synthesized Okazaki fragment, the distribution of Okazaki fragment sizes should be unaffected by alterations to the system. However, when the concentration of deoxynucleoside triphosphates was decreased during a T_4 DNA replication reaction, the distribution of Okazaki fragments was observed to shift to smaller average sizes. In addition, altering the concentration of ribonucleoside triphosphates also changed the size distribution in Okazaki fragments [6]. This laboratory has made similar observations with replication forks assembled from the DNA Pol III HE and the primosomal proteins [Mok and Marians, unpublished data].

This observation has led Alberts and co-workers to propose that a "timing" mechanism may govern the size distribution of Okazaki fragments. This is based on the reasoning that if the only effect of lowering the dNTP concentration is to slow the speed of polymerization, then a polymerase that remains bound to the primer terminus for a fixed interval and is unaffected by this reaction perturbation will consequently synthesize smaller products. On the other hand, if Okazaki fragment size were set only by the size of the penultimate fragment, then altering the velocity of movement of the replication fork should not change the size of the Okazaki fragments produced. Of course, one cannot easily eliminate the possibility that the observed effect is a consequence of a change in concentration of a conformational effector (i.e., a dNTP) acting on a protein involved in the Okazaki fragment synthetic cycle and not a result of altering the velocity of replication fork movement.

How does one picture such a clock? In a simple sense, the k_{off} of a protein bound to DNA serves as a clock measuring the time the protein is, in fact, on the DNA. In the protein machine at the replication fork, an effective clock might be the half-life of a particular conformational change of a key component in the machinery that recycles the lagging-strand DNA polymerase from primer to primer.

If the size of Okazaki fragments is governed by an internal clock, what components of the replication fork constitute the clock? The trombone model proposed that an RNA primer should be synthesized as the lagging-strand loop is reset. Likely candidates for the internal clock, therefore, are the components of the fork involved in the formation of the loop. In the T_4 DNA replication system, these could be the DNA polymerase accessory proteins.

Although potential initiation sites are spaced roughly 50 to 60 nt apart, the T_4 primosome does not necessarily have to synthesize an RNA primer at each site. It would be useful to vary the concentration of the gene 41/61 primase/helicase in the T_4 replication reaction in vitro in order to determine the minimum possible size of an Okazaki fragment (under conditions of both limiting and saturating concentrations of ribonucleoside triphosphates). The size of the Okazaki fragments may be an intrinsic property of the mechanism of the primase itself.

In addition, when one considers the maintenance of Okazaki fragment size, the processivity of each of the replication components must be taken into consideration. It has been shown that leading-strand DNA synthesis in vitro in both bacteriophage T_7 [40] and E. coli [97] replication systems proceeds by a highly processive mechanism. When coupled to a helicase, both the bacteriophage T_7 and E. coli DNA polymerase holoenzymes are capable of synthesizing contiguous stretches of DNA in excess of 40 kb per binding event [40,97]. However, it has been shown that lagging-strand DNA synthesis during the replication of DNA in vitro by T_7 replication proteins is distributive [40]. The distributive nature of lagging-strand DNA synthesis in the T_7 system appears to reflect the properties of the DNA primase. When competitor single-stranded DNA was added to a T_7 DNA replication reaction, lagging-strand DNA synthesis was preferentially inhibited, and primer synthesis occurred predominantly on the competitor DNA. This result implied that the leading-strand DNA polymerase and helicase were processive in nature, whereas the DNA primase (and possibly the lagging-strand DNA polymerase) was distributive in nature.

If the primase acts distributively in the T_4 system, i.e., dissociates from the fork after each priming event, and if one makes the assumption that the primase utilizes the next available priming site when it does associate with the fork, then average Okazaki fragments size could be governed by the association constant between the primase and helicase. Therefore, the priming frequency on the lagging-strand DNA template could be affected by a number of factors, including primase concentration and ribonucleoside triphosphate concentration. The absolute amount of the primase/helicase complex present on replicating DNA molecules would be determined by the concentration of primase, whereas, once formed, the priming efficiency of this complex would be affected by the ribonucleoside triphosphate concentration.

We have recently determined that the DNA G primase also appears to act in a distributive manner in replication forks formed with the DNA Pol III HE and the primosomal proteins [Mok and Marians, unpublished data]. If reaction mixtures containing such active replication forks were diluted shortly after the initiation of DNA synthesis, pulse labeling revealed that the mean of the size distribution of the Okazaki fragments was immediately increased. If the concentration of primase was maintained in the dilution media at the original value, the size of the Okazaki fragments remained relatively constant.

Caution should be used when the term "distributive mode of action" is applied to a replication fork. Clearly the primase must be dissociated from the 3' end of the newly synthesized primer. It is conceivable that the primase also dissociates from the lagging-strand DNA template during this event. However, one can envision that even if this were to occur, the primase could remain associated with the replication fork by protein-protein interactions. These interactions might be overcome by challenge with excess single-strand DNA, giving the appearance that the enzyme did not remain continuously associated with the replication fork.

If the DNA primase itself is the internal clock in the T_4 DNA replication system, then there may be some type of mechanism to stabilize the primer on the template. This would be necessary, since the primase may synthesize a primer before the termination of the previous Okazaki fragment and resetting of the lagging-strand loop.

It has been proposed that the T_4 DNA polymerase accessory proteins (the T_4 gene 44/62 and gene 45 proteins) act as a sliding clamp and confer processivity on the T_4 DNA polymerase [15]. These proteins also appear to have a single-stranded DNA terminus-dependent ATPase activity [111]. It is possible that these proteins may be functionally equivalent to the $\gamma \cdot \delta$ complex and β subunit of the DNA Pol II HE in the *E. coli* DNA replication system. Perhaps these proteins interact with the primer, stabilize it, and promote the interaction of primer with the T_4 DNA polymerase. The DNA terminus-dependent ATPase of the gene 44/62 and gene 45 DNA polymerase accessory proteins was not significantly stimulated by the presence of short DNA primers; however, RNA primers, and specifically primers synthesized by the T_4 primase, were not examined.

The concept of a coupled replication fork has also been exposed for *E. coli* replication fork movement [1]. Several properties of the *E. coli* replication proteins make the trombone model for replication fork coupling attractive in *E. coli* replication fork propagation.

The existence of two distinct species of DNA Pol III HE proposed by Johanson and McHenry [109] and Maki and Kornberg [19] fits nicely into a model in which one species of the DNA Pol III HE would function in leading-strand DNA synthesis and the other in lagging-strand DNA synthesis. In addition, there is evidence that DNA polymerase III may function as a dimer. DNA polymerase III', a subassembly of the holoenzyme containing the DNA polymerase III core enzyme and τ subunit, can exist as a dimer in solution. This has led McHenry to propose that the τ subunit may act as the glue between two holoenzyme monomers [23]. Recently, Maki and Kornberg [19] have been able to isolate dimeric forms of DNA polymerase III* and the DNA polymerase III core enzyme. The dimerization of the core enzyme presumably occurs through the θ subunit, since reconstituted core containing only the α and ϵ subunits failed to dimerize. It is difficult to assess the significance of this finding, however, because dimerization of the core required very high protein concentrations. Mass action effects, occurring at these high protein concentrations, may result in dimerization of the polymerase, an event that normally would be catalyzed by the τ subunit.

A major caveat in the experimental dissection of *E. coli* replication fork movement lies in the complex nature of the protein-protein and protein-DNA interactions that occur at the fork. The replication fork, as an entity, consists of a helicase/primase coupled with a DNA polymerase(s). In addition, a requirement for structural proteins, hitherto undetected, that are necessary for proper replication fork function cannot be excluded.

To establish the mechanism of replication fork movement, one must specifically perturb one of these activities and observe the subsequent effect; however, this specificity is not easy to obtain. For example, one way to slow replication fork movement would be to lower the concentration of ribonucleoside triphosphates in the reaction. Since the primosomal helicase is fueled by nucleoside triphosphate hydrolysis, lowering their concentration should slow the unwinding of the parental duplex. Unfortunately, this will also affect the DNA Pol III HE and DNA G primase. The holoenzyme requires ATP for the efficient recognition of primer termini. In addition, since the DNA G protein utilizes ribonucleoside triphosphates for primer synthesis [52], altering the concentration may affect primer site selection. Slowing replication fork movement by altering the deoxynucleoside triphosphate concentration is equally nonspecific in its effect, since both the DNA G protein and primosomal helicases can also utilize deoxynucleoside triphosphates. One key aspect of a coupled replication fork that does seem demonstrable is that the sites of primase action and lagging-strand DNA polymerase action are both, in fact, at the replication fork and not nonspecifically distributed distally along the nascent leading strand.

Summary and Perspectives

The coupling of leading- and lagging-strand DNA synthesis has been proposed to explain how a protein of low abundance can rapidly and efficiently synthesize the entire 4.5×10^6 bp *E. coli* genome. However, it is unclear whether DNA synthesis occurring at the replication fork need be coupled.

An essential feature of the current model of replication fork coupling is the presence of a dimeric form of the DNA Pol III HE. Although the holoenzyme and various subassemblies of the enzyme appear to be able to dimerize, these forms of the polymerase appear to be very unstable. The purification of a dimeric form of the holoenzyme requires that both Mg^{2+} and ATP be present in the column buffers and that the protein concentration be kept very high throughout purification [19]. In addition, the purification of intact holoenzyme requires the use of relatively weak ion exchange resins in order to maintain its integrity.

The concentration of the holoenzyme in the cell has been estimated at 40 nM. At this low protein concentration, none of the purified subassemblies of DNA polymerase III, or the holoenzyme itself, exists as a dimer in vitro. In fact, at this protein concentration, the holoenzyme breaks down into multiple subassemblies [19].

It has been estimated that at any particular time during replication, there may be eight or more RNA primers associated with replication forks (in addition to the original bidirectional replication fork, the two daughter molecules initiate bidirectional DNA synthesis before cell division occurs). Therefore, the intracellular concentration of primers approaches that of the holoenzyme itself. The likelihood of one holoenzyme molecule locating an RNA primer is just as probable as the enzyme locating another holoenzyme molecule. The location of a primer, in fact, may be more probable, since it has been shown that the $\gamma \cdot \delta$ complex and β subunit of the holoenzyme can actively recruit the holoenzyme to a primer terminus [33,106,107]. Of course compartmentalization cannot be ruled out, and it is possible that the actual concentration of some components, in the presence of the replication fork, is very high. In addition, all studies on the action of the DNA Pol III HE to date have focused on the synthesis of a single strand of DNA. It is certainly possible that studies in the future that examine the simultaneous synthesis of both the leading and lagging strands may reveal the requirement for an additional subunit necessary to maintain the integrity of a dimeric DNA Pol III HE at the replication fork.

The existence of the coupling of leading- and lagging-strand DNA synthesis can be supported by physical biochemical studies. For example, it is essential to determine a dimerization constant for the DNA Pol III HE and an association constant for the holoenzyme with a preinitiation complex (the primer terminus in association with the $\gamma \cdot \delta$ complex and β subunit of holoenzyme). An understanding of these parameters will help determine whether the coupling of leading- and lagging-strand DNA synthesis is necessary for efficient DNA synthesis. If the association constant for DNA Pol HE with the preinitiation complex is significantly higher than the dimerization constant for the holoenzyme, an argument may be made that a dimeric polymerase need not be necessary for DNA synthesis. In addition, if the association constant between the DNA Pol III HE and the preinitiation complex is extremely high, one may not have to evoke the concept of coupling for efficient DNA synthesis.

Although the concept of a dimeric holoenzyme is crucial to the coupling of leading- and lagging-strand DNA synthesis, the replication of both strands may not require the presence of two different species of holoenzyme. The asymmetric nature of DNA synthesis at the fork may be dictated by the fork itself. Perhaps the signal for the holoenzyme to recycle to the next primer terminus occurs when the enzyme encounters a long stretch of double-stranded DNA. Leading-strand DNA synthesis could continue unimpeded, since the leading-strand DNA polymerase would only encounter single-stranded DNA. The lagging-strand DNA polymerase, however, would encounter approximately 2,000 nucleotides of single-stranded DNA before encountering the previously synthesized Okazaki fragment. Once the DNA polymerase encounters the double-stranded DNA, a conformational change may occur, signaling the polymerase to dissociate from the template.

Little is known about the interaction between the DNA Pol III HE and double-stranded DNA. O'Donnell and Kornberg [112] have shown that the DNA Pol III HE is capable of diffusing along double-stranded regions of DNA (in the elongation direction) as long as 2 kb in search of a primer terminus. The holoenzyme was capable of efficiently diffusing over short double-stranded regions of DNA; however, the efficiency of transfer across a 2-kb stretch of DNA was significantly decreased. This may reflect a lower affinity for double-stranded DNA by the DNA Pol III HE.

The study of prokaryotic DNA replication offers an unparalleled opportunity to study both protein-protein and protein-DNA interactions. The replication of the *E. coli* chromosome may require the coordination and activities of as many as 20 different proteins. Unlike eukaryotic systems, the majority of these proteins have been purified to near homogeneity and have been well characterized biochemically. Thus, the study of prokaryotic DNA replication is entering a new era where more sophisticated problems may be addressed. How does a replication fork form? Is the replication of leading- and lagging-strand DNA coupled or uncoupled? How is a replication fork terminated? What roles do topo-isomerases have in the replication of DNA?

The amenability of *E. coli* to combined biochemical and genetic manipulations offers a distinct advantage over eukaryotic systems. In addition, the insight gained from the study of *E. coli* DNA replication appears to be applicable to eukaryotic systems. Except for minor differences, analogous DNA replication activities appear to be present in eukaryotes (a notable exception may be DNA gyrase). A greater understanding of the mechanisms of DNA replication in *E. coli* will invariably give more insight into the nature of eukaryotic DNA replication and perhaps into the nature of protein-protein and protein-DNA interactions in general.

Acknowledgments. Studies from the authors' laboratory were supported by grants from the NIH (GM34557), ACS (FRA261), and the Irma T. Hirschl Charitable Trust. R.J.D. was supported by a National Institutes of Health postdoctoral fellowship (F32GM11381).

References

1. Kornberg, A. (1980) DNA Replication. W.H. Freeman, San Francisco.
2. Kornberg, A. (1982) Supplement to DNA Replication. W.H. Freeman, San Francisco.
3. Alberts, B.M., Barry, J., Bedinger, P., Burke, R.L., Hibner, U., Liu, C.-C., and Sheridan, R. (1980) In ICN-UCLA Symposia on Molecular Biology. Academic Press, New York.
4. Alberts, B., Barry, J., Bedinger, P., Formosa, C., Jongeneel, C., and Kreuzer, K. (1983) Cold Spring Harbor Symp. Quant. Biol. *47*, 655–688.
5. Alberts, B.M. (1985) Cold Spring Harbor Symp. Quant. Biol. *49*, 1–12.
6. Selick, H.E., Barry, J., Cha, T.-A., Munn, M., Nakanishi, M., Wong, M.L., and Alberts, B.M. (1987) In DNA Replication and Recombination. McMacken, R., and Kelly, T.J., eds. Alan R. Liss, New York.
7. Richardson, C.C. (1983) Cell *33*, 315–318.
8. Richardson, C.C., Beauchamp, B.B., Huber, H.E., Ikeda, R.A., Myers, J.A., Nakai, H., Rabkin, S.D., Tabor, S., and White, J. (1987) In DNA Replication and Recombination. McMacken, R., and Kelly, T.J., eds. Alan R. Liss, New York.
9. Wu, Y.H., Franden, M.A., Hawker, J.R., and McHenry, C.S. (1984) J. Biol. Chem. *259*, 12117–12122.
10. Modrich, P., and Richardson, C.C. (1975) J. Biol. Chem. *250*, 5508–5514.
11. Goulian, M., Lucas, Z.J., and Kornberg, A. (1968) J. Biol. Chem. *243*, 627–638.
12. Hori, K., Mark, R.F., and Richardson, C.C. (1979) J. Biol. Chem. *254*, 11591–11597.
13. Adler, S., and Modrich, P. (1979) J. Biol. Chem. *254*, 11605–11614.
14. Tabor, S., Huber, H.E., and Richardson, C.C. (1987) J. Biol. Chem. *262*, 16212–16223.
15. Huang, C.-C., Hearst, J., and Alberts, B. (1981) J. Biol. Chem. *256*, 4087–4094.
16. Gefter, M., Hirota, Y., Kornberg, T., Wechsler, J., and Barnoux, C. (1971) Proc. Natl. Acad. Sci. USA *68*, 3150–3153.
17. McHenry, C.S., and Crow, W. (1979) J. Biol. Chem. *254*, 1748–1753.
18. McHenry, C.S., and Kornberg, A. (1977) J. Biol. Chem. *252*, 6478–6484.
19. Maki, S., and Kornberg, A. (1988) J. Biol. Chem. *263*, 6570–6578.
20. Maki, H., and Kornberg, A. (1985) J. Biol. Chem. *260*, 12987–12992.
21. Schevermann, R.H., and Echols, H. (1984) Proc. Natl. Acad. Sci. USA *81*, 7747–7751.

22. Maki, H., and Kornberg, A. (1987) Proc. Natl. Acad. Sci. USA *84*, 4389–4392.
23. McHenry, C.S. (1982) J. Biol. Chem. *257*, 2657–2663.
24. Maki, S., and Kornberg, A. (1988) J. Biol. Chem. *263*, 6561–6569.
25. Burgers, P., Kornberg, A., and Sakakibara, Y. (1981) Proc. Natl. Acad. Sci. USA *78*, 5391–5395.
26. Hurwitz, J., and Wickner, S. (1974) Proc. Natl. Acad. Sci. USA *71*, 6–10.
27. Maki, S., and Kornberg, A. (1988) J. Biol. Chem. *263*, 6547–6554.
28. Kodiara, M., Biswas, S.B., and Kornberg, A. (1983) Mol. Gen. Genet. *192*, 80–86.
29. Mullin, D.A., Woldringh, C.L., Henson, J.M., and Walker, J.R. (1983) Mol. Gen. Genet. *192*, 73–79.
30. Flower, A., and McHenry, C.S. (1986) Nucleic Acids Res. *14*, 8091–8101.
31. Yin, K.C., Blinkowa, A., and Walker, J.R. (1986) Nucleic Acids Res. *14*, 6541–6549.
32. Johanson, K.O., and McHenry, C.S. (1980) J. Biol. Chem. *255*, 10984–10990.
33. Maki, S., and Kornberg, A. (1988) J. Biol. Chem. *263*, 6555–6560.
34. Scherzinger, E., Lanka, E., Molelli, G., Seifert, D., and Yuki, A. (1977) Eur. J. Biochem. *72*, 543–558.
35. Matson, S., Tabor, S., and Richardson, C.C. (1983) J. Biol. Chem. *258*, 14017–14024.
36. Kilodner, R., Masamune, Y., LeClerc, J., and Richardson, C.C. (1978) J. Biol. Chem. *253*, 566–573.
37. Fischer, H., and Hinkle, D. (1980) J. Biol. Chem. *255*, 7956–7964.
38. Dunn, J.J., and Studier, F.W. (1983) J. Mol. Biol. *166*, 477–535.
39. Bernstein, J., and Richardson, C.C. (1988) Proc. Natl. Acad. Sci. USA *85*, 396–400.
40. Nakai, H., and Richardson, C.C. (1988) J. Biol. Chem. *263*, 9818–9830.
41. Venkatesan, M., Silver, L.L., and Nossal, N.C. (1982) J. Biol. Chem. *257*, 12426–12434.
42. Hinton, D.M., Silver, L.L., and Nossal, N.C. (1985) J. Biol. Chem. *260*, 12851–12857.
43. Morris, C., Moran, L., and Alberts, B. (1979) J. Biol. Chem. *254*, 6797–6802.
44. Liu, C.-C., and Alberts, B.M. (1981) J. Biol. Chem. *256*, 2813–2820.
45. Nossal, N.C. (1980) J. Biol. Chem. *255*, 2176–2182.
46. Silver, L.L., and Nossal, N.C. (1982) J. Biol. Chem. *257*, 11696–11703.
47. Hinton, D., and Nossal, N.C. (1985) J. Biol. Chem. *260*, 11696–11705.
48. Liu, C.-C., and Alberts, B.M. (1981) J. Biol. Chem. *256*, 2821–2829.
49. Cha, T.-A., and Alberts, B.M. (1986) J. Biol. Chem. *261*, 7001–7010.
50. Kaguni, J.M., and Kornberg, A. (1984) Cell *38*, 183–190.
51. LeBowitz, J.H., and McMacken, R. (1986) J. Biol. Chem. *261*, 4738–4748.
52. Rowan, L., and Kornberg, A. (1978) J. Biol. Chem. *253*, 758–764.
53. Wickner, S., and Hurwitz, J. (1975) Proc. Natl. Acad. Sci. USA *72*, 3342–3346.
54. Shlomai, J., and Kornberg, A. (1980) Proc. Natl. Acad. Sci. USA *77*, 799–803.
55. Shlomai, J., and Kornberg, A. (1980) Proc. Natl. Acad. Sci. USA *77*, 6521–6525.

56. Wickner, S., and Hurwitz, J. (1974) Proc. Natl. Acad. Sci. USA *71*, 4120–4124.
57. Shlomai, J., Polder, L., Arai, K.-I., and Kornberg, A. (1981) J. Biol. Chem. *256*, 5233–5238.
58. Arai, K.-I., and Kornberg, A. (1981) Proc. Natl. Acad. Sci. USA *78*, 69–73.
59. Lee, M.S., and Marians, K.J. (1987) Proc. Natl. Acad. Sci. USA *84*, 8345–8349.
60. Lasken, R., and Kornberg, A. (1988) J. Biol. Chem. *263*, 5512–5518.
61. Meyer, R.R., Glassberg, J., and Kornberg, A. (1979) Proc. Natl. Acad. Sci. USA *76*, 1702–1705.
62. Alberts, B.M., and Frey, L. (1970) Nature *227*, 1313–1318.
63. Reuben, R.C., and Gefter, M.L. (1974) J. Biol. Chem. *249*, 3843–3850.
64. Molineux, I.J., Friedman, S., and Gefter, M.L. (1974) J. Biol. Chem. *249*, 6090–6098.
65. Fuller, C.W., and Richardson, C.C. (1985) J. Biol. Chem. *260*, 3197–3206.
66. Nakai, H., and Richardson, C.C. (1988) J. Biol. Chem. *263*, 9831–9839.
67. Kohara, Y., Akiyama, K., and Isono, K. (1987) Cell *50*, 495–508.
68. Orr, E., Fairweather, N.F., Holland, I.B., and Pritchard, R.H. (1979) Mol. Gen. Genet. *177*, 103–112.
69. Steck, T., and Drlica, K. (1984) Cell *36*, 1081–1088.
70. Kreuzer, K., and Cozzarelli, N.R. (1979) J. Bacteriol. *140*, 424–435.
71. Drlica, K., and Snyder, M. (1978) J. Mol. Biol. *120*, 145–154.
72. Bliska, J.B., and Cozzarelli, N.R. (1987) J. Mol. Biol. *194*, 205–218.
73. Sugino, A., Peebles, C.L., Kreuzer, K.N., and Cozzarelli, N.R. (1977) Proc. Natl. Acad. Sci. USA *74*, 4767–4771.
74. Mizuuchi, K., O'Dea, M.H., and Gellert, M. (1978) Proc. Natl. Acad. Sci. USA *75*, 5960–5963.
75. Higgins, N.P., Peebles, C.L., and Cozzarelli, N.R. (1979) Proc. Natl. Acad. Sci. USA *75*, 1773–1777.
76. Gellert, M., Mizuuchi, K., O'Dea, M., and Nash, H. (1976) Proc. Natl. Acad. Sci. USA *73*, 3872–3876.
77. Kreuzer, K., and Cozzarelli, N.R. (1980) Cell *20*, 245–254.
78. Krasnow, M.A., and Cozzarelli, N.R. (1982) J. Biol. Chem. *257*, 2687–2693.
79. Mizuuchi, K., Fisher, L.M., O'Dea, M.H., and Gellert, M. (1980) Proc. Natl. Acad. Sci. USA *77*, 1847–1851.
80. Marians, K.J. (1987) J.Biol. Chem. *262*, 10361–10368.
81. Filutowicz, M., and Jonczyk, P. (1983) Mol. Gen. Genet. *191*, 282–287.
82. Minden, J.S., and Marians, K.J. (1985) J. Biol. Chem. *260*, 9316–9325.
83. DiGate, R.J., and Marians, K.J. (1988) J. Biol. Chem. *263*, 13366–13373.
84. Itoh, T., and Tomizawa, J. (1980) Proc. Natl. Acad. Sci. USA *77*, 2450–2459.
85. Itoh, T., and Tomizawa, J. (1982) Nucleic Acids Res. *10*, 5949–5965.
86. Zipursky, S.L., and Marians, K.J. (1980) Proc. Natl. Acad. Sci. USA *77*, 6521–6525.
87. Chakraborty, T., Yoshinaga, K., Lother, H., and Messer, W. (1982) EMBO J. *1*, 1545–1549.
88. Fuller, R.S., Funnell, B.E., and Kornberg, A. (1984) Cell *38*, 889–900.
89. Funnell, B.E., Baker, T.A., and Kornberg, A. (1986) J. Biol. Chem. *261*, 5616–5624.

90. Funnell, B.E., Baker, T.A., and Kornberg, A. (1987) J. Biol. Chem. *262*, 10327–10334.
91. Van der Ende, A., Baker, T.A., Ogawa, T., and Kornberg, A. (1985) Proc. Natl. Acad. Sci. USA *82*, 3954–3958.
92. Lark, K.D. (1972) J. Mol. Biol. *64*, 47–60.
93. Hirose, S., Hiraya, S., and Okazaki, T. (1983) Mol. Gen. Genet. *189*, 422–431.
94. Kohara, Y., Tohdoh, N., Jiang, X., and Okazaki, T. (1985) Nucleic Acids Res. *13*, 6847–6866.
95. Junker, D.E., Rokeach, L.A., Ganea, D., Chiaramello, A., and Zyskind, J.W. (1986) Mol. Gen. Genet. *203*, 101–109.
96. Schauzu, M.-A., Kucherer, C., Kolling, R., Messer, W., and Lother, H. (1987) Nucleic Acids Res. *15*, 2479–2497.
97. Mok, M., and Marians, K.J. (1987) J. Biol. Chem. *262*, 16644–16654.
98. Seufert, W., and Messer, W. (1986) EMBO J. *5*, 3401–3406.
99. Fay, P.J., Johanson, K.O., McHenry, C.S., and Bambara, R.A. (1981) J. Biol. Chem. *256*, 976–983.
100. Fay, P.J., Johanson, K.O., McHenry, C.S., and Bambara, R.A. (1982) J. Biol. Chem. *257*, 5692–5699.
101. Mok, M., and Marians, K.J. (1987) J. Biol. Chem. *262*, 2304–2309.
102. Lechner, R.L., and Richardson, C.C. (1983) J. Biol. Chem. *258*, 11185–11196.
103. LaDuca, R.J., Fay, P.J., Chuang, C., McHenry, C.S., and Bambara, R.A. (1983) Biochemistry *22*, 5177–5188.
104. Wickner, W., Schekman, R., Geider, K., and Kornberg, A. (1973) Proc. Natl. Acad. Sci. USA *70*, 1764–1767.
105. Hurwitz, J., and Wickner, S. (1974) Proc. Natl. Acad. Sci. USA *71*, 6–10.
106. O'Donnell, M.E. (1987) J. Biol. Chem. *262*, 16558–16565.
107. Wickner, S. (1976) Proc. Natl. Acad. Sci. USA *73*, 3511–3515.
108. Lasken, R., and Kornberg, A. (1987) J. Biol. Chem. *262*, 1720–1724.
109. Johanson, K.O., and McHenry, C.S. (1984) J. Biol. Chem. *259*, 4589–4595.
110. Liu, C.-C., and Alberts, B.M. (1980) Proc. Natl. Acad. Sci. USA *77*, 5698–5702.
111. Piperno, J.R., and Alberts, B.M. (1978) J. Biol. Chem. *253*, 5174–5179.
112. O'Donnell, M.E., and Kornberg, A. (1985) J. Biol. Chem. *260*, 12875–12883.

Chapter 3
The Role of RecA Protein in Homologous Genetic Recombination

Michael M. Cox

Homologous genetic recombination is the process by which genetic information is rearranged between chromosomes that possess similar sequences. This type of recombination is responsible, in prokaryotes, for DNA rearrangements that occur during transformation or conjugation, postreplication repair of DNA, and a number of other processes [1]. Additional roles have been identified in eukaryotic cells [1,2]. About 30 homologous recombination events occur somewhere in the human genome during each meiosis, scattered more or less randomly about the chromosomes [3]. In bacteria, the probability of recombination occurring in a given length of DNA is increased about 500-fold relative to human meiotic chromosomes. A recombination event takes at least 35 minutes in bacteria [4] and perhaps 8 to 10 times longer in fungi [5].

A variety of models for homologous genetic recombination have been proposed, based primarily on genetic observations in fungi [2,5,6]. Although these models differ in important details, in most cases they share a set of fundamental steps. These are outlined in Figure 3.1, and include (1) cutting or creation of a gap in the DNA molecules to be recombined; (2) alignment or pairing of homologous sequences in two different DNA molecules; (3) the creation of a crossover or Holliday junction between the molecules in which DNA strands from each are paired, creating short regions of heteroduplex DNA; (4) extension of heteroduplex DNA (branch migration); and (5) resolution of the crossover junction to yield products [2,5].

This is an extremely complicated reaction that requires the action of several different proteins. Moreover, the reaction must occur in a cellular environment filled with structural, topological, and energetic barriers. Recombination proteins must compete for binding sites with a plethora of other cellular DNA-binding proteins. The helical nature of DNA places topological constraints on the reaction. Energy is required to confer order and a driving force in a reaction where products and substrates can be chemically indistinguishable or nearly so.

Proteins that carry out all of the proposed steps in recombination have

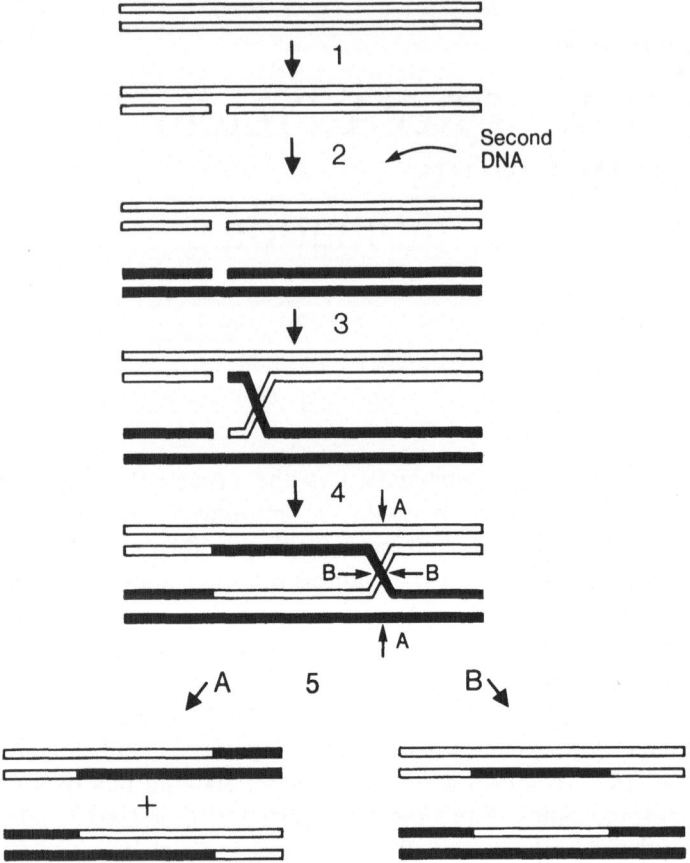

Figure 3.1. A model for homologous genetic recombination. The model is arranged to highlight the contribution of recA protein. Steps are (1) creation of a nick or gap in at least one strand of a duplex DNA; (2) alignment of two DNAs cut in this way; (3) creation of a crossover structure or Holliday junction; (4) branch migration; and (5) resolution of the Holliday junction. A ligation step sealing the original DNA nicks is included in step 4 but could occur at another point. Step 5 is a combination of cleavage by a specialized resolution nuclease and religation. The two sets of products are those generated by cleavage at the points labeled A and B, respectively. Additional steps, likely to occur, that require additional nucleases and DNA polymerase are not shown. [For variations, see references 3 and 6.]

been purified from many sources and studied in vitro [7]. These studies are quite advanced in the case of bacteria. This chapter will focus on the recA protein of *E. coli* as a model system for key events in homologous recombination.

The recA protein promotes the reactions depicted in steps 2 through 4 of Figure 3.1. These steps are the heart of a recombination event. Steps 1 and 5 and the depicted ligation reactions can be rationalized as the activities of specialized versions of familiar enzyme classes such as helicases, nucleases, and ligases. Several nucleases have been purified with the unusual and interesting substrate specificities required for these steps in recombination [7]. Steps 2 through 4 could similarly be rationalized as an unusual combination of helicase and strand-annealing activities. In this case, however, the rationalization falls far short of the biochemical mark. The recA protein appears to be a prototype of a completely distinct class of enzyme.

The activity of recA protein is not easily categorized. An in vitro reaction that mimics recombination steps 2 through 4 and that has been widely employed in recA studies is illustrated in Figure 3.2. The reaction involves an exchange of strands between linear duplex DNAs and homologous circular single strands. RecA protein will also promote strand exchanges between two duplex DNAs (a four-strand exchange as opposed to the three-strand exchange described here) if one of the duplexes is partially single stranded [8–10]. The substrates are generally derived from bacteriophages. The mechanism by which recA protein carries out this reaction is currently a focus of research in many laboratories. The three reaction phases illustrated in Figure 3.2 are easily distinguished experimentally. The system offers a number of biochemical problems that can be most easily described in a summary of these reaction phases.

The first phase is the formation of a stoichiometric, filamentous complex of recA protein on the ssDNA [11,12]. This complex contains approximately one recA monomer per three or four nucleotides of DNA, arranged in a right-handed helical array [13–17]. An example is illustrated in Figure 3.3. Formation of this complex is facilitated by a second protein, the single-stranded DNA-binding protein of *E. coli* (SSB). The role of SSB in this reaction is discussed at length elsewhere [7,18,19] and will not be considered here. During strand exchange, the SSB is transferred to the displaced linear "+" strand, as shown in Figure 3.2 (B.C. Schutte and M.M. Cox, unpublished).

In the second phase, the ssDNA in the complex is aligned with homologous sequences in the duplex. Several steps have been identified in this process [20,21]. Initial interactions with nonhomologous sequences are manifested by the formation of aggregates of complexes and duplex DNA molecules that may facilitate the search for homology [21]. Additional studies initiated by the Radding group have defined two types of homologous interaction. The first is a paranemic joint in which the ssDNA and duplex DNA are aligned homologously but no strand interwinding to produce heteroduplex DNA has occurred [20,22,23]. Recent evidence has demonstrated that paranemic joints can be thousands of base pairs long [24–26]. The structure of this paranemic joint is a central issue in understanding the overall reaction. Paranemic joints are formed fast enough to be considered intermediates in the pathway leading to the

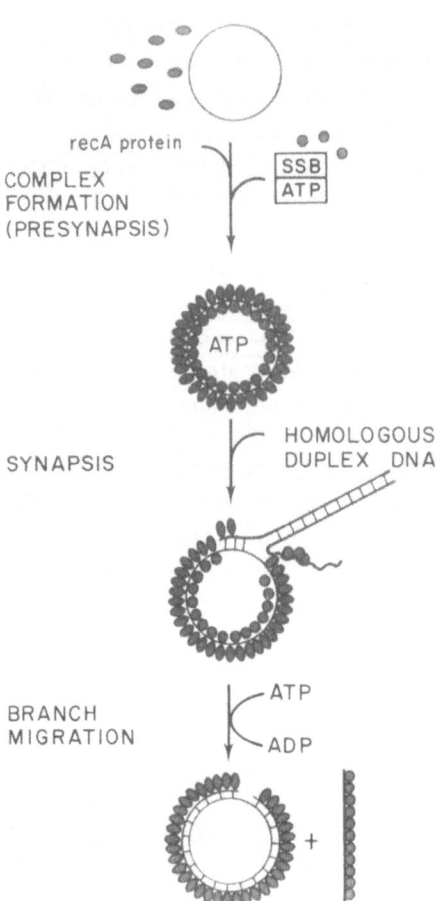

Figure 3.2. RecA protein-promoted DNA strand exchange. The three phases of the reaction are described in the text. RecA protein and SSB are denoted by ovals and circles, respectively.

second type of homologous interaction in which some interwinding of strands from the two DNA molecules occurs [20,22,23]. This species is a plectonemic joint [22]. The formation of plectonemic joints is limited only by topological considerations [27]. RecA protein is a DNA-dependent ATPase, and ATP is required for the pairing process. Paranemic joints, however, can be formed in the presence of ATP analogs that are not hydrolyzed by recA protein [23,28,29].

ATP hydrolysis is required only in the final phase illustrated in Figure 3.2. This is a branch migration reaction in which the linear "+" strand of the duplex is progressively replaced by the circular "+" strand until products are formed. This is probably the least well understood phase of the DNA strand exchange reaction. As pointed out elsewhere [7], a major contribution of recA protein to this phase of the reaction is the absolute polarity it exhibits. Branch migration proceeds only in the 5' → 3' direc-

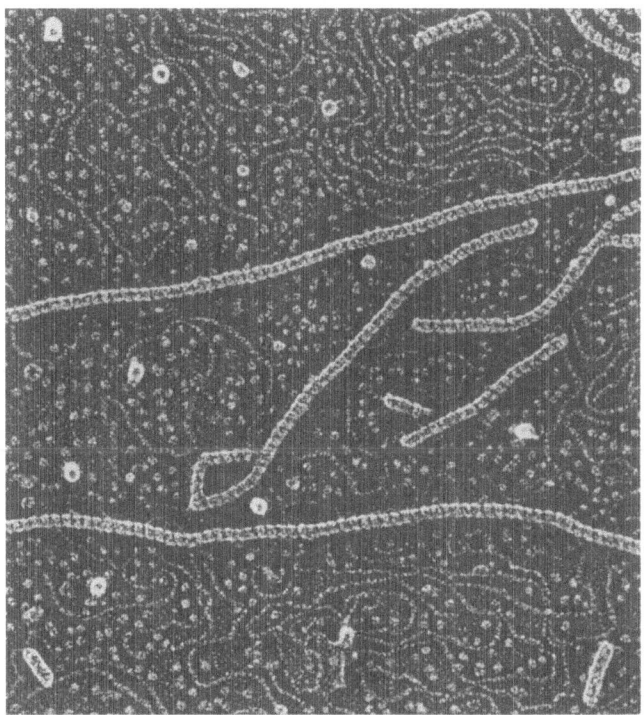

Figure 3.3. RecA protein helical filaments formed on duplex DNA in the presence of ATPγS. The filaments are right-handed helices. Unbound DNA is evident in the background. (Courtesy of J. Heuser, Washington University.)

tion relative to the incoming ssDNA within the complex [30–33]. The process is much slower than spontaneous branch migration (5 to 20 bp sec^{-1} vs. 6,000 bp sec^{-1}) [7]. By giving the process a direction, however, the efficiency of an extensive branch migration reaction is greatly increased relative to the faster, but random, uncatalyzed process. This raises a classical biochemical question concerning the coupling of chemical energy to a unidirectional process. The problem is analogous, in this respect, to muscle contraction and ATP-driven ion pumps. The problem is made more complicated by the fact that the active species is a long nucleoprotein filament.

A complete understanding of this reaction has proved elusive. What follows is a short overview of the fundamental properties and activities of recA protein. This is followed by an attempt to rationalize a wealth of data in terms of some possible mechanisms for strand exchange. Of necessity, this treatment will be somewhat selective, and the reader is encouraged to seek out several recent reviews [7,18,19] for different perspectives.

General Properties of recA Protein

Binding to Single-Stranded DNA

RecA protein will bind readily to single-stranded DNA at pHs ranging from 5 to 9. The binding stoichiometry appears to be 1 recA monomer per 4 nucleotides (nt) when solution methods are used for measurement [7]. A somewhat smaller binding site is seen with electron microscopic measurements of complexes on duplex DNA [15] (see below). The reason for the apparent discrepancy is not clear. ATP is not required for binding to ssDNA. The protein is bound along the phosphate backbone of the DNA [34,35], with the bases exposed in a large helical groove evident in image reconstructions derived from electron micrographs of filaments formed in the presence of ATPγS [36]. An inherent polarity in the filament is evident both from image reconstructions and from the experimental observation that the filament is assembled in a 5' → 3' direction along the DNA [37,38].

ATP Hydrolysis

At physiological salt concentrations recA protein is a DNA-dependent ATPase, exhibiting a turnover number (per monomer) varying between 20 and 35 min^{-1} depending on conditions and the DNA cofactor [7]. A key characteristic of this reaction is that it occurs throughout recA nucleoprotein filaments [7,19,39]. There is no relationship between the location of a recA monomer with respect to the filament ends and its rate of ATP hydrolysis.

This reaction (and many other recA activities) is inhibited by low concentrations (20 to 100 mM) of NaCl [7]. The reaction is much less sensitive to glutamate, recently shown to be the physiologically relevant anion in *E. coli* [40]. The ATPase activity is activated in the absence of DNA by very high concentrations (2 M) of a variety of salts [41]. The turnover number under high salt conditions is at least as high as that observed in the presence of DNA at low salt concentrations.

ATP hydrolysis is inhibited by ADP and ATPγS [7]. Both act as competitive inhibitors, but the inhibition patterns are complicated by cooperativity between adjacent recA monomers in the presence of both high salt and DNA. A slow conformation change mediated by ADP leads to recA protein dissociation when the ADP/ATP ratio exceeds 1.0 [42; J. Lee and M. Cox, unpublished].

Binding to Duplex DNA

Recent findings [43–48] have clarified many aspects of this interaction. RecA protein binds to duplex DNA via a multisegment pathway (Fig.

3.4). This association exhibits an absolute requirement for ATP or an ATP analog. The first segment is probably a weak binding of recA monomers or small oligomeric units that does not alter the structure of the DNA. This step is rapid, reversible, and dependent on recA protein concentration [44,45]. In the second reaction segment, a tighter association of a recA monomer or small oligomeric unit occurs that results in the underwinding of a corresponding region of DNA. This nucleation event is rate limiting overall [43–45]. Nucleation is followed by a very rapid propagation segment in which the DNA is completely covered with recA protein. On nicked circular duplex DNA the final complex is stable, and ATP hydrolysis occurs throughout with a k_{cat} (monomer) of 20 to 25 min^{-1} [44,45]. The recA protein is arranged in a right-handed helical structure, following the phosphate backbone of the DNA. There is a net uptake of two protons in the first two reaction segments of Figure 3.4, resulting in a very slow rate of de novo association at pHs above 7.0 [44,45].

The DNA within this complex is underwound by 39.6% in the presence of ATP [48]. This corresponds to an underwinding of 13.1°/bp and a helical periodicity of 17.6 bp per turn. This underwinding increases slightly, to 43% or 18.6 bp per turn, when ATPγS is substituted for ATP [48,49]. Underwinding of DNA in the recA-dsDNA complexes in the presence of ATPγS is manifested in the electron microscope by a 1.6-fold extension of the complexes relative to B-form DNA [13–17,49]. There is an 8 to 10-nm repeat in the filament structure that corresponds to the ~18 bp per turn helical repeat of the DNA. There are approximately six recA monomers per repeat in the filament [15,49].

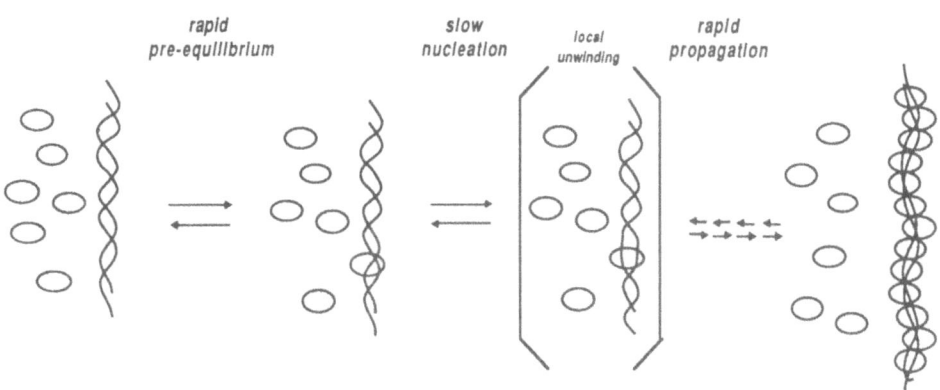

Figure 3.4. Pathway for recA protein binding to duplex DNA. The second step (nucleation) is generally rate limiting. ATP is hydrolyzed only after the complex is formed.

As with ssDNA, the recA-dsDNA nucleoprotein filament has an inherent polarity. The polarity reflects an asymmetry in binding to the two DNA strands. One of the two bound DNA strands exhibits a greater resistance to nuclease treatments than the other [46,50,51]. The more resistant strand can be referred to as the initiating strand. RecA protein filaments are assembled in the 5' → 3' direction relative to this strand [46,47,52]. It is reasonable to infer that the initiating strand occupies the same binding site occupied by ssDNA in recA-ssDNA complexes.

At pHs above 7.0, recA protein binds more rapidly to DNA with single-stranded gaps or tails, damaged DNA, DNA rich in A-T base pairs, and left-handed Z-form DNA than to B-form DNA [43,45–47,52,53]. This can be attributed to an enhancement in the rate of nucleation brought about by these structural features. Binding to Z-form DNA is much faster than to B-form DNA [53]. The final complex on Z DNA, however, is arranged in a right-handed helix structurally indistinguishable from that formed on B DNA [J. Kim, M. Cox, and J. Heuser, unpublished results]. The complex on Z DNA is also less stable than the complex formed on B DNA. When B and Z DNA are both present and compete for a limiting pool of recA protein, binding is ultimately restricted to the B DNA.

Once formed on duplex DNA, the recA nucleoprotein filament is stable at pHs between 5 and 9 as long as ATP is regenerated and the DNA substrate is circular (i.e., form II or nicked circular DNA). Complexes formed on linear duplex DNA are unstable at pHs above 6.8. This instability reflects a progressive dissociation of recA protein monomers (or other small units) from one end of the linear filament [46]. The maximum rate of this dissociation (observed at pHs above 8.0) is about 200 monomers min^{-1} per complex [46]. Relative to the initiating strand, dissociation occurs in the 5' → 3' direction. This is the same direction as assembly, so association and dissociation are largely restricted to opposite ends of the filament [46].

DNA Strand Exchange

Formation of a recA-DNA nucleoprotein filament that hydrolyzes ATP serves no useful end in itself. This complex is designed to find a second homologous DNA molecule, align it with the DNA already bound, and promote an exchange of DNA strands between these two DNA molecules. The mechanism of this reaction is not yet understood, but a wealth of experimental data is available to guide speculation. Approaching an understanding of this reaction requires a consideration of the cellular environment in which it takes place, the cellular problems it addresses, the structure of DNA, and the structure and kinetic properties of the recA nucleoprotein filament.

The Biological Problem

An analysis of the biological problem can begin with the question, Why would the cell employ a long protein filament to promote DNA pairing and branch migration? This strategy is expensive in two respects. A considerable energetic investment is made simply to assemble the 352 amino acids in each recA monomer in the filament. If each of the monomers then hydrolyzes ATP, the expense increases. At first glance, this investment seems unnecessary. Rates of uncatalyzed branch migration are sufficient to explain many of the genetic observations that require branch migration [54]. If directed pairing and branch migration are required, then it is easy to envision an enzyme with appropriate helicase and reannealing activities that could do the job. The in vitro evidence that the active species is a long filamentous complex, however, is extensive [7], and it is complemented by evidence that a similar structure is utilized in vivo [55] and that sufficient recA protein exists in the average cell to form a long filament [56]. To rationalize this, it is necessary to assume that extensive branch migration is an essential cellular process and that the filament is required to bring it about.

The importance of strand exchange can be made apparent by examining one role of homologous recombination in *E. coli*—i.e., postreplication repair. Evidence suggests that the replication apparatus is halted by unrepaired DNA lesions such as pyrimidine dimers [57,58]. If replication continues further downstream, the pyrimidine dimer is left in a region of ssDNA as illustrated in Figure 3.5, inaccessible to the normal cellular DNA repair systems. Cleavage of this single strand is a potentially lethal event. Repair of this lesion requires conversion of this DNA to duplex, and this is a recA-dependent process in *E. coli* [59]. It can be accomplished by utilizing recA protein to recruit a complementary strand from the other side of the replication fork. This is paired over a long region of

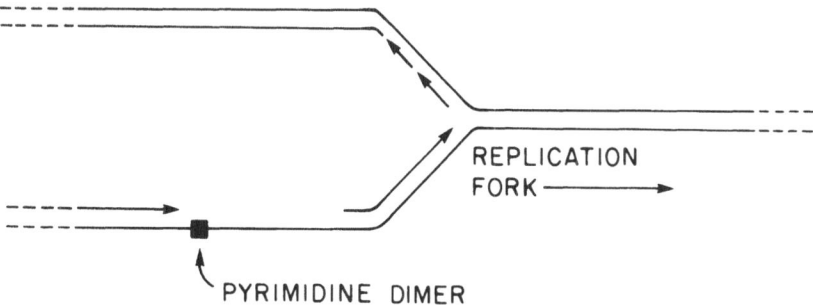

Figure 3.5. Postreplication repair. A pyrimidine dimer is positioned in a region of ssDNA after the replication fork has passed. Repair of this lesion requires recA protein as described in the text.

the damaged strand employing the pairing and branch migration activities of recA protein, as described in a model proposed by Howard-Flanders and colleagues [59]. RecA protein-promoted branch migration in vitro proceeds efficiently past pyrimidine dimers [60] and short mismatches [61]. This suggests that extensive branch migration is essential under some conditions for cell survival.

Once the importance of strand exchange is accepted, employing a filament can be rationalized as a logical cellular strategy and potentially the only strategy that can ensure that extensive branch migration will occur. A homologous interaction between two DNA molecules is likely to involve an unusual and unstable DNA structure (see "Paranemic Joints," below). A filament may be required to stabilize this structure. Also, in the cell there are a large number of barriers to branch migration that do not exist in the test tube. These include (1) nucleases, which may resolve crossover or Holliday junctions before branch migration has proceeded for any distance; (2) a variety of DNA-binding proteins that would represent impassable obstacles to a reaction of this kind; and (3) the DNA lesions themselves, which could represent a significant kinetic barrier to uncatalyzed branch migration. If extended branch migration is to occur, a mechanism must exist to protect the branch point, exclude other proteins from the path of the migrating branch, and provide the energy required to overcome short regions of heterology and DNA lesions. A filament meets these requirements.

Paranemic Joints

Before an exchange of strands begins, recA protein aligns two DNA molecules homologously without Watson-Crick interwinding of strands from the two DNA molecules. The resulting structure is very unstable, dissociating immediately if recA protein is removed [23]. Little is known about the architecture of this paranemic complex. The structure of paranemic joints is a critical parameter for understanding strand exchange. This can be seen by the fact that different models for strand exchange generally envision different types of paranemic joints. The structure also has interesting implications as a precedent for structures formed in synaptonemal complexes in eukaryotic cells during meiosis.

There are several different proposals for paranemic joints (Fig. 3.6). It is important to remember in each case that the DNA within the nucleoprotein filament is extended and underwound by 39.6% when ATP is present. The second DNA must be equally underwound over the region in which it is homologously aligned. The first model, proposed by Howard-Flanders and colleagues [62], envisions a three-stranded (or four-stranded in the case of four-strand exchanges) DNA structure such as that proposed by McGavin [63] and Wilson [64], with homology-dependent hydrogen-bonding between duplexes in the major groove. In a second

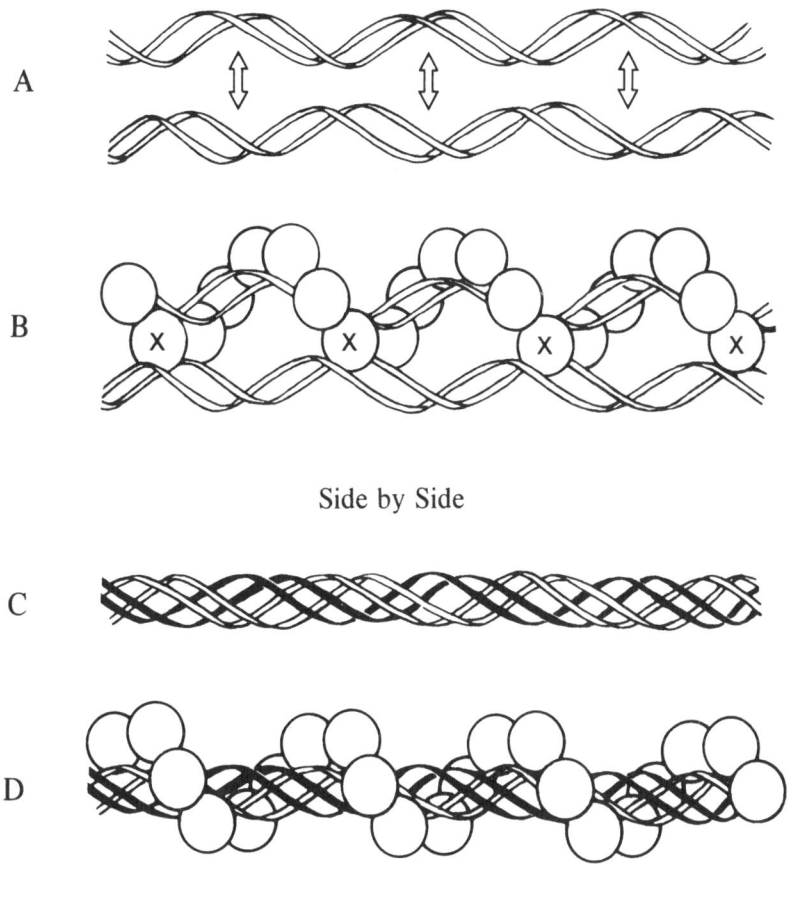

Side by Side

4-Stranded

Figure 3.6. Some possible structures for paranemic joints. The side-by-side and four-stranded structures are shown with and without recA protein. In the side-by-side configuration, open arrows (panel A) indicate points where periodic homologous interactions between the two DNAs might occur. In panel B the second DNA molecule is shown bound to every sixth recA monomer (denoted by the symbol "X"). In the four-stranded structure, homologous contacts between the two DNAs in their major grooves are continuous. The two DNAs are still bound asymmetrically, however. The dark strands denote the second DNA to be bound in the paranemic joint. This is the DNA that remains DNase sensitive in studies of paranemic joining.

model, the two DNAs are aligned side by side [24,65]. Periodic homologous contacts would occur in the major grooves of the DNAs, with perhaps 2 or 3 bp homologously paired for each 18-bp turn of the underwound helices. A third model, in which the two DNAs are aligned with alternating right- and left-handed helical regions (not shown), seems unlikely in view of the fact that recA protein always forms regular right-handed helical structures on DNA, even when the bound DNA is in the left-handed Z conformation prior to binding.

Because of the inherent instability of the paranemic joint, available evidence concerning the structure of these complexes is indirect. RecA protein in recA-ssDNA complexes hydrolyzes ATP with a turnover number of approximately 25 min^{-1}. Upon addition of homologous dsDNA, a 33% drop in the rate of ATP hydrolysis is observed that has been shown to be associated with the formation of paranemic joints [24]. The decrease is complete in less than 2 minutes, is completely homology dependent, and occurs even when homologous pairing is topologically restricted to paranemic complexes. More important, the extent of the decrease is linearly proportional to the *length* of available homology between the two DNA molecules [24]. These and other results imply that paranemic joints are formed in this reaction that are much longer than had been previously suspected, possibly extending for as much as 7,000 bp.

Extensive homologous pairing was detected in this study between circular DNA molecules in which interwinding over the entire homologous region was topologically prohibited [24]. At the moment this extended paranemic alignment is completed, the ssDNA is found to be completely protected from DNase digestion, while the incoming duplex DNA is still almost completely sensitive to digestion [66]. This indicates that in the paranemic complex the ssDNA is bound within the recA filament and the incoming duplex is bound in a position accessible to nuclease.

As expected, the incoming duplex DNA is underwound as a result of alignment with the DNA in the nucleoprotein filament [25]. This work employed nicked circular duplexes, and underwinding that occurred as a result of paranemic joint formation was trapped by sealing the nick with DNA ligase. As with the drop in ATP hydrolysis, the underwinding is homology dependent and kinetically competent to reflect formation of paranemic complexes. In addition, the underwinding occurs in circular duplexes that do not permit plectonemic joint formation, and the degree of underwinding exhibits a linear dependence on the fraction of the duplex homologous to the DNA within the filament [25]. This suggests again that all available homology is detected in paranemic joints.

The observed degree of underwinding, however, is insufficient for complete alignment throughout the homologous region. There is a broad distribution in the observed degree of underwinding, consistent with an average of 30% and a maximum of 45% of the available homologous duplex DNA aligned at any instant [25]. This indicates that in a 7,200-bp homologous duplex up to 2,900 base pairs can be paired simultaneously in a paranemic complex. Since the relationship between underwinding and length of homology suggests that all homology is detected, the paranemic complex may be noncontiguous, or it may be dynamic and move back and forth rapidly within the region of homology.

Pairing two circular DNA molecules, as in the experiment above, precludes net interwinding of the two DNAs. This topological constraint is removed if the duplex DNA is linear. The ends of the linear DNA must be heterologous to prevent plectonemic joint formation. Underwinding of the duplex resulting from paranemic joint formation can be trapped by ligation or by using a site-specific recombination reaction. The underwinding of duplex DNA in such an experiment is similar to that observed with nicked circular DNAs. More important, the duplex, after it is circularized, is interwound (catenated) with the circular single strand [B.S. Schutte and M.M. Cox, unpublished results]. This result is consistent with the formation of a three-stranded DNA helix over at least a short distance in the paranemic complex. Alternatively, the interwinding could be a fortuitous outcome of the formation of long paranemic joints. Determining whether this catenane reflects a three-stranded paranemic DNA structure will require a quantitative comparison of the degree of DNA interwinding and the degree of duplex underwinding.

Measurements of paranemic joints observed in the electron microscope by Griffith and colleagues [26] also indicate they can extend for more than 1,000 bp. Notably, the incoming duplex disappears into the filament in these complexes without significantly altering the ultrastructure of the complex [26,67].

Most of these results can be rationalized by either of the models in Figure 3.6. The two are not necessarily mutually exclusive. A paranemic joint could, for example, involve a region of four-stranded DNA flanked on one or both sides by DNA aligned side by side.

Properties and Kinetics of the DNA Strand Exchange Reaction

Whereas formation of paranemic joints is rapid, the subsequent branch migration phase is slow and rate limiting. About 10 to 20 minutes is required to cover 7,000 bp [7], yielding an optimal rate of at least 350 bp min^{-1}. As it proceeds, one strand of the duplex is progressively paired with the ssDNA in the complex and thus gradually becomes partially resistant to DNase treatment [66].

A successful model for this reaction must explain a number of observations related to this process:

1. As demonstrated by Brenner et al. [39], ATP hydrolysis occurs throughout the filament, rather than being restricted to filament ends or to monomers at the branch point. This is true prior to and during strand exchange [39].
2. ATP hydrolysis is *not* tightly coupled to association and/or dissociation of recA monomers with or from complexes [46,68,69].

3. After the 33% decrease in the rate of ATP hydrolysis that follows rapidly after the addition of homologous duplex DNA, ATP hydrolysis then continues at a constant level throughout and subsequent to the branch migration phase of strand exchange [24].
4. RecA protein remains on the heteroduplex product after the reaction is complete as determined by measuring underwinding of the nicked circular DNA products. Note that observation 3 also requires continued recA protein binding [66].
5. As seen in the electron microscope, the ultrastructure of the complex is unaffected by strand exchange, although dissociation of recA protein is sometimes observed in these studies well after strand exchange has occurred [18,26,67].
6. ATP hydrolysis in this system is apparently inefficient, with over 100 ATPs hydrolyzed per base pair of heteroduplex DNA formed [42].
7. In a three-strand exchange, branch migration can proceed not only past mismatches and pyrimidine dimers, but also past insertions or deletions involving hundreds of base pairs at a significant frequency [70]. This does not occur as readily in four-strand exchanges [71].
8. Both three- and four-strand exchange reactions can traverse a double-strand break in the incoming duplex, but not in the DNA within the nucleoprotein filament [10,24,71; B.C. Schutte and M.M. Cox, unpublished].

Models for Strand Exchange: Constraints

Models for strand exchange can be considered on several levels. The first obvious question involves the branch point where exchange is taking place. Is the recA protein filament stable and contiguous across the branch point, or is branch migration linked to association/dissociation of recA protein at this point? An obvious possibility is a treadmilling reaction such as that observed with actin and tubulin filaments [72]. A variation of treadmilling, called polar polymerization, has also been suggested in which strand exchange is coupled to addition of recA protein at a growing filament end [19]. Both assembly and disassembly of recA protein filaments proceed in the same direction as branch migration is driven [37,46,47]. Assembly, however, occurs at least 3- to 10-fold faster than branch migrations [44,46]. Dissociation occurs under some conditions at rates comparable to strand exchange [46], although strand exchange proceeds efficiently under conditions in which no net dissociation is observed (observations 3 and 4 above). Nuclease protection experiments and observations in the electron microscope provide additional evidence that the recA filament spans the branch point [26,50,51] without a structural alteration or interruption.

If branch migration were linked to a treadmilling process, ATP hydrolysis would be restricted to the ends of the filament. For this reason, observations 1 and 2 effectively eliminate a classical treadmilling reaction as a mechanism in this system. Observation 1 echoes the structural data above, indicating that there is nothing special about the branch point with respect to ATP hydrolysis. Since ATP hydrolysis is required to drive branch migration, we infer that the important molecular events are not restricted to filament ends. Observations 1 through 5 together suggest that the nucleoprotein filament remains essentially intact throughout the reaction. This indicates that models must be considered that do not require association or dissociation of recA protein to drive branch migration. Griffith and Harris [18] refer to these as stable filament models.

To develop models for strand exchange in which the nucleoprotein filament is static, it is necessary to consider energetics, structure, and DNA topology. Observation 6 is especially problematic. ATP hydrolysis clearly occurs even before duplex DNA is introduced. This and other results may indicate that much of this ATP hydrolysis is uncoupled. This view has most recently been advanced by Roman and Kowalczykowski [73]. At first glance, an efficiency of 1 ATP per base pair appears logical. This estimate can be reduced by considering the fact that one recA monomer binds three to four base pairs, and reduced further by considering the fact that the activation energy for branch migration is negligible. Strand exchange itself, in fact, may not require ATP hydrolysis, as suggested by numerous reports of eukaryotic proteins that promote strand exchange without an ATP requirement [74,75]. The apparent excess of ATP hydrolysis, however, is an experimental fact that may be an important mechanistic clue. The reaction as a whole accomplishes something of importance to the cell, and an understanding of the use of ATP in this system requires that a complete assessment of the biological and physical properties of the reaction be added to the chemical equation. The use of ATP can, in fact, be readily explained in a simple model (see below). Notably, on a cellular level, the expense is not great. The turnover number for ATP hydrolysis per recA monomer is about 25 min^{-1}. The 1,000 recA monomers present in the average cell [56] are therefore capable of hydrolyzing about 25,000 ATPs per minute. This is less than 0.1% of the potential energy consumption of the 10,000 to 20,000 ribosomes present in the same cell. In spite of the apparent inefficiency, the price may not be exorbitant when the alternative is cell death. The potential return on this investment will be described below.

Observation 7 is also difficult to explain. Bianchi and Radding [70] observed branch migration through long insertions at varying efficiencies. When the insertion was in the duplex DNA, the efficiency dropped off rapidly with increasing insertion size. When the insertion was in the ssDNA, these workers observed a low, but constant and significant, reaction efficiency that was relatively independent of insertion size. Insertions of hundreds of base pairs were bypassed in both cases.

Observation 7 suggests that an asymmetry exists in the recA nucleo-protein filament with respect to the binding sites of the ssDNA and the duplex DNA. This asymmetry is also reflected in observation 8 and in the nuclease sensitivity of the two DNAs in the paranemic joint. As suggested by West and Howard-Flanders [10], this is evidence for two nonidentical binding sites for DNA on each recA monomer.

In addition to these observations, a successful model must provide a viable role for a recA filament and a mechanism for coupling ATP hydrolysis to a molecular event that would result in unidirectional branch migration. If complete coupling is assumed, a direct relationship between the rate of ATP hydrolysis and the observed rate of strand exchange in vitro should be evident. The right-handed DNA helix also imposes topological constraints on a workable model that must be considered.

Models

Only two models have been outlined in the literature in sufficient detail to consider at length. For some additional mechanistic suggestions and viewpoints, the reader is encouraged to consult two recent articles [18,19]. The first model is that proposed by Howard-Flanders and colleagues [62]. This model develops directly from the four-stranded DNA paranemic joint illustrated in Figure 3.6. At one end of this joint the two homoduplexes enter and become intertwined. Within the complex there is a rotation of bases to effect strand exchange, and two heterodu-plexes exit at the other end of the complex. As originally proposed, the model links strand exchange with recA dissociation, although this is not

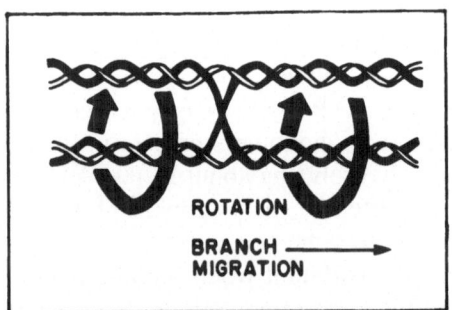

Figure 3.7. Model for DNA strand exchange. The proposed rotation of DNA molecules about the other is illustrated above. In each of the succeeding panels (next page), the dark strand corresponds to the "−" strand as described in the text. The arrow labeled "B" denotes the position of the branch point in panels 3 to 6. The balls represent individual recA monomers in a helical filament. The monomers to which the outer duplex DNA molecule is bound are denoted by "X" symbols in each panel. See text for details.

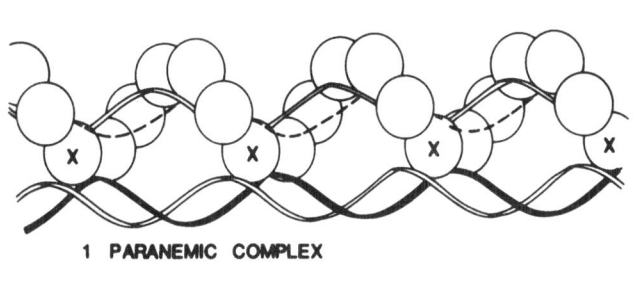

1 PARANEMIC COMPLEX

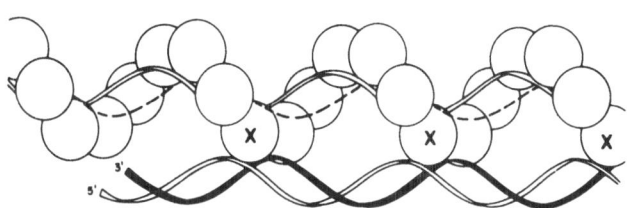

2 FREE END / PARANEMIC COMPLEX

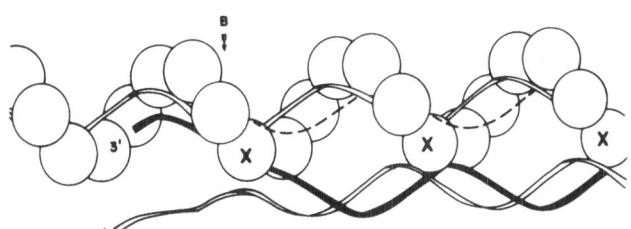

3 STRAND SWITCH

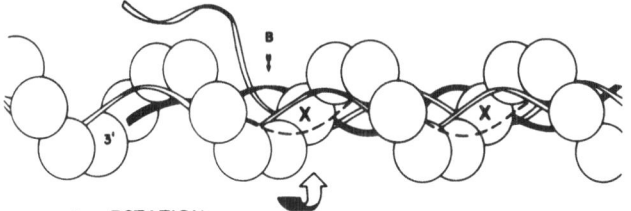

4 ROTATION

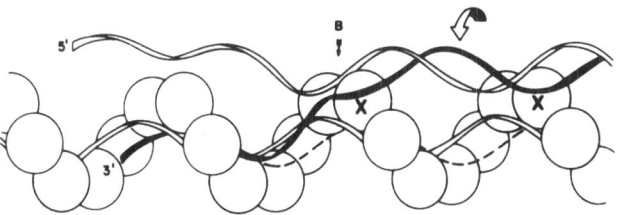

5 CONTINUED ROTATION

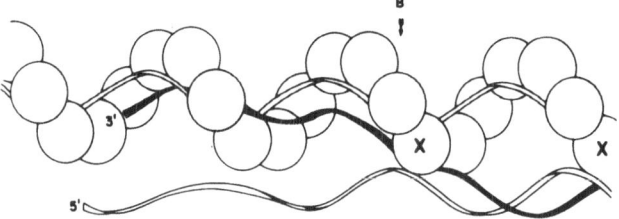

6 ROTATION / BRANCH MIGRATION

essential. This model has been outlined in much detail [62,71] and will not be reviewed extensively here, although a modified version will be illustrated below. The model is consistent with most of the data outlined above, and the four-stranded intermediate makes it easy to visualize homologous paranemic interactions. The model is less successful at explaining the energetics of the system, unless the assumption is made that much of the ATP hydrolysis is uncoupled.

The second model, the "rotation model" [65], will be expanded here and is presented in Figure 3.7. We begin with the paranemic complex, in which DNAs are aligned side by side. The paranemic complex includes a filament of recA protein enclosing the ssDNA. The protein is arranged in the right-handed helix with six recA monomers per turn as already described. The bases of the ssDNA (DNA 1) are exposed continuously in the helical groove of this filament. A similar complex can be formed on duplex DNA, and we infer that binding sites for two DNA strands exist within this complex.

The binding site for the second DNA strand is indicated by a dashed line. A partially underwound homologous duplex (DNA 2) is bound to the outside of this complex as described above, aligned side by side with the ssDNA. As drawn, this duplex is bound to every sixth recA monomer in the filament. We will arbitrarily designate the set of monomers to which DNA 2 is bound to be 6n. These recA monomers bind the outer duplex along the phosphate-deoxyribose backbone, and at these points the bases of the two DNA molecules are oriented away from each other. Between each periodic point of protein contact the duplex spans the groove of the filament. The twist of the DNA helices will cause the bases of the two separate DNA molecules to juxtapose in the groove, and at these points homologous contacts between the incoming duplex and the ssDNA are made.

This paranemic complex, with periodic homologous contacts, can span thousands of base pairs. The side-by-side alignment facilitates the formation of extended paranemic joints by minimizing topological constraints at this stage of the reaction. As shown in panels 2 and 3 (Fig. 3.7), strand exchange will be initiated readily if a free DNA end occurs anywhere within this paranemic complex. The melting of the end of this duplex and the pairing of the "−" strand with its complement within the complex to initiate the exchange will be facilitated by the intramolecular nature of the reaction, partial underwinding of the external duplex, base-pairing inter- actions, and possible favorable protein contacts within the unoccupied DNA strand binding site within the complex.

Once initiated, propagation of the exchange reaction encounters the topological requirement for rotation of the DNA molecules. This con- straint can be met by rotating both DNA molecules in the same direction (longitudinal rotation [65]) or, similarly, by rotating one DNA molecule around the other as illustrated in the Figure 3.7 inset. The central feature of this model is the controlled rotation of the duplex DNA molecule relative to the nucleoprotein filament, as illustrated in Figure 3.8. The rotation does not result in movement of either DNA relative to the filament along its length, but it does result in movement of the branch point. Extensive paranemic contact between the two DNAs is maintained at all times. This is accomplished by having the set of recA monomers to which the duplex is bound transfer the duplex unidirectionally to the neighboring set of recA monomers. The duplex will again be bound to every sixth recA monomer in the filament, but now this involves the monomer set $6n + 1$. This transfer is coupled to ATP hydrolysis via a set of conformation changes. These recA conformation changes could be analogous to the activity of the myosin ATPase during muscle contraction. The result is a unidirectional rotation of the duplex DNA around the ssDNA that leads directly to a unidirectional movement of the branch point.

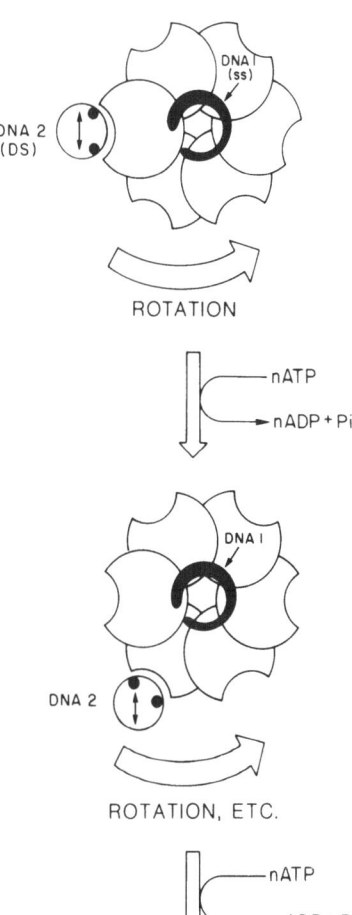

Figure 3.8. Proposed rotation of duplex DNA about the recA nucleoprotein filament. This is an end view of the recA nucleoprotein filament, as seen looking down the axis of the filament from right to left as illustrated in Figure 3.7. See text for details.

The rotation of the duplex DNA is better considered from the vantage point of a cross section of the complex shown in Figure 3.8. The complex may be considered fixed relative to an arbitrary plane bisecting it from top to bottom as drawn (not shown). The duplex DNA (DNA 2) is drawn as a cylinder viewed end-on and bound to one of the six recA monomers visible. ATP hydrolysis results in transfer of the duplex to the neighboring recA monomer, and repetition of this process causes rotation of the duplex around the outside of the complex in the direction indicated. The investment of chemical energy is used in part to ensure that this rotation, and the resulting branch migration, proceeds in only one direction. The system must therefore represent a coupled vectorial reaction as defined by Jencks [76]. The specific direction of branch movement will be predetermined and invariant relative to the asymmetry inherent in the filament.

As the rotation occurs, an arbitrary plane drawn through the axis of the duplex from top to bottom as shown in Figure 3.9 must be maintained parallel to the plane drawn through the complex. This ensures that each succeeding set of recA monomers will be bound in an identical fashion to the duplex and that homologous DNA base contacts in the complex groove occur similarly at each point in the rotation. Each monomer will

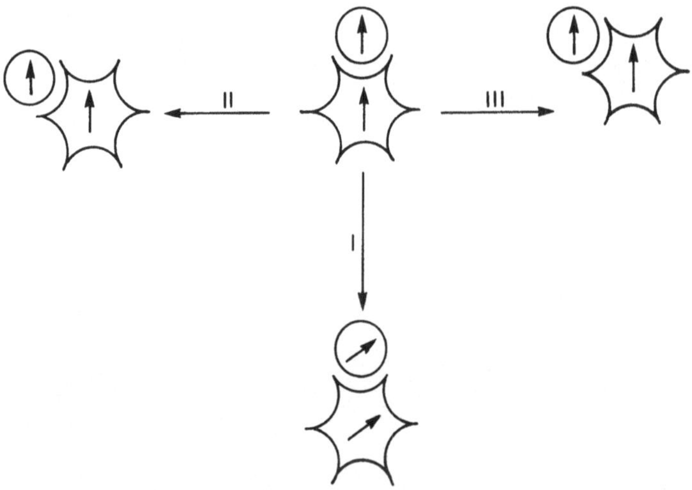

Figure 3.9. Three different types of rotation. The complex is viewed from one end, as in Figure 3.8. The planes bisecting the DNA within the complex (DNA 1) and the outer DNA (DNA 2—circle) are shown. Rotation I involves rotation of both DNAs about their longitudinal axis. In rotation II, DNA 2 rotates around DNA 1, and in rotation III, DNA 1 rotates around DNA 2. All three are compatible with the coupling mechanism outlined in Figures 3.7 and 3.8. The product of a single rotational step is structurally the same regardless of the rotation mechanism, although the positions in space of the two DNAs differ.

bind to the duplex DNA at the same point each time the duplex rotates 360° around the complex. Thus, each set of recA monomers will each have a unique and mutually exclusive set of DNA contact points for DNA 2.

Panels 4 through 6 in Figure 3.7 are related successively to each other by rotation through two of the transfer steps described above. The position of the branch point in each case is indicated by an arrow, and the rotation results in branch movement from left to right as drawn. The branch point is maintained within the helical groove of the filament at all times. In the cell, this rotation would cause topological strain elsewhere in the DNA that would require the action of topoisomerases. This is true, however, for any mechanism one might propose for branch migration.

The model is drawn in Figures 3.7 and 3.8 with the duplex rotating around the filament for ease of illustration. The same mechanism is compatible, however, with rotation of the filament around the duplex, or with the rotation of both the filament and the duplex around their rotational axes. The three possibilities are illustrated in Figure 3.9. The important feature is not the type of rotation itself, but the requirement that the planes through the two DNAs (shown with arrows) remain parallel at all points in the rotation. Any transfer mechanism that meets this condition is compatible with all three types of rotation. Strand exchange proceeds efficiently in this system when rotation is restricted to the rotational axes of the two DNAs [77].

This model satisfies all of the constraints listed above. A mechanistic role for the filament is provided that employs the important structural features of the filament and the paranemic complex. The filament groove becomes not only the site for homologous contacts between the two DNA molecules but an extended active site for migration of the branch point as well. There is no need for association or dissociation of recA protein. No special conformation or reaction of recA monomers at the branch point is required.

The model also provides a mechanistic role for ATP hydrolysis. The important molecular events that bring about branch migration are occurring throughout the filament rather than uniquely at the branch point. Each transfer event involves the making and breaking of hundreds of contacts between DNA 2 and recA monomers along the complex. It is not necessary, therefore, to invoke the idea of uncoupling to explain the observed reaction efficiency. In fact, both the *efficiency* and the *rate* of branch migration observed in vitro can be explained by assuming that ATP hydrolysis and branch migration are tightly coupled, with each recA monomer in the complex hydrolyzing one ATP to bring about a 360° rotation. Each rotation will result in an 18-bp (approximately) migration of the branch point. If the filament contains 2,000 monomers (as in a typical in vitro experiment), the expense will be nearly 100 ATPs per base pair—very close to the efficiency actually observed [42]. If the recA turnover number for ATP hydrolysis during strand exchange (18 to 20 min^{-1}) is translated directly into 18 to 20 rotations min^{-1}, the rate of branch migration will be 320 to 360 bp min^{-1}—again very close to the rate observed.

The system is easy to rationalize in vivo. The filament would serve to protect the crossover junction from nucleases and to exclude other DNA-binding proteins from the region. In addition to driving the reaction, a role of ATP hydrolysis in this system would be to maintain homologous alignment throughout the filament at all times and to avoid topological complications in the region covered by the filament. The extent of branch migration in a given recombination event may even be defined or regulated by the length of the filament. Resolution of the branch would occur as soon as the branch migrated beyond the end of the filament.

The system, in a sense, reflects what is probably its primary day-to-day function—postreplication repair. The strand exchange reaction itself does not require a large infusion of chemical energy, as already noted. ATP hydrolysis may be required, however, for extensive, unidirectional strand exchange through barriers such as sequence heterologies and pyrimidine dimers. By making the reaction unidirectional and insensitive to small barriers in DNA structure, this expenditure of chemical energy can be viewed as a device to ensure that life-or-death repair events are promoted efficiently.

The maintenance of homologous alignment throughout the complex helps explain the capability to branch-migrate through mismatches and pyrimidine dimers. If homologous alignment were maintained only through a short region near the branch point, a pyrimidine dimer or a short stretch of heterology could destabilize these interactions and produce a significant barrier to branch migration. By stabilizing homologous contact through thousands of base pairs, however, the effect of pyrimidine dimers or short mismatches is rendered insignificant. The energy invested in rotating the duplex all along the filament will overcome most barriers of this kind.

The observation that branch migration proceeds through large insertions can also be explained. This is accomplished without a requirement for halting branch migration and reinitiation on the other side of the insertion. The solution to this problem will be different depending on whether the insertion is located in the ssDNA or the duplex. The key factor in both cases is the maintenance of homologous paranemic contacts on both sides of the insertion. An insertion in the ssDNA would probably require dissociation of recA monomers bound to the insertion sequences. We speculate that this might be induced by strain if the complex is bent enough to permit paranemic contacts on both sides of the insertion. In the case of an insertion in the duplex, paranemic contacts would cause a looping out of the insertion sequences. A possible structure for the final paranemic complex formed in both cases is illustrated in Figure 3.10.

For the ssDNA insertion, the problem of branch migration is then

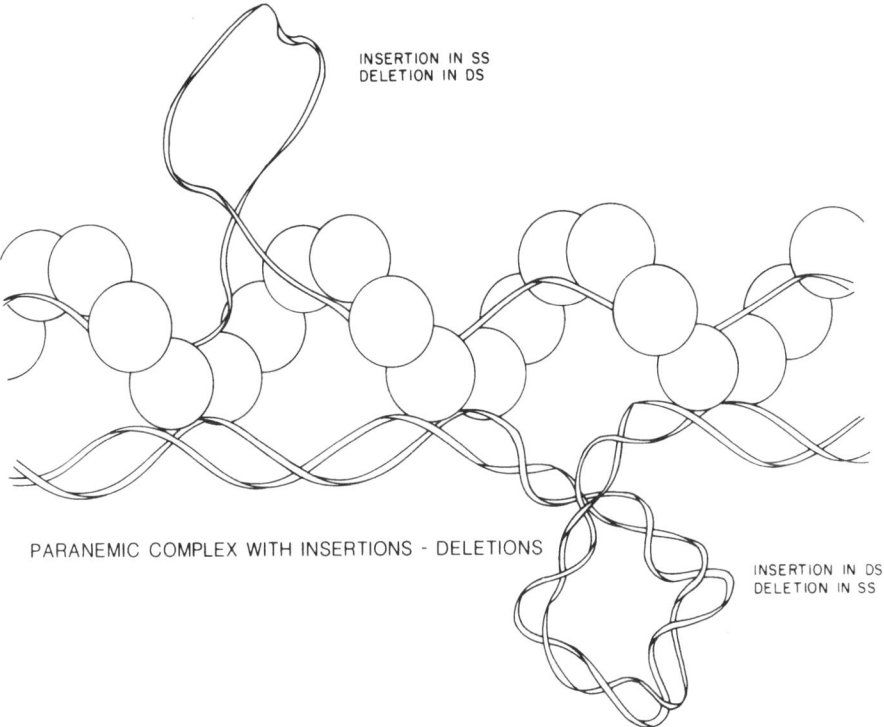

INSERTION IN SS
DELETION IN DS

PARANEMIC COMPLEX WITH INSERTIONS - DELETIONS

INSERTION IN DS
DELETION IN SS

Figure 3.10. Paranemic complex with insertions in the ssDNA or the duplex DNA. DNA corresponding to insertion sequences is extruded from the complex, with homologous contacts maintained on both sides of each insertion. See text for details.

reduced to a steric restriction on the rotation of the duplex caused by the extruded single strand. This steric problem could be exacerbated by any protein that remains associated with this extruded DNA. Rotation of the duplex would require the extruded DNA to be wrapped tightly around the complex. The topological problem is different and possibly greater for an insertion in the duplex. As shown in Figure 3.11, the proposed rotation would result in the unwinding of the duplex DNA in the insertion. The "+" strand of the duplex will be extruded and displaced. The "−" strand will be wrapped around the incoming "+" strand in the complex without base pairing as shown. The number of non-base-paired wraps will be equivalent to the number of helical DNA turns within the heterologous region of the original duplex. The number of these wraps will increase with insertion size. Since the wrapping would be constrained to a small region of the complex, an inverse relationship between reaction efficiency and insertion size might be expected here, as observed. Upon dissociation of recA protein, this structure would relax spontaneously, as indicated in Figure 11.

The model as drawn requires that each recA monomer have two DNA

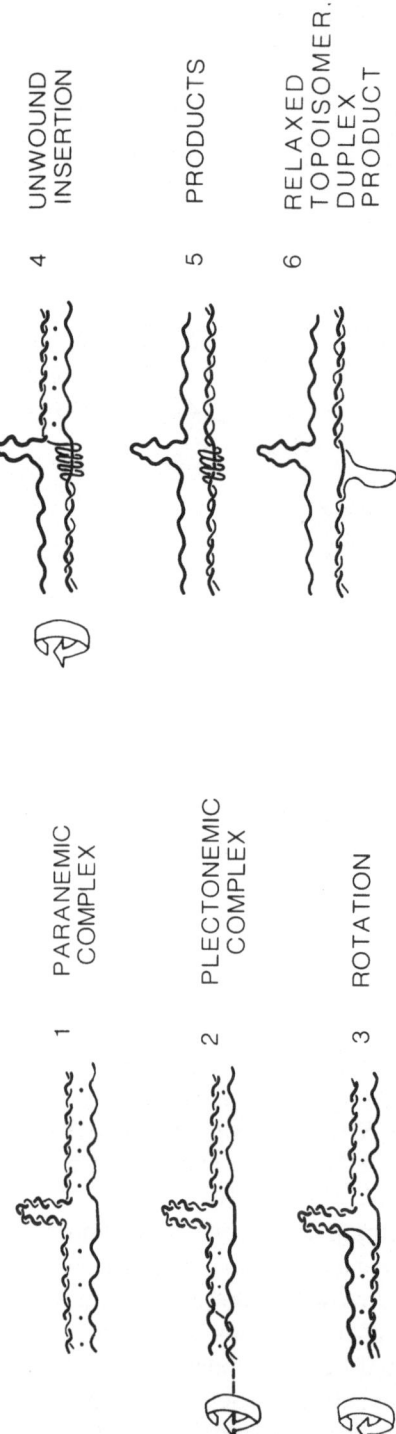

1 PARANEMIC COMPLEX

2 PLECTONEMIC COMPLEX

3 ROTATION

4 UNWOUND INSERTION

5 PRODUCTS

6 RELAXED TOPOISOMER. DUPLEX PRODUCT

Figure 3.11. Topological consequences of branch migration through an insertion in the duplex DNA substrate. The lighter of the three DNA strands corresponds to the "−" strand as described in the text. Panel descriptions: (1) The paranemic complex is illustrated, with homologous contacts spanning the insertion in the duplex. The recA filament (not shown) encloses the ssDNA at the bottom of the panel. (2) A nascent plectonemic complex results from the first rotation of the duplex around the ssDNA as shown. (3) Continued rotation brings the branch to the boundary of the insertion sequences. (4) The next four rotations unwind the duplex within the insertion without advancing the branch. The "−" strand in this region becomes wrapped around the incoming "+" strand without base-pairing. Once the insertion is unwound, branch migration proceeds normally. (5) Strand exchange is complete. (6) The duplex DNA will relax spontaneously to this form if the wrapped structure is not maintained by recA protein.

Figure 3.12. A composite model for recA protein-mediated DNA strand exchange. The four-stranded structure proposed by Howard-Flanders and colleagues is seen in the center of this complex. Rotation of DNA 2 in regions flanking the four-stranded region, in the direction and by the mechanism illustrated in Figure 3.7, will result in branch migration left to right.

binding sites as proposed by Howard-Flanders and co-workers [62], although here they must be positioned on opposite sides of the protein. The model predicts the pronounced asymmetry, observed in several studies, in the interaction of the nucleoprotein filament with the two DNA molecules participating in the exchange. The observation that strand exchange is blocked by a double strand break in a gapped duplex [10,71] is readily explained. In a four-strand exchange reaction, the gapped duplex will be within the filament, and a break in this duplex will result in a break in the entire complex. On the other hand, a break in the linear duplex (which will be outside the filament) will simply result in the formation of two contiguous paranemic complexes. Strand exchange should traverse the break in this outer duplex, as observed. In the case of a three-strand exchange, the model predicts that essentially any number of linear duplex fragments with nonoverlapping homology should be able to participate in strand exchange simultaneously on a single recA nucleoprotein filament. Results consistent with this prediction have already been obtained [24; B.C. Schutte, unpublished].

Several factors distinguish the rotation model from the four-strand model. The most important potential problem with the rotation model is the structure of the paranemic joint itself. The side-by-side alignment is difficult to envision, and no molecular models for it are available. The four-stranded structure proposed by Howard-Flanders, in contrast, is based on published models [63,64] and is accepted as a reasonable possibility. The interwinding of DNAs in paranemic complexes may be an experimental manifestation of this structure. A principal advantage of the rotation model is the rationalization it provides for energy use in this system. The two models differ in several other respects. For example, if the very long paranemic joints detected in several studies are completely three or four-stranded, considerable rotation of the two DNAs must occur during paranemic joint formation. For the model described above, rotation is not needed until branch migration commences.

For all of this, the two models are not necessarily incompatible. A composite model is illustrated in Figure 3.12. In this version, a rotation of the two DNAs as outlined in Figures 3.7 through 3.9 is coupled to formation and resolution of a four-stranded DNA intermediate, the four-stranded region has a significant length, and rotation leads to extension on one end and resolution on the other. The position of the four-stranded region would change as the reaction proceeded, and conversion of homoduplex to heteroduplex could occur at any point within it. This model is flexible in that the four-stranded region could be very short or extend for thousands of base pairs. The rotation at either end provides the same coupling and energetic advantages as the rotation model. Rotation could be driven with or without homologous interactions in the DNA flanking the four-stranded region.

Summary

Several models for recA-protein-mediated strand exchange are discussed in light of available evidence. A final resolution of the structure of paired DNA complexes bound by recA protein and the mechanism of DNA strand exchange continues to be elusive. Continued study of this problem should contribute to an understanding of recombination, cellular energetics, and chromosome-chromosome interactions.

Acknowledgments. Work cited from the author's laboratory was supported by National Institutes of Health grant GM 32335. The author is supported by N.I.H. Research Career Development Award AI 00599. This chapter is dedicated to the memory of Paul Howard-Flanders.

References

1. Radding, C.M. (1973) Annu. Rev. Genet. *7*, 87–111.
2. Radding, C.M. (1978) Annu. Rev. Biochem. *47*, 847–880.
3. Low, K.B. (1988) In The Recombination of Genetic Material, 1–21, Low, K.B., ed. Academic Press, San Diego.
4. Porter, R.D., McLaughlin, T., and Low, B. (1979) Cold Spring Harbor Symp. Quant. Biol. *43*, 1043–1047.
5. Holliday, R. (1971) Nature *232*, 233–236.
6. Szostak, J.M., Orr-Weaver, T.L., Rothstein, R.J., and Stahl, F.W. (1983) Cell *33*, 25–35.
7. Cox, M.M., and Lehman, I.R. (1987) Annu. Rev. Biochem. *56*, 229–262.
8. Das Gupta, C., Wu, A.M., Kahn, R., Cunningham, R.P., and Radding, C.M. (1981) Cell *25*, 507–516.
9. West, S.C., Cassuto, E., and Howard-Flanders, P. (1981) Proc. Natl. Acad. Sci. USA *78*, 2100–2104.
10. West, S.C., and Howard-Flanders, P. (1984) Cell *37*, 683–691.
11. Cox, M.M., and Lehman, I.R. (1982) J. Biol. Chem. *257*, 8523–8532.
12. Flory, J., Tsang, S.S., and Muniyappa, K. (1984) Proc. Natl. Acad. Sci. USA *81*, 7026–7030.
13. Flory, J., and Radding, C.M. (1982) Cell *28*, 747–756.
14. Dunn, K., Chrysogelos, S., and Griffith, J. (1982) Cell *28*, 757–765.
15. Di Capua, E., Engel, A., Stasiak, A., and Koller, T. (1982) J. Mol. Biol. *157*, 87–103.
16. Williams, R.C., and Spengler, S.J. (1986) J. Mol. Biol. *187*, 109–118.
17. Stasiak, A., and Egelman, E.H. (1987) UCLA Symp. Mol. Cell. Biol., New Ser. *47*, 619–628.
18. Griffith, J.D., and Harris, L.D. (1988) CRC Crit. Rev. Biochem. (Suppl.) *23*, 543–586.
19. Kowalczykowski, S.C. (1987) Trends Biochem. Sci. *12*, 141–145.
20. Bianchi, M., Das Gupta, C., and Radding, C.M. (1983) Cell *34*, 931–939.

21. Chow, S.A., and Radding, C.M. (1985) Proc. Natl. Acad. Sci. USA *82*, 5646–5650.
22. Flory, S.S., Tsang, J., Muniyappa, K., Bianchi, M., Gordon, D., Kahn, R., Azhderian, E., Egner, C., Shaner, S., and Radding, C.M. (1984) Cold Spring Harbor Symp. Quant. Biol. *49*, 513–523.
23. Riddles, P.W., and Lehman, I.R. (1985) J. Biol. Chem. *260*, 165–169.
24. Schutte, B.C., and Cox, M.M. (1987) Biochemistry *26*, 5616–5625.
25. Schutte, B.C., and Cox, M.M. (1988) Biochemistry *27*, 7886–7894.
26. Register, J.C., Christiansen, G., and Griffith, J. (1987) J. Biol. Chem. *262*, 12812–12820.
27. Das Gupta, C., Shibata, T., Cunningham, R.P., and Radding, C.M. (1980) Cell *22*, 437–446.
28. Cox, M.M., and Lehman, I.R. (1981) Proc. Natl. Acad. Sci. USA *78*, 3433–3437.
29. Honigberg, S.M., Gonda, D.K., Flory, J., and Radding, C.M. (1985) J. Biol. Chem. *260*, 11845–11851.
30. West, S.C., Cassuto, E., and Howard-Flanders, P. (1981) Proc. Natl. Acad. Sci. USA *78*, 6149–6153.
31. Kahn, R., Cunningham, R.P., Das Gupta, C., and Radding, C.M. (1981) Proc. Natl. Acad. Sci. USA *78*, 4786–4790.
32. Cox, M.M., and Lehman, I.R. (1981) Proc. Natl. Acad. Sci. USA *78*, 6018–6022.
33. Wu, A.W., Kahn, R., Das Gupta, C., and Radding, C.M. (1982) Cell *30*, 37–44.
34. Leahy, M.C., and Radding, C.M. (1986) J. Biol. Chem. *261*, 6954–6960.
35. Di Capua, E.D., and Müller, B. (1987) EMBO J. *6*, 2493–2498.
36. Egelman, E.H., and Stasiak, A. (1986) J. Mol. Biol. *191*, 677–697.
37. Register, J.C., and Griffith, J. (1985) J. Biol. Chem. *260*, 12308–12312.
38. Stasiak, A., Egelman, E.H., and Howard-Flanders, P. (1988) J. Mol. Biol. *202*, 659–662.
39. Brenner, S.L., Mitchell, R.S., Morrical, S.W., Neuendorf, S.K., Schutte, B.C., and Cox, M.M. (1987) J. Biol. Chem. *262*, 4011–4016.
40. Leirmo, S., Harrison, C., Cayley, D.S., Burgess, R.R., and Record, M.T. Jr. (1987) Biochemistry *26*, 2095–2101.
41. Pugh, B.F., and Cox, M.M. (1988) J. Biol. Chem. *263*, 76–83.
42. Cox, M.M., Soltis, D.A., Livneh, Z., and Lehman, I.R. (1983) J. Biol. Chem. *258*, 2577–2585.
43. Kowalczykowski, S.C., Clow, J., and Krupp, R.A. (1987) Proc. Natl. Acad. Sci. USA *84*, 3127–3131.
44. Pugh, B.F., and Cox, M.M. (1987) J. Biol. Chem. *262*, 1326–1336.
45. Pugh, B.F., and Cox, M.M. (1988) J. Mol. Biol. *203*, 479–493.
46. Lindsley, J.E., and Cox, M.M. (1989) J. Mol. Biol. *205*, 695–711.
47. Shaner, S.C., and Radding, C.M. (1987) J. Biol. Chem. *262*, 9211–9219.
48. Pugh, B.F., and Cox, M.M. (1989) J. Mol. Biol. *205*, 487–492.
49. Stasiak, A., and Di Capua, E. (1982) Nature *299*, 185–186.
50. Chow, S.A., Honigberg, S.M., Bainton, R.J., and Radding, C.W. (1986) J. Biol. Chem. *261*, 6961–6971.
51. Chow, S.A., Honigberg, S.M., and Radding, C.M. (1988) J. Biol. Chem. *263*, 3335–3347.

52. Cassuto, E., and Howard-Flanders, P. (1986) Nucleic Acids Res. *14*, 1149–1157.
53. Blaho, J.A., and Wells, R.D. (1987) J. Biol. Chem. *262*, 6082–6088.
54. Warner, R.C., and Tessman, I.T. (1978) In The Single-Stranded DNA Phages, 417, Denhardt, D.T., et al. (eds.) Cold Spring Harbor Laboratory, Cold Spring Harbor, New York.
55. Yancey, S.D., and Porter, R.D. (1984) Mol. Gen. Genet. *193*, 53–57.
56. Salles, B., and Paoletti, C. (1983) Proc. Natl. Acad. Sci. USA *80*, 65–69.
57. Ganesan, A.K. (1974) J. Mol. Biol. *87*, 103–119.
58. Rupp, W.D., Wilde, C.E. III, Reno, D.L., and Howard-Flanders, P. (1971) J. Mol. Biol. *61*, 25–44.
59. West, S.C., Cassuto, E., and Howard-Flanders, P. (1981) Nature *294*, 659–662.
60. Livneh, Z., and Lehman, I.R. (1982) Proc. Natl. Acad. Sci. USA *79*, 3171–3175.
61. Das Gupta, C., and Radding, C.M. (1982) Proc. Natl. Acad. Sci. USA *79*, 762–766.
62. Howard-Flanders, P., West, S.C., and Stasiak, A. (1984) Nature *309*, 215–219.
63. McGavin, S.J. (1971) J. Mol. Biol. *55*, 293–298.
64. Wilson, J.H. (1979) Proc. Natl. Acad. Sci. USA *76*, 3641–3645.
65. Cox, M.M., Pugh, B.F., Schutte, B.C., Lindsley, J.E., Lee, J., and Morrical, S.W. (1987) UCLA Symp. Mol. Cell. Biol., New Ser., *47*, 597–607.
66. Pugh, B.F., and Cox, M.M. (1987) J. Biol. Chem. *262*, 1337–1343.
67. Christiansen, G., and Griffith, J. (1986) Proc. Natl. Acad. Sci. USA *83*, 2066–2070.
68. Neuendorf, S.K., and Cox, M.M. (1986) J. Biol. Chem. *261*, 8276–8282.
69. Menetski, J.P., and Kowalczykowski, S.C. (1987) J. Biol. Chem. *262*, 2093–2100.
70. Bianchi, M.E., and Radding, C.M. (1983) Cell *35*, 511–520.
71. Howard-Flanders, P., West, S.C., Cassuto, E., Hahn, T.R., and Egelman, E. (1987) UCLA Symp. Mol. Cell. Biol., New Ser. *47*, 609–617.
72. Cox, M.M., Morrical, S.W., and Neuendorf, S.K. (1984) Cold Spring Harbor Symp. Quant. Biol. *49*, 525–533.
73. Roman, L.J., and Kowalczykowski, S.C. (1986) Biochemistry *25*, 7375–7385.
74. Kolodner, R., Evans, D.H., and Morrison, P.T. (1987) Proc. Natl. Acad. Sci. USA *84*, 5560–5564.
75. Hsieh, P., Meyn, M.S., and Camerini-Otero, R.D. (1986) Cell *44*, 885–894.
76. Jencks, W.P. (1980) Adv. Enzymol. *51*, 75–106.
77. Honigberg, S.M., and Radding, C.M. (1988) Cell *54*, 525–532.

Chapter 4
The SV40 Large Tumor Antigen

Arnold Jay Levine

The purpose of this review is to bring together a large number of observations and facts that have been assembled during the study of the large tumor antigen protein (T antigen) encoded by simian virus 40 (SV40). This protein and its functions have been one the most intensively studied gene systems derived from a eukaryotic environment. The reasons for this are several:

1. The SV40 T antigen is a protein with multiple functions that correspond to multiple structural and functional domains of the protein. Among these functions are the initiation and propagation of DNA replication and the positive and negative regulation of transcription.
2. The SV40 T antigen can act to alter the growth potential and regulation of the host cell expressing this protein. The T antigen can form oligomeric protein complexes with at least two oncogene products (p53 and the retinoblastoma susceptibility gene product, or RB), a transcriptional factor (AP-2), and an enzyme involved in DNA replication (DNA polymerase-alpha).
3. Viruses like SV40 may well be under considerable evolutionary constraints to keep the size of their genome small, and this, in turn, may have provided a selection for the development of a single protein with multiple structural and functional domains. This protein is the product of a viral encoded oncogene, able to cause cancer in animals and alter the growth of cells in culture (transformation), and these experimental studies are a most reasonable way to understand, at a molecular level, how oncogenes may function.

A number of diverse approaches have been employed in the study of the SV40 tumor antigen. Molecular biology and genetics have led to the construction of hundreds of mutants across the gene for this protein. Immunologists have developed more than 80 independently derived monoclonal antibodies and cell lines of killer T cells that recognize diverse epitopes in this protein. Biochemists, cancer biologists, cell biologists, and virologists have developed dozens of functional assays, both in vitro and in vivo, to study the protein and its mutant forms or to examine the role of an epitope in a particular gene function. All of this provides an intensity of study and an information base that is unique. By collecting and examining this wide body of information, we can hope for a synthesis that provides a real understanding of the processes of DNA replication and transcription and, through this, the nature of oncogene function and growth regulation [for additional recent reviews see references 1–6].

Simian Virus 40

SV40 was first isolated as a passenger of primary rhesus monkey kidney cells in culture that were being used to produce the poliovirus vaccine [7]. The virus is composed of DNA (12.5% by weight) and protein and is constructed of an icosahedral coat, 45 nm in diameter, which encloses a nuclear protein core [1]. The icosahedron is assembled from 72 capsomeres that are constructed from three viral proteins: VP-1 (45₃ kD), VP-2 (42 kD), and VP-3 (30 kD). The DNA is a double-stranded closed circular molecule composed of 5,243 base pairs (in the SV40 776 strain) [8,9] wrapped around 24 or 25 nucleosomes which are formed from the four cellular histone proteins—H2A, H2B, H3, and H4—in an octameric structure [1]. The genome of this virus is organized (Fig. 4.1) in two halves encoding six protein gene products. The BglI site (at nucleotides 5243/0) serves to divide the two major categories of viral genes that are expressed at early or later times after a virus infection of cells. The two early gene products, the SV40 large tumor (T) and small tumor (t) antigens, are produced by alternate splicing of a common precursor RNA transcribed across this early region [10,11]. The large T antigen mRNA is translated into a protein of 708 amino acids, while the small t antigen is composed of an N-terminal domain of 82 amino acids that is identical with the N terminus of the large T antigen and a unique C-terminal domain of 97 residues derived from the intron of the large T antigen gene (Fig. 4.1). The three major structural proteins of the virus or late genes encoding VP-1, VP-2, and VP-3 are similarly derived from spliced mRNAs transcribed from the opposite DNA template strand [1]. The leader sequence (nucleotides 325 and 526) of the 16S mRNA for VP-1 (a bicistronic mRNA) encodes the sixth virus protein termed the agnoprotein. This mRNA produces a 71–amino acid basic protein, not found in the virion, which is thought to aid cell-to-cell spread of the infectious virus [12].

When SV40 was inoculated into a newborn hamster, tumors were produced at the site of injection within 6 to 12 months [13]. These tumors did not contain detectable infectious virus or the late structural

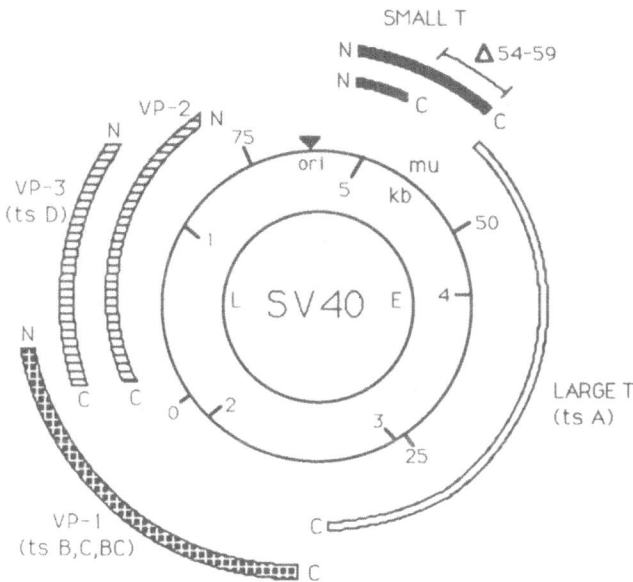

Figure 4.1. Map of the SV40 genome. The circular map of the SV40 genome divided into kilobases (kb) or map units (mu; 100 units) shows the early region (E) and late region (L) with the position of the large T antigen (ts A) and small t antigen (Δ54 to 59) genes indicated in the N- to C-terminal regions of the protein and VP-1 (ts B, C, BC), VP-2, and VP-3 (ts D). The origin (ori) of DNA replication separates the early and late genes.

proteins—VP-1, VP-2, or VP-3—of this virus [1]. The tumor cells did contain a copy of the viral genome integrated into a host cell chromosome, and this DNA was transcribed across the early region genes producing the large T and small t antigen proteins [1]. The host animal recognized these viral proteins as foreign and responded by producing antibodies (and sometimes a killer T cell response) against these proteins. These antisera became the major reagent employed to detect the SV40 T antigens; hence the name for these proteins [14]. When a protein expressed by a tumor elicits an antibody or humoral response (B cell mediated), it has come to be called a tumor antigen (T antigen). When such a protein elicits a cellular mediated tumor rejection (cytotoxic T cell) response, the protein is termed a tumor-specific transplantation antigen (TSTA). The SV40 large T antigen is both a T antigen and a TSTA [15].

Life Cycle of the Virus

The extent of expression of the SV40 genome and viral proteins during infection depends on the cell type and species of the host cell. The virus

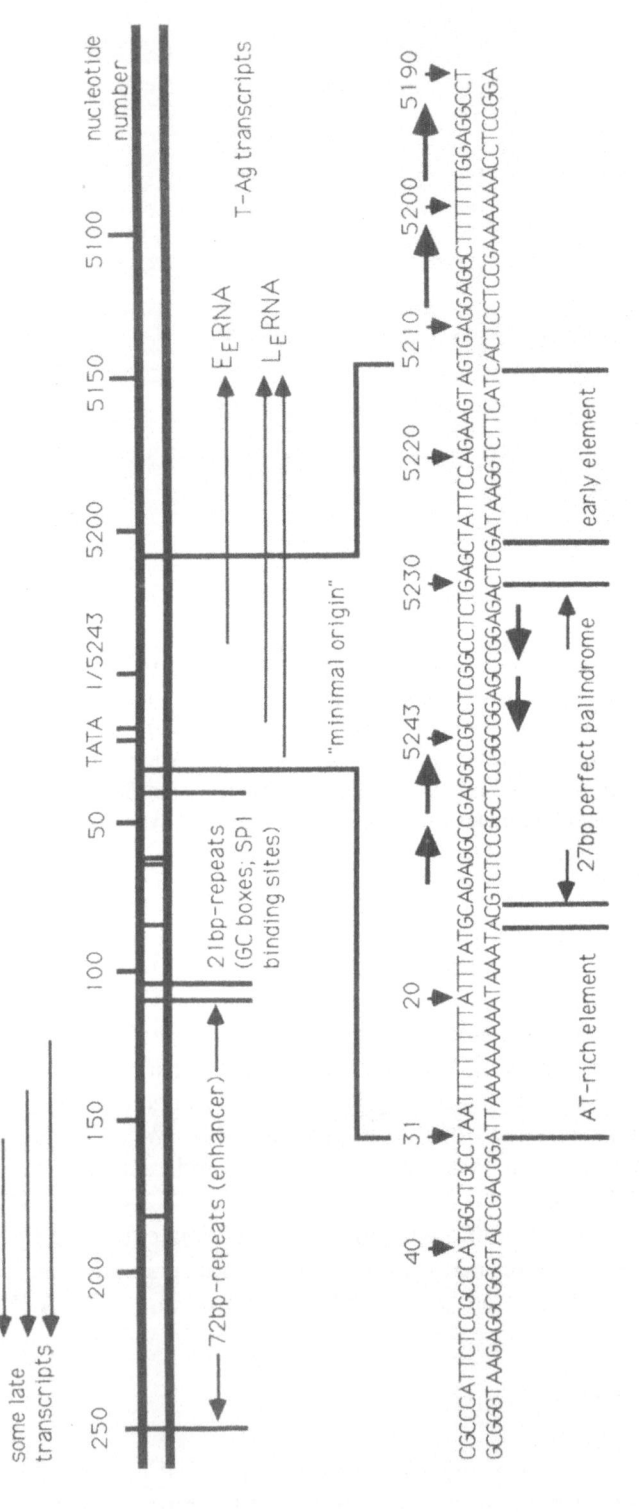

Figure 4.2. Detailed map of the SV40 T antigen-binding sites. The map is given in nucleotide numbers (1 is the Bgl-I site) showing the starts of early (E$_E$, early-early; and L$_E$, late-early) transcripts of T antigen, minimal origin, site 1, and site 2 for T antigen binding (arrows over the GAGGC sites). The enhancer (72-bp repeats) and promoter-G-C boxes and TATA regions are shown with the nucleotide sequence of this region.

duplicates itself efficiently in African green monkey kidney cells (permissive host) but poorly in human cells (semipermissive), and it fails to replicate in murine cells (nonpermissive). In mouse cells, the virus expresses its early proteins, T and t antigen, but does not replicate the viral DNA, nor does it synthesize the late viral proteins [1]. Mouse-monkey cell hybrids are permissive for virus production, suggesting the absence of some factor critical to virus DNA replication in mouse cells or the failure of a murine factor to interact with a viral protein employed in DNA replication [16].

In the permissive cells, the virus attaches to a receptor on the cell surface and penetrates into the cell in pinocytic vesicles which then ferry the intact virus into the nucleus [1]. Uncoating exposes the histone-DNA viral core to the transcription machinery of the cell nucleus. Several *cis*-acting signals encoded in the enhancer (72-bp repeat) [1] and promoter region (21-bp repeat and TATA box; see Fig. 4.2) are recognized by specific transcription factors that bind to this DNA in order to promote transcription. The SP-1 transcription factor binds to several GGGCGC sites in the 21-bp repeat region of the promoter, and a TATA-box binding factor also recognizes this element of the early promoter region (Fig. 4.2). AP-1 binds to two sites (TGACTCA) in the 72-bp repeat and one site located on the late gene side of the 72-bp repeat [17]. AP-2 binds to CCCCAGGC sequences in the 72-bp repeat, and AP-3 recognizes a sequence, TGTGGAAAGT, adjacent to an AP-2 site in the 72-bp enhancer [18]. RNA polymerase II presumably interacts with these factors (and others yet to be found), initiating transcription at the early-early site (E_ERNA), and synthesizes the primary transcript for the T and t antigen spliced mRNAs (Fig. 4.2).

The SV40 T antigen, once synthesized, can bind to at least two sites on the SV40 genome in this critical control region of the chromosome (Fig. 4.2). The T antigen recognizes several pentanucleotide sequences 5'-(G > T) (A > G)GGC-3' that are located in different patterns or pairs at nucleotides 5210 to 5190 (site I) adjacent to the early region gene for T antigen, and nucleotides 12 to 5230 (site II) where two pairs of inverted binding sites are found within a 27-bp perfect palindrome (Fig. 4.2) (20). T antigen binding to site I results in the negative regulation of early gene transcription, and so this protein autoregulates its own levels [21]. T antigen binding at site II helps to initiate a new round of viral DNA replication [22] by unwinding the DNA duplex helix at this site [23]. To do this, the T antigen employs its ATPase and helicase activities [24].

Initiation of DNA replication proceeds by the synthesis of multiple initiation RNAs (7 to 10 bp) synthesized by a DNA primase activity using 3' dPuT sites preferentially (which occur at multiple sites on both strands) [25]. DNA polymerase alpha (which binds to T antigen) can then utilize these RNA primers to produce Okazaki fragments. The leading strand of newly made DNA is probably made in a continuous fashion (5' to 3') while the lagging strand is synthesized by multiple RNA initiations and Okazaki fragments which are then connected by gap filling and ligase reactions [26]. Replication around the chromosome proceeds in a bi-directional fashion with the two forks meeting at 180° from the origin [27]. The SV40 large T antigen is apparently utilized at each replication fork to continue to unwind the double helix [28] at that site (helicase activity). Both viral DNA replication [1] and the positive, *trans*-acting function of the T antigen [29] are next required to begin transcription of the late genes. This process also utilizes AP-1 and AP-4 (CCAGCTGTGG), which have binding sites on the late side of the 72-bp repeated enhancer sequences [19]. The late transcript is spliced into mRNAs that are then translated to produce VP-1, VP-2, VP-3, and the agnoprotein [1]. The virus is assembled in the nucleus, and the agnoprotein helps (in an unknown way) to spread the infectious virus to new cells [12]. This entire process takes 50 to 70 hours in cell culture, depending on the cell type [1].

During SV40 infection of a nonpermissive host, such as mouse cells in culture, all of the early events appear to occur normally, but viral DNA replication does not occur. The late mRNAs and proteins are not made in detectable amounts, and no infectious virus is produced [1]. SV40 infection of cells in G_o arrest results in the stimulation of net RNA synthesis including ribosomal RNAs [30] and increased levels of a variety of enzymes involved in the deoxypyrimidine biosynthetic pathway [31]. The SV40 large T antigen is required for this induction of cellular functions [30,31] and the stimulation of cellular DNA synthesis that follows these events [32]. Although some murine cells are killed by the virus infection, others go on to divide. A small percentage of the cells in the culture (depending on the multiplicity of infection, cell type, and species [1]) contain integrated SV40 DNA in their genome [33] expressing the large T and small t antigen.

A variable percentage of these cells produce cloned cell lines that express one or more properties of the transformed phenotype [34]: (1) growth in medium containing low serum concentrations; (2) loss of contact inhibition and foci formation; (3) ability to form colonies in agar; and (4) the ability to produce tumors in isogenic hosts or immunocompromised hosts. The SV40 large T antigen is required both to initiate and to maintain the transformed phenotype in these cells [1]. Temperature-sensitive (tsA) mutations in the T antigen gene have been employed to demonstrate that the transformed phenotypes described above are temperature conditional in cell lines transformed by SV40 tsA mutants of this type [1]. Clearly, the large T antigen regulates these phenotypes.

Finally, it should be pointed out that all oncogenes and their products are defined and tested only in the context of a particular cell type. Some viral or cellular oncogenes can immortalize primary cells in culture but do not have dramatic effects on permanent cell lines. The SV40 large T antigen both has immortalizing capabilities [35] and can fully induce the transformed phenotype in primary and permanent cell lines [1]. As we shall see in the next sections, different domains of the SV40 T antigen can be identified and separated based on their abilities to immortalize cells in culture [35] and transform some cell lines but not others [36]. In studies of this type, both the phenotype of transformation and the cell line employed are important variables that need to be recognized.

The SV40 Large T Antigen

Primary Sequence and Structure Predictions

The SV40 large T antigen is composed of 708 amino acid residues with a molecular weight of about 81,500 daltons. When this protein is analyzed by SDS polyacrylamide gel electrophoresis, it has an apparent molecular weight of 88–94,000 daltons, but a more accurate molecular weight is obtained by Sepharose chromatography in 6 M guanidine HCl [37]. The primary sequence of T antigen is presented in Figure 4.3. Based on this sequence, it is possible to postulate the presence of different structural features over the length of the protein. Employing these postulates [38,39] one may suggest a conformation for regions of the protein with alpha helix, turns, random coils, or extended portions as well as local regions of hydrophobicity. These data are presented in Figure 4.4 and are, at best, a guide to look for patterns in the protein rather than a true reflection of reality.

Protein Modifications

The population of T antigen molecules as isolated from a cell is very heterogeneous with many molecules having several different types of modifications. While all of the T antigen molecules have a blocked (acetylated) N-terminal methionine residue [40], about 5% of the molecules are acetylated with a fatty acid [41]. This subfraction of T antigen appears to be preferentially associated with the plasma membrane of the cell [41]. A small percentage of the T antigen moieties are modified with galactose or glucosamine [42], while some molecules are poly-ADP ribosylated [43]. A larger percentage of the T antigen molecules are adenylated with the AMP linked to a serine residue via a phosphodiester bond [44]. There are at least two distinct serine-AMP residues per T antigen, and these modifications have been mapped between amino acids 412 and 528 [44].

Finally, there are at least 10 amino acid residues that can be phosphorylated—eight on a serine and two at a threonine site. These 10 residues are localized on T antigen in two clusters as follows: ser-106, ser-111, ser-112, ser-123, and thr-124; and ser-639, ser-676, ser-677, ser-679, and

1	2	3	4	5	6	7	8	9	10	11	12	13	14	15
MET	ASP	LYS	VAL	LEU	ASN	ARG	GLU	GLU	SER	LEU	GLN	LEU	MET	ASP

16	17	18	19	20	21	22	23	24	25	26	27	28	29	30
LEU	LEU	GLY	LEU	GLU	ARG	SER	ALA	TRP	GLY	ASN	ILE	PRO	LEU	MET

31	32	33	34	35	36	37	38	39	40	41	42	43	44	45
ARG	LYS	ALA	TYR	LEU	LYS	LYS	CYS	LYS	GLU	PHE	HIS	PRO	ASP	LYS

46	47	48	49	50	51	52	53	54	55	56	57	58	59	60
GLY	GLY	ASP	GLU	GLU	LYS	MET	LYS	LYS	MET	ASN	THR	LEU	TYR	LYS

61	62	63	64	65	66	67	68	69	70	71	72	73	74	75
LYS	MET	GLU	ASP	GLY	VAL	LYS	TYR	ALA	HIS	GLN	PRO	ASP	PHE	GLY

76	77	78	79	80	81	82	83	84	85	86	87	88	89	90
GLY	PHE	TRP	ASP	ALA	THR	GLU	ILE	PRO	THR	TYR	GLY	THR	ASP	GLU

91	92	93	94	95	96	97	98	99	100	101	102	103	104	105
TRP	GLU	GLN	TRP	TRP	ASN	ALA	PHE	ASN	GLU	GLU	ASN	LEU	PHE	CYS

106	107	108	109	110	111	112	113	114	115	116	117	118	119	120
SER	GLU	GLU	MET	PRO	SER	SER	ASP	ASP	GLU	ALA	THR	ALA	ASP	SER

121	122	123	124	125	126	127	128	129	130	131	132	133	134	135
GLN	HIS	SER	THR	PRO	PRO	LYS	LYS	LYS	ARG	LYS	VAL	GLU	ASP	PRO

136	137	138	139	140	141	142	143	144	145	146	147	148	149	150
LYS	ASP	PHE	PRO	SER	GLU	LEU	LEU	SER	PHE	LEU	SER	HIS	ALA	VAL

151	152	153	154	155	156	157	158	159	160	161	162	163	164	165
PHE	SER	ASN	ARG	THR	LEU	ALA	CYS	PHE	ALA	ILE	TYR	THR	THR	LYS

166	167	168	169	170	171	172	173	174	175	176	177	178	179	180
GLU	LYS	ALA	ALA	LEU	LEU	TYR	LYS	LYS	ILE	MET	GLU	LYS	TYR	SER

181	182	183	184	185	186	187	188	189	190	191	192	193	194	195
VAL	THR	PHE	ILE	SER	ARG	HIS	ASN	SER	TYR	ASN	HIS	ASN	ILE	LEU

196	197	198	199	200	201	202	203	204	205	206	207	208	209	210
PHE	PHE	LEU	THR	PRO	HIS	ARG	HIS	ARG	VAL	SER	ALA	ILE	ASN	ASN

211	212	213	214	215	216	217	218	219	220	221	222	223	224	225
TYR	ALA	GLN	LYS	LEU	CYS	THR	PHE	SER	PHE	LEU	ILE	CYS	LYS	GLY

226	227	228	229	230	231	232	233	234	235	236	237	238	239	240
VAL	ASN	LYS	GLU	TYR	LEU	MET	TYR	SER	ALA	LEU	THR	ARG	ASP	PRO

241	242	243	244	245	246	247	248	249	250	251	252	253	254	255
PHE	SER	VAL	ILE	GLU	GLU	SER	LEU	PRO	GLY	GLY	LEU	LYS	GLU	HIS

256	257	258	259	260	261	262	263	264	265	266	267	268	269	270
ASP	PHE	ASN	PRO	GLU	GLU	ALA	GLU	GLU	THR	LYS	GLN	VAL	SER	TRP

Figure 4.3. The amino acid sequence of the SV40 T antigen (*continued*).

271 272 273 274 275 276 277 278 279 280 281 282 283 284 285
LYS LEU VAL THR GLU TYR ALA MET GLU THR LYS CYS ASP ASP VAL

286 287 288 289 290 291 292 293 294 295 296 297 298 299 300
LEU LEU LEU LEU GLY MET TYR LEU GLU PHE GLN TYR SER PHE GLU

301 302 303 304 305 306 307 308 309 310 311 312 313 314 315
MET CYS LEU LYS CYS ILE LYS LYS GLU GLN PRO SER HIS TYR LYS

316 317 318 319 320 321 322 323 324 325 326 327 328 329 330
TYR HIS GLU LYS HIS TYR ALA ASN ALA ALA ILE PHE ALA ASP SER

331 332 333 334 335 336 337 338 339 340 341 342 343 344 345
LYS ASN GLN LYS THR ILE CYS GLN GLN ALA VAL ASP THR VAL LEU

346 347 348 349 350 351 352 353 354 355 356 357 358 359 360
ALA LYS LYS ARG VAL ASP SER LEU GLN LEU THR ARG GLU GLN MET

361 362 363 364 365 366 367 368 369 370 371 372 373 374 375
LEU THR ASN ARG PHE ASN ASP LEU LEU ASP ARG MET ASP ILE MET

376 377 378 379 380 381 382 383 384 385 386 387 388 389 390
PHE GLY SER THR GLY SER ALA ASP ILE GLU GLU TRP MET ALA GLY

391 392 393 394 395 396 397 398 399 400 401 402 403 404 405
VAL ALA TRP LEU HIS CYS LEU LEU PRO LYS MET ASP SER VAL VAL

406 407 408 409 410 411 412 413 414 415 416 417 418 419 420
TYR ASP PHE LEU LYS CYS MET VAL TYR ASN ILE PRO LYS LYS ARG

421 422 423 424 425 426 427 428 429 430 431 432 433 434 435
TYR TRP LEU PHE LYS GLY PRO ILE ASP SER GLY LYS THR THR LEU

436 437 438 439 440 441 442 443 444 445 446 447 448 449 450
ALA ALA ALA LEU LEU GLU LEU CYS GLY GLY LYS ALA LEU ASN VAL

451 452 453 454 455 456 457 458 459 460 461 462 463 464 465
ASN LEU PRO LEU ASP ARG LEU ASN PHE GLU LEU GLY VAL ALA ILE

466 467 468 469 470 471 472 473 474 475 476 477 478 479 480
ASP GLN PHE LEU VAL VAL PHE GLU ASP VAL LYS GLY THR GLY GLY

481 482 483 484 485 486 487 488 489 490 491 492 493 494 495
GLU SER ARG ASP LEU PRO SER GLY GLN GLY ILE ASN ASN LEU ASP

496 497 498 499 500 501 502 503 504 505 506 507 508 509 510
ASN LEU ARG ASP TYR LEU ASP GLY SER VAL LYS VAL ASN LEU GLU

511 512 513 514 515 516 517 518 519 520 521 522 523 524 525
LYS LYS HIS LEU ASN LYS ARG THR GLN ILE PHE PRO PRO GLY ILE

526 527 528 529 530 531 532 533 534 535 536 537 538 539 540
VAL THR MET ASN GLU TYR SER VAL PRO LYS THR LEU GLN ALA ARG

Figure 4.3. Continued

541	542	543	544	545	546	547	548	549	550	551	552	553	554	555
PHE	VAL	LYS	GLN	ILE	ASP	THE	ARG	PRO	LYS	ASP	TYR	LEU	LYS	HIS

556	557	558	559	560	561	562	563	564	565	566	567	568	569	570
CYS	LEU	GLY	ARG	SER	GLU	PHE	LEU	LEU	GLU	LYS	ARG	ILE	ILE	GLN

571	572	573	574	575	576	577	578	579	580	581	582	583	584	585
SER	GLY	ILE	ALA	LEU	LEU	LEU	MET	LEU	ILE	TRP	TYR	ARG	PRO	VAL

586	587	588	589	590	591	592	593	594	595	596	597	598	599	600
ALA	GLU	PHE	ALA	GLN	SER	ILE	GLN	SER	ARG	ILE	VAL	GLU	TRP	LYN

601	602	603	604	605	606	607	608	609	610	611	612	613	614	615
GLU	ARG	LEU	ASP	LYS	GLU	PHE	SER	LEU	SER	VAL	TYR	GLN	LYS	MET

616	617	618	619	620	621	622	623	624	625	626	627	628	629	630
LYS	PHE	ASN	VAL	ALA	MET	GLY	ILE	GLY	VAL	LEU	ASP	TRP	LEU	ARG

631	632	633	634	635	636	637	638	639	640	641	642	643	644	645
ASN	SER	ASP	ASP	ASP	ASP	GLU	ASP	SER	GLN	GLU	ASN	ALA	ASP	LYS

646	647	648	649	650	651	652	653	654	655	656	657	658	659	660
ASN	GLU	ASP	GLY	GLY	GLU	LYS	ASN	MET	GLU	ASP	SER	GLY	HIS	GLU

661	662	663	664	665	666	667	668	669	670	671	672	673	674	675
THR	GLY	ILE	ASP	SER	GLN	SER	GLN	GLY	SER	PHE	GLN	ALA	PRO	GLN

676	677	678	679	680	681	682	683	684	685	686	687	688	689	690
SER	SER	GLN	SER	VAL	HIS	ASP	HIS	ASN	GLN	PRO	THY	HIS	ILE	CYS

691	692	693	694	695	696	697	698	699	700	701	702	703	704	705
ARG	GLY	PHE	THR	CYS	PHY	LYS	LYS	PRO	PRO	THR	PRO	PRO	PRO	GLU

706	707	708
PRO	GLU	THR

Figure 4.3. Continued

thr-701 [45]. Experiments with T antigens that fail to enter the nucleus (cytoplasmic mutants) suggest that the thr-124 and thr-701 sites are phosphorylated in the cytoplasm of the cell shortly after synthesis while the serine sites are phosphorylated at later times in the nucleus [45]. Phosphorylation at thr-124 appears to be essential for efficient interaction of T antigen with the origin DNA sequences [46], while the presence of phosphate on the serine residues appears to decrease the affinity of the protein for origin DNA [46]. Removal of the phosphate residues from serine increases the efficiency of T antigen for initiation of each round of DNA synthesis in vitro [47]. Thus, it appears that phosphorylation modification is essential for T antigen binding at site II (Fig. 4.2) [46], and thr-124 is involved in this interaction. Thr-124 can be changed to ile or ala, and the T antigen produced fails to replicate SV40 DNA [48]. Interest-

Hydropathic profile of SU4CG5
Mean hydropathic index: -0.06

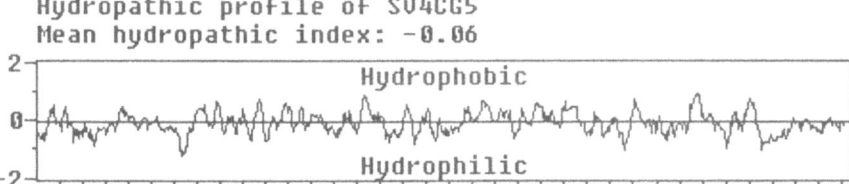

Predicted secondary structure of SU4CG5

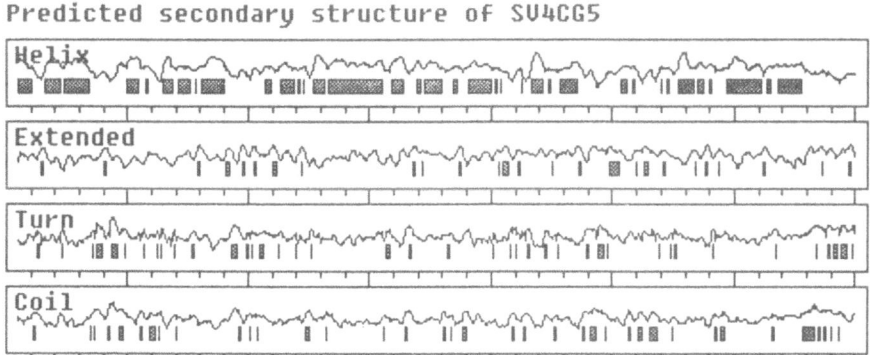

Figure 4.4. Structural predictions of the SV40 T antigen. The hydropathic profile of T antigen is shown based on the primary sequence [39]. The predicted helix, extended, turn, and coil regions of the protein (the solid rectangle indicates a probable conformation) are deduced from the sequence of amino acids [38].

ingly, these modified thr or ser residues are not in the DNA binding domain of T antigen (found in residues 135 to 249 [49,50], see below), so these phosphate groups are not directly involved in DNA-protein interactions but instead alter protein-protein interactions or conformational changes of the T antigen.

The T antigen recognition sites, GAGGC, are arranged in a different pattern in site I and site II (Fig. 4.2). The origin binding site II has two pairs of recognition sequences (inverted), while site I has a single tandem pair. A monomer or dimer of T appears to bind at each pentanucleotide, so phosphorylation could affect protein-protein interactions differently at sites I and II. This suggestion is confirmed by the observation [46] that the extent of phosphorylation of a T antigen population does not affect the affinity of T antigen binding at site I but does at site II [46,47]. T antigens accumulate increasing levels of phosphate moieties on the serine residues with time after synthesis, making these molecules less efficient for initiation of DNA replication with time. Whether this process activates other functions of T antigen remains to be explored.

Oligomeric Forms of T Antigen

SV40 T antigen from virus-infected cells sediments in three broad species: 5–7S, 15–16S, and 21–23S. The 5–7S species are labeled first with ^{35}S-methionine in an experiment measuring the kinetics of synthesis of T antigen, and these appear to be the monomer (5.5S) and dimer (7S) forms [51] of this protein. These molecules can be chased into a tetrameric form (15–16S), which then can associate with one or more cellular proteins (p53, the RB gene product) to produce the 21–23S species. The 7S form of T antigen has ATPase and DNA-binding activities (the 5.5S form appears not to have these activities [51]) and a much higher affinity for binding to origin DNA [51] than the larger forms of T antigen. This is in good agreement with the observation that the 15–16S and 21–23S forms of T antigen contain more phosphate residues (on serine) per molecule (bind poorly to Site II, see above) than the 5–7S forms. Clearly, some of the T antigen activities can be modulated by the state of oligomerization and protein modification.

Localization of T Antigen in the Cell

The great majority of the SV40 large T antigen is localized in the nucleus of the cell. A five amino acid sequence between residues 127 and 131, lys–lys–lys–arg–lys, is required for T antigen accumulation in the nucleus [52]. This peptide has been fused to other proteins, and the majority of the fused product is localized in the nucleus. The lys-128 residue appeared to be the most significant amino acid of the five examined (lys-128 mutated to thr lost all activity of nuclear localization) [52]. Interestingly, in some T antigen mutants a defective nuclear signal is dominant over the wild type, suggesting that the altered signal inhibits the transport mechanism [53]. Cytoplasmic mutants of SV40 T antigen can transform established mouse or rat cell lines but fail to transform primary cells in culture [54]. These results suggest that higher concentrations of nuclear T antigen are required for immortalization of cells. These data do not prove that cytoplasmic T antigen can transform cells, because some percentage of this protein is in the nucleus (and replicates SV40 DNA inefficiently) even with these cytoplasmic mutant forms of T antigen.

About 1% to 5% of the T antigen appears at the plasma membrane, and this form is associated with a fatty acid acetylated T antigen [41] (see above). There has always been considerable debate about the possibility that T antigen could stick to membranes in an artifactual manner. Cytotoxic T cell clones have been shown to kill SV40-transformed cells expressing T antigen (in an H-2 restricted fashion), and the mechanism is presumably via T antigen epitope recognition at the cell surface [15,55] (see below).

Association of T Antigen with Cellular Proteins

T antigen has been shown to be associated with several cellular proteins: p53 [56,57], the retinoblastoma sensitivity gene product or RB [58], the transcription factor AP-2 [18], and the DNA polymerase alpha enzyme activity [59]. In most cases, the evidence for a specific association rests with criteria such as copurification, coimmunoprecipitation, and, in some cases, in vitro association of purified components.

The retinoblastoma susceptibility gene (RB gene), when deleted or inactivated, predisposes the host to malignant retinoblastoma, and this gene has been recently identified [60,61]. The protein (RB) encoded by this gene is a nuclear phosphoprotein that binds to DNA. Because both alleles of this gene must be inactivated to predispose the host to cancer, it has been termed a suppressor oncogene product or a recessive oncogene [62]. The binding of the T antigen to RB can be disrupted by mutations localized between residues 104 and 114 in the T antigen protein [58], and the same mutations eliminate the ability of T antigen to transform some cells [35,36,58]. These mutations map in the N-terminal transforming domain of T antigen (see below).

p53 is also a nuclear phosphoprotein that binds to DNA. Wild-type p53 cDNA clones or genomic clones expressing the p53 protein fail to transform primary rat cells when cotransfected with an activated *ras* oncogene, while mutant p53 cDNA and genomic clones will transform such cells [63,64]. Mutations at many different residues [63] in the p53 gene activate this gene product for transformation (mutants between amino acids 130 and 230 out of 390 amino acids), suggesting that the inactivation of a function may be important here. The mutant p53 protein is always produced in large amounts in transformed cells, and it complexes with a heat shock protein (hsc70). The p53-hsc70 complex then sequesters the rat wild-type p53 protein produced by the host cell [63,64]. This "mutant p53-hsc70/wild-type p53" complex may then lead to the inactivation of wild-type p53 function in such cells (a transdominant p53 mutation).

The deletion or rearrangement of p53 genes and altered p53 gene products (loss of wild-type function) have been observed in erythroleukemias induced by the Friend virus complex in mice [65,66]. These experiments suggest that p53 is an antioncogene, a recessive oncogene, or a suppressor of oncogenic activity that must be altered or deleted in cancer cells. The available evidence suggests that p53 binds to the SV40 large T antigen between amino acid residues 371 and 625 [50] in a domain that is also required for the transformation of some cell types [67] (see below). This domain also contains the ATPase and helicase activity [23,24,28] and the DNA polymerase alpha binding site [59]. Indeed, the p53 protein and DNA polymerase α can compete for binding to the SV40

T antigen [68] and the p53 protein can inhibit SV40 DNA replication in monkey cells [69].

It is probably more than coincidence that two nuclear phosphoproteins that appear to be recessive oncogenes (p53 and RB) both bind to the SV40 T antigen. RB binding can be localized to the T antigen transforming domain between amino acid residues 1 and 120 while p53 interactions occur in a second transforming domain at residues 371 to 625. These two domains can be distinguished by the cell type transformed by each of these regions of the SV40 T antigen; residues 1 to 120 will immortalize primary rat cells in culture [35] and induce foci production in C3H 10T½ cells but not in REF52 cells in culture [67]. T antigen containing the 371 to 625 region of the protein can transform these REF52 cells, indicating the additional transformation activity of this domain [67]. The role of T antigen here may then be to bind to p53 and RB and inactivate or prevent their normal function in a cell (i.e., eliminate the antioncogene function). The failure of p53 and RB to function could start the cell on the pathway to transformation. A similar suggestion for the T-AP-2 interaction has been made by Mitchell et al. [18]. AP-2 is a transcription factor that is required for expression of some cellular gene products but is not essential for SV40 transcription [18]. The possibility that p53 and RB are also transcription factors (negative or positive activities) must now be considered.

The Immune Response to SV40 T Antigen

Animals containing SV40-induced tumors contain cells expressing the viral-encoded T antigens. These proteins are recognized as foreign by the immune system, and both a humoral- and cellular-mediated response is mounted. The humoral response produces antibodies directed against a variety of epitopes [1,5]. Lane and his colleagues have mapped the epitopes on T antigen detected by a series of 53 monoclonal antibodies that were derived as independent events in a BALB/c mouse. Figure 4.5 presents a distribution of antibodies mapped to various regions of T antigen. It is clear from this distribution that there is a 3- to 16-fold higher incidence of epitope recognition at the N and C termini (amino acids 1 to 82 and 682 to 708) of the molecule. This is undoubtedly a reflection of several variables:

1. The ends of proteins are often immunodominant sites.
2. The ends of T antigen (amino acids 1 to 120 and 615 to 708; see Fig. 4.4) are among the most hydrophilic regions of the molecule and are probably on the outside surface of the antigen.
3. The N terminus has a strong alpha-helical content while the C terminus is rich in a coil-turn motif (Fig. 4.4) that provides structure to the epitopes.

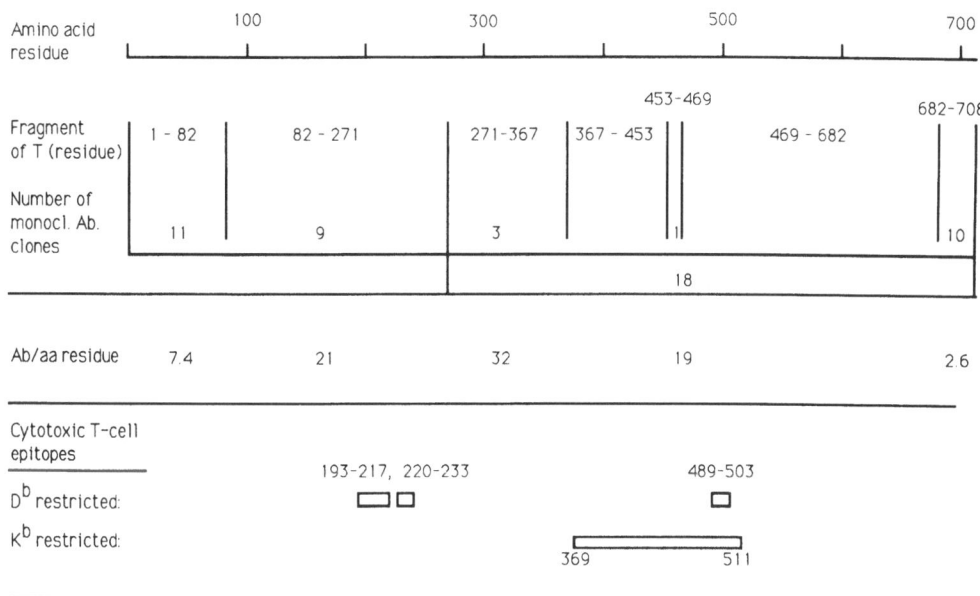

Figure 4.5. Distribution of epitopes recognized by the B and T cell responses to SV40 large T antigen. The number of independently derived monoclonal antibodies per amino acid residue (Ab/aa residue) demonstrates the density of epitopes recognized by antibodies. Cytotoxic T cells recognize different epitopes.

While these structures are likely important for immunogenic activity (antigen presentation), the majority of monoclonal antibodies that react with the ends of T antigen are conformation independent and can bind to T antigen denatured in sodium dodecyl sulfate [5].

When animals are immunized with killed SV40-transformed cells or purified T antigen, they reject (no tumor formation) a subsequent injection of live tumorigenic cells expressing SV40 T antigen [1]. A variety of experiments have demonstrated that the tumor-specific transplantation antigen is T antigen itself and the rejection is mediated by killer or cytotoxic T-cells (CD-8 cells) [1]. The T cells are probably recognizing the SV40 T antigen on the cell surface (see above). CTL (cytotoxic lymphocyte) clones of cells have been developed in a C57B1/6J mouse background. These T cell clones kill SV40-transformed cells in culture, and this killing is H-2 or histocompatibility restricted (i.e., both the CTL and target must be the same H-2 type) [70]. Employing deletion mutants of SV40 T antigen, different clones of CTL cells were shown to recognize several discrete epitopes on the T antigen molecule (Fig. 4.5). Clones that were H-2D^b restricted recognized amino acid residues 193 to 217, 220 to 233, and 489 to 503 (based on deletion mutant analysis) [70]. Employing peptides as recognition sites for this epitope, the 193- to 217-site could be confirmed by this additional approach.

An H-2K^b-restricted epitope (at the K gene instead of the D gene) was also found between residues 369 and 511 on the T antigen molecule. Clearly, the major epitopes recognized by CTL clones (residues 193 to 233; Fig. 4.5) do not correspond to the predominant immunogenic regions eliciting a B cell response (residues 1 to 82 and 682 to 708). While this is expected because T cells and B cells have differences in recognition of immunogenes, a drawback of this comparison is that the monoclonal antibodies were produced in a BALB/c (H-2^d) background and the CTL clones were produced in a C57Bl/6(H-2^b) background. In spite of this, the major epitopes recognized by T cell clones (193–233; see Fig. 4.4) have structural features (no helix, rich in extended regions and turns) and a hydropathy index that may be consistent with a membrane insertional region of T antigen as a recognition site for CTLs. Residues 489 to 503 also have no alpha helical regions but are not notably hydrophobic.

SV40 T Antigen: DNA Replication Functions

Binding to the SV40 Origin, Site II

SV40 DNA replicates bidirectionally from a unique origin of replication [1,26,27]. The interaction of T antigen with this origin DNA (site II; Fig. 4.2) was proved by the requirement for a second site mutation in the SV40 T antigen gene (at amino acid 157 or 166 in the ori-DNA binding domain) to suppress a primary site mutation in the origin DNA region (at *cis*-acting site II) [22]. The minimal origin of SV40 DNA replication, a *cis*-acting site, was defined by cloning different DNA fragments into a plasmid and observing whether the plasmid replicated in monkey cells expressing T antigen (COS cells) or by an extensive deletion and point mutation analysis around this origin region [reviewed in 26]. The minimal origin is composed of 65 base pairs between nucleotides 31 and 5209 (Fig. 2) [26]. The origin is constructed from (1) a 17-bp AT-rich region, (2) a 27-bp perfect palindrome containing two pairs of T antigen-binding sites (GAGGC) in inverted tandem arrays, and (3) a so-called early element (located on the early side of the origin) which also contains a palindromic sequence which can be folded into a second hairpin loop (ATAAGG-TCTT(C)AT).

Binding of T antigen to site II is weaker than to site I [20,26], and even though T antigen recognizes the same core sequence, CAGGC in each case, the pattern or arrangement of these sequences in site I and II differs (Fig. 4.2). The functions of site I (negative regulation of early transcription) and site II (DNA replication initiation) also differ [20]. Binding of T antigen to site II is apparently independent of site I binding (not cooperative) [20]. All of these interactions have been demonstrated by DNA-protein protection experiments (nuclease, methylation), immunoprecipitation of DNA-protein complexes with antibody to T antigen, and genetic analyses using second-site suppressor mutations to indicate these interactions [reviewed in 2,4,6,26].

An extensive genetic analysis of the SV40 T antigen gene [2,4,6,22,26,36,47–49,67] has localized a domain in this protein, between amino acid residues 135 and 249, that is required for sequence specific binding to SV40 DNA. In addition, fragments of SV40 T antigen (fusion proteins) have been employed to demonstrate that amino acids 1 to 271 bind to SV40 DNA, but the C-terminal two-thirds of the protein is not required and does not bind to this DNA [5,50]. Almost all of the mutations in this region of the T antigen gene (residues 135 to 249) are DNA replication defective and cellular transformation competent [36,48,49,67].

The SV40 Zinc Finger

Klug and Rhodes [71] have pointed out that many proteins that recognize DNA sequences in a specific fashion utilize a structural element termed a zinc finger, which is composed of the sequence motif cys-$X_{(2-5)}$-cys-$X_{(2-12)}$-his or cys-$X_{(2-5)}$-cys, where X may be a variety of amino acids. The histidines and cysteines are thought to interact to hold a zinc ion in a core structure with two minor loops (cys-$X_{(2-5)}$-cys or his-$X_{(2-5)}$-his) and a major loop (cys-$X_{(2-12)}$-his) which forms the finger. The SV40 T antigen has a single zinc finger motif between residues 302 and 320 just outside (139 to 249 residues) the region of the protein required for specific DNA binding (a fragment of amino acids 1 to 271 binds specifically to sites I and II [5]). An extensive mutational analysis of the amino acids that compose this region (302 to 320) demonstrates that many amino acids in the proposed zinc finger are required for DNA replication [72]. Some of these mutants were replication poor and transformation competent, while others were replication negative and transformation inefficient [72]. This latter class of mutants could result from large structural changes in the protein and as such may be less informative about discrete domain structure-function relationships.

ATPase and Helicase Activity of T Antigen

The SV40 T antigen has an intrinsic phosphohydrolase activity that is stimulated by the addition of poly dT [73]. Monoclonal antibodies directed against defined epitopes in T antigen block this ATPase activity (PAb204). This antibody-binding site has been mapped to an epitope between amino acids 453 and 469 in a domain (amino acids 371 to 625) required for DNA replication [5]. This ATPase works in association with the helicase activity to unwind the DNA at the origin of replication (initiation events) [4,23,24] and at each replication fork (propagation events) [4,23,24,28]. Monoclonal antibody PAb204 also inhibits the helicase activity of T antigen and blocks in vitro DNA replication [5]. A number of mutants in the T antigen domain localized between residues 371 and 625 fail to replicate DNA, and some are ATPase negative/helicase negative while others are ATPase positive/helicase negative [67]. Similarly, proteolytic fragments of T antigen and fusion proteins that contain this region of the protein (371 to 625) have ATPase activity [5,50].

T Antigen Binding of the Alpha DNA Polymerase and p53

Monoclonal antibodies that bind to epitopes on T antigen localized between residues 271 to 628 (PAb205, 414, 413) block the binding of alpha DNA polymerase and p53 to T antigen in vitro. Other antibodies such as PAb204 (epitope 453 to 469 residues) fail to block p53 or alpha DNA polymerase binding to T even though the ATPase and helicase activities are inhibited [5]. Similarly, fusion proteins of T antigen [5] and proteolytic fragments of T [50] bind to p53 or alpha DNA polymerase so long as the residues between 271 and 708 are present. A variety of mutants in T antigen localized in the 371- to 625-domain fail to bind to p53 or the alpha DNA polymerase [36,40,48,49,67,74].

Thus, the functional domain localized in T antigen between residues 371 and 625 contains several activities essential for DNA replication: (1) ATPase-helicase, and (2) the alpha DNA polymerase-binding site (this enzyme is likely the major polymerase employed during SV40 DNA replication [26]). The binding of murine p53 to T antigen, which appears to be localized in a site similar to the alpha polymerase, blocks alpha polymerase binding and DNA replication in monkey cells [68,69]. Whether or not this is employed as a mechanism to regulate DNA replication in the normal infected cell remains unclear.

Transcriptional Regulation by T Antigen

The monomeric or dimeric T antigen molecule binds to SV40 DNA site I (Fig. 4.2) with a higher affinity than site II. While binding to site II appears to be weakened by phosphorylation modifications at serine residues (see above), site I binding is not affected by this. As RNA polymerase II prepares to transcribe the early region T antigen genes, AP-1 and AP-3 must bind to the enhancer DNA [17,18], SP-1 binds to the CG boxes in the 21-bp repeats (Fig. 4.2), and the TATA box is recognized by its factor. This collection of nucleoproteins presumably guides the RNA polymerase to the correct site on the DNA. The initiation site for these transcripts to produce T antigen (before DNA replication, early-early or E_ERNA) and the TATA box that determines the initiation site (20 to 24 bp upstream) are localized in site II within the minimal origin region. Similarly the L_E or late-early RNA (made after DNA replication) initiates in this region

(Fig. 4.2). Because the binding of T antigen to site I is known to modulate down the rate of transcription across this gene [21], it is felt that T antigen binding to site I blocks the progress of RNA polymerase II and so it acts as a repressor at that site.

T antigen also acts as a positive effector, able to transactivate and increase the level of the late set of viral transcripts [29,75]. Two discrete sequences in the origin region appear to be critical for this T antigen–mediated positive regulation of late transcription: (1) within the 72-bp enhancer region, and (2) within the minimal origin sequence (Fig. 4.2) [29,75]. It is of some interest that the spacing between these elements is critical for this positive regulation [76], suggesting that transcription factors, interacting with DNA and T antigen in a complex that spans this region, may be responsible for such positive regulation [76]. This organization of a protein, T antigen, binding or interacting with two adjacent sites, one for positive and one for negative regulation of transcription, oriented on two different DNA strands, is a familiar set of observations first described in prokaryotes [77].

T antigen also has a positive effect on the level of late mRNA synthesis by virtue of its ability to initiate DNA replication. At least one or several rounds of DNA replication are required for optimal late gene transcription [1], and the reasons for this requirement are unclear. Increased numbers of viral DNA templates contribute to this increased level of late mRNA, but this does not fully explain the high rate of late gene transcription. Similarly, T antigen is required over and above these events to produce maximum levels of late mRNA. Binding of T antigen to site III on the DNA (Fig. 4.2; GGGGC, TGGGC on the late side of the minimal origin DNA) is the weakest binding site in vitro and is not involved in T antigen–mediated gene regulation [20].

While it is clear that binding to site I (negative regulation) requires the sequence-specific DNA binding domain of T antigen (residues 135 to 249), the positive regulation of late transcription has not been mapped to a functional domain of this protein. Indeed, it is not even clear if this positive regulation requires T antigen binding to DNA, although one of the T antigen mutants that failed to bind to DNA also failed to transactivate late virus functions [29,75,76].

Just as T antigen can regulate viral RNA synthesis in a positive and/or negative manner, it has often been suggested that this protein may regulate cellular gene transcripts in a similar or analogous fashion [1–4,6]. No clear-cut examples of this have been demonstrated at the gene or transcription level. T antigen can mediate an increased level of several cellular enzyme activities such as thymidine kinase or deoxycytidine kinase [1,31]. The mechanism by which T antigen does this (transcription, protein stability, indirect effects) is not clear. Such effects may be a consequence of T antigen's ability to stimulate cells from G_0 arrest into the S phase of the cycle. The role of T antigen modulation of transcription in transformation remains to be explored. The binding of T antigen to AP-2 [18], p53 [56,57], and the RB protein [58] may well suggest one mechanism for the way in which T antigen could regulate transcription (see above).

The Host Range Domain of SV40 T Antigen

Mutants of T antigen lacking the last 26 amino acid residues (dl 2459 has residues 1 to 682) [78,79] replicate the virus normally in some monkey cell lines (CV-1P cells) but not in other monkey kidney cell lines (BSC-1 or Vero cells). These mutants synthesize normal levels of late mRNA, but little or no late proteins are made, nor are they assembled into virions. This mutant (dl 2459) can be complemented (intragenic complementation) by some SV40 T antigen gene mutants [2], suggesting a different set of functions in the same molecule and different genetic domains.

This same domain (residues 676 to 708) has a second phenotype—i.e., when present it is able to permit human adenovirus to replicate and produce infectious virus in monkey cells in culture [80]. The human adenoviruses normally produce infectious virus in human cell cultures but fail to synthesize infectious viruses in monkey cells [1,80]. The block to adenovirus growth in monkey cells is after viral DNA replication, which is normal. Some late mRNAs (especially the mRNA for fiber antigen) are made in much reduced amounts (splicing defect, attenuation effects), and several late proteins are also produced at very low levels (translational defect), leading to the failure to make virions in monkey cells [81]. An adenovirus-encoded mutation in the adenovirus 72-kD single-stranded DNA binding protein (also an early viral protein) can compensate for this defect and produce virions in monkey cells [82]. Similarly, a fusion protein with only the amino acids from 676 to 708 of SV40 T antigen can permit adenovirus to replicate and produce virions in monkey cells [79]. Thus, the last 32 or 33 amino acids of T antigen (carboxyterminal domain) can act independently to help the synthesis and assembly of SV40 virion proteins or adenovirion proteins in certain host cells. This has been called the host cell or helper function. Interestingly, the N-terminal portion of this host range domain contains some of the protein modification (ser-676, ser-677, ser-679, and thr-701) sites for phosphorylation (see above).

Stimulation of Cellular DNA Synthesis

Infection of resting cells with SV40 often results in the stimulation of cellular DNA synthesis in these cells [1]. Microinjection of purified T-antigen into serum-deprived resting cells also results in the synthesis of cellular DNA [32]. A detailed genetic analysis employing deletion mutants of T antigen localized the property of induction of cellular DNA synthesis between amino acid residues 20 and 275 [83]. The mechanism of this induction of cellular entry into S phase is unclear, but it does not require the ability of SV40 T antigen to bind to specific viral DNA sequences. There are many examples of viral DNA replication-defective, viral DNA binding-defective T antigen mutants, which stimulate cellular DNA synthesis and transform cells in culture [35,36,40,48,49,74,83]. Thus, the mechanism by which T antigen induces a cell to leave G_o and enter S phase is not by initiating cellular DNA synthesis at origin sequences that resemble viral DNA sequences.

Summary: The Functional Domains of SV40 Large T Antigen

Figure 4.6 presents a representation of the experimental data reviewed here that identifies nine properties of SV40 large T antigen localized into the functional domains of this protein. It is clear from this summary that there are two well-separated domains (1 to 120 and 371 to 625 residues) involved in the transformation of cells in culture and for which the nature of the host cell type or assays employed permits the identification of different regions of the protein. The two transformation-associated domains each correspond to a region of the protein that binds to or interacts with a different recessive oncogene; 1 to 120 amino acids (RB) and 371 to 625 residues (p53), and one of these domains overlaps the ability of a T antigen fragment (20 to 275 residues) to act as a mitogen and stimulate G_o-arrested cells into S phase. The nuclear localization signal (127 to 131 residues), the adjacent sequence-specific DNA binding domain (135 to 249 residues), and the adjacent zinc finger (302 to 320 residues) all lie outside the cellular transforming functions (and therefore are not required for this). Instead, these elements are the core of the viral DNA-protein interaction machinery.

The adjacent protein modifications (thr 124 and ser-106, 111, 112, and 123) can affect DNA-protein interactions. The sequence-specific recognition element (135 to 249 residues) binds to DNA (GAGGC sites) and brings the adjacent domain (371 to 625 residues) onto the origin of DNA replication where its ATPase helicase activity unwinds the DNA and initiates primase function—i.e., RNA primer synthesis. The same domain (371 to 625 residues) binds to alpha DNA polymerase, placing it adjacent to an RNA primer, extending the 3′ ends, and producing Okazaki fragments or continuous-strand DNA synthesis. This combination of T-ATPase-helicase-alpha DNA polymerase acts not only at the origin of DNA replication but at each replication fork (propagation step).

There is a certain symmetry in these domain interactions: (1) the two transforming domains (1 to 120—RB with 371 to 625—p53) cooperate to

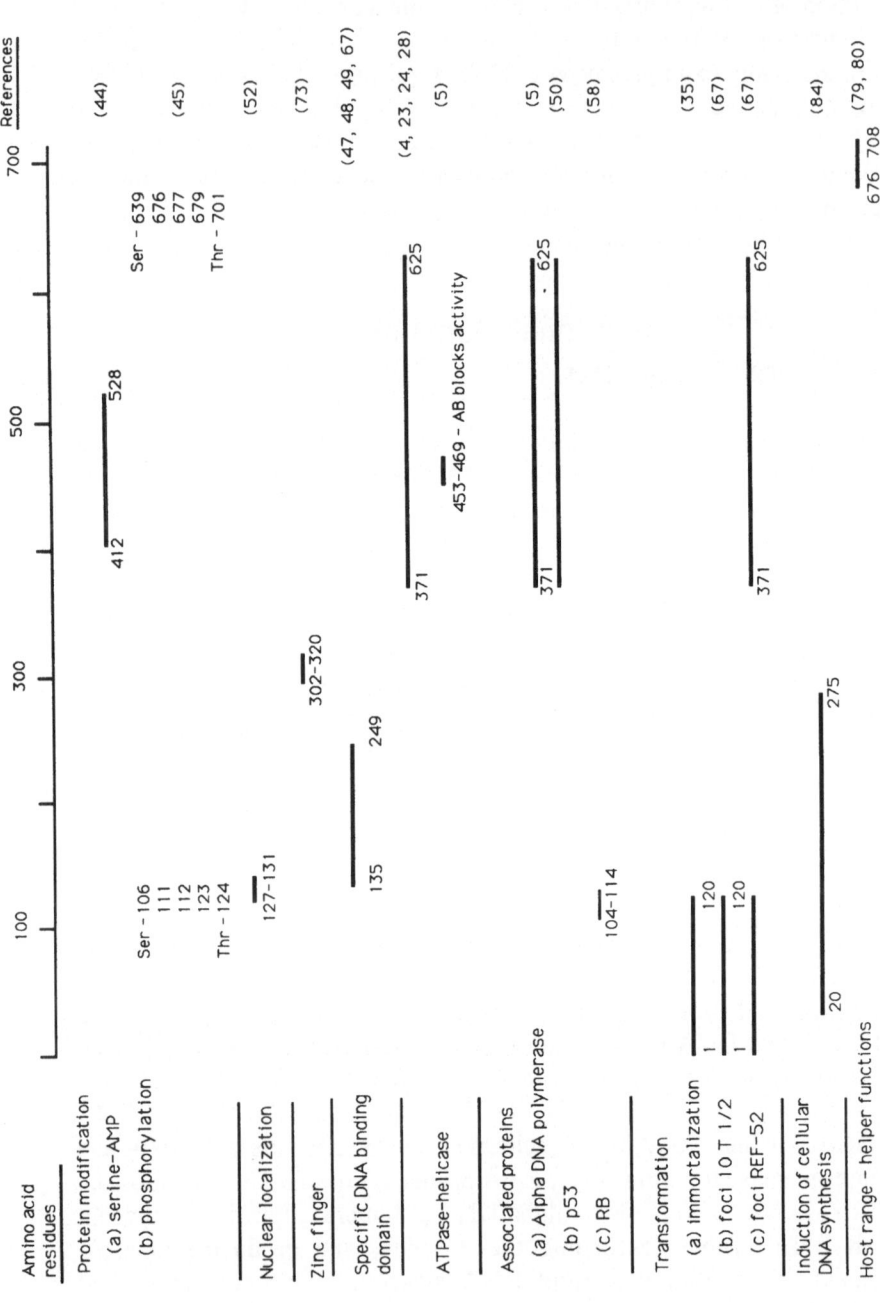

Figure 4.6. A map of the functional domains of the SV40 T antigen. A list of nine protein modifications, structural features, functional activities, regions associated with other proteins, and regions responsible for transforming activities are indicated on a map of the T

transform cells, and (2) the two replication domains (135 to 249—DNA binding; 302 to 320—zinc finger with 371 to 625—ATPase-helicase-alpha DNA polymerase domain) cooperate to initiate and propagate a replicating molecule. These two types of domain interactions share a common domain (371 to 625) which functions at a distinct region of the protein. Furthermore, the first transforming domain is separated from the first DNA-binding domain by a cluster of phosphorylation sites that modulates binding activity at the origin site. The second cluster of such protein modification sites are adjacent to the common domain (371 to 625 is followed by ser-639 to thr-701 sites). These phosphorylation sites can affect (or be affected by) the oligomeric state of a molecule, p53 binding, or binding to DNA, suggesting a role for protein modification in conformation and function of this protein. Finally, the C-terminal fragment (676 to 708 residues) of T antigen can somehow regulate synthesis, stability, and/or assembly of late structural virion proteins in specific host cells. It does so for SV40 and can even act on behalf of the human adenoviruses in monkey cells to permit a new host range extension. Remarkably, the 676 to 708-fragment can act alone as a fusion protein, attached to a heterologous carrier; it behaves like an independent functional domain.

The temperature-sensitive mutations of the SV40 T antigen gene (tsA mutations [1]) all map (with the exception of tsA.7) between residues 250 and 550. Most of these mutants produce proteins that are unstable (do not form oligomeric proteins or complexes) and are degraded at the nonpermissive temperature. As such, these mutants regulate both viral and transforming functions in a temperature-conditional fashion. These mutants also map in the domain common to DNA replicative and transforming activities.

Presumably, the SV40 large T antigen gene has evolved to package multiple functions in a single protein so as to conserve coding information and minimize the size of the viral genome. The gene organization, in both the early and late regions of the genome, has common sequences shared by different proteins (T and t antigens) and overlapping sequences (VP-2, VP-3, and VP-1) (Fig. 4.1). It is perhaps the need for a compact genome that such a protein as the SV40 T antigen evolved with multiple functions and enormous regulatory abilities. It will be interesting to determine if cellular encoded proteins will also evolve to cluster so many functional units in one protein or, in fact, that the SV40 T antigen will be more unique in that it will have many lessons to teach us in a single protein.

Acknowledgments. The author thanks K. James and D. Welsh for help with the manuscript. This research is supported by grants MV47-I from the American Cancer Society and CA38757-04 and CA41086-03 from the National Institutes of Health.

References

1. Tooze, J. (ed.) (1982) Molecular Biology of Tumor Viruses, Part 2, DNA Tumor Viruses, Chaps. 2–5, Cold Spring Harbor Laboratory, Cold Spring Harbor, New York.
2. Livingston, D., and Bradley, M. (1987) Mol. Biol. Med. *4*, 63–80.
3. Rigby, P., and Lane, D. (1983) Adv. Viral Oncol. *3*, 31–57.
4. Stahl, H., and Knippers, R. (1987) Biochim. Biophys. Acta *910*, 1–10.
5. Mole, S.E., Gannon, J.V., Ford, M.J., and Lane, D.P. (1987) Phil. Trans. R. Soc. Lond. *317*, 455–469.
6. Salzman, N. (ed.) (1986) The Papovaviridae, Vol. 1, Chap. 3, Plenum Publishing Corp., New York.
7. Sweet, B.H., and Hilleman, N.R. (1960) Proc. Soc. Exp. Biol. Med. *105*, 420–426.
8. Fiers, W., Contreras, R., Haegeman, G., Rogiers, R., Van de Voorde, A., Van Heuversyrun, H., Van Herreweghe, J., Volckaert, G., and Ysebaert, M. (1978) Nature *273*, 113–125.
9. Reddy, V.B., Thimmappaya, B., Dahr, R., Subramanian, K., Zain, B.S., Pan, J., Ghosh, P.K., Celma, M.L., and Weissman, S.M. (1978) Science *200*, 494–502.
10. Berk, C.J., and Sharp, P. (1978) Proc. Natl. Acad. Sci. USA *75*, 1274–1278.
11. Crawford, L.V., Cole, C.N., Smith, A.E., Paucha, E., Tegtmeyer, P., Rundell, K., and Berg, P. (1978) Proc. Natl. Acad. Sci. USA *75*, 117–121.
12. Resnick, J., and Shenk, T. (1986) J. Virol. *60*, 1098–1106.
13. Girarde, A.J., Sweet, B.H., Slotnick, V.B., and Hilleman, M. (1962) Proc. Soc. Exp. Biol. Med. *109*, 649–654.
14. Black, P.H., Rowe, W.P., Turner, H.C., and Hubner, R.J. (1963) Proc. Natl. Acad. Sci. USA *50*, 1148–1152.
15. Levine, A.J. (1982) Adv. Cancer Res. *37*, 75–109.
16. Watkins, J.F., and Dulbecco, R. (1967) Proc. Natl. Acad. Sci. USA *58*, 1396–1401.
17. Lee, W., Mitchell, P., and Tjian, R. (1987) Cell *49*, 741–752.
18. Mitchell, P., Wang, C., and Tjian, R. (1987) Cell *50*, 847–861.
19. Mermod, N., Williams, T.J., and Tjian, T. (1988) Nature *332*, 557–561.
20. DeLucia, A.L., Lewton, B.A., Tjian, R., and Tegtmeyer, P. (1983) J. Virol. *46*, 143–150.
21. Rio, D., Robbins, A., Meyers, R., and Tjian, R. (1980) Proc. Natl. Acad. Sci. USA *77*, 5706–5710.
22. Shortle, D.R., Margolskee, R., and Nathans, D. (1979) Proc. Natl. Acad. Sci. USA *76*, 6128–6131.
23. Stahl, H., Scheffner, M., Wiekowski, M., and Knippers, R. (1988) In Cancer Cells, 6, Kelly, T., and Stillman, B. (eds.) Cold Spring Harbor Laboratory, Cold Spring Harbor, New York.
24. Stahl, H., Droge, P., and Knippers, R. (1986) EMBO J. *5*, 1939–1944.
25. Hay, R.T., Hendrickson, E.A., and DePamphilis, M.L. (1984) J. Mol. Biol. *175*, 131–144.
26. DePamphilis, M.L., and Bradley, M. (1986) In The Papovaviridae, Vol. 1, Chap. 3, N.P. Salzman (ed.). Plenum Publishing Corp., New York.
27. Dana, K.J., and Nathans, D. (1972) Proc. Natl. Acad. Sci. USA *69*, 3097–3101.

28. Wiekowski, M., Droge, P., and Stahl, H. (1987) J. Virol. *61*, 411–418.
29. Brody, J., Bolen, J.B., Radonovich, M., Salzman, N., and Khoury, G. (1984) Proc. Natl. Acad. Sci. USA *81*, 2040–2044.
30. Galanti, N., Jonak, G.J., Saprano, K.J., Floros, J., Kaczmarech, L., Weissman, S., Reddy, V.B., Tilghman, S.M., and Baserga, R. (1981) J. Biol. Chem. *256*, 6469–6476.
31. Postel, E.H., and Levine, A.J. (1976) Virology *73*, 206–216.
32. Tjian, R., Fey, G., and Graessmann, A. (1978) Proc. Natl. Acad. Sci. USA *75*, 1279–1284.
33. Sambrook, J., Westphal, H., Srinivasan, P.R., and Dulbecco, R. (1968) Proc. Natl. Acad. Sci. USA *60*, 1288–1293.
34. Risser, R., and Pollack, R. (1974) Virology *59*, 477–483.
35. Colby, W.N., and Shenk, T. (1982) Proc. Natl. Acad. Sci. USA *79*, 5189–5194.
36. Peden, K.W.C., and Pipas, J.M. (1985) J. Virol. *55*, 1–9.
37. Griffin, J.D., Light, S., and Livingston, D.M. (1978) J. Virol. *27*, 218–226.
38. Garnier, J., Osguthorpe, D.J., and Robson, B. (1978) J. Mol. Biol. *120*, 97–120.
39. Kyte, J., and Doolittle, R.F. (1982) J. Mol. Biol. *157*, 105–132.
40. Paucha, E., Mellon, A., Harvey, R., Smith, A.E., Hewick, R.M., and Waterfield, M.D. (1978) Proc. Natl. Acad. Sci. USA *75*, 2165–2169.
41. Klockman, U., and Deppert, W. (1983) EMBO J. *2*, 1151–1157.
42. Jarvis, D.L., and Butel, J.S. (1985) Virology *141*, 173–189.
43. Goldman, N.D., Brown, M., and Khoury, G. (1981) Cell *24*, 567–572.
44. Bradley, M.K., Hudson, J., Villaneuva, M.S., and Livingston, D.M. (1984) Proc. Natl. Acad. Sci. USA *81*, 6574–6578.
45. Scheidtmann, K.H., Echle, B., and Walter, G. (1982) J. Virol. *44*, 2165–2169.
46. Klausing, K., and Knippers, R. (1988) In Current Topics in Microbiology and Immunology, Knippers, R., and Levine, A.J. (eds.). (In press.)
47. Mohr, I.J., Stillman, B., and Gluzman, Y. (1987) EMBO J. *6*, 153–160.
48. Kalderon, D., and Smith, A.E. (1984) Virology *139*, 109–134.
49. Paucha, E., Kalderon, D., Harvey, R.W., and Smith, A.E. (1986) J. Virol. *57*, 50–64.
50. Simmons, D.T. (1988) Proc. Natl. Acad. Sci. USA *85*, 2086–2090.
51. Bradley, M.K., Griffin, J.D., and Livingston, D.M. (1982) Cell *28*, 125–134.
52. Kalderon, D., Richardson, W.D., Markham, A.F., and Smith, A.E. (1984) Nature *311*, 33–35.
53. Lanford, R.E., and Butel, J. (1980) Virology *105*, 314–327.
54. Lanford, R.E., Wong, C., and Butel, J.S. (1985) J. Virol. *5*, 1043–1050.
55. Tevethia, S., Tevethia, M., Lewis, A., Reddy, V., and Weissman, S. (1983) Virology *128*, 319–330.
56. Lane, D.P., and Crawford, L.V. (1979) Nature *278*, 261–263.
57. Linzer, D.I.H., and Levine, A.J. (1979) Cell *17*, 43–52.
58. DeCaprio, J.A., Ludlow, J.W., Figge, J., Shew, J.Y., Huang, C.M., Lee, W.H., Marsilio, E., Paucha, E., and Livingston, D.M. (1988) Cell *54*, 275–283.
59. Smale, S.T., and Tjian, R. (1986) Mol. Cell. Biol. *6*, 4077–4087.
60. Friend, S.H., Horowitz, J.M., Gerber, M.R., Wang, X.F., Bagenmann, E., Li, F.R., and Weinberg, R.A. (1987) Proc. Natl. Acad. Sci. USA *84*, 9059–9063.

61. Lee, W.H., Bookstein, R., Hong, F., Young, L.J., Shew, J.Y., and Lee, E.Y.H.P. (1987) Science *235*, 1394–1399.
62. Knudson, A.G. (1971) Proc. Natl. Acad. Sci. USA *68*, 820–823.
63. Finlay, C.A., Hinds, P.W., Tan, T.-H., Eliyahu, D., Oren, M., and Levine, A.J. (1988) Mol. Cell. Biol. *8*, 531–539.
64. Hinds, P., Finlay, C., and Levine, A.J. (1988) J. Virol *63*, 739–746.
65. Mowat, M., Cheng, A., Kimura, N., Bernstein, A., and Benchinol, W. (1985) Nature *315*, 633–636.
66. Munroe, D.G., Rovinski, B., Bernstein, A., and Benchinol, S. (1988) Oncogene *2*, 621–624.
67. Srinivasan, A., Peden, K.W.C., and Pipas, J.M. (1988) In Current Topics in Microbiology and Immunology, Knippers, R., and Levine, A.J. (eds.). (In press.)
68. Gannon, J.V, and Lane, D.P. (1987) Nature *329*, 456–468.
69. Braithwaite, A.W., Sturzbecher, H.W., Addison, C., Palmer, C., Rudge, K., and Jenkins, J.R. (1987) Nature *329*, 458–460.
70. Anderson, R.W., Tevethia, M., Kalderon, D., Smith, A.E., and Tevethia, S.S. (1988) J. Virol. *62*, 285–296.
72. Klug, A., and Rhodes, D. (1987) Science *12*, 464–469.
73. Loeber, G., Parsons, R., and Tegtmeyer, P. (1988) In Current Topics in Microbiology and Immunology, Knippers, R., and Levine, A.J. (eds.). (In press.)
73. Tjian, R., and Robbins, A. (1979) Proc. Natl. Acad. Sci. USA *76*, 610–615.
74. Manos, M.M, and Gluzman, Y. (1985) J. Virol. *53*, 120–127.
75. Keller, J., and Alwine, J. (1984) Cell *36*, 381–389.
76. Brady, J., Loeken, M.R., and Khoury, G. (1985) Proc. Natl. Acad. Sci. USA *82*, 7299–7303.
77. Ptashne, M. (1987) In A Genetic Switch, Gene Control and Phage Lambda. Blackwell Science Publishers, London.
78. Tornow, J., and Cole, C. (1983) J. Virol. *47*, 487–494.
79. Tornow, J., Polvino-Bodnan, M., Santangelo, G., and Cole, C.N. (1985) J. Virol. *53*, 415–424.
80. Rabson, A., O'Connor, G.T., Berezesky, I.K., and Paul, F.J. (1964) Proc. Soc. Exp. Biol. Med. *116*, 187–194.
81. Klessig, D.F. (1984) In The Adenoviruses, 426–435, Ginsberg, H.S. (ed.). Plenum Press, New York.
82. Klessig, D.F., and Grodzicker, T. (1979) Cell *17*, 957–966.
83. Saprano, K., Galanti, N., Janok, G., McKercher, S., Pipas, J., Peden, K., and Basilico, R. (1983) Mol. Cell. Biol. *3*, 214–219.

Chapter 5
Replication of the Human Immunodeficiency Virus

Charles S. McHenry

Human immunodeficiency virus (HIV), the primary etiologic agent of the acquired immunodeficiency syndrome (AIDS), is among a group of human retroviruses specific for T4$^+$ lymphocytes [1]. Several independent isolates of HIV* have been obtained from AIDS patients. These have been specified human T cell lymphotropic virus III (HTLV III [2]), lymphadenopathy-associated virus (LAV [3]), and AIDS-associated retrovirus (ARV [4]). Molecular clones of all have been described and show a high level of identity in sequence and products produced [5–7]. The sequence indicates similarities with other retroviruses, including an overall *gag–pol–env* genome organization for the genes that encode the viral core *gag* (group antigen) proteins; a *pol* polyprotein that is processed to produce a protease, reverse transcriptase, and integrase; and a group of viral envelope (*env*) proteins[†]. Other genes unique to HIV are present, some of which are thought to be involved in the regulation of HIV gene expression and activation [7–10].

This article reviews the current state of knowledge of the replication of the human immunodeficiency virus from viral RNA up to its integration into the human genome. An emphasis is placed upon in vitro studies as they relate to protein–nucleic acid interactions and enzymology. In spite of the explosive growth of HIV research, much of our knowledge of the replication of this virus lags behind that of other retroviruses. Thus, I often relate what is known of other systems with a comparison of likely

---

* Abbreviations: HIV, human immunodeficiency virus; AIDS, acquired immunodeficiency syndrome; LTR, long terminal repeat, MoMLV, Moloney murine leukemia virus; MMTV, murine mammary tumor virus.
† The nomenclature used in this review is that recommended by Leis et al. (156). The full names are generally used. Retroviral proteins resulting from the *gag* and *gag–pol* precursors are referred to as MA, matrix proteins; CA, capsid protein; NC, nucleocapsid protein; PR, protease; RT, reverse transcriptase; IN, integrase. Where necessary, distinctions are made between the p51 and p66 forms of HIV RT.

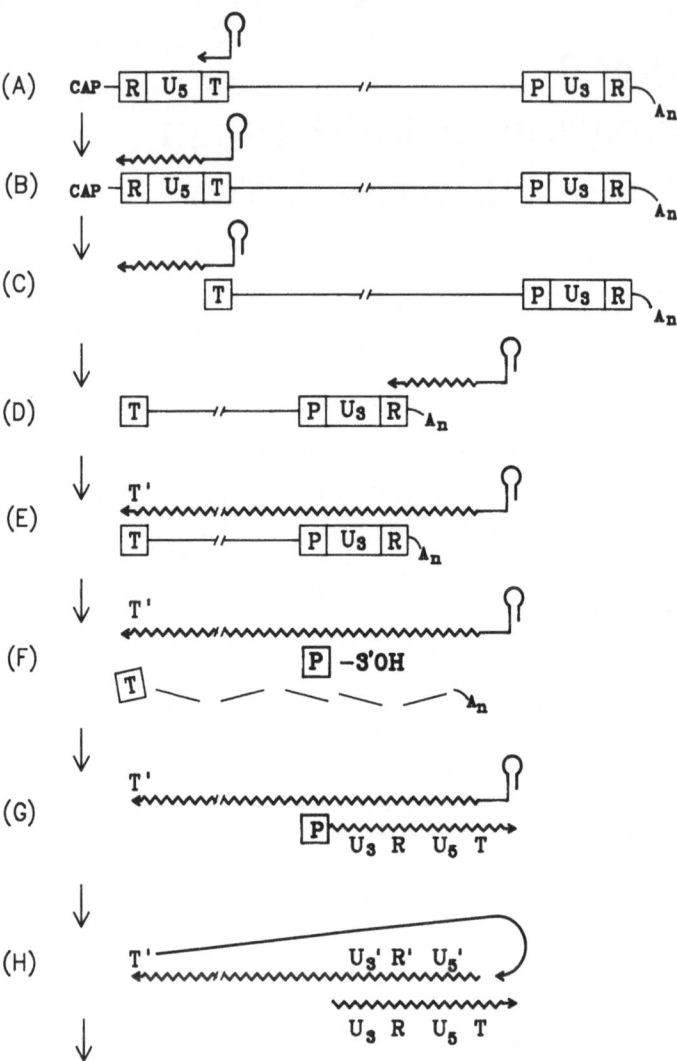

Figure 5.1. General scheme for retroviral replication. The arrows are included at the ends of nucleic acid strands to simplify keeping track of termini potentially active for further DNA elongation. T = priming tRNA binding site. R = redundancy common to 5′ and 3′ ends. U_5 and U_3 = regions unique to 5′ and 3′ LTRs, respectively. P = primer for (+) strand synthesis created by RNase H. The viral strand has the (+) strand designation.

similarities and differences for HIV. In doing this, I will cite only representative references that illustrate a point since many excellent reviews have been written on retroviral replication [11–17]. I apologize in advance to the authors of any work that I may have slighted by this approach.

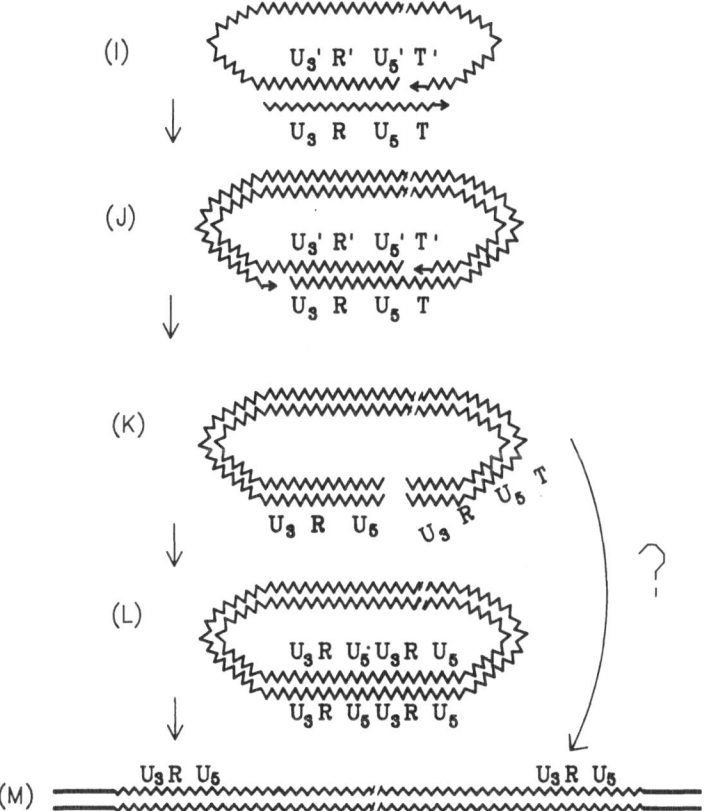

Figure 5.1. Continued

Overview of Retroviral Replication

Through the elegant work of multiple laboratories over the past 18 years a consensus of the general mechanistic features of retroviral replication has arisen. This scheme, summarized in Figure 5.1, will be used as a framework for discussion throughout this review. The details are reviewed in a later section. Upon entering the cell, cellular tRNA annealed to the primer binding site near the 5' end of the viral RNA is elongated by viral reverse transcriptase to generate the strong stop (−) strand intermediate (step A → B). The 5'-RNA end of this RNA-DNA duplex is hydrolyzed by the RNase H activity of the reverse transcriptase to expose the 3' terminus of the nascent (−) strand DNA product (step B → C). This exposed DNA terminus is complementary to the terminal redundancies (R) present at both the 3' and 5' termini of the viral RNA. This permits hybridization of the nascent (−) strand DNA intermediate to the 3' end of the viral template (step C → D) and elongation to synthesize full length (−) strand DNA (step D → E). At a specific site (P) in the 3' region of the viral RNA template, RNase H creates a primer for (+) strand synthesis (step E → F). This primer is elongated on a DNA template to synthesize (+) strand strong stop DNA intermediate (step F → G). Note that the final stage of strong stop (+) strand synthesis includes reverse transcription of a 3' portion of the priming tRNA (step G). This leads to exposure of the 3' end of the nascent (+) strand DNA through the action of RNase H (step G → H), permitting hybridization to the complement of the tRNA primer binding site on the (−) strand DNA (step H → I) and completion of full-length (+) strand DNA (step I → J). Final extension to form duplex DNA with LTRs containing both 5' and 3' viral RNA sequences at both ends takes place by strand displacement (step J → K). These molecules, upon entering the nucleus, circularize to form molecules with one or two copies of the LTRs (2 LTR form shown in Fig. 5.1L). Integration of duplex viral DNA into the host chromosome completes the initial stage of viral replication (step L → M). The integrated copy of proviral DNA (M) serves as a transcriptional template for generation of more viral RNA by the host RNA polymerase II.

Reverse Transcriptase

From the above scheme, it can be seen that a polymerase with incredible versatility is required. The viral reverse transcriptase must be able to utilize efficiently both RNA and DNA templates in both single-stranded and duplex configuration and templates coated with viral and perhaps cellular nucleic acid–binding proteins. The intrinsic RNase H must serve as both a general RNase, facilitating the displacement of all the replicated viral RNA, and a nuclease with the specificity required to generate a (+) strand primer with precise 5' and 3' termini. This enzyme also serves in the unusual role of making precise scissions at RNA/DNA junctions during primer removal.

Polymerase Properties of Reverse Transcriptase

The sequence of all HIV isolates reveals a long open reading frame colinear with the *pol* genes of other retroviruses [5,7,18,19]. The template preferences of HIV reverse transcriptase released or purified from viral particles are listed in Table 5.1. The variability of the results obtained is due, at least in part, to differences in assay conditions. In all studies, poly A primed with oligo dT is an efficiently used template. On most templates, except for 2'-O methylated polymers, the enzyme prefers Mg^{2+} over Mn^{2+} [20,21]. Although reverse transcriptase uses both DNA and RNA templates in its natural replicative reaction, the enzyme exhibits a marked preference for an RNA template in vitro. Perhaps additional viral or cellular proteins facilitate this reaction in vivo.

Table 5.1. Template specificity for HIV reverse transcriptase.

	Reference				
Template	157	158	22	159	21
$(A)_n(dT)_{10}$	100^a	100	100	100	100
$(C)_n(dG)_{12-18}$	136	18	31.6	16	1,820
$(G)_n(dC)_{12-18}$	0				
$(U)_n(dA)_{12-18}$	4				
$(Cm)_n(dG)_{12-18}$		1	3.1		253
$(dA)_n(dT)_{12-18}$		14	2.9	7	
$(dT)_n(A)_{12-18}$		0			

a All values are normalized to poly A–oligo dT within each study.

Reverse transcriptase, isolated from viral particles, contains two different polypeptides with apparent molecular weights of 51,000 and 66,000 (p51 and p66). Both proteins exhibit identical amino terminal sequences [22,23] and are the products of cleavage from a *gag–pol* precursor protein (Table 5.2). Two forms of HIV reverse transcriptase have been separated by preparative isoelectric focusing [21]. Whether these two forms relate to the two molecular weight forms has not been determined.

Only trace amounts of reverse transcriptase can be obtained from infected cells or viral particles. Detailed biochemical studies require larger quantities. Particularly encouraging is the recent success of multiple labs in producing HIV reverse transcriptase in bacteria and yeast. Goff and co-workers [24] were the first to succeed in expressing active reverse transcriptase in *E. coli*. This enzyme exhibited properties similar to the viral enzyme. Others [25–30] have subsequently expressed the HIV

Table 5.2. Products of the *gag* and *gag–pol* precursor protein.

Protein	Amino terminus	Carboxyl terminus	References
gag reading frame			
MA p17[gag]	Myr^a–Gly–Ala–Arg	Ser–Gln–Asn–Tyr	6,7,38
CA p25[gag]	Pro–Ile–Val–Gln	Ala–Arg–Val–Leu	5,19
NC p7[gag]	Met–Gln–Arg–Gly	Arg–Gln–Ala–Asn	5,38
p6[gag]	Leu–Gln–Ser–Arg		38
pol reading frame			
PR	Pro–Gln–Ile–Thr	Thr–Leu–Asn–Phe	102,38
RT p51	Pro–Ile–Ser–Pro	N.D.	23,22
RT p66	Pro–Ile–Ser–Pro	N.D.	23,22
RNaseH p13.3	N.D.	N.D.	38,37
IN p31	Phe–Leu–Asp–Gly	N.D.	23

a Myristyl group.

pol gene in *E. coli, B. subtilis,* and yeast to yield proteins closely related to the authentic viral enzyme.

HIV reverse transcriptase arises from proteolysis of a precursor protein; the native enzyme contains a proline at the amino terminus. Therefore, expression systems must add amino acid(s) to the terminus to permit translational initiation. The approach taken by Goff and co-workers has been to make fusion proteins that contain the *trp*E gene at the amino terminus [24,31]. Initially, a segment containing all reverse transcriptase sequences and nearly all of the *pol* open reading frame was fused in frame [24]. Reverse transcriptase activity appeared in extracts, but the resulting protein was unstable. In spite of these fusion proteins containing the entire coding sequence for the HIV protease, significant quantities of appropriately processed p51 or p66 forms of reverse transcriptase were not detected. To permit the preparation of larger quantities of enzyme with increased activity, a series of deletions were made in the *pol* gene. Optimal levels of protein were recovered which contained *trp*E fused to the entire p66 sequence plus minimal sequences from the adjacent protease and integrase segments (30 and 20 base pairs, respectively). Internal portions of the *trp*E gene were required for stability. Using a similar approach, Farmerie et al. [27] fused a segment of HIV *pol* containing the complete HIV protease coding sequence with the poly-linker of a *lac* expression vector resulting in a 29 amino acid amino terminal extension.

Using sensitive immunoblotting procedures, proteins identical or similar to the p51 and p66 forms of reverse transcriptase were detected. Apparently, the level of expression of processed reverse transcriptase was very low. It was concluded that the HIV protease was responsible for both the amino and carboxyl-terminal processing of p51 and p66, since deletion mutants that lacked protease were not processed. As pointed out by the authors, this interpretation was complicated by the use of a large deletion to inactivate the HIV protease, perhaps generating a protein with a very different conformation that might not be susceptible to cleavage by a cellular protease or that might be a more favorable substrate for competing cellular proteases. This reservation has been overcome with the recent observation that the protease activity required to generate p51 and p66 efficiently from a defective protease-*pol* fusion can be supplied in *trans* [32].

Recently, with an emphasis on generating HIV reverse transcriptase with an amino terminus as close as possible to the authentic viral enzyme, several laboratories have succeeded in producing nearly natural reverse transcriptase in large quantity. Barr and co-workers, through the clever design of oligonucleotides, generated a yeast vector that produces HIV reverse transcriptase with only the initiating methionine preceding the amino terminal proline [25]. This sequence is a known substrate for the yeast methionine aminopeptidase [33]. That the amino terminal methionine was completely removed was verified by direct sequencing of the purified protein [25]. In addition to an approximately full-length p66, a p51-size form of the reverse transcriptase is generated, presumably through the action of a yeast protease [25]. Methods have been developed for purification of this recombinant reverse transcriptase to homogeneity with preservation of activity [Moen, Bathurst, Santi, and Barr, personal communication]. This yeast clone terminates 22 residues before the integrase processing site. With more recent information indicating that some or all of these missing carboxyl-terminal residues are part of the p66 reverse transcriptase [34–36], the missing region is currently being replaced with an oligonucleotide to permit generation of full-length reverse transcriptase with authentic amino and carboxyl termini [Moen, Bathurst, Santi, and Barr, personal communication].

In *E. coli,* Larder et al. [26] generated a vector that produces a protein containing extra Met–Asn–Ser residues on the amino terminus of p66. A termination codon was inserted in place of the amino-terminal residue of the downstream integrase protein. This recombinant gene, under control of the *tac* promoter, produces HIV reverse transcriptase to a level of 10% the total *E. coli* protein. The resulting active enzyme can be purified to near homogeneity by two chromatographic steps. Taking a similar approach, Hizi et al. [36] constructed a vector that produces a protein with the same carboxyl terminus and a Met–Val dipeptide extension in place of the tripeptide used by Larder et al. [26]. In both studies [26,36], a deletion of carboxyl-terminal sequences up to the innermost KpnI site generated a protein slightly larger that the p51 form of reverse transcriptase found in viral particles and infected cells. Although produced in high quantity, it was found to be inert on the poly A template used to assay it.

Mölling and colleagues [28], using *trp–pol* fusions similar to the initial Goff plasmids, found expression of processed p66 and p51 to 0.01% of the *E. coli* protein. In spite of this low-level expression, these investigators developed an effective three-column purification of the reverse transcriptase, finding that both forms cosedimented as a single peak of approximately 120,000 daltons in a glycerol gradient. Both reverse transcriptase and RNase H activity sedimented in parallel with these polypeptides. Some correctly processed p51 and p66 were observed even in the absence of HIV protease.

RNase H Activity of Reverse Transcriptase

HIV reverse transcriptase produced in *E. coli* was found to copurify with RNase H activity as expected from earlier studies with other retroviruses [28]. This exonuclease activity was classified as processive since exogenous nonradioactive RNA-DNA hybrid did not compete with the digestion of radioactive hybrid in ongoing reactions [37]. The enzyme is free of general RNase H endonucleolytic activity. A second HIV-encoded RNase H activity of 15,000 daltons has been purified free of reverse transcriptase [37]. Monoclonal antibody specifically directed against the carboxyl terminus of p66 also reacts with the p15 RNase H. Thus, p15 RNase H probably arises by proteolysis of p66. This enzyme is a distributive exonuclease [37]. It is not certain whether this lower-molecular-weight form of RNase H serves a role in normal viral replication, although a protein of similar size and composition has been detected in viral particles [38].

Both the avian retroviruses and, apparently, HIV contain two forms of reverse transcriptase. The α and β forms of the avian enzymes both contain full polymerase and RNase H regions [39]. They differ by the presence or absence of the integrase protein; only the larger β subunit contains this sequence [16,17]. With HIV, the integrase sequences are lacking from both the p51 and p66 forms. These forms differ by whether or not they contain RNase H domain as defined by Johnson et al. [34].

However, even though the α and β subunits of the avian reverse transcriptases contain RNase H sequences, the activities of the two subunits are very different. The ratio of RNase H activity to polymerase activity of the α subunit is sixfold lower than the $\alpha\beta$ form of the enzyme [40]. The $\alpha\beta$ form of the enzyme is a processive nuclease; the monomeric α form is distributive [41]. The β subunit, dissociated from the $\alpha\beta$ complex, appears to undergo a loss of polymerase activity while preserving nucleic acid–binding activity [42]. Thus, the properties of the individual subunits are very different from when they are associated. This situation is not unique to retroviruses. In the *E. coli* replicative complex, both the polymerase and $3' \rightarrow 5'$ exonuclease subunits are activated by association; neither is efficient alone [43]. This type of synergistic interaction could be used by a cell or virus to prevent one component from acting before it enters a complete, properly assembled complex. These lessons should be recalled when evaluating the roles of the p51 and p66 forms of the HIV reverse transcriptase. The shorter p51 form, when examined by activity gel analysis or through production from deletion mutants, has been found to be inactive to date. These findings could be expected if the p51 form refolds with greater difficulty or if the unnatural carboxyl-terminal sequences prevent the artificial "p51" protein from assuming its proper conformation. These points aside, even if the isolated protein is potentially active, it may require the p66 form to assemble into a heterodimeric complex.

Although both p51 and p66 forms exist in viral particles, it is not yet clear whether they form a well-defined p51–p66 heterodimer. Keeping in mind the established existence of the avian heterodimeric reverse transcriptase and assuming the existence of a HIV heterodimer, what might the role of such an asymmetric enzyme be? In *E. coli*, considerable evidence has been gathered that an asymmetric replicative complex exists

[44–50]. Interestingly, this asymmetry may arise in part by a difference in the carboxyl-termini of two proteins with identical amino termini, similar to the retroviral situations [46,48–49]. This asymmetry is thought to exist in *E. coli* to permit solution of the problem of the asymmetric roles of leading and lagging strand polymerases [49]. In retroviruses, the roles of plus and minus strand polymerases are very different. The (−) strand polymerase must use an RNA template and interact with viral core RNA-binding proteins. The (+) strand polymerase replicates a DNA template except while making a limited and accurate copy of the first 18 bases of the priming tRNA; it may also have to displace RNA sequences that remain hybridized to the DNA template. One of the two polymerases must catalyze a strand displacement reaction at the termini of the nearly completed duplex product, and both enzymes may have to interact with unique RNA- and DNA-binding proteins present in the cytoplasmic milieu. The requirements for RNase H activity are very asymmetric for (+) and (−) strand synthesis. During viral-length (−) strand synthesis, all that is required is the limited removal of the 5′-terminal redundancy sequences to permit template transfer. Considerable RNase H activity is required during (+) strand synthesis. Thus, the principle of an asymmetric dimeric replicative complex may be universal. This possibility should be kept in mind as the properties of heterodimeric reverse transcriptases are examined on natural templates.

Domain Structure of Reverse Transcriptase

Doolittle and colleagues [34], in their computer-assisted studies of retroviral sequence relationships, revealed a striking similarity between the carboxyl-terminal sequences of retroviral reverse transcriptases and RNase H of *E. coli*. A 30% identity was observed between *E. coli* RNase H and MoMLV and a 26% similarity to the HIV reverse transcriptase carboxyl-terminal region. Similarities were also observed between the amino-terminal regions of the reverse transcriptase and other polymerases including the α subunit of the *E. coli* RNA polymerase. A key region of similarity in other polymerases are two adjacent aspartate residues surrounded by nonpolar amino acids. Such aspartates are found at residues 185 and 186 of HIV reverse transcriptase (numbering from the amino terminus). A less conserved region, designated the "tether," is located between the polymerase and RNase H domains. Thus, the domain ordering in HIV reverse transcriptase from amino to carboxyl terminus was proposed to be reverse transcriptase-tether-RNase H.

This model, initially based on sequence comparisons, has since been experimentally verified in several ways. Tanese and Goff [51] generated a series of linker insertion mutants in MoMLV reverse transcriptase and found that those in the 5′ proximal two thirds of the gene affected reverse transcriptase activity while those in the 3' one third abolished RNase H activity. Using this information, plasmids were constructed that expressed fusion proteins with each activity independent of the other. Mölling and co-workers [37] demonstrated that a 15,000-dalton protein that possesses RNase H activity can be cleaved from the carboxyl-terminal sequence of HIV p66 reverse transcriptase. However, the two domains are not completely independent. Hizi et al. [36] demonstrated that deletion of the codons for the 23 carboxyl-terminal residues of HIV p66 reverse transcriptase inactivate the polymerase.

Integrase

An essential stage in the propagation of HIV, like all retroviruses, is integration of the provirus into the host genome [16,17]. At least one viral protein is required for this event, the integrase protein encoded by the 3' end of the *pol* open reading frame. The existence of this protein in HIV was inferred through similarity between the 3' end of the *pol* open reading frame and the integrase protein of other retroviruses [5–7,19]. Recombinant proteins containing sequences from the putative integrase competed for binding of antibodies present in sera from AIDS patients to a 31,000-dalton protein present in viral particles [52,53]. The amino terminus of the integrase was determined and found to coincide with sequences from the expected region of the HIV genome (Table 5.2). The carboxyl terminus has not been experimentally determined.

The enzymatic activity of the HIV integrase has not yet been examined. A protein with the complete integrase sequence has not been produced in high level in microorganisms. Doolittle has observed a "zinc finger" that may bind to DNA and be important for integrase function. This sequence, two histidines separated by 20 to 30 residues from two closely spaced cysteines, is conserved in all retroviral integrase sequences. In HIV, it occurs between residues 12 and 43. The 23–amino acid loop between the Zn-binding residues can be drawn in a β structure [34]. In computer-assisted searches for similarities to other proteins, an interesting resemblance of the HIV integrase to an *E. coli* transposase (Tn3) has been detected [34]. Recent information has been obtained that increases the confidence in this similarity [R. Doolittle, personal communication].

Hints of potential HIV integrase activities can be gleaned from other retroviral systems. Integrases have been observed to have both nonspecific and specific DNA binding, and endonucleolytic activities. The integrase of avian retroviruses has been the most revealing in terms of its reactions. This integrase exists in two states—as a free protein of roughly 32,000 daltons, and as the carboxyl-terminal domain of the large β subunit of the avian heterodimeric α-β reverse transcriptase [54–56]. Both the free integrase and the β subunit are phosphorylated [57,58]. The importance of this modification to integrase function is not yet established.

The avian retroviral integrase exhibits specific DNA-binding properties, being able to retain on nitrocellulose filters viral LTR sequences and a subset of bacterial plasmid sequences, including the inverted repeat sequences of transposon Tn3 [59]. DNase I protection experiments revealed a nucleosomelike pattern of alternating protected and hypersensitive sites over a 170-nucleotide region from the U_3 region extending downstream about 135 nucleotides through U_5 to the unique $U_5 : U_3$ junction and 34 nucleotides downstream into the second copy of U_3 [59] (see Fig. 5.1L for LTR circle junction arrangement). The affinity of the integrase appears to slightly favor supercoiled DNA (twofold) over linear molecules. Integrase exhibited a similar preference for single-stranded DNA over double-stranded [60]. Circles containing two tandem LTRs (Fig. 5.1L), as found in the nuclei of infected cells, are specifically cleaved by the integrase at the unique $U_5 : U_3$ circle junction [61–65]. Cleavage by the p32 form of the enzyme is immediately 3' to the CA dinucleotide pair that defines the border with cellular sequences during integration [62–64]. The *pol–int* form (β subunit of avian reverse transcriptase) cleaves between the C and the A [65]. This cleavage reaction is inefficient. At ratios of 44 integrase molecules per DNA molecule, only 25% of the LTR junctions are cleaved. The specificity of the *pol*-associated integrase differs for single- and double-stranded endonuclease reactions [63]. The reason for this difference is not clear, but the sequences that support viral production correlate most closely with the duplex DNA cleavage requirements. In both cases, the sequences required for integrase cleavage are more limited than the regions of integrase interaction, defined by nuclease protection.

The activity of other retroviral integrases is less clear. For integrases isolated from murine leukemia viral particles or as fusion proteins in *E. coli,* DNA binding was the only function revealed. DNA binding is nonspecific, both in terms of sequence and for duplex vs. single-stranded DNA. In vitro mutagenesis of a *trp*E fusion protein has permitted determination that the first eight residues are not essential for nonspecific binding nor is the Cys in the putative Zn finger [66]. However, the latter mutation, when placed into full-length viral clones, was found to be lethal for virus production. Thus, the Zn finger may be required for a specific DNA interaction or for an unrevealed catalytic function of the integrase. Luk et al. [67] observed similar properties with a *lac*-spleen necrosis virus integrase fusion. In this study, a hint of specific binding to sequences that contained an intact U_5 region was revealed by the formation of slowly moving, ethidium-bromide-staining species on agarose gels. Whether these complexes contain more than one DNA or are due to aggregation of multiple integrase proteins at a site in one circle is not clear. Incubation at

60°C is required for formation of these aggregates. Whether this is due to overcoming a kinetic barrier with a high energy of activation or due to local denaturation of a domain of the integrase is not yet certain.

Endonucleolytic activity can be detected in MoMLV particles [68–71], but recent genetic tests suggest that the activity may be of cellular origin [72]. In spite of the inability to find activities that are unequivocally associated with integrase function, it is certain from genetic tests that the protein is required for integration into the host genome with functions that are distinct from DNA synthesis [73–75].

Whether the HIV integrase catalyzes LTR : LTR cleavage at the unique HIV U_5:U_3 circle junction must await isolation of the active enzyme from viral particles or cells carrying molecular clones. It should be possible to generate an HIV integrase with, at most, a methionine extension preceding the amino-terminal phenylalanine of the natural protein.

Core Proteins Involved in Nucleic Acid Interactions

All retroviruses have multiple core proteins, most associated to some extent with the viral RNA [76,77]. Because of this association, these proteins must influence the reaction catalyzed by reverse transcriptase. Their roles may be positive—helping the polymerase to overcome structural barriers, strengthening the association of the polymerase through specific protein-protein interactions, or participating in one or more of the specific stages of retroviral replication. Alternatively, these associations may present barriers to the replicative reaction that must be overcome. In either case, a complete understanding of retroviral replication cannot be achieved without understanding the components, structure, and role of the core proteins. It is of interest that the DNA replication systems already characterized depend extensively upon the formation of protein-coated templates (see Chapter 2, this volume).

The *gag* gene of HIV encodes three proteins: p18 (matrix protein), p24 (capsid protein), and p7 (NC, nucleocapsid protein) [5–7,19,38]. The correspondence of these proteins with other retroviral *gag* products is given in Table 5.3. The HIV matrix protein is derived from the amino terminus of the *gag* precursor. Typical of retroviral matrix proteins, it

Table 5.3. Equivalence of retroviral core proteins.

Protein	Avian (RSV)	Murine (MoMLV)	HIV	Function
MA	p19^{gag}	p15^{gag}	p18^{gag}	Core-membrane associated
	p10^{gag}	p12^{gag}	Not present	
CA	p27^{gag}	p30^{gag}	p24^{gag}	Major core structural component
NC	p12^{gag}	p10^{gag}	p7^{gag}	NC, RNA-binding protein

contains a myristyl group covalently attached to the penultimate residue of the processed protein [38,76,78]. This protein is typically membrane associated, although a subpopulation has been reported to be associated with specific structures of the viral RNA [76]. The p24 capsid protein is the major structural component of the core. It has a potential asparagine-linked glycosylation site [19]. This protein, isolated from viral particles, contains phosphorylated serine and threonine residues [38]. The protein derived from near the carboxyl terminus of the gag precursor, p7, is the basic core-RNA-binding nucleocapsid protein. HIV lacks a protein in the position corresponding to p10 and p12 of avian retroviruses and MoMLV, respectively, but an extra protein, p6, is located at the carboxyl terminus of the gag precursor. Whether these proteins share similar functions is not known. Given the shortage of functional information on HIV core proteins, examples from other systems that are likely to be pertinent to HIV will be discussed.

All replication-competent retroviruses contain a nucleic acid–binding protein that is tightly associated with the viral RNA in the core. These proteins are recognized by their basic composition and the existence of a sequence $Cys-X_2-Cys-X_3-Gly-His-X_4-Cys$ (X refers to intervening amino acids) [79]. A sequence very similar to this, involved in the binding of Zn^{2+} and DNA, is also present (inverted amino to carboxyl polarity) in the single-stranded DNA-binding protein of bacteriophage T4 [80–82]. Nucleocapsid proteins interact strongly with single-stranded RNAs [83a,83b,84–85] and DNA [86]. Recent site-directed mutagenesis of cysteine codons in the structural gene for MoMLV nucleocapsid protein results in virus unable to specifically recognize and package viral RNA [83b]. A function of these zinc-fingerlike sequences in sequence-specific recognition of RNA has been proposed [83b].

Leis and co-workers have conducted an elegant series of experiments that correlate the structural, biochemical, and biological properties of the avian retrovirus nucleocapsid protein. The avian pp12 protein is phosphorylated in viral particles. Dephosphorylation results in a 100-fold decrease in the apparent affinity for RNA [87]. Strong binding can be restored by phosphorylation with protease-activated kinase I from rabbit reticulocytes [88]. Avian retrovirus p12 has been produced in *E. coli* with an extra methionine at the amino terminus [89]. This protein is indistinguishable from the uphosphorylated viral protein. High-affinity binding can be induced by phosphorylation of the recombinant protein in vitro [89]. Site-directed mutagenesis of the target serine led to an unexpected result [90]. Mutant viruses (Ser→Ala-40), containing an alanine in place of phosphorylated serine, were viable and produced NC with nearly the same affinity as the phosphorylated protein. Surprisingly, the Ala_{40} NC is locked in the high-affinity conformation normally induced by phosphorylation. This indicates that low-affinity NC is not essential for any stage of the viral life cycle, such as conversion to a dissociable form during replication. The mutant protein is still phosphorylated at an alternate site (Ser-76, 77) without an accompanying change in binding affinity [90]. Double mutants lacking both phosphorylation target sites (Ser-40, 76, 77) will have to be constructed to test whether phosphorylation is essential for a function distinct from modulating affinity for nucleic acids. Consistent with the hypothesis that the high-affinity conformation is required is the finding that mutations made elsewhere (Ile replacements for both Lys-36 and Lys-37) generate a protein that is incapable of assuming the high-affinity conformation upon phosphorylation even though it has a slightly higher affinity for RNA than the wild-type unphosphorylated form [91]. Mutant virus containing this mutation produce particles lacking RNA. Continued use of this approach should prove valuable for determining the functions of the core proteins and for testing biochemical hypotheses in a biologically relevant context.

The murine leukemia virus nucleocapsid proteins also bind tightly to single-stranded RNA and DNA [79,83a,86,92]. The binding ratio is approximately 1 protein: 6 nucleotides, similar to the value for avian retroviral NCs [83a]. Apparently, the murine protein is not phosphorylated. Its binding constant, in the form isolated from the viral particle, approximates that of the phosphorylated avian nucleocapsid protein [83a]. Thus, the murine leukemia retroviral NCs may be locked in a high-affinity conformation, even when unphosphorylated, like the Ala_{40} mutants of avian retroviruses [90]. Measurement of the affinity of the $Pr65^{gag}$ precursor indicates a higher affinity for RNA than the processed nucleocapsid protein.

Multiple assays are available for assessing nucleic acid binding by retroviral nucleocapsid proteins. These include measuring the hypochromicity resulting from disruption of secondary structure in single-stranded nucleic acids [85], extrinsic fluorescence resulting from fewer fluorescamine-sensitive lysines being available upon binding of the NC to nucleic acids [84], quenching of tryptophan fluorescence of the NC upon interacting with nucleic acids [83a], an increase in the fluorescence of poly (εA) upon NC binding [83a], and retention of bound nucleic acid on nitrocellulose filters. The latter method gives only an apparent binding constant that varies with polymer size unless corrected for the number of binding sites [83a].

How do nucleocapsid protein-RNA interactions influence the efficiency of the replication process? Native ribonucleoprotein is an efficient transcriptional template [93]. This complex, fixed to maintain protein-protein interactions, appears as a circular polynucleosomelike complex in electron micrographs. These structures may facilitate the strand transfer reactions. Avian retroviral ribonucleoprotein particles contain the α and β subunits of reverse transcriptase, higher levels of the cleaved integrase (pp32), the p27 capsid protein, and large quantities of the p12 NC and an unidentified protein that has the mobility of the phosphorylated form of matrix protein (93). Unphosphorylated matrix protein and protease are absent from these complexes; bands corresponding to $p10^{gag}$ (not discovered at the time of this study) are not evident in the gel system used. Thus, the proteins associated with the efficiently replicating core (reverse transcriptase, integrase, nucleocapsid protein, capsid protein, and a protein of unknown origin) are candidates for functions that might facilitate the natural retroviral replicative reaction. The corresponding proteins in HIV would be the p51 and p66 reverse transcriptase, integrase, capsid ($p24^{gag}$) and nucleocapsid ($p7^{gag}$) proteins. Determining the necessity of each of these proteins will require learning how to appropriately reconstitute these complexes in a functionally active state. The efficient synthesis of full-length (−) strand DNA may provide an appropriate assay for this function.

Adding avian nucleocapsid protein to reactions containing avian myeloblastosis virus reverse transcriptase results in up to a sixfold stimulation of DNA synthesis, depending on the template used [85]. It also causes an inhibition of RNase H activity when measured on exogenous templates [85]. It will be interesting to investigate whether this inhibition occurs in RNA hydrolytic reactions that are coupled to replication. Supporting this reasoning, the exonucleolytic activity of T4 bacteriophage polymerase is inhibited by the T4 single-stranded DNA-binding protein when measured alone, but the processivity of hydrolysis is enhanced when measured with a properly assembled replicative complex [94].

Besides the nucleocapsid protein, the capsid protein of MoMLV, $p30^{gag}$, has been reported to associate with polymerase in a high-molecular-weight complex and to concomitantly stimulate the polymerase activity severalfold [95]. Supporting the notion that this association is functionally relevant, weaker associations and lower levels of stimulation were observed in interspecies mixing experiments [96].

The $p19^{gag}$ matrix protein, normally found associated with the membrane, has also been observed to associate at low extent with RNA. This association is to regions of RNA with secondary structure. Treatment of viral RNA with single-strand-specific RNases does not influence binding; however, *E. coli* RNase III, which only cleaves in duplex regions, diminishes binding significantly [97]. Binding of $p19^{gag}$ has been proposed to influence splicing reactions [97,98]. It could influence the course of the replication reaction as well.

Thus, in HIV, all three gag proteins may play a role in the replicative reaction. The participation of each individually or together as complexes can be tested biochemically by addition of purified proteins to purified viral RNA and observing whether they stimulate the reaction or alter its course to one that more closely mimics the authentic viral reaction. For example, a protein that only binds to duplex regions might stimulate the strand transfer reaction. A protein that inhibits uncoupled RNase H activity might permit the enzyme to function more efficiently on regions undergoing active replication. Only through detailed biochemical and structural studies, exploiting complete and natural replicative assays, will a solution to these problems be achieved.

Biosynthesis of HIV Replicative Proteins

The HIV *pol* gene products are synthesized from a gag-pol precursor protein of 160,000 daltons (Fig. 5.2) [19]. The pol sequences are out of frame with the *gag* gene. The large polyprotein arises by translational frame shifting [99]. This was first established by expression of *gag–pol* mRNA in an in vitro system where the mRNA was produced by a bacterial RNA polymerase, precluding complications of splicing. A rabbit reticulocyte lysate produces both the normal Pr55^{gag} precursor and the Pr160$^{gag-pol}$ precursor in a ratio of 1 : 8.

By analogy to the RSV frameshift, Jacks and co-workers predicted that HIV would frame-shift (-1 relative to *gag*) by a slippage of leucyl tRNA at the sequence U UUA [99]. This position of frameshifting was verified through determination of the amino-terminal sequence of a fusion protein generated by using a synthetic oligonucleotide to cover the relevant HIV region (first 50 nucleotides of the *gag–pol* reading frame overlap). Supporting the notion that the leucyl tRNA is responsible for the translational slippage leading to a frameshift, C residues in place of the U

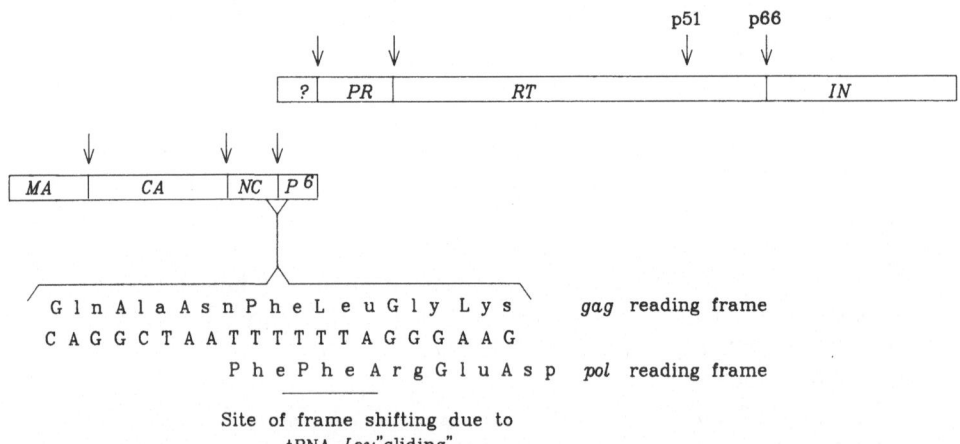

Figure 5.2. Diagram of the *gag–pol* region of the HIV genome. The *pol* reading frame overlaps with that of *gag*, but in a different frame. Translation of *pol* is the result of occasional frame shifting near the beginning of the overlapping region [99]. Vertical arrows signify regions of proteolytic processing of the translated protein catalyzed by the HIV protease.

opposite to the 3'-A in the anticodon loop block this process when placed at either the 5' end of the correct codon or −1 codon (U UUA → either U CUA or C UUA). Stem and loop structures downstream from the site of frame shifting are thought to be important [99], perhaps by causing the ribosome to pause for sufficient time to favor leucyl tRNA slippage.

The Pr160$^{gag-pol}$ precursor is processed proteolytically to generate the p18^{gag} matrix protein, p24^{gag} capsid protein, p7^{gag} nucleocapsid protein, and the pol gene products: HIV protease, p51 and p66 reverse transcriptase subunits, and integrase. Similarly, the Pr55^{gag} precursor is cleaved to generate matrix, capsid, and nucleocapsid proteins [100]. The HIV protease is involved in some or all of these proteolytic cleavages.

HIV protease generated in *E. coli* or yeast has been found to function in processing gag and pol precursors whether supplied as part of the same polyprotein precursor or as a separate polypeptide [27,28,30,32,35,101–103]. Recombinant HIV protease is capable of autocatalytic cleavage from its higher-molecular-weight precursors [102,103]. Another cleavage catalyzed by HIV protease that occurs within *E. coli* leads to generation of p24^{gag} capsid protein from the recombinant HIV gag precursor protein. Taken together these three established sites suggest the consensus cleavage site Asn–Y–Pro, where Y is an aromatic residue. These cleavage sites are roughly within the consensus for other retroviral proteases that are specific for cleavage between proline and an aromatic or bulky hydrophobic residue [104]. Therefore, it came as a surprise when Darke et al. [35] found that chemically synthesized HIV protease could cleave sequences preceding residues other than proline. Synthetic peptides representing all the established processing sites in HIV gag and pol could be cut with this enzyme. These include the Val Leu/Phe Leu junction between the p66 form of HIV reverse transcriptase and the integrase. In addition to the p66-integrase cleavage, evidence has been obtained that the HIV protease cleaves p66 to generate p51 [Moen, Bathurst, Santi, and Barr, unpublished]. Forms resembling p51 can be generated in both yeast and *E. coli* without the HIV protease [25,28,36], but these could be the result of fortuitous processing of an exposed hinge region between the reverse transcriptase and RNase H domains, generating a protein of approximately the correct molecular weight. Carboxyl-terminal sequencing of the p51 form of the enzyme from infected cells or viral particles is needed to permit solution of this problem and construction of vectors that produce authentic p51. Similarly, sequencing of p66 and the integrase needs to be conducted to ensure that secondary carboxyl-terminal processing does not occur. This is especially important for the integrase, since there is precedent for carboxyl-terminal processing. Avian retroviral pp32 integrase and the β subunit of reverse transcriptase are processed to remove a 37-residue polypeptide that is not required for viral replication in tissue culture [105,106].

Details of the Replication Process

Minus Strand Synthesis

tRNA Priming

All retroviruses examined to date use a specific host tRNA as a primer. HIV contains a primer binding site complementary to 18 nucleotides on the 3' acceptor stem of tRNA$^{Lys}_3$. That this site serves as the primer binding site has been confirmed by examining the (−) strand strong stop product of the endogenous reaction in viral particles [6]. For another retrovirus, MMTV, the use of tRNA$^{Lys}_3$ has been confirmed by the observation of radioactivity being added to the tRNA's terminus during primer extension in disrupted viral particles [107].

The function of the tRNA primer has been best studied with the avian retroviruses. tRNA^{Trp} serves as a primer in these systems [108–110]. Eighteen nucleotides of the 3' acceptor stem hybridize to the viral RNA template. Apparently, the reverse transcriptase functions to melt the acceptor stem, facilitating annealing at biologically relevant temperatures [111]. tRNA^{Trp} binds specifically to the avian reverse transcriptase [112]. Reverse transcriptase only binds the priming species from a mixture of tRNAs. Apparently, this interaction is critical for packaging the priming tRNA in the viral particle. Mutants with defective polymerase do not package priming tRNAs; mutants lacking viral RNA but containing functional polymerase produce particles containing tRNA^{Trp} [113,114]. The sequence and structural determinants are more stringent than the requirements for priming. Only the portion of the tRNA complementary to the viral RNA is necessary for priming. The entire primary structure of tRNA^{Trp} and native conformation are required for strong binding to reverse transcriptase [109,110]. These observations are consistent with the tRNA-enzyme contacts identified by Litvak and co-workers. Photocrosslinks occur between tRNA and reverse transcriptase in two regions—in the anticodon loop and stem and in the 3'-terminal residues.

Comparative studies of avian reverse transcriptase forms indicate a strong asymmetry in function of the two subunits when monitored by this physiologically relevant binding assay. Both the β and α-β form of the enzyme bind tRNA^{Trp} strongly. However, the α form by itself does not bind tRNA strongly enough to be detected by cosedimentation in glycerol gradients or comigration during gel filtration [109,115]. The α subunit is ineffective in transcribing the AMV 70S RNA complex [109]. Interestingly, interactions of tRNA with the α subunit can be detected in the heterodimeric α-β complex. Photocrosslinking results in crosslinks of tRNA to both the α and the β subunits [116]. However, when interactions with the 3' primer terminus are probed with periodate oxidized tRNA, forming a Schiff's base adduct with lysine side chains, only the α subunit becomes linked to tRNA. Perhaps β is important for initial interactions, followed by relevant priming interactions with α.

Most of the tRNA in cells is amino-acylated. Litvak and co-workers discovered that avian retroviral reverse transcriptase could use amino-acylated tRNA with nearly the same efficiency as that with a free 3'-OH group. Only an initial lag in deoxynucleoside incorporation distinguished the two primers. Apparently the basis for this finding is that reverse transcriptase facilitates the hydrolytic removal of the amino acid. Under the assay conditions used, the charged tRNA by itself is unstable; the reverse transcriptase only accelerates the reaction approximately three-fold. The primer tRNA must be annealed to the template for this reaction to occur. Neither template nor reverse transcriptase alone facilitates hydrolysis. The specificity of this reaction relative to other tRNAs annealed to complementary templates has not yet been examined.

In MoMLV, tRNA[Pro] serves as the primer [117]. The tRNA–reverse transcriptase interaction is not as strong [109], and apparently this interaction does not play such an important role in tRNA packaging in viral particles [118]. Binding of the tRNA[Pro] to MoMLV reverse transcriptase is characterized by a shift in apparent molecular weight from 70,000 to 150,000. Whether the enzyme dimerizes under these conditions is not certain.

SYNTHESIS OF MINUS STRAND DNA

The priming tRNA is elongated until the 5' end of the viral RNA is reached, a distance of 100 nucleotides (nt) in avian retroviruses and 135 nucleotides in MoMLV [119]. The length of this DNA product is 181 to 182 nt in HIV [5,6,19,120]. This intermediate is designated as the (−) strand strong stop. The RNase H activity of reverse transcriptase can destroy the template only after full-length (−) strand strong stop DNA is synthesized. Intermediates not fully extended are inert to RNase action [121]. This provides a mechanism to prevent premature template destruction. Incisions are at discrete points. With avian retrovirus, 60% of the ribonucleotides released reside in a single oligonucleotide (12 to 15 mer) that contains the 5'-7-MeG cap [121]. The redundancies present at both the 3' and 5' termini of the viral RNA are exploited to permit a switch in templates enabling full-length complementary DNA to be synthesized [122–124]. Annealing of the exposed DNA complementary to the terminal redundancy to the 3' end of the viral template provides a template for synthesis of full-length (−) strand DNA.

In in vitro reconstructed reactions, hairpin formation, mediated by an inverted repeat present within the 5' LTR, competes with the template switch [125]. These hairpins do not form in reactions proceeding in partially disrupted viral particles [123]. The competing reaction may be prevented by association of the nascent DNA with specific viral proteins that either facilitate transfer to the 3' LTR or specifically prevent hairpins from forming.

Since two copies of full-length 35S RNA are present in the viral particle, the question arises of whether transfer from the 5' to a 3' LTR (−) strand template occurs as an intramolecular transfer within the same RNA molecule or as an intermolecular transfer between molecules. To permit resolution of this issue, Panganiban and Fiore [126] recently generated spleen necrosis virus stocks that contain two different viral RNAs differentially marked with restriction enzyme recognition sequence polymorphisms within their termini. By analysis of the restriction endo-nuclease cleavage pattern of the integrated provirus, the origin of the (−) and (+) strands of DNA could be determined. Minus-strand synthesis occurred exclusively by intramolecular template transfer, consistent with the expected function of a virus that contains two identical templates.

Proviral DNA is slightly longer than the viral RNA. This results from a duplication of terminal sequences, a consequence of the template trans-fers that occur during viral replication [16,17]. An extra copy of U3 occurs 5' to the 5' LTR, and an extra copy of U5 is found distal to the 3' LTR. Generation of 7.7-kb genome length cDNA in melittin-disrupted avian retroviral particles occurs in the presence of actinomycin, an inhibitor of DNA template-directed synthesis by reverse transcriptase. Extension to the 8-kb (−) strand species is inhibited by actinomycin, indicating that a DNA template is used for this final reaction phase [127]. This finding supports the model that the RNA template for extensive (−) strand synthesis is processed by RNase H and extends only to the primer binding site (Fig. 1E).

Plus Strand Synthesis

Shortly after synthesis of DNA complementary to the 3'-terminal region of the viral RNA, prior to synthesis of the full-length (−) strand transcript, the synthesis of a (+) DNA strand is initiated [128-129]. This occurs through the action of RNase H generating a unique RNA primer in the region of the conserved polypurine stretch adjacent to the U3 region [130,132]. The polypurine stretch of avian retrovirus has been demon-strated to be essential for a cis-acting function in viral replication [133]. The (+) strand RNA primer is elongated using the nascent (−) strand DNA as a template. Elongation occurs until the end of the template is reached. Thus, the initial DNA is composed of the LTR U3-R-U5 and primer binding site (P) [134]. In the avian retroviruses, the (+) strand primer is 11 to 12 nt long [130,131]. This primer is extended to generate a 340-nt strong stop (+) strand DNA molecule for avian retrovirus. Secondary priming may also occur. Structures containing displaced single strands can be seen as products resulting from replication of melittinpermeabilized virus [135]. These products are exclusively of (+) strand polarity, presumably resulting from initiation at more than one site [136].

The mechanism of primer generation and selection is not yet clear. Specificity of priming can be reconstituted with reactions that contain viral RNA hybridized to complementary DNA [130,131,137]. This specificity is lost if the RNA is not prehybridized before introduction of reverse transcriptase [130]. The specific primer hybridizes much more readily with the DNA template (−) strand than other oligonucleotides generated. The hybrid formed is particularly stable (T_m = 51°C in low salt) [138]. It has been proposed that this may be due in part to a structure formed by the single-stranded polypurine that facilitates hybridization [138]. This polypurine-polypyrimidine structure may be in part responsible for the high specificity of cleavage within this region. But the specificity cannot simply be due to difficulty in cleaving within polypurine stretches. The primer for MoMLV results from cleavage preceding two adenosine residues after a stretch of purines. It is interesting that avian and MoMLV reverse transcriptases both generate the same primers when assayed on identical substrates [132,138]. This suggests a recognition of a specific structure rather than a specific sequence. For the determinants of primer creation and utilization to be more fully understood, a systematic investigation of the in vitro cleavage of model substrates correlated with the ability of similarly altered retroviruses to support replication is required.

Once extended, RNase H can remove the (+) strand RNA primer, generating a DNA polymer with a 5′-phosphorylated end, an appropriate substrate for DNA ligase following completion of synthesis [130,132,137]. Apparently, there is no specificity for (+) strand primer removal, since RNA primers are removed efficiently from nonspecifically primed molecules [130].

A polypurine stretch of 16 nucleotides may provide the primer for HIV plus strand synthesis. The sequence AAAAGAAAAGGGGGGA immediately precedes the U3 region of HIV [139]. It remains to be experimentally verified whether this sequence provides the initial primer for (+) strand HIV DNA synthesis. An identical polypurine stretch also appears within the integrase coding sequences starting with base 4,338 [5]. Whether this site serves as a secondary initiation site for HIV (+) strand replication has not been determined. From the sequence [7,139] one would expect the HIV (+) strand strong stop product to be 557 deoxynucleotides. This (+) strand intermediate would contain portions of the transcribed tRNA. Only the first 18 nucleotides should be copied for the product to represent accurately the sequence of the mature product— reverse transcription of additional sequences would result in insertion mutants.

What induces the reverse transcriptase to stop at this position? It has been noted [140] that all priming tRNAs contain a 1-Me A position in the

19 position from the 3' end. This modification would be expected to interfere with base pairing, perhaps causing the reverse transcriptase to stall for sufficient time for the tRNA to be removed so that it can no longer serve as a template. This mechanism has never been experimentally tested, but this should now be possible with the capability of synthesizing tRNAs with T7 RNA polymerase in vitro [141]. If this mechanism is operative, another route to antiviral chemotherapy may be opened that, to my knowledge, has not yet been addressed. Drugs that block the action of the tRNA methylase responsible for this reaction may interfere with normal viral replication. This assumes that this enzyme could be blocked without lethal effect on normal cells. Blockage of tRNA modifications is possible in bacteria without harmful effect on cell growth.

Upon transcription of tRNA sequences, the tRNA is removed by an endonucleolytic clip at or near the RNA-DNA junction [142]. This reaction requires most or all of the first 18 bases of the tRNA acceptor stem to be transcribed. Products containing only six to eight bases complementary to the 3' end of the priming tRNA are not substrates for RNase H cleavage [142]. This reaction exhibits similarities to the (+) strand primer cleavage reaction where a scission at an RNA:DNA junction also occurs. The (+) strand primer is also removed intact without secondary degradation [132]. The free (+) strand 3' terminus sequence containing the tRNA primer binding site can then hybridize to its complement in the (−) strand transcript, permitting the (+) strand to be elongated to full length by a mechanism similar to the (−) strand template jumping reaction. Using methods described in the (−) strand synthesis section, Panganiban and Fiore [126] have demonstrated that this transfer is intramolecular.

Final stages of the reaction require a strand displacement reaction to generate the mature LTRs of the duplex viral DNA double strand [140]. Whether this strand displacement occurs preferentially by (+) or (−) strand elongation is not certain. All stages involving replication up to the linear duplex occur in the cytoplasm [143].

The linear duplex migrates to the nucleus where a fraction of it becomes circular. In the nucleus circles exist with both one and two copies of the LTRs. It has been proposed that the circles containing only one copy of the LTR may arise by intramolecular homologous recombination between the repeated LTRs [144]. Alternatively, I propose that single LTR circles could result from the premature action of DNA ligase on the circular intermediate that precedes strand displacement (Fig. 5.1J).

Integration

Once in the nucleus, duplex DNA is integrated into the host genome. There it can serve as a template for further viral RNA production. For

HIV, the virus remains dormant until activated, a process that can take years in the human host.

The site of retroviral insertion into the host genome is apparently random relative to sequence [145], but a bias toward insertion at DNase-I-hypersensitive sites has been detected [146]. In vitro studies (see below) indicate that there is no requirement for chromatin structure, but this could be reconciled with the above observation if certain chromatin structures inhibited insertion. Varmus et al. [147] have demonstrated that efficient integration requires concomitant host DNA synthesis. Whether this generates a specific structure or is the result of a requirement for host replication enzymes is not known. Insertion is characterized by a duplication of a small block of host genome sequence flanking the site of insertion. The size of this sequence is characteristic of the inserting virus. More HIV-host junctions need to be published before the host insertion duplication size can be accurately determined.

During the insertion event, a fixed number of bases characteristic of individual retroviruses are lost from the unique circle junction that occurs upon blunt end ligation of the completely replicated viral DNA. For example, with MoMLV, the circle junction sequence

$$U_5 \qquad \text{T C T T T C A T T : A A T G A A A G A} \qquad U_3$$
$$\text{A G A A A G T A A : T T A C T T T C T}$$

must be cleaved 3′ to the CA dinucleotide resulting in the donor sequence

$$\text{T C T T T C A}$$
$$\text{A G A A A G T A A T T}$$

Ligation of the 3′ terminus to host DNA and the ensuing repair reaction results in loss of AATT from the final integrated sequence (see below). This mechanism assumes that a viral DNA circle is the precursor for integration. Panganiban and Temin [148] demonstrated that an LTR-LTR circle junction can serve as an integration site by showing that LTR-LTR junctions inserted within internal viral sequences could serve as an integration site for host DNA. Alternative pathways are still possible (see below).

For HIV, the probable circle junction sequences of various isolates are shown in Table 5.4. By analogy to other retroviral systems, cleavage probably occurs 3′ to the CA dinucleotide near the junction. HIV-1 has only a two-nucleotide spacing between the probable cleavage points and forms an imperfect palindrome. This would lead to loss of only one nucleotide from each of the termini of HIV upon integration into the human genome. HIV-2 has another two nucleotides inserted, creating a probable asymmetric cleavage. Colicelli and Goff [149] have made insertion mutants in the LTR-LTR junction of MoMLV that create asymmetry without loss of function. These authors argue that the

Table 5.4. Sequence of HIV circle junction.

Circle junction sequence (U₅:U₃)		Reference	Isolate
GTGGAAAAATC T CTAGCAG	: C TGGAAGGG CTAATT T G G T	5	HIV-1 (ARV 2)
GTGG_AAAATC T CTAGCAG	: c TGGAAGGG CTAATT C A C T	7	HIV-1 (BH 10)
GTGG_AAAATC T CTAGCAG	: C TGGAAGGG CTAATT C A C T	6	HIV-1 (l J19)
GTGG_AAAATC T CTAGCAG	: C TGGAAGGG CT A ATT C A C T	19	HIV-1 (H9c.7)
C AGG_AAAATC C CTAGCAGGT: C TGGAAGGG ATG TTT T A C A		160	HIV-2

The *c* is shown for consistency. Although not assigned as part of U₃ in reference 7, it was present in the sequence. The difference arose from the assignment of the polypurine stretch-U₃ junction, not in determination of the sequence. All of the above assigned junctions are subject to revision when the exact placement of the polypurine stretch has been experimentally determined. The presence of CCA on the end of all tRNA primers permits unambiguous assignment of the U₅ portion of the junction.

cleavage apparatus measures from sequences external to the site, rather than measuring a precise spacing from the junction. Even with imprecision in the cleavage apparatus, the preference for scissions after the CA dinucleotide is retained.

Resolution of the mechanism of retroviral integration has been hampered by the lack of in vitro systems that permit identification of the protein components, the substrates, and the reaction intermediates for the integration reaction. Striking progress has been made in this area for MoMLV. Brown et al [150] were the first to set up an in vitro reaction. The starting material was a protein–nucleic acid complex isolated from infected cells at a stage of infection where integration is known to occur. The virus used for these studies was marked genetically to permit assay of low-efficiency events. A replication-competent MoMLV derivative was used that had the *E. coli* amber suppressor *sup*F inserted in the U3 region. λ DNA that contained amber mutations in multiple genes required for lytic growth was used as an insertional target. Phage plated on wild-type *E. coli* could only form plaques if MoMLV carrying the amber suppressor had integrated. Mixing cell-free nuclear or cytoplasmic extracts with purified λ DNA resulted in a substantial increase in infectious phage after packaging. It was estimated that about 0.3% of the retroviral DNA integrated into λ. Examination of the sites of insertion indicated no specificity. Random integration occurred in regions not required to support λ infection. Examination of the λ-MoMLV junctions indicated normal integration structures. Each integrated provirus lacked the 2 bp thought to be present in the unintegrated precursor. Four base pairs of the λ target site were duplicated, characteristic of normal MoMLV integration. Surprisingly, when cytoplasmic and nuclear extracts were compared, the former proved to be more active. This was not due to the higher concentration of DNA or the presence of specific factors, since mixing experiments proved to give additive results. Circular viral duplex DNA containing two tandem copies of the viral LTRs that form a unique U3:U5 junction (att site) are thought to be the precursors for integration.

None of these molecules could be detected in the cytoplasmic extracts. Extract-specific activity could be increased 100-fold by gel filtration of the viral DNA-protein complexes, indicating that all required components remain tightly bound. Most of the reaction is complete with a 15-minute incubation at 30°C to 37°C. The reaction is dependent on Mg^{2+}. Treatment of extracts with protease K or N-ethylmaleimide abolished activity; treatment with RNase A was without effect. Neither ATP nor any other nucleoside or deoxynucleoside triphosphate was required. This does not preclude the prior formation of the isolated complexes requiring ATP or the presence of a factor carrying ATP in a nondissociable form.

Fujiwara and Mizuuchi [151] have extended the finding of Brown and co-workers [150] to develop a physical assay to detect the integration intermediates. This permits direct assay of integration without coupling detection with the requirement for repair and other subsequent steps. The use of gel assays permitted quantitating, by viral LTR sequence hybridization, the formation of higher-molecular-weight viral sequences that resulted from insertion of MoMLV DNA into a homogeneous population of exogenous DNA acceptor molecules. The basic findings were consistent with those of Brown et al. The reaction did not require the addition of an external energy source, it was dependent upon Mg^{2+} (or Mn^{2+} or Co^{2+}), and the sites of integration were random. The major advantage of this new approach is the improved efficiency of integration achieved and the resulting ability to examine the integration intermediates directly. Between 10% and 25% of the input linear viral DNA was integrated. Hybridization with specific probes for each strand of the two LTRs on denaturing gels indicated that only the 3' ends of the viral DNA become covalently bound to λ DNA. The 5' ends remain free. Cleavage upstream with a restriction endonuclease and determination of fragment size indicated their terminal sequence was 5'-AATG. This sequence is expected from a linear viral integration precursor and not the 5'-TTAATG anticipated if the viral DNA had first been circularized and then cleaved to form staggered ends. Thus, given that the integrating donor was a linear molecule, how were the two nucleotides removed from the 3' end of the viral sequences? Either the terminal two nucleotides from the 3' end of the retroviral inserting DNA had to be cleaved from a linear molecule or the molecules that lacked the final two bases from flush ends were selected. Colicelli and Goff [149], providing an explanation for genetic findings, have proposed that the integrase may function on unjoined LTR sequences that are stabilized in close proximity by integrase protein-protein interactions. This explanation is consistent with the findings of Fujiwara and Mizuuchi [151].

Fujiwara and Mizuuchi [151] point out the striking parallels between this integration reaction and that of bacteriophage mu. In both cases, only the 3' ends of cleaved mu DNA are joined to the 5' ends of host DNA at the integration site. Normally, the mu A protein cleaves mu DNA as part of the integration reaction. However, this event can be uncoupled from integration [152,153]. Mu DNA can be efficiently supplied for the integration reaction in a precisely cleaved form. Even the CA-3'-terminal nucleotide sequence is conserved between mu and retroviruses. Temin [154] has proposed that retroviruses have evolved from transposable elements.

The breakthroughs of Brown et al. [150] and Fujiwara and Mizuuchi [151] provide a pathway to understanding the components, structures, and mechanisms operational in this intriguing step of retroviral replication. Given the stability of the viral DNA-protein complexes, they should be amenable to further purification and characterization of their components and structural arrangements. With cloning and overproduction of authentic replicas of the viral and host proteins, it should be possible to obtain sufficient material to learn how to reconstitute these complexes. True, this may prove difficult considering the inability to provide viral DNA in *trans* as an integration substrate even in the presence of helper virus [155]. This difficulty is not due solely to proper assembly of the integrase, since this function can be provided in *trans* [75]. Perhaps the assembly of specific viral core proteins or host factors plays a critical role. The use of these novel integration assays coupled with the replicative assays described above to provide substrates for the integration step should provide appropriate avenues to the solution of this difficult but critical problem.

Future Perspectives

In spite of the productive and imaginative contributions of multiple laboratories over the past 18 years, many central questions remain regarding retroviral replication. These questions begin with our lack of understanding of the structure of the viral RNA template. What is the nucleoprotein structure of the template? What are the critical protein components, and where are they associated? Can these structures be reconstituted from purified components using their efficient replication as an assay?

Concerning the replication process itself, what host factors are involved? Cellular DNA ligase would be expected to be involved in circular duplex formation—what associated proteins are necessary for its recruitment and effective action? Are other host proteins involved in earlier stages of retroviral replication? Candidates could include the 7-MeG cap-binding factors, poly A–binding proteins, and general cytoplasmic RNA-binding proteins. Appropriate assays may not be found in the ability to influence gross DNA synthesis but perhaps in the specificity or efficiency of the discrete steps delineated in Figure 5.1. In latter stages, are cellular DNA polymerases or accessory factors involved in DNA strand displacement, RNA displacement, or the coordination of these processes?

Concerning the reverse transcriptase itself, does it form an asymmetric dimeric structure with distinct plus and minus strand polymerases? Do the halves of this complex communicate to coordinate $(+)$ and $(-)$ strand replication? Do the halves of the polymerase remain associated with the ends of replicative intermediate K (Fig. 5.1) in association with integrase and perhaps other factors to stabilize a structure that is active in subsequent integration events? How are the activities of RNase H coordinated with polymerase action? Is there a separate role for the low-molecular-weight form of RNase H created concomitant with p51 generation?

In the final stages of the replicative reactions, questions remain concerning the favored precursor for the integration event: Is it a linear or covalently closed circular molecule? If linear, do the strands contain blunt ends or staggered ends resulting from controlled incomplete replication, generating the appropriate termini for the integration event? What is the nucleoprotein structure of the DNA product? Do any of the capsid proteins associated with viral RNA remain in the final product? Are they important for the subsequent integration reaction? What other factors are involved in this process.

These and many other fundamental questions remain. Among the contributions required for their solution, many will continue to be derived from nucleic acid enzymology. A worthy goal is to establish a reconstituted reaction, containing viral RNA, a tRNA primer, and necessary purified viral and host factors, that supports the orderly and efficient synthesis of viral duplex DNA competent for integration into exogenous DNA. Such an approach will permit the examination of intermediates formed in a state amenable to study—free of nucleases and interfering substances. Careful examination of the enzymatic mechanisms of such reactions will reveal important new information concerning cellular and viral nucleic acid metabolism, new strategies for the use of retroviruses as tools for the molecular biologist, novel targets for antiviral chemotherapy, and superior systems for the preliminary screening of antiviral agents.

References

1. Broder, S., and Gallo, R.C. (1985) Annu. Rev. Immunol. 3, 321–336.
2. Popovic, M., Sarngadharan, M., Read, E., and Gallo, R. (1984) Science 224, 497–500.
3. Barre-Sinoussi, F., Chermann, J.C., Rey, F., Nugeyre, M.T., Chamaret, S., Gruest, J., Dauguet, C., Axler-Blin, C., Vezinet-Brun, F., Rouzioux, C., Rozenbaum, W., and Montagnier, L. (1983) Science 220, 868–871.
4. Levy, J., Hoffman, A., Kramer, S., Landis, J., Shimabukuro, J., and Oshiro, L. (1984) Science 225, 840–842.
5. Sanchez-Pescador, R., Power, M.D., Barr, P.J., Steimer, K.S., Stempien,

M.M., Brown-Shimer, S.L., Gee, W.W., Renard, A., Randolph, A., Levy, J.A., Dina, D., and Luciw, P.A. (1985) Science *227*, 484–492.

6. Wain-Hobson, S., Sonigo, P., Danos, O., Cole, S., and Alizon, M. (1985) Cell *40*, 9–17.

7. Ratner, L., Haseltine, W., Patarca, R., Livak, K.J., Starcich, B., Josephs, S.F., Doran, E.R., Rafalski, J.A., Whitehorn, E.A., Baumeister, K., Ivanoff, L., Petteway, S.R., Pearson, M.L., Lautenberger, J.A., Papas, T.S., Ghrayeb, J., Chang, N.T., Gallo, R.C., and Wong-Staal, F. (1985) Nature *313*, 277–284.

8. Luciw, P.A., Cheng Mayer, C., and Levy, J.A. (1987) Proc. Natl. Acad. Sci. USA *84*, 1434–1438.

9. Seigel, L.J., Ratner, L., Josephs, S.F., Derse, D., Feinberg, M.B., Reyes, G.R., O'Brien, S.J., and Wong-Staal, F. (1986) Virology *148*, 226–231.

10. Dayton, A.I., Sodroski, J.G., Rosen, C.A., Goh, W.C., and Haseltine, W.A. (1986) Cell *44*, 941–947.

11. Gerard, G., and Grandgenett, D. (1980) In Molecular Biology of RNA Tumor Viruses, 345–394, Stephenson, J.R. (ed.). Academic Press, New York.

12. Gerard, G. (1981) In Enzymes of Nucleic Acid Synthesis and Identification, Vol. 1, 1–38, Jacob, S.T. (ed.). CRC Press, Boca Raton, Florida.

13. Coffin, J. (1979) J. Gen. Virol. *42*, 1–26.

14. Temin, H., and Baltimore, D. (1972) Adv. Virus Res. *17*, 129–186.

15. Mason, W., Taylor, J., and Hull, R. (1987) Adv. Virus Res. *32*, 35–96.

16. Varmus, H., Swanston, R., and Coffin, J. (1984) In RNA Tumor Viruses, Molecular Biology of Tumor Viruses, 369–512, Weiss, R., Teich, N., Varmus, H., and Coffin, J. (eds.). Cold Spring Harbor Laboratory, Cold Spring Harbor, New York.

17. Varmus, H., and Swanstrom, R. (1985) In RNA Tumor Viruses, Molecular Biology of Tumor Viruses, 75–134, Weiss, R., Teich, N., Varmus, H., and Coffin, J. (eds.). Cold Spring Harbor Laboratory, Cold Spring Harbor, New York.

18. Alizon, M., Wain-Hobson, S., Montagnier, L., and Sonigo, P. (1986) Cell *46*, 63–74.

19. Muesing, M.A., Smith, D.H., Cabradilla, C.D., Benton, C.V., Lasky, L.A., and Capon, D.J. (1985) Nature *313*, 450–458.

20. Casareale, D., Sinangil, F., Hedeskog, M., Ward, W., Volsky, D.J., and Sonnabend, J. (1983) AIDS Res. *1*, 253–270.

21. Chandra, A., Gerber, T., and Chandra, P. (1986) FEBS Lett. *197*, 84–88.

22. Veronese, F.D.M., Copeland, T.D., Devico, A.L., Rahman, R., Oroszlan, S., Gallo, R.C., and Sarngadharan, M.G. (1986) Science *231*, 1289–1291.

23. Lightfoote, M.M., Coligan, J.E., Folks, T.M., Fauci, A.S., Martin, M.A., and Venkatesan, S. (1986) J. Virol. *60*, 771–775.

24. Tanese, N., Sodroski, J., Haseltine, W.A., and Goff, H.L. (1986) J. Virol. *59*, 743–745.

25. Barr, P.J., Power, M.D., Lee-Ng, C.T., Gibson, H.L., and Luciw, P.A. (1987) Bio/Technology *5*, 486–489.

26. Larder, B., Purifoy, D., Powell, K., and Darby, G. (1987) EMBO J. *6*, 3133–3138.

27. Farmerie, W.G., Loeb, D.D., Casavant, N.C., Hutchison, C.A, Edgell, M.H., and Swanstrom, R. (1987) Science *236*, 305–308.

28. Hansen, J., Schulze, T., and Mölling, K. (1987) J. Biol. Chem. *262*, 12393–12396.
29. Moore, R., Dixon, M., Smith, R., Peters, G., and Dickson, C. (1987) J. Virol. *61*, 480–490.
30. Le Grice, F., Beuck, V., and Mous, J. (1987) Gene *55*, 95–103.
31. Tanese, N., Prasad, V.R., and Goff, S. (1988) DNA *7*, 407–416.
32. Le Grice, S., Mills, J., and Mous, J. (1988) EMBO J. *7*, 2547–2553.
33. Sherman, F., and Stewart, J. (1982) In The Molecular Biology of the Yeast Saccharomyces, 301–333, Cold Spring Harbor Laboratory, Cold Spring Harbor, New York.
34. Johnson, M.S., McClure, M.A., Feng, D.F., Gray, J., and Doolittle, R.F. (1986) Proc. Natl. Acad. Sci. USA *83*, 7648–7652.
35. Darke, P., Nutt, R., Brady, S., Garsky, V., Ciccarone, T., Leu, C., Lumma, P., Freidinger, R., Veber, D., and Sigal, I. (1988) Biochem. Biophys. Res Commun. *156*, 297–303.
36. Hizi, A., McGill, C., and Hughes, S.H. (1988) Proc. Natl. Acad. Sci. USA *85*, 1218–1222.
37. Hansen, J., Schulze, T., Mellert, W., and Mölling, K. (1988) EMBO. J. *7*, 239–243.
38. Henderson, L.E., Copeland, T.D., Sowder, R.C., Schultz, A.M., Oroszlan, S., and Bolognesi, D. (1988) In UCLA Symposia on Molecular and Cellular Biology New Series, Human Retroviruses, Cancer, and AIDS: Approaches to Prevention and Therapy, Vol. 71, 135–148. Alan R. Liss, New York.
39. Grandgennet, D.P., Gerard, G.F., and Green, M. (1973) Proc. Natl. Acad. Sci. USA *70*, 230–234.
40. Hizi, A., and Joklik, W. (1977) J. Biol. Chem. *252*, 2281–2289.
41. Grandgenett, D., and Green, M. (1974) J. Biol. Chem *249*, 5148–5152.
42. Grandgenett, D. (1976) J. Virol. *17*, 950–961.
43. Maki, H., and Kornberg, A. (1987) Proc. Natl. Acad. Sci. USA *84*, 4389–4392.
44. Johanson, K.O., and McHenry, C.S. (1984) J. Biol. Chem. *259*, 4589–4595.
45. McHenry, C.S. (1985) Mol. Cell Biochem. *66*, 71–85.
46. Hawker, J.R., and McHenry, C.S. (1987) J. Biol. Chem. *262*, 12711–12727.
47. McHenry, C.S., and Johanson, K.O. (1984) In Proteins Involved in DNA Replication, 315–319, Hübscher, U., and Spadari S. (eds.). Plenum, New York.
48. McHenry, C.S., Flower, A.M., and Hawker, J.R. (1987) In Eukaryotic DNA Replication, 35–41, Kelly, T., and Stillman, B. (eds.). Cold Spring Harbor Laboratory, Cold Spring Harbor, New York.
49. McHenry, C. (1988) Annu. Rev. Biochem. *57*, 519–550.
50. Maki, H., Maki, S., and Kornberg, A. (1988) J. Biol. Chem. *263*, 6570–6578.
51. Tanese, N., and Goff, S.P. (1988) Proc. Natl. Acad. Sci. USA *85*, 1777–1781.
52. Steimer, K.S., Higgins, K.W., Powers, M.A., Stephans, J.C., Gyenes, A., George Nascimento, C., Luciw, P.A., Barr, P.J., Hallewell, R.A., and Sanchez-Pescador, R. (1986) J. Virol. *58*, 9–16.
53. Chang, N., Huang, J., Ghrayeb, J., McKinney, S., Chanda, P., Chang, T., Putney, S., Sarngadharan, M., Wong-Staal, F., and Gallo, R. (1985) Nature *315*, 151–154.
54. Golomb, M., and Grandgenett, D. (1979) J. Biol. Chem. *254*, 1606–1613.

55. Golomb, M., Grandgenett, D.P., and Mason, W. (1981) J. Virol. *38*, 548–555.
56. Schiff, R.D., and Grandgenett, D.P. (1978) J. Virol. *28*, 279–291.
57. Schiff, R.D., and Grandgenett, D.P. (1980) J. Virol. *36*, 889–893.
58. Hizi, A., and Joklik, W. (1977) Virology *78*, 571–575.
59. Misra, T., Grandgenett, D.P., and Parsons, J.T. (1982) J. Virol. *44*, 330–343.
60. Knaus, R.J., Hippenmeyer, P.J., Misra, T.K., Grandgenett, D.P., Mueller, U.R., and Fitch, W.M. (1984) Biochemistry *23*, 350–359.
61. Duyk, G., Longiaru, M., Cobrinik, D., Kowal, R., Dehaseth, P., Skalka, A., and Leis, J. (1985) J. Virol. *56*, 589–599.
62. Grandgenett, D.P., Vora, A.C., Swanstrom, R., and Olsen, J.C. (1986) J. Virol *58*, 970–974.
63. Cobrinik, D., Katz, R., Terry, R., Skalka, A., and Leis, J. (1987) J. Virol. *61*, 1999–2008.
64. Grandgenett, D., and Vora, A. (1985) Nucleic Acids Res. *13*, 6205–6221.
65. Duyk, G., Leis, J., Longiaru, M., and Skalka, A. (1963) Proc. Natl. Acad. Sci. USA *80*, 6745–6749.
66. Roth, M., Tanese, N., Schwartzberg, P., and Goff, S. (in press).
67. Luk, K.C., Gilmore, T.D., and Panganiban, A.T. (1987) Virology *157*, 127–136.
68. Nissen-Meyer, J., Raae, A., and Nes, I. (1981) J. Biol. Chem *256*, 7985–7989.
69. Nissen-Meyer, J., and Nes, I. (1980) Biochim. Biophys. Acta *609*, 148–157.
70. Nissen-Meyer, J., and Nes, I. (1980) Nucleic Acids Res. *8*, 5043–5055.
71. Kopchick, J.J., Harless, J., Geisser, B.S., Killam, R., Hewitt, R.R., and Arlinghaus, R.B. (1981) J. Virol. *37*, 274–283.
72. Panet, A., and Baltimore, D. (1987) J. Virol. *61*, 1756–1760.
73. Donehower, L.A., and Varmus, H.E. (1984) Proc. Natl. Acad. Sci. USA *81*, 6461–6465.
74. Schwartzberg, P., Colicelli, J., and Goff, S.P. (1984) Cell *37*, 1043–1052.
75. Panganiban, A.T., and Temin, H.B. (1984) Proc. Natl. Acad. Sci. USA *81*, 7885–7889.
76. Dickson, C., Eisenman, R., Fan, H., Hunter, E., and Teich, N. (1984) In RNA Tumor Viruses, Molecular Biology of Tumor Viruses, Vol. 1, 2d Ed., 513–648, Weiss, R., Teich, N., Varmus, H., and Coffin, J. (eds). Cold Spring Harbor Laboratory, Cold Spring Harbor, New York.
77. Dickson, C., Eisenman, R., and Fan, H. (1985) In RNA Tumor Viruses, Molecular Biology of Tumor Viruses, Vol. 2, 2d Ed., 135–146, Weiss, R., Teich, N., Varmus, H., and Coffin, J. (eds.). Cold Spring Harbor Laboratory, Cold Spring Harbor, New York.
78. Oroszlan, S., and Copeland, T.D. (1985) Curr. Topics Microbiol. Immunol. *115*, 221–233.
79. Henderson, L., Copeland, T., Sowder, R., Smythers, G., and Oroszlan, S. (1981) J. Biol. Chem. *256*, 8400–8406.
80. Williams, K., Lopresti, M., and Setoguchi, M. (1981) J. Biol. Chem. *256*, 1754–1762.
81. Gauss, P., Krassa, K.B., McPheeters, D.S., Nelson, M.A., and Gold, L. (1987) Proc. Natl. Acad. Sci. USA *84*, 8515–8519.
82. Giedroc, D.P., Keating, K.M., Williams, K.R., Konigsberg, W.H., and Coleman, J.E. (1986) Proc. Natl. Acad. Sci. USA *83*, 8452–8456.

83a. Karpel, R., Henderson, L., and Oroszlan, S. (1987) J. Biol. Chem. *262*, 4961–4967.

83b. Gorelick, R., Henderson, L., Hanser, J., and Rein, A. (1988) Proc. Natl. Acad. Sci. USA *85*, 8420–8424.

84. Smith, V., and Bailey, J. (1979) Nucleic Acids Res. *7*, 2055–2072.

85. Sykora, K., and Mölling, K. (1981) J. Gen Virol. *55*, 379–391.

86. Schulein, M., Burnette, W., and August, J. (1978) J. Virol. *26*, 54–60.

87. Leis, J., and Jentoft, J. (1983) J. Virol. *48*, 361–369.

88. Leis, J., and Johnson, S. (1984) J. Biol. Chem. *259*, 7726–7732.

89. Katz, R.A., Fu, X.D., Skalka, A.M., and Leis, J. (1986) Gene *50*, 361–370.

90. Fu, X., Tuazon, P.T., Traugh, J.A., and Leis, J. (1988) J. Biol. Chem. *263*, 2134–2139.

91. Fu, X., Katz, R.A., Skalka, A.M., and Leis, J. (1988) J. Biol. Chem. *263*, 2140–2145.

92. Davis, J., Scherer, M., Tsai, W., and Long, C. (1976) J. Virol. 18, 709–719.

93. Chen, M., Garon, C., and Papas, T. (1980) Proc. Natl. Acad. Sci. USA *77*, 1296–1300.

94. Bedinger, P., and Alberts, B.M. (1983) J. Biol. Chem. *258*, 9649–9656.

95. Bandyopadhyay, A. (1977) J. Biol. Chem. *252*, 5883–5887.

96. Bandyopadhyay, A., and Levy, C. (1978) J. Biol. Chem. *253*, 8285–8290.

97. Leis, J., McGinnis, J., and Green, R. (1978) Virology *84*, 87–98.

98. Leis, J., Scheible, P., and Smith, R. (1980) J. Virol. *35*, 722–731.

99. Jacks, T., Power, M. Masiarz, F., Luciw, P., Barr, P., and Varmus, H. (1988) Nature *331*, 280–283.

100. Veronese, S., Copeland, T., Oroszlan, S., Gallo, R.C., and Sarngadharan, M.G. (1988) J. Virol. *62*, 795–801.

101. Kramer, R., Schaber, M.D., Skalka, A.M., Ganguly, K., Wong-Staal, F., and Reddy, E.P. (1986) Science *231*, 1580–1584.

102. Debouck, C., Gorniak, J., Strickler, J., Meek, T., Metcalf, B., and Rosenberg, M. (1987) Proc. Natl. Acad. Sci. USA *84*, 8903–8906.

103. Graves, M.C., Lim, J.J., Heimer, E.P., and Kramer, R.A. (1988) Proc. Natl. Acad. Sci. USA *85*, 2449–2453.

104. Pearl, L., and Taylor, W. (1987) Nature *328*, 482.

105. Katz, R.A., and Skalka, A.M. (1988) J. Virol. *62*, 528–533.

106. Grandgenett, D., Quinn, T., Hippenmeyer, P., and Oroszlan, S. (1985) J. Biol. Chem. *260*, 8243–8249.

107. Peters, G., and Glover, C. (1980) J. Virol. *35*, 31–40.

108. Harada, F., Sawyer, R., and Dahlberg, J. (1975) J. Biol. Chem. *250*, 3487–3497.

109. Haseltine, W.A., Panet, A., Smoler, D., Baltimore, D., Peters, G., Harada, F., and Dahlberg, J.E. (1977) Biochemistry *16*, 3625–3632.

110. Cordell, B., Swanstrom, R., Goodman, H.G., and Bishop, J.M. (1979) J. Biol. Chem. *254*, 1866–1874.

111. Sarih, L., Araya, A., and Litvak, S. (1988) FEBS Lett. *230*, 61–66.

112. Panet, A., Haseltine, W.A., Baltimore, D., Peters, G., Harada, F., and Dahlberg, J.E. (1975) Proc. Natl. Acad. Sci. USA *72*, 2535–2539.

113. Sawyer, R.C., and Hanafusa, H. (1979) J. Virol. *29*, 863–871.

114. Peters, G., and Hu, J. (1980) J. Virol. *36*, 692–700.

115. Grandgenett, D., and Vora, A. (1976) Virology *75*, 26–32.
116. Araya, A., Keith, G., Fournier, M., Gandar, J., Labouesse, J., and Litvak, S. (1980) Arch. Biochem. Biophys. *205*, 437–448.
117. Harada, F., Peters, G.G., and Dahlberg, J.E. (1979) J. Biol. Chem. *254*, 10979–10985.
118. Levin, J., and Seidman, J. (1981) J. Virol. *38*, 403–408.
119. Haseltine, W.A., Kleid, D.G., Panet, A., Rothenberg, E., and Baltimore, D. (1976) J. Mol. Biol. *106*, 109–131.
120. Starcich, B., Ratner, L., Josephs, S., Okamoto, T., Gallo, R., and Wong-Staal, F. (1985) Science *227*, 538–540.
121. Collett, M., Dierks, P., and Parsons, J. (1978) Nature *272*, 181–184.
122. Haseltine, W.A., Coffin, J.M., and Hageman, T.C. (1979) J. Virol. *30*, 375–383.
123. Swanstrom, R., Varmus, H.E., and Bishop, J.M. (1981) J. Biol. Chem. *256*, 1115–1121.
124. Omer, C., Parsons, J., and Faras, A. (1981) J. Virol. *38*, 398–402.
125. Collett, M.S., and Faras, A.J. (1978) Virology *86*, 297–311.
126. Panganiban, A., and Fiore, D. Science *241*, 1064–1069.
127. Boone, L., and Skalka, M. (1981) J. Virol. *37*, 109–116.
128. Mitra, S., Goff, S., Gilboa, E., and Baltimore, D. (1979) Proc. Natl. Acad. Sci. USA *76*, 4355–4359.
129. Varmus, H.E., Heasley, S., Kung, H.J., Oppermann, H., Smith, V.C., Bishop, J.M., and Shank, P.R. (1978) J. Mol. Biol. *120*, 55–82.
130. Smith, J.K., Cywinski, A., and Taylor, J.M. (1984) J. Virol. *52*, 314–319.
131. Resnick, R., Omer, C.A., and Faras, A.J. (1984) J. Virol. *51*, 813–821.
132. Champoux, J.J., Gilboa, E., and Baltimore, D. (1984) J. Virol. *49*, 686–691.
133. Sorge, J., and Hughes, S.H. (1982) J. Virol. *43*, 482–488.
134. Swanstrom, R., Bishop, J.M., and Varmus, H.E. (1982) J. Virol. *42*, 337–341.
135. Junghans, R.P., Boone, L.R., and Skalka, A.M. (1982) J. Virol. *43*, 544–554.
136. Boone, L.R., and Skalka, A.M. (1981) J. Virol. *37*, 117–126.
137. Finston, W.I., and Champoux, J.J. (1984) J. Virol. *51*, 26–33.
138. Taylor, J., and Sharmeen, L. (1987) J. Cell Sci. (Suppl.) *7*, 189–195.
139. Holck, A., Lossius, I., Aasland, R., and Kleppe, K. (1987) Biochim. Biophys. Acta *914*, 49–54.
140. Gilboa, E., Mitra, S., Goff, S., and Baltimore, D. (1979) Cell *18*, 93–100.
141. Sampson, J.R., and Uhlenbeck, O.C. (1988) Proc. Natl. Acad. Sci. USA *85*, 1033–1037.
142. Omer, C.A., and Faras, A.J. (1982) Cell *30*, 797–805.
143. Varmus, H., Guntaka, R., Fan, W., Heasley, S., and Bishop, J. (1974) Proc. Natl. Acad. Sci. USA *71*, 3874–3878.
144. Shank, P.R., Hughes, S.H., Kung, H.J., Majors, J.E., Quintrell, N., Guntaka, R.V., Bishop, J.M., and Varmus, H.E. (1978) Cell *15*, 1383–1395.
145. Hughes, S., Shank, P., Spector, D., Kung, H., Bishop, J., Varmus, H., Vogt, P., and Breitman, M. (1978) Cell *15*, 1397–1410.
146. Vijaya, S., Steffen, D.L., and Robinson, H.L. (1986) J. Virol. *60*, 683–692.
147. Varmus, H., Padgett, T., Heasley, S., Simon, G., and Bishop, M. (1977) Cell *11*, 307–319.
148. Panganiban, A.T., and Temin, H.M. (1984) Cell *36*, 673–680.

149. Colicelli, J., and Goff, S. (1988) J. Mol. Biol. *199*, 47–59.
150. Brown, P.O., Bowerman, B., Varmus, H.E., and Bishop, J.M. (1987) Cell *49*, 347–356.
151. Fujiwara, T., and Mizuuchi, K. (1988) Cell *54*, 497–504.
152. Craigie, R., and Mizuuchi, K. (1987) Cell *51*, 493–501.
153. Surette, M.G., Buch, S.J., and Chaconas, G. (1987) Cell *49*, 253–262.
154. Temin, H.M. (1980) Cell *21*, 599–600.
155. Kriegler, M., and Botchan, M. (1983) Mol. Cell Biol. *3*, 325–339.
156. Leis, J., Baltimore, D., Bishop, J.M., Coffin, J., Flessiner, E., Goff, S.P., Oroszlan, S., Robinson, H., Skalka, A.M., Temin, H., and Vogt, V. (1988) J. Virol. *62*, 1808–1809.
157. Cheng, Y.C., Dutschman, G.E., Bastow, K.F., Sarngadharan, M.G., and Ting, R.Y.C. (1987) J. Biochem. *262*, 2187–2189.
158. Hoffman, A.D., Banapour, B., and Levy, J.A. (1985) Virology *147*, 326–335.
159. Rey, M.A., Spire, B., Dormont, D., and Barre-Sinouss, F. (1984) Biochem. Biophys. Res. Commun. *121*, 126–133.
160. Guyader, M., Emerman, M., Sonigo, P., Clavel, F., Montagnier, L., and Alizon, M. (1987) Nature *326*, 662–670.

Section II Gene Transcription

Replication sustains the genome, but transcription is the reason for its existence. The synthesis of messenger RNA transcripts from the chromosomal DNA template and the translation of the mRNA into proteins are primary events in the functioning of the cell. Topics concerning different aspects of gene transcription are therefore discussed in this section. Studies in this research area are at a sophisticated stage because of the availability of advanced techniques in biochemistry, molecular biology, and molecular genetics. As with investigations of DNA replication, in vitro experiments allow the roles of specific proteins and DNA sequences to be examined. The basic occurrence in the synthesis of mRNA involves the activity of RNA polymerase, and properties of the RNA polymerases receive considerable attention in the following chapters. But equally significant events come before and after the polymerization of the mRNA polynucleotide chain. Regulation of the initiation of transcription by the binding of specific proteins controls the spectrum of proteins that are produced in a given cell at a given time. Following mRNA synthesis, splicing of pre-mRNA is required to generate processed transcripts that are functional in translation. And, during transcription and splicing, pre-mRNA appears to be packaged in regular ribonucleoprotein particles. Chapters concerning these subjects are, on account of their importance, included in the section.

Highly refined answers can be given to the fundamental questions concerning gene transcription because of the experimental tools available for modern biology research. The ability to clone specific gene segments and to determine their sequences has allowed answers to be given at the level of individual nucleotides. Advances in this field of research have also been due to the use of organisms such as *E. coli* that can be genetically manipulated. The application of molecular genetics along with molecular biology has proved to be a powerful combination. The techniques of biochemistry, particularly protein purification and gel electrophoresis procedures, have further permitted the proteins involved in transcription to be identified and characterized. Eukaryotic systems can now be tackled, so that the data should shed light directly on the process in human cells. A future challenge will be to extend the understanding of gene transcription derived from in vitro studies to the level of the intact, living cell.

Chapter 6
Escherichia coli Repressor Proteins

Kyle L. Wick and Kathleen S. Matthews

Genetic regulation is an essential function in all living organisms. In prokaryotes genetic control provides responsivity to a constantly changing external milieu, and bacterial systems have proved extremely valuable in elucidating the variety of potential mechanisms. Both positive and negative strategies exist for transcriptional control; this format for regulation at the initiation of mRNA synthesis allows maximum conservation of cellular energy. Positive control over a wide range of systems in *E. coli* is exerted by catabolite repression mediated by the CAP-cAMP system. Negative control in bacteria involves a series of repressor proteins, each specific for a particular pathway. Several of these systems and the details of their mechanisms of action will be discussed in this chapter.* The basic regulatory scheme involves the interaction of a repressor protein with its target DNA sequence in an inducibile or repressible manner (Fig. 6.1). Inducible mechanisms generally govern catabolic pathways where expression is necessary only in the presence of substrate. These genes are turned "off" through a repressor protein-operator DNA interaction which interferes with transcription initiation by RNA polymerase. Derepression or induction is accomplished through destabilization of the repressor-operator complex upon binding of substrate, or a closely related inducer molecule, to the repressor protein. In contrast, repressible mechanisms most. often apply to biosynthetic pathways where gene expression is modulated by levels of the end product. To avoid overaccumulation, repression is established when the end product or related compound, acting as corepressor, binds to and activates an aporepressor protein. The resulting complex recognizes its cognate DNA sequence and thereby blocks transcription initiation.

---

* Because of constraints many references could not be included. The citation for a specific observation may contain references to earlier work in related areas.

INDUCIBLE SYSTEM

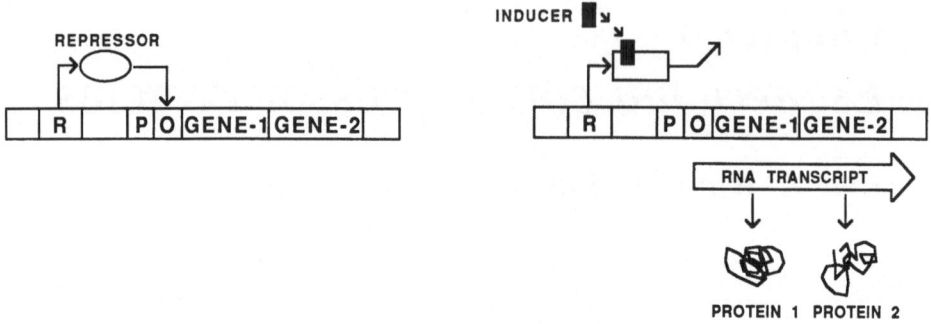

REPRESSIBLE SYSTEM

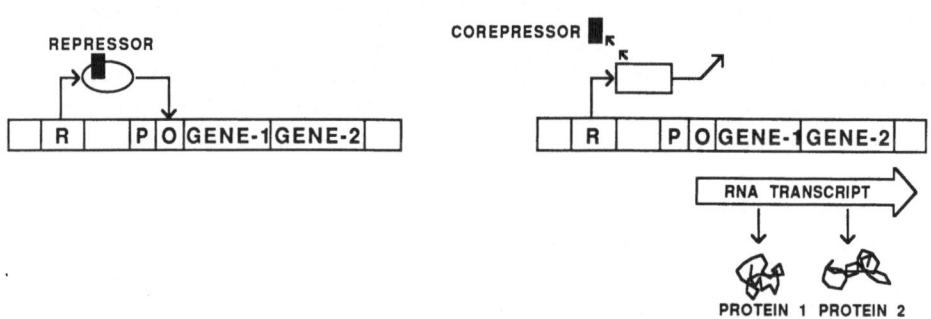

Figure 6.1. Schematic representation of regulatory interactions operative in inducible and repressible negative control systems. The genes or sequences for repressor protein (R), promoter (P), operator (O), and the structural genes 1 and 2 are shown. Active repressor is depicted by ◯, while ▭ represents the induced or inactive form of the protein. The inducer or corepressor molecule is shown as ■.

A feature common to many of these repressors is the presence of regions homologous to a secondary structural motif found in DNA-binding domains of proteins with known X-ray crystallographic structures [1–4]. In this general model for DNA binding, a pair of helices from each of two subunits of a DNA-binding protein makes contact with adjacent major grooves in right-handed B-DNA. Within the protein, the α helices are connected by a β turn. Sequence homologies among these proteins and structurally undefined DNA-binding proteins suggest the use of a prototypic structure for sequence-specific complex formation [5]. Another common feature that has emerged, at least for inducible systems, is the apposition of distant DNA-binding sites mediated by direct protein binding or by protein-protein contacts to form a looped DNA structure [6,7]. Many aspects of repressor function in prokaryotes find analogies in more complex systems, and detailed understanding in these simpler systems may inform exploration of eukaryotic genetic control.

Inducible Systems

Transcription of mRNA for inducible systems occurs in the presence of the substrate for the pathway and, in the case of sugar pathways, in simultaneous absence of glucose. Thus, the enzymes necessary for catabolism are synthesized only under conditions that require their activity (e.g., for generation of a carbon source or for protection from antibiotic). The operons that produce enzymes involved in sugar metabolism are also under positive control of the catabolite activator protein (CAP). The CAP protein ($M_r \sim 45,000$) functions as a dimer of identical subunits. The crystal structure of the CAP-cAMP complex has been solved and refined to a resolution of 2.5 Å. [1,8]. Each subunit of CAP binds a cAMP molecule in its NH_2-terminal domain and contains the helix-turn-helix DNA-binding structure within its COOH-terminal domain [8]. CAP binds to sites within the vicinity of many *E. coli* promoters when activated by complex formation with cAMP. The presence of this CAP-cAMP complex stimulates transcription of the associated structural genes by RNA polymerase. CAP-cAMP binding induces a conformational change that has properties consistent with a bend in the DNA [9–11]. cAMP levels are modulated by the influence of glucose availability on adenylate cyclase activity; thus, alternate carbon sources are utilized only in the absence of glucose. Although an activator rather than a repressor protein, the intimate involvement of CAP-cAMP in regulating production of mRNA for sugar-metabolizing enzyme systems dictates its inclusion in the discussion of several of these systems.

Lactose Operon

The lactose operon of *Escherichia coli* represents the classic inducible negative control system [12]. The operon consists of the genes encoding enzymes responsible for the transport and metabolism of lactose as well as the *lac*I gene coding for the repressor protein (Fig. 6.2). The lactose structural genes are subject to both positive and negative coordinate transcriptional control [for a review, see 13]. In the absence of lactose, the *lac*I gene product represses initiation of *lac*ZYA transcription by binding with high affinity to its recognition site within the control region, the operator. Derepression of the structural genes is accomplished through interaction of the repressor protein with 1,6-allolactose, a metabolite of lactose produced by β galactosidase, or other β galactosides. The protein responds to this inducer binding with a conformational change which translates to a destabilization of the operator binding interaction [14–16]. In its inducer-bound state, the repressor exhibits diminished affinity for operator and binds with comparable affinity to nonspecific DNA sequences present in the genome [17]. With the operator site open transcription can proceed, provided RNA polymerase and the CAP-

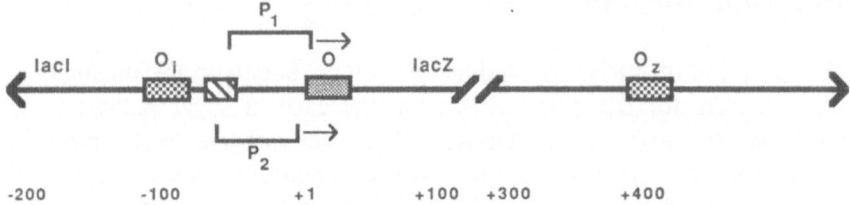

Figure 6.2. Genetic organization of the regulatory elements of the *lac* operon [13]. Centrally positioned, the primary operator (O) represents the target sequence for *lac* repressor. The two overlapping promoters, P_1 and P_2, are shown along with the position of the CAP-cAMP binding site ( ). Flanking the central regulatory region, the pseudooperators, which act as secondary repressor recognition sites, are located upstream within the *lac*I gene (O_i) and downstream within the *lac*Z gene (O_z). The small arrows indicate transcriptional polarity from P_1 and P_2.

cAMP complex are bound at their neighboring recognition sites. Transcription and translation of the structural genes continue until the lactose and consequently 1,6-allolactose pools are depleted or glucose becomes available. With dissociation of bound inducer, the repressor resumes its high-affinity binding conformation to interact at the operator site and to shut down expression from the *lac* operon. In addition to its role in repression, Straney and Crothers [18] have shown that *lac* repressor bound at the operator site increases the binding of RNA polymerase to the promoter. Thus, the repressor also serves to facilitate the first round of transcription once inducer is bound and therefore promotes a rapid induction response.

Modulation of the negative control exerted by the *lac* repressor is effected not only by inducer binding but also by environmental conditions [see 19 for a summary]. The binding of repressor to nonspecific DNA is highly electrostatic in nature and therefore sensitive to ionic strength [19]. The influence of ionic strength on repressor-operator interaction has indicated that in addition to ionic components, apolar interactions contribute significantly to the free energy of binding [19]. Kinetic analyses of repressor-operator binding suggest a two-step mechanism of association with a search of decreased dimensions once the DNA is bound nonspecifically [19].

The primary structure of the lactose repressor protein is known both from its gene sequence and from analysis of the protein itself [20,21]. In solution the repressor exists as a tetramer of identical subunits ($M_r \sim 150{,}000$) [22] and has four inducer sites and two operator DNA-binding sites [15,23–25]. Cooperativity in inducer binding is observed in the presence of operator DNA or at elevated pH in the absence of operator [15,16]. The protein can be separated into two structural domains through mild proteolytic digestion: a tetrameric core consisting of amino acids 60–360 and four amino-terminal domains composed of the first 59 amino acids [26,27]. Amino acid sequence homology between proteins of known structure and *lac* repressor NH₂ terminus strongly suggests the existence of a helix-turn-helix DNA-binding motif [5]. Further support has been derived from NMR analysis of the isolated *lac* repressor NH₂-terminal domain in which the folding patterns obtained are consistent with the postulated two-helix structure [28,29].

The arrangement of the core and NH₂-terminal domains within the tetrameric protein has been deduced from both powder X-ray diffraction analysis of microcrystals [30] and low-angle X-ray and neutron diffraction studies of the native and core repressors [31–33]. These studies suggest an elongated shape for the protein with the NH₂-terminal domains situated at the end of the tetrameric core. These end domains exhibit differential mobility and appear motionally flexible relative to the core region [34]. A definitive structural description awaits X-ray crystallographic analysis, which has been hampered by the difficulty in obtaining crystals.

Mutational and physical analyses have made possible assignment of function to the two domains. Mutational studies suggested structural independence of inducer and DNA-binding activities [35–40]. Mutations that affect DNA binding cluster in the NH₂-terminal region, with a lesser number scattered throughout the core domain. In contrast, amino acid changes that alter the inducer-binding capability of *lac* repressor are found entirely within the COOH-terminal 300 amino acids of the protein. Mutations affecting subunit aggregation cluster within the COOH-terminal 100 amino acids of the protein. Genetic methods have been combined with physical characterization to demonstrate effects of specific repressor amino acid substitution and/or operator base alteration on the interaction between the two [41,42]. These studies have demonstrated the importance of specific residues within the helix-turn-helix region for operator recognition and identified the bases contacted within the operator sequence.

Physical and chemical methods have complemented information from genetic analysis. The tetrameric core protein produced by proteolytic digestion binds inducer with wild-type affinity [26]; although this species does not bind to nonspecific DNA sequences [43], it may contain determinants for operator binding [44,45]. Examination of DNA binding of the isolated NH₂-terminal domain indicates nonspecific binding as well as interaction with operator sequences [46–52]. Results from chemical studies are also consistent with the placement of the subunit interface at the COOH-terminal region of the lactose repressor monomer [53,54]. Sequence homology with periplasmic sugar-binding proteins [55] has formed the basis for postulation of residues which form an inducer-binding site [56]; this model is consistent with effects of variation of inducer structure on inducer binding [57].

As the specific target for repression control, the operator element was hypothesized by Jacob and Monod [12] on the basis of *cis*-dominant mutations that resulted in operon constitutivity in the presence of active repressor. Many point mutations within the operator sequence have been identified and characterized [58,59]. These mutational analyses together with experiments probing the repressor-operator interaction through protection of the DNA against nuclease digestion, methylation, and UV-induced breakage defined the operator locus as a specific sequence spanning up to 40 nucleotides and displaying a high degree of dyad symmetry [45,47,59–62]. The repressor appears to favor a single face of the operator and binds in an asymmetric fashion with some of the most critical contacts located within the promoter-proximal half of the sequence [45,59–63]. Characteristics of the operator base sequence essential for binding have been identified by production of synthetic substituted operators [63,64], and the importance of spacing between symmetrically disposed sequences has been indicated by modified operators with increased affinity [65,66]. Sequences homologous to the operator have been found both ~100 bp upstream (O_i) and ~400 bp downstream (O_z) of the regulatory region [67,68]. These sequences, termed pseudooperators, have been shown to bind repressor, albeit with decreased affinities compared to the primary operator, and have been implicated in the overall regulatory mechanism. Although a second overlapping promoter site has been indicated by sequence analysis and mutational studies, this second promoter appears to play a minimal role in activation of the primary promoter by CAP-cAMP [69].

The pseudooperator sequences in the *lacI* and *lacZ* genes influence binding to operator both in vitro and in vivo. Although these sequences individually bind repressor with affinities 20-fold (O_z) and 100-fold (O_i) lower than primary operator, their presence in linear operator-containing DNA results in modest stabilization of the repressor-DNA complex [25,70,71]. Ternary complex formation with occupation of both operator sites in the protein has been suggested to explain this effect of sequence context on binding affinity [25]. Significant enhancement of binding affinity (>1,000-fold) by supercoiling of dual operator-containing DNAs is consistent with ternary complex and consequent DNA loop formation [72,73]. Such DNA looping is influenced by the spacing between operators in linear DNA [74], and, as anticipated based on mechanism, the effect of spacing is diminished and altered by supercoiling [72,73]. Direct electron microscopic examination has confirmed the presence of DNA loop formation via operator-repressor-operator association [73,74]. In vivo measurements of the effect of multiple operators on expression are consistent with these in vitro results and with DNA loop formation [75–78]. The role of looped DNA structures in genetic regulation is evident in other prokaryotic systems; the *ara, gal,* and *deo* operons all require occupation of multiple operators for full repression, and DNA loop formation offers a probable explanation for the observations in all of these inducible systems [7,79].

Galactose Operon

Expression of the galactose operon of *E. coli* is negatively regulated by the product of an unlinked gene (*gal*R) which encodes the galactose repressor protein. The genetic organizaiton of the regulatory region of the *gal*ETK operon is illustrated in Figure 6.3. Structurally unlike the *lac* operon, this region suggests a potentially complex mode of transcriptional regulation with a central role for the two overlapping promoters, P_1 and P_2. Catabolite repression is intimately involved in the selection of one promoter over the other with CAP-cAMP binding just upstream of the RNA polymerase sites to trigger activation of P_1 and simultaneously inhibit transcription initiation from P_2 [80]. Further control of the *gal* operon is mediated through *gal* repressor-binding interactions at each of two operator sequences. Located upstream from the promoter region, O_1 is considered to be the "classical" *gal* operator [81,82]. Positioned downstream from O_1 and within the *gal*E structural gene, the second operator, O_2, is also essential for transcriptional control [83,84].

The *gal*R gene product is responsible for negative regulation of the *gal*ETK operon through its specific binding to operator sequences O_1 and O_2 [85,86]. Von Wilcken-Bergmann and Müller-Hill [87] calculated the size of the *gal* polypeptide to be $M_r \sim 38,000$ based on its nucleotide sequence, a value confirmed by $NaDodSO_4$/PAGE of the purified protein [86]. The *gal* repressor sediments as a $M_r \sim 72,000$ protein and presumably exists as a dimer in solution [86]. A chimeric gene, composed of the sequences for the NH_2-terminal 72 amino acids from *gal* repressor and all but the first five amino acids from β-galactosidase, was constructed and expressed by Von Wilcken-Bergmann et al. [85]; this design was based on the high degree of sequence homology in both the NH_2-terminal and COOH-terminal domains between the *gal* and *lac* repressor proteins. The resulting fusion protein was a tetramer in solution due to its β-galactosi-

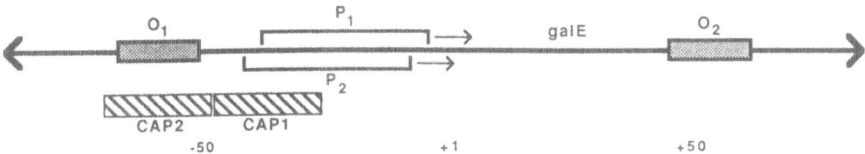

Figure 6.3. Regulatory determinants within the *gal*ETK operon [93]. The two overlapping promoter regions, P_1 and P_2, are shown with arrows indicating the direction of transcription. Two binding sites for CAP-cAMP (▨) are located just upstream. The operators are shown with O_1 upstream of the promoter region and O_2 within the downstream *gal*E gene.

dase core and actively repressed galactokinase (galE) synthesis in a noninducible manner. DNaseI footprint analysis using this engineered protein established its specific recognition and protection of a DNA sequence corresponding to the O_1 operator [85]. Thus, strong evidence exists for structural homology with the lac repressor where a portion of DNA-binding determinants are in the NH_2-terminal domain and the inducer-binding domain is localized in the COOH-terminal core [26,35]. The structural relatedness of these two repressor proteins has been shown to extend to their DNA-binding activities. Indeed, their respective operators share sequence similarities to the extent that the gal repressor has been shown to bind to lac operator DNA [87]. No details on the binding of inducer to isolated repressor or on conformational effects of this binding are available.

The two operator sequences of the gal operon, defined by positions of O^c point mutations and sequence homology, bind gal repressor with relatively equivalent affinities ($K_d \simeq 100$ nM) [81–83,86]. The ability of the gal repressor to bind its two operators simultaneously was suggested based on electrophoretic mobility shift assay results with DNA fragments containing various combinations of wild-type and mutant gal operators in the presence and absence of the inducer D-fucose [86]. With DNA fragments bearing both wild-type O_1 and O_2, inducer-sensitive gal repressor binding was observed as two complexed bands, presumed to correspond to repressor occupation of a single operator or both operators.

DNaseI footprinting experiments established the regions of protection provided by the gal repressor [88]. For each of the operator sequences, the results indicated that the repressor covers a region that extends about three bases beyond either side of the palindromic 16-bp conserved control sequence. In methylation protection experiments, symmetrically positioned guanine residues were protected by gal repressor, and residues on the opposite face were also protected [88]. The near-perfect dyad symmetry of the O_1 and O_2 sequences, along with the symmetrical protection pattern generated by bound repressor, suggests that each subunit of the dimeric protein contacts an operator half-site. From these studies, a model was proposed for the repressor-operator complex where two adjacent major grooves on one face of the DNA are simultaneously covered, while the intermediate major groove on the opposite face is also protected by the protein. This model would require flexibility within the protein to accommodate contacts on the opposite face. An alternative explanation is that the protein induces structural changes in the DNA that render the intermediate major groove residues nonreactive [88].

Regulation of expression within the gal operon is not a simple case of inducer-modulated negative control. Transcription initiation from P_1 and P_2 is positively controlled by the CAP-cAMP complex [89]. Without cAMP, transcription is initiated from P_2, while in its presence the CAP-cAMP complex binds just upstream of the promoter region, blocking P_2 and activating the P_1 promoter [90]. Repression of P_1 and P_2 activation has been investigated through promoter and regulatory site mutations [91–95]. The results indicated that occupation of the upstream operator (O_1) is sufficient for repression of the cAMP-dependent promoter (P_1) [93]. The second promoter (P_2) is also sensitive to repressor bound at O_1 but requires occupancy of the downstream O_2 site for complete repression [93]. These results support a model proposed by Di Lauro et al. [82] and modified by Shanblatt and Revzin [96] for repression of P_1. Under noninducing conditions, the binding of *gal* repressor to O_1 prevents initiation from P_1 by potentially interfering with CAP-cAMP binding. This model is further substantiated by the finding that two CAP-cAMP molecules are required for P_1 activation [96]. With overlapping binding sites, competition between the *gal* repressor and the second molecule of CAP-cAMP could impede the protein-protein interactions necessary for P_1 activation.

The role of the two operators in the control of P_2 transcription has been more difficult to determine. One model involves the formation of DNA loops via interactions between repressors bound at O_1 and O_2 [86]. Consequent stabilization of the repressor-operator complex or unfavorable DNA conformational changes could result in transcriptional repression. Alternatively, the second operator could serve the purpose of increasing the local concentration of *gal* repressor and thereby improve the efficiency of repression [93]. Clarification of these complex interactions will require further experimentation.

Arabinose Operon

The *ara*C gene product controls expression of the proteins involved in L-arabinose utilization in *Escherichia coli* through differential occupancy of its various DNA-binding sites [for a review see 97]. Figure 6.4 illustrates the organization of the *ara*BAD operon and includes the divergently transcribed *ara*C region. Along with the CAP-cAMP complex, the AraC protein in the presence of arabinose stimulates transcription of *ara*BAD, *ara*E, and *ara*FGH operons, which encode proteins for arabinose catabolism and transport. Without available arabinose, the AraC protein represses the *ara*BAD operon. This protein also serves an autoregulatory function in controlling its own levels through repression of the *ara*C gene in both the presence and absence of arabinose.

The functional AraC protein has been purified and shown to exist as a dimer ($M_r \sim 60,000$) of identical subunits [97]. Attempts to define functional domains of the protein have included mutational analyses and DNA-binding studies. The results of these studies on the AraC protein have provided evidence for DNA-binding determinants within the COOH-terminal region [98,99]. Properties of some mutants have suggested that the same region on the protein is essential for both its positive

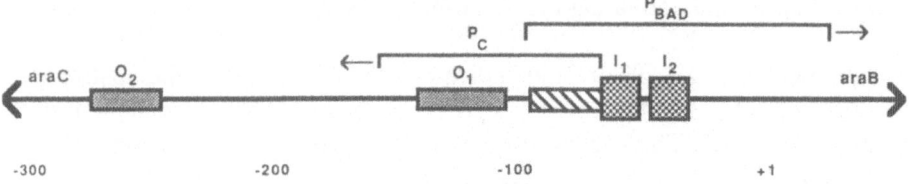

Figure 6.4. Organization of the control sequences for *ara*C and *ara*BAD expression [105]. Divergent transcription of *ara*C and *ara*BAD is initiated from the overlapping promoters P_C and P_{BAD}, respectively. The AraC protein recognition sequences are shown: O_1 and O_2 function in autoregulation and *ara*BAD repression, while I_1 and I_2 are involved in autoregulation of *ara*C and *ara*BAD activation. The CAP-cAMP binding site (▨) is just upstream and adjacent to I_1.

and negative regulatory properties [99]. A mutant P_{BAD} promoter was used to select for suppressor mutations in *ara*C generating altered DNA specificities; these mutations also mapped within the COOH terminus of the AraC protein [100]. Thus, evidence gathered to date convincingly defines the COOH terminus as the DNA-binding domain of the protein. The amino acid sequence in this region of AraC has been analyzed for homology to DNA-binding domains of other transcriptional regulatory proteins. Two regions near the COOH terminus of this protein possess homology to the known and proposed helix-turn-helix motifs of such proteins as CAP and lambda *trp* and *lac* repressors [100].

Transition between repressor and activator states of the AraC protein is modulated by L-arabinose. Wilcox [101] used spectrofluorometric methods to monitor the AraC-inducer interactions ($K_d \sim 3 \times 10^{-3}$ M) and proposed that L-arabinose, and not the antiinducer D-fucose, activates AraC through an induced conformational change in the protein. With altered DNA specificities relative to the nonliganded form, the AraC-inducer complex recognizes new sequences within the *ara*I region and stimulates *ara*BAD expression.

DNaseI footprint analyses of AraC binding were performed in the absence and presence of L-arabinose using both wild-type sequences and DNA containing a promoter-null *ara*I mutation [102]. In the absence of arabinose, AraC protein protected two turns of the DNA (-54 to -73) against DNaseI digestion. The protected sequence spanned one face of the DNA and extended to include four adjacent major grooves when arabinose was bound (-35 to -73). These data indicate that the *ara*I region is composed of two AraC binding sites, *ara*I$_1$ and *ara*I$_2$. The two sites are equivalent in size, exist on the same face of the DNA, and show some sequence homology. Further evidence for this division of *ara*I into two binding sites was derived from DNaseI protection studies using a promoter-null mutant *ara*I sequence [102]. Binding of AraC to premethylated and preethylated DNAs showed contact with only three helical turns within the *ara*I region [103]. The absence of guanine and phosphate contact within *ara*I$_2$ suggested a weaker binding interaction of this region with AraC-arabinose. AraC protein binds *ara*I$_1$ with a relatively high affinity ($\sim$2 $\times$ 10^{-12} *M*, equivalent to O$_1$) in both its liganded and unliganded forms. *ara*O$_2$ and *ara*I$_2$, though unable to bind AraC with high affinity, bind AraC protein cooperatively with *ara*I$_1$ [104,105]. In the absence of arabinose, *ara*BAD repression is accomplished through the cooperative binding of AraC at *ara*I$_1$ and *ara*O$_2$ [106]. Under inducing conditions, the occupancy within *ara*I is shifted such that cooperative binding of AraC occurs at the adjacent sites *ara*I$_1$ and *ara*I$_2$ [105].

AraC performs its autoregulatory function both in its ligand-bound and ligand-free states [105]. Although the CAP-cAMP site also overlaps the *ara*C promoter, expression is dominated by autoregulation [107]. Early studies suggested that *ara*C autoregulation occurred through the competitive binding of AraC protein with RNA polymerase at *ara*O$_1$, a site within the *ara*C promoter. To test whether this was the only site involved in autoregulation, an *ara*O$_1^-$ mutant was constructed such that the AraC binding consensus sequence was altered at four positions while the -10 and -35 regions of the *ara*C promoter remained intact [105]. This mutant operator showed a loss of autoregulation in the presence of the inducer, L-arabinose. Further experiments demonstrated that intact *ara*I$_1$ and *ara*O$_2$ are required for autoregulation of *ara*C expression in the absence of L-arabinose. Under inducing conditions, however, the binding site requirement is shifted to the *ara*O$_1$ and *ara*O$_2$ sites. Because of the lower affinity of the *ara*O$_2$ site for AraC ($\sim$5 $\times$ 10^{-10} *M*) [108], the cooperative interactions through *ara*I$_1$ and *ara*O$_1$ are presumed essential for AraC binding to *ara*O$_2$ and the maintenance of autoregulation. Although only two of these three sites are bound at any one time, all three of the sites are requisite for *ara*C repression in the presence and absence of arabinose. Thus, this protein appears to function through different states of cooperative binding to specific DNA sequences, and these states are influenced by the presence or absence of arabinose.

DNA loop formation in systems where a protein makes cooperative interactions with widely spaced DNA sequences has been proposed as a means by which separate sites can interact to exert observed effects [6,7,104,106]. DNA looping has been suggested to function in the repression of *ara*BAD expression in the absence of inducer [104,106]. Under these conditions, the repressor protein occupies *ara*I and *ara*O$_2$. Looping is presumed to involve dimer-dimer interaction to form a tetramer with each of its two potential DNA-binding sites occupied. Dunn et al. [106] tested the looping hypothesis by varying the spacing between the two sites and measuring the repression response. Their results supported the DNA loop model by showing that the phasing of the two sites is critical. Similar experimental evidence for the formation of loops between *ara*O$_1$ and *ara*O$_2$ is not yet available. The intervening distance, however, measures 158 bp or 15 helical turns, indicating that the phasing of *araO*$_1$ and *araO*$_2$ could likewise facilitate a DNA loop [105]. The apparent use of DNA looping in this and other systems allows for communication of regulatory interactions over relatively long distances on the DNA.

Tetracycline Resistance Operon

The Tn10-encoded tetracycline resistance unit consists of *tet*A and *tet*R, the resistance and repressor genes, respectively. Expression of these genes is regulated at the level of transcription through interactions within the common regulatory region. Figure 6.5 illustrates the spatial orientation of the two genes and the regulatory determinants. The complexity of this system is exemplified by three overlapping promoters [109], under the regulatory control of two operator sequences [110] and responsible for transcriptional initiation of two divergently expressed genes. Negative regulation of these genes is exerted by the *tet*R gene product, the *tet* repressor protein [110]. Induction of *tet*R and *tet*A transcription occurs by tetracycline binding to the repressor protein to elicit a decreased affinity for both of the operator sequences [111,112]. Subsequent expression of the *tet*A gene yields the *tet* protein, an inner membrane protein of $M_r \sim$ 36,000 that confers drug resistance [110]. This Tn10-encoded resistance, rather than depending on inactivation of the drug, results from the active export of tetracycline from the cell [113,114]. With the antibiotic itself serving as inducer, this mechanism of drug resistance requires an efficient switch between its repressed and induced states to assure survival of the organism. Indeed, the low equilibrium dissociation constant exhibited by the repressor for tetracycline (6×10^{-10} M) results in expression of the resistance gene at low antibiotic concentrations [112]. Because of its high affinity for tetracycline, it has been suggested that the *tet* repressor may also function in the resistance mechanism by titrating incoming tetracycline until the active export mechanism is functioning.

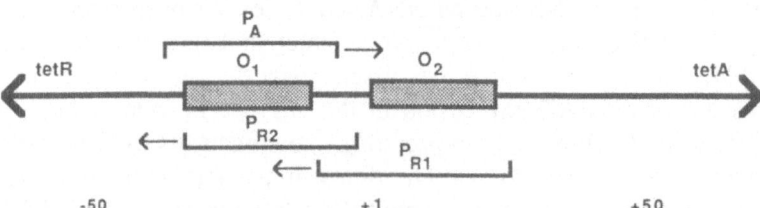

Figure 6.5. Regulatory region for *tet*R and *tet*A expression [117]. Transcription of *tet*A initiates at P$_A$, while transcripts of *tet*R originate from P$_{R1}$ and P$_{R2}$. The operator sequences, O$_1$ and O$_2$, are specific for the *tet*R protein and function in the *tet*R and *tet*A repression response.

The *tet* repressor protein has been shown by a variety of techniques to function as a dimer ($M_r \sim 47,000$) of identical subunits [110,111]. Based on studies using intrinsic tryptophan fluorescence, the binding of tetracycline appears to induce a conformational change in the protein that shields the DNA-binding determinants such that the operator sequence cannot be recognized [112]. The results of site-specific mutagenesis, binding measurements, and spectroscopic studies [112] support the structural separation of inducer and DNA-binding domains as seen in *lac* and other repressor proteins.

Sequence analysis of the *tet*R gene detected a region, from amino acids 26–47, which exhibits homology to the helix-turn-helix structures of λCro, λCI repressor, and the CAP protein [115]. A number of dominant *tet*R mutations mapping within this region have been characterized and show ~1,000-fold decreased operator-binding affinities with no effect on tetracycline binding [115]. Operator-binding affinities of Trp→Phe mutants at each of the two intrinsic sites were measured [116]. Substitution at Trp 75 had no effect on the equilibrium association constant, while Trp→Phe mutation at position 43 resulted in a 1,000-fold decrease in affinity for operator. Fluorescence experiments further implicated Trp 43 in the binding of operator by showing that its fluorescence was completely quenched when operator bound. These results led to the proposal that Trp 43 makes a sequence-specific contact with operator DNA [116]. The homology previously observed with conserved helix-turn-helix motifs of other DNA binding proteins would place Trp 43 within helix 3 of the proposed DNA binding domain.

The *tet* repressor-operator interaction has received much attention due to the unique arrangement of the regulatory sequences and the high degree of sequence homology between the two palindromic operator sequences, O_1 and O_2. Wray and Reznikoff [110] used an in vivo repressor-binding assay to demonstrate the specific binding of this region by *tet* repressor and thereby defined the symmetrical sites as operators. DNaseI footprint analysis illustrated that repressor protects each of the palindromic sequences and that the two operators are occupied simultaneously in a noncooperative manner [117]. In terms of binding affinities, Wissmann et al. [118] recently showed that repressor-O_2 interactions are twofold tighter than those with O_1 (2 pM *vs.* 4 pM, a difference previously suggested by Wray and Reznikoff [110]). A recent report from Meier and co-workers [117] offers evidence for differential regulation of *tet*R and *tet*A expression by repressor binding to the two operators. Repression of P_A was equally efficient from either O_1 or O_2, but complete repression of P_R required occupancy of O_1. This differential regulation leads to strict control of *tet*A expression, whereas the autoregulated *tet*R repression is more relaxed. The *tet* system offers a unique example of coordinate regulation of divergent genes from a common control region on the DNA.

Repressible Systems

Repressible systems of negative control require the presence of the product of an anabolic pathway, primarily those involved in amino acid biosynthesis, for effective repressor binding to control sequences. Although catabolite repression is not observed in these systems, attenuation mechanisms that couple premature transcription termination to translational state are frequently operative. This discussion will be confined to aspects of repressor-mediated control of transcription initiation.

Tryptophan Operon

The tryptophan repressor exerts its regulatory influence in response to L-tryptophan levels on three separate operons: the *trp*EDCBA operon, including genes for the final enzymes involved in tryptophan biosynthesis; the *aro*H operon, encoding one of three isoenzymes responsible for the first step in aromatic amino acid synthesis; and, in an autoregulatory mode, the *trp*R operon, containing the gene (*trp*R) for the tryptophan aporepressor. A schematic representation of the genetic organization found within the *trp*EDCBA operon is shown in Figure 6.6. The *trp* holorepressor protein binds operator and thereby blocks binding of RNA polymerase to the encompassing promoter region [119]. A second promoter, P_2, located within the distal translated portion of *trp*D, has been identified [120], although its role in regulation is unclear. The additional level of attenuation control common among amino acid biosynthetic operons couples transcription and translation. This mechanism of transcriptional termination is independent of the repressor protein and relies on the influence of tryptophanyl-tRNA levels on ribosome position during translation within the leader region (*trp*L) and on the formation of

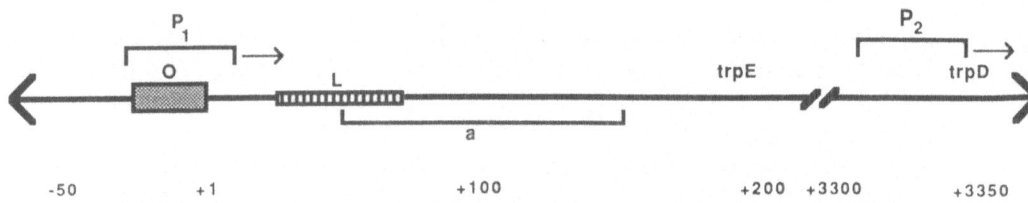

Figure 6.6. Genetic organization of transcription and translation control elements of the *trp*EDCBA operon [119,120]. The two promoters are shown with small arrows indicating the transcriptional polarity. P_1 completely encompasses the operator sequence (O), while P_2 is located about 3,300 bp downstream and within the translated region of *trp*D. The *trp* leader region (L) and the attenuator sequence (a) are responsible for premature transcriptional termination via an attenuation mechanism.

particular secondary structures by transcripts in the attenuator region [121].

The *trp* aporepressor product of the *trp*R gene is a dimer ($M_r \sim 24,000$) of identical subunits of known amino acid sequence [119,122]. Prior to crystallization, NMR and circular dichroism measurements indicated that the protein exhibited high α-helical content with about 20% β-sheet and turns [123]. X-ray crystallographic analysis of *trp* holorepressor has defined the three-dimensional structure [4]: six α-helices (A–F) separated by short turns are found in each subunit; the structure is extensively interlocked such that five of the six helices (all but D) from each monomer make contacts with the opposing subunit. These interactions result in the formation of a rigid central core domain which is flanked by flexible structures formed by helices D and E from each subunit [124]. The flexible helices D and E from each subunit form helix-turn-helix domains so that the *trp* repressor appears to share this common structural motif for its sequence-specific DNA interactions [4,125]. There are two tryptophan-binding sites, each of which contains amino acid residues from the two component subunits (4). Zhang et al. (125) recently solved the crystal diffraction pattern of the *trp* aporepressor to 1.8 Å resolution. Comparison of the structure for the holorepressor [4] with that of the unliganded aporepressor yielded details of the tryptophan-induced conformation change demonstrated earlier by Tsapakos et al. [126]. Residues in the central core (interlocking helices A, B, C, and F from both subunits) are unaffected by ligand binding, while tertiary structural changes occur in the presumed DNA-binding domains (helices D and E) to bring them into proper alignment for operator recognition [125]. L-tryptophan binding appears to activate the repressor by forcing these critical helices out and away from the molecular dyad and into their proper orientations for a snug fit within the contacted major grooves [125]. Once activated in this fashion, the *trp* repressor is able to recognize and bind the *trp*EDCBA, *aro*H, and *trp*R operators.

L-Tryptophan binding to aporepressor has been studied using a variety of spectral techniques as well as equilibrium and flow dialysis [127–130]. Results indicate that the indole ring and the α-carboxyl group of L-tryptophan contribute substantially to the stability of binding. The affinity of L-tryptophan for aporepressor is $\sim 2 \times 10^{-5}$ M, while *trp*EDCBA operator DNA binding has been measured to be ~ 0.5 nM to 5 nM [129–132]. Binding to tryptophan exhibits minimal cooperativity, and nonspecific affinity of holorepressor or aporepressor for DNA sequences is approximately 100-fold weaker than operator DNA binding [129–131]. Studies of mutant *trp* repressors confirm a role for residues in the D and E helices in operator recognition [133,134]. Kumamoto and co-workers [135] have shown using methylation protection and DNaseI footprinting experiments that the *trp* repressor contacts each of its operators differently. Their results indicate that two neighboring major grooves are contacted within the *trp*R operator, three for *aro*H and four for the *trp*EDCBA operator sequence. Not only does the *trp* repressor appear to recognize its individual operators differently, but Bass and co-workers [136] presented evidence suggesting that the repressor uses its DNA-binding motif in a manner distinct from that employed by λ repressor, Cro protein, or 434 repressor. The level of understanding of the *trp* system coupled with the availability of crystallographic information makes this system ideal for continued study of repression mechanisms.

Methionine Operon

The *met*J gene product of both *E. coli* and *S. typhimurium* is responsible for negative transcriptional control of seven unlinked genes. These include the five genes encoding enzymes involved in methionine biosynthesis, the gene for S-adenosylmethionine synthesis (*met*K), and transcription from its own gene, *met*J [137,138]. While *met* aporepressor activity is responsive to L-methionine levels, the effects of *met*K mutations on expression of the *met* genes as well as in vivo studies suggest that S-adenosylmethionine (SAM) acts as the corepressor for repression of the methionine regulon [139,140]. Equilibrium dialysis experiments showed that two molecules of SAM are bound by one molecule of dimeric aporepressor ($M_r \sim 24,000$) [138] with a dissociation constant of 200 μM [141]. The lack of cooperativity of binding the two molecules of SAM indicated that the two sites for corepressor are identical and independent. This interaction with corepressor has been shown to enhance the binding of aporepressor to its specific *met*F recognition site by an order of magnitude (aporepressor-DNA, $K_d \sim 10$ n*M*; holorepressor-DNA, $K_d \sim 1$ n*M*) [141].

Information on the mechanism of DNA recognition employed by the *met* repressor has been derived primarily from assessment of the nucleotide sequences for the regulatory regions of *met* genes. Figure 6.7 shows the sequence of the *met*F regulatory region, illustrating the promoter and mutationally defined operator region. Various alterations and repeats of a consensus palindrome, AGACGTCT, termed the "*met* box," appear before each of the genes [142,143]. Two repeats are found for *met*C and

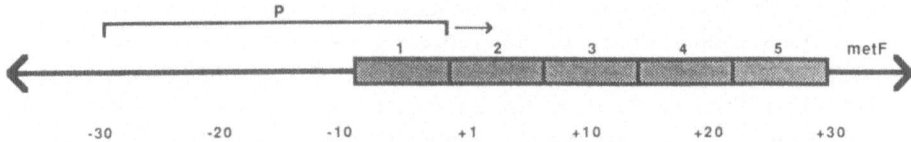

Figure 6.7. The *met*F control region [145]. The promoter (P) is shown along with the five binding sites for the *met* repressor protein known as "*met* boxes" (▬). The 8-bp "*met* box" consensus sequence is AGACGTCT.

four for *met*A, while five such repetitive sequences precede the *met*F and *met*B genes [144]. Balfaiza and co-workers [145] described the generation and analysis of five distinct operator mutations of the *met*F gene. Upon sequencing, the positions of the mutations were defined within the proposed *met* repressor recognition sequence. Any mutation that decreased the homology of a repeat with the consensus octamer resulted in derepression of *met*F. This result suggests that all five repeats are necessary for full repression of *met*F expression. There is some evidence for a correlation between the number of homologous repeats and the level of repression conferred by *met* repressor [137].

Although the sites of DNA recognition have been identified, the actual mechanism of repressor-DNA binding remains unknown. Primary structure analysis suggests that the protein is primarily α-helical, but no strong homology to the relatively common helix-turn-helix DNA-binding motif has been found [146]. The *E. coli met* aporepressor was recently crystallized, and further structural information may soon be available [146]. The *met* aporepressor crystals obtained thus far diffract to beyond 1.5 Å and are quite resistant to X-ray damage. Any attempt at introducing the corepressor, SAM, into the crystals resulted in cracking and a reduction in resolution, consistent with a conformational change induced by ligand binding. Attempts at *met* repressor-operator cocrystallization are under way using a synthetic 16-bp oligonucleotide composed of two 8-bp consensus units [146], and additional insight into the mechanism of action should derive from these studies.

Arginine Operon

Arginine modulates the formation of the enzymes involved in its biosynthesis via the arginine repressor protein, encoded by the *arg*R gene [147]. The arginine-ArgR protein complex negatively regulates transcription of the genes within the arginine regulon. The regulatory region for the divergently transcribed *arg*ECBH operon is illustrated in Figure 6.8. Each of the regulatory regions for *arg* genes contains two repeats of a conserved 18-bp sequence which was postulated as the arginine operator sequence and termed the "*arg* box" [148]. Comparisons between the repression response of each of the *arg* genes and the sequence of their respective *arg* boxes revealed a correlation between the stringency of repression and the degree of dyad symmetry within the *arg* box sequence [148]. Mutational studies also showed the importance of sequence composition apart from any symmetry [149]. The amplitude of repression was likewise related to the relative positions of the *arg* boxes and promoter sequences. The most efficient repression was observed for those genes transcribed from a promoter maximally overlapped by an *arg* box, suggesting that repression is exerted at the level of transcription initiation [148].

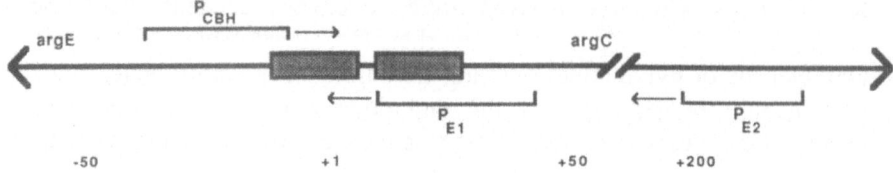

Figure 6.8. Regulatory organization of the divergently transcribed *arg*ECBH region [147]. A single promoter (P_{CBH}) is responsible for *arg*CBH expression, while *arg*E transcription is controlled from P_{E1} and P_{E2}. The neighboring 18-bp *arg*R protein recognition sites (▓) are termed *"arg* boxes."

The *arg*R gene has recently been sequenced, and its product has been purified and analyzed [150]. The *arg*R polypeptide is composed of 156 amino acids and aggregates to form a hexamer ($M_r \sim 98,000$). This protein is interesting, therefore, in its contrast to the *trp* and *met* repressors, which function as dimers. Also, whereas the *trp* repressor binds to a single symmetrical sequence (119), the arginine repressor appears to contact two neighboring *arg* boxes. Two such sequences have also been found within the *arg*R regulatory region, suggesting that ArgR maintains an autoregulatory function [150]. Autoregulation, as well as transcriptional control of *arg*F, shows an absolute requirement for arginine and is dependent on the concentration of the repressor protein [150]. Inspection of the control sequences for *arg*R reveals two promoters; P_1 is proximal to the transcription start site and within the *arg* boxes, while P_2 is distally located and uninfluenced by ArgR [150]. Therefore, *arg*R, though autoregulated, is expressed under conditions of repression (from P_2) and derepression (from P_1). Using their purified repressor protein and the DNaseI footprinting technique, Lim and co-workers [150] clearly demonstrated arginine-dependent binding of ArgR within the *arg* boxes of both *arg*R and *arg*F. The arginine system is the least defined of the repressible systems, and its unique aspects suggest that expansion of our understanding of regulatory mechanisms will be derived from further examination of this regulon.

Conclusion

These systems illustrate the complexity that can be generated from the single fundamental theme of protein recognition of specific DNA sites modulated by ligand binding. Overlapping sites for different proteins and cooperative interactions at adjacent sites or between distant sites with consequent DNA looping provide responsivity to multiple states within the cells. Signal molecule availability, whether from the environment or internal synthesis, and local ionic environment determine occupation of sites at the first level, but the topology of the DNA can become a significant influence in bridging of sites where loop formation occurs [72–78]. Results from eukaryotic systems indicate similar types of basic mechanisms, and the functionality of prokaryotic control in mammalian cells testifies to the viability of these regulatory strategies in higher organisms. The *lac* repressor has been shown to regulate expression in a variety of mammalian cells [151–154], an indication that recognition of operator sequences occurs in intact chromatin as well as isolated nucleosomal structure [155]. The interaction of distant sites on a single DNA observed in these bacterial control systems also parallels the effects of widely spaced sequences found in eukaryotic cells. (e.g., enhancers) and demonstrates that protein interaction at multiple sites and protein-protein contacts are common genetic regulatory themes throughout living organisms. It is becoming less possible to make a clear distinction between modes of eukaryotic and prokaryotic transcriptional control, and the availability of detailed mechanistic analysis of these readily obtained prokaryotic repressor proteins informs our understanding of the range and type of interactions that may exert transcriptional control in higher organisms.

Acknowledgments. Work from this laboratory was supported in part by grants from the National Institutes of Health (GM 22441, GM 35133) and the Robert A. Welch Foundation (C-576). We appreciate the efforts of Marlene Ross, Marie Monroe, and Jennifer Binford in assisting in the preparation of this manuscript.

References

1. McKay, D.B., and Steitz, T.A. (1981) Nature *290*, 744–749.
2. Anderson, J., Ptashne, M., and Harrison, S.C. (1984) Proc. Natl. Acad. Sci. USA *81*, 1307–1311.
3. Anderson, W.F., Ohlendorf, D.H., Takeda, Y., and Matthews, B.W. (1981) Nature *290*, 754–758.
4. Schevitz, R.W., Otwinowski, Z., Joachimiak, A., Lawson, C.L., and Sigler, P.B. (1985) Nature *317*, 782–786.
5. Pabo, C.O., and Sauer, R.T. (1984) Annu. Rev. Biochem. *53*, 293–321.
6. Ptashne, M. (1986) Nature *322*, 697–701.
7. Schleif, R. (1988) Science *240*, 127–128.
8. Weber, I.T., and Steitz, T.A. (1987) J. Mol. Biol. *198*, 311–326.
9. Kolb, K., Spassky, A., Chapon, C., Blazy, B., and Buc, H. (1983) Nucleic Acids Res. *11*, 7833–7852.
10. Liu-Johnson, H.N., Gartenberg, M.R., and Crothers, D.M. (1986) Cell *47*, 995–1005.
11. Dripps, D., and Wartell, R.M. (1987) J. Biomol. Struct. Dynam. *5*, 1–14.
12. Jacob, F., and Monod, J. (1961) J. Mol. Biol. *3*, 318–356.

13. Miller, J.H., and Reznikoff, W.S. (eds.) (1980) The Operon. Cold Spring Harbor Laboratory, Cold Spring Harbor, New York.

14. Barkley, M.D., Riggs, A.D., Jobe, A., and Bourgeois, S. (1975) Biochemistry *14*, 1700–1712.

15. O'Gorman, R.B., Rosenberg, J.M., Kallai, O.B., Dickerson, R.E., Itakura, K., Riggs, A.D., and Matthews, K.S. (1980) J. Biol. Chem. *255*, 10107–10114.

16. Daly, T.J., and Matthews, K.S. (1986) Biochemistry *25*, 5479–5484.

17. Lin, S.-Y., and Riggs, A.D. (1975) Cell *4*, 107–111.

18. Straney, S.B., and Crothers, D.M. (1987) Cell *51*, 699–707.

19. Lohman, T.M. (1985) CRC Crit. Rev. Biochem. *19*, 191–245.

20. Beyreuther, K., Adler, K., Fanning, E., Murray, C., Klemm, A., and Geisler, N. (1975) Eur. J. Biochem. *59*, 491–509.

21. Farabaugh, P.J. (1978) Nature *274*, 765–769.

22. Riggs, A.D., and Bourgeois, S. (1968) J. Mol. Biol. *34*, 361–364.

23. Butler, A.P., Revzin, A., and Von Hippel, P.H. (1977) Biochemistry *16*, 4757–4768.

24. Culard, F., and Maurizot, J.C. (1981) Nucleic Acids Res. *9*, 5175–5184.

25. Whitson, P.A., and Matthews, K.S. (1986) Biochemistry *25*, 3845–3852.

26. Platt, T., Files, J.G., and Weber, K. (1973) J. Biol. Chem. *248*, 110–121.

27. Jovin, T.M., Geisler, N., and Weber, K. (1977) Nature *269*, 668–672.

28. Kaptein, R., Zuiderweg, E.R.P., Scheek, R.M., Boelens, R., and Van Gunsteren, W.F. (1985) J. Mol. Biol. *182*, 179–182.

29. Zuiderweg, E.R.P., Scheek, R.M., and Kaptein, R. (1985) Biopolymers *24*, 2257–2277.

30. Steitz, T.A., Richmond, T.J., Wise, D., and Engelman, D. (1974) Proc. Natl. Acad. Sci. USA *71*, 593–597.

31. Charlier, M., Maurizot, J.C., and Zaccai, G. (1981) J. Mol. Biol. *153*, 177–182.

32. McKay, D.B., Pickover, C.A., and Steitz, T.A. (1982) J. Mol. Biol. *156*, 175–183.

33. Pilz, I., Goral, K., Kratky, O., Bray, R.P., Wade-Jardetzky, N.G., and Jardetzky, O. (1980) Biochemistry *19*, 4087–4090.

34. Wade-Jardetzky, N., Bray, R.P., Conover, W.W., Jardetzky, O., Geisler, N., and Weber, K. (1979) J. Mol. Biol. *128*, 259–264.

35. Adler, K., Beyreuther, K., Fanning, E., Geisler, N., Gronenborn, B., Klemm, A., Müller-Hill, B., Pfahl, M., and Schmitz, A. (1972) Nature *237*, 322–327.

36. Pfahl, M., Stockter, C., and Gronenborn, B. (1974) Genetics *76*, 669–679.

37. Müller-Hill, B., Fanning, T., Geisler, N., Gho, D., Kania, J., Kathmann, P., Meissner, H., Schlotmann, M., Schmitz, A., Triesch, I., and Beyreuther, K. (1975) In Protein-Ligand Interactions, 211–227, Sund, H., and Blauer, G. (eds.). Walter de Gruyter and Co., Berlin.

38. Miller, J.H. (1979) J. Mol. Biol. *131*, 249–258.

39. Miller, J.H. (1984) J. Mol. Biol. *180*, 205–212.

40. Gordon, A.J.E., Burns, P.A., Fix, D.F., Yatagai, F., Allen, F.L., Horsfall, M.J., Halliday, J.A., Gray, J., Bernelot-Moens, C., and Glickman, B.W. (1988) J. Mol. Biol. *200*, 239–251.

41. Ebright, R.H. (1986) Proc. Natl. Acad. Sci. USA *83*, 303–307.

42. Lehming, N., Sartorius, J., Niemöller, M., Genenger, G., Von Wilcken-Bergmann, B., and Müller-Hill, B. (1987) EMBO J. *6*, 3145–3153.
43. Friedman, B.E., and Matthews, K.S. (1978) Biochem. Biophys. Res. Commun. *85*, 497–505.
44. Matthews, K.S. (1979) J. Biol. Chem *254*, 3348–3353.
45. Manly, S.P., Bennett, G.N., and Matthews, K.S. (1984) J. Mol. Biol. *179*, 335–350.
46. Geisler, N., and Weber, K. (1977) Biochemistry *16*, 938–943.
47. Ogata, R.T., and Gilbert, W. (1979) J. Mol. Biol. *132*, 709–728.
48. Culard, F., Schnarr, M., and Maurizot, J.C. (1982) EMBO J. *1*, 1405–1409.
49. Nick, H., Arndt, K., Boschelli, F., Jarema, M.A.C., Lillis, M., Sadler, J., Caruthers, M., and Lu, P. (1982) Proc. Natl. Acad. Sci. USA *79*, 218–222.
50. Schnarr, M., Durand, M., and Maurizot, J.C. (1983) Biochemistry *22*, 3563–3570.
51. Barbier, B., Charlier, M., and Maurizot, J.C. (1984) Biochemistry *23*, 2933–2939.
52. Boelens, R., Scheek, R.M., Van Boom, J.H., and Kaptein, R. (1987) J. Mol. Biol. *193*, 213–216.
53. Daly, T.J., and Matthews, K.S. (1986) Biochemistry *25*, 5474–5478.
54. Sams, C.F., Hemelt, V.B., Pinkerton, F.D., Schroepfer, G.J.S. Jr., and Matthews, K.S. (1985) J. Biol. Chem. *260*, 1185–1191.
55. Müller-Hill, B. (1983) Nature *302*, 163–164.
56. Sams, C.F., Vyas, N.K., Quiocho, F.A., and Matthews, K.S. (1984) Nature *310*, 429–430.
57. Chakerian, A.E., Olson, J.S., and Matthews, K.S. (1987) Biochemistry *26*, 7250–7255.
58. Smith, T.F., and Sadler, J.R. (1971) J. Mol. Biol. *59*, 273–305.
59. Gilbert, W., Gralla, J., Majors, J., and Maxam, A. (1975) In Protein-Ligand Interactions, 193–210, Sund, H., and Blauer, G. (eds.). Walter de Gruyter and Co., Berlin.
60. Ogata, R., and Gilbert, W. (1977) Proc. Natl. Acad. Sci. USA *74*, 4973–4976.
61. Schmitz, A., and Galas, D.J. (1979) Nucleic Acids. Res. *6*, 111–137.
62. Manly, S.P., and Matthews, K.S. (1984) J. Mol. Biol. *179*, 315–333.
63. Goeddel, D.V., Yansura, D.G., and Caruthers, M.H. (1978) Proc. Natl. Acad. Sci. USA *75*, 3578–3582.
64. Caruthers, M.H., Beaucage, S.L., Efcavitch, J.W., Fisher, E.F., Goldman, R.A., De Haseth, P.L., Mandecki, W., Matteucci, M.D., Rosendahl, M.S., and Stabinsky, Y. (1982) Cold Spring Harbor Symp. Quant Biol. *47*, 411–418.
65. Sadler, J.R., Sasmor, H., and Betz, J.L. (1983) Proc. Natl. Acad. Sci. USA *80*, 6785–6789.
66. Simons, A., Tils, D., Von Wilcken-Bergmann, B., and Müller-Hill, B. (1984) Proc. Natl. Acad. Sci. USA *81*, 1624–1628.
67. Reznikoff, W.S., Winter, R.B., and Hurley, C.K. (1974) Proc. Natl. Acad. Sci. USA *71*, 2314–2318.
68. Pfahl, M., Gulde, V., and Bourgeois, S. (1979) J. Mol. Biol. *127*, 339–344.
69. Donnelly, C.E., and Reznikoff, W.S. (1987) J. Bacteriol. *169*, 1812–1817.
70. Winter, R.B., and Von Hippel, P.H. (1981) Biochemistry *20*, 6948–6960.
71. Hsieh, W.-T., Whitson, P.A., Matthews, K.S., and Wells, R.D. (1987) J. Biol. Chem *262*, 14583–14591.

72. Whitson, P.A., Hsieh, W.-T., Wells, R.D., and Matthews, K.S. (1987) J. Biol. Chem. *262*, 14592–14599.
73. Krämer, H., Amouyal, M., Nordheim, A., and Müller-Hill, B. (1988) EMBO J. *7*, 547–556.
74. Krämer, H., Niemöller, M., Amouyal, M., Revet, B., Von Wilcken-Bergmann, B., and Müller-Hill, B. (1987) EMBO J. *6*, 1481–1491.
75. Mossing, M.C., and Record, M.T. Jr. (1986) Science *233*, 889–892.
76. Besse, M., Von Wilcken-Bergmann, B., and Müller-Hill, B. (1986) EMBO J. *5*, 1377–1381.
77. Borowiec, J.A., Zhang, L., Sasse-Dwight, S., and Gralla, J.D. (1987) J. Mol. Biol. *196*, 101–111.
78. Eismann, E., Von Wilcken-Bergmann, B., and Müller-Hill, B. (1987) J. Mol. Biol. *195*, 949–952.
79. Dandanell, G., Valentin-Hansen, P., Larsen, J.E.L., and Hammer, K. (1987) Nature *325*, 823–826.
80. Busby, S., Aiba, H., and De Crombrugghe, B. (1982) J. Mol. Biol. *154*, 211–227.
81. Adhya, S., and Miller, W. (1979) Nature *279*, 492–494.
82. Di Lauro, R., Taniguchi, T., Musso, R., and De Crombrugghe, B. (1979) Nature *279*, 494–500.
83. Fritz, H.-J., Bicknäse, H., Gleumes, B., Heibach, C., Rosahl, S., and Ehring, R. (1983) EMBO J. *2*, 2129–2135.
84. Irani, M.H., Orosz, L., and Adhya, S. (1983) Cell *32*, 783–788.
85. Von Wilcken-Bergmann, B., Koenen, M., Griesser, H.W., and Müller-Hill, B. (1983) EMBO J. *2*, 1271–1274.
86. Majumdar, A., and Adhya, S. (1984) Proc. Natl. Acad. Sci. USA *81*, 6100–6104.
87. Von Wilcken-Bergmann, B., and Müller-Hill, B. (1982) Proc. Natl. Acad. Sci. USA *79*, 2427–2431.
88. Majumdar, A., and Adhya, S. (1987) J. Biol. Chem *262*, 13258–13262.
89. Musso, R., Di Lauro, R., Rosenberg, M., and De Crombrugghe, B. (1977) Proc. Natl. Acad. Sci. USA *74*, 106–110.
90. Aiba, H., Adhya, S., and De Crombrugghe, B. (1981) J. Biol. Chem. *256*, 11905–11910.
91. Irani, M.H., Orosz, L., Busby, S., Taniguchi, T., and Adhya, S. (1983) Proc. Natl. Acad. Sci. USA *80*, 4775–4779.
92. Busby, S., Spassky, A., and Chan, B. (1987) Gene *53*, 145–152.
93. Kuhnke, G., Krause, A., Heibach, C., Gieske, U., Fritz, H.-J., and Ehring, R. (1986) EMBO J. *5*, 167–173.
94. Kuhnke, G., Fritz, H.-J., and Ehring, R. (1987) EMBO J. *6*, 507–513.
95. Ponnambalam, S., Spassky, A., and Busby, S. (1987) FEBS Lett. *219*, 189–196.
96. Shanblatt, S.H., and Revzin, A. (1983) Proc. Natl. Acad. Sci. USA *80*, 1594–1598.
97. Lee, N. (1980) In The Operon, 389–409, Miller, J.H., and Reznikoff, W.S. (eds.). Cold Spring Harbor Laboratory, Cold Spring Harbor, New York.
98. Brunelle, A., Hendrickson, W., and Schleif, R. (1985) Nucleic Acids Res. *13*, 5019–5026.
99. Cass, L.G., and Wilcox, G. (1986) J. Bacteriol. *166*, 892–900.
100. Francklyn, C.S., and Lee, N. (1988) J. Biol. Chem. *263*, 4400–4407.

101. Wilcox, G. (1974) J. Biol. Chem. *249*, 6892–6894.
102. Lee, N., Francklyn, C., and Hamilton, E.P. (1987) Proc. Natl. Acad. Sci. USA *84*, 8814–8818.
103. Hendrickson, W., and Schleif, R. (1985) Proc. Natl. Acad. Sci. USA *82*, 3129–3133.
104. Martin, K., Huo, L., and Schleif, R.F. (1986) Proc. Natl. Acad. Sci. USA *83*, 3654–3658.
105. Hamilton, E.P., and Lee, N. (1988) Proc. Natl. Acad. Sci. USA *85*, 1749–1753.
106. Dunn, T.M., Hahn, S., Ogden, S., and Schleif, R.F. (1984) Proc. Natl. Acad. Sci. USA *81*, 5017–5020.
107. Miyada, C.G., Stoltzfus, L., and Wilcox, G. (1984) Proc. Natl. Acad. Sci. USA *81*, 4120–4124.
108. Hendrickson, W., and Schleif, R.F. (1984) J. Mol. Biol. *174*, 611–628.
109. Bertrand, K.P., Postle, K., Wray, L.V. Jr., and Reznikoff, W.S. (1983) Gene *23*, 149–156.
110. Wray, L.V. Jr., and Reznikoff, W.S. (1983) J. Bacteriol. *156*, 1188–1191.
111. Hillen, W., Gatz, C., Altschmied, L., Schollmeier, K., and Meier, I. (1983) J. Mol. Biol. *169*, 707–721.
112. Hansen, D., Altschmied, L., and Hillen, W. (1987) J. Biol. Chem. *262*, 14030–14035.
113. Ball, P.R., Shales, S.W., and Chopra, I. (1980) Biochem. Biophys. Res. Commun. *93*, 74–81.
114. McMurray, C., Petrucci, R.E. Jr., and Levy, S.B. (1980) Proc. Natl. Acad. Sci. USA *77*, 3974–3977.
115. Isackson, P.J., and Bertrand, K.P. (1985) Proc. Natl. Acad. Sci. USA *82*, 6226–6230.
116. Hansen, D., and Hillen, W. (1987) J. Biol. Chem. *262*, 12269–12274.
117. Meier, I., Wray, L.V. Jr., and Hillen, W. (1988) EMBO J. *7*, 567–572.
118. Wissmann, A., Meier, I., Wray, L.V. Jr., Geissendörfer, M., and Hillen, W. (1986) Nucleic Acids Res. *14*, 4253–4266.
119. Gunsalus, R.P., and Yanofsky, C. (1980) Proc. Natl. Acad. Sci. USA *77*, 7117–7121.
120. Horowitz, H., and Platt, T. (1982) J. Mol. Biol. *156*, 257–267.
121. Yanofsky, C. (1981) Nature *289*, 751–758.
122. Singleton, C.K., Roeder, W.D., Bogosian, G., Somerville, R.L., and Weith, H.L. (1980) Nucleic Acids Res. *8*, 1551–1560.
123. Lane, A.N., and Jardetzky, O. (1985) Eur. J. Biochem. *152*, 405–409.
124. Lawson, C.L., Zhang, R., Schevitz, R.W., Otwinowski, Z., Joachimiak, A., and Sigler, P.B. (1988) Proteins *3*, 18–31.
125. Zhang, R.-G., Joachimiak, A., Lawson, C.L., Schevitz, R.W., Otwinowski, Z., and Sigler, P.B. (1987) Nature *327*, 591–597.
126. Tsapakos, M.J., Haydock, P.V., Hermodson, M., and Somerville, R.L. (1985) J. Biol. Chem. *260*, 16383–16394.
127. Lane, A.N. (1986) Eur. J. Biochem. *157*, 405–413.
128. Marmorstein, R.Q., Joachimiak, A., Sprinzl, M., and Sigler, P.B. (1987) J. Biol. Chem. *262*, 4922–4927.
129. Arvidson, D.N., Bruce, C., and Gunsalus, R.P. (1986) J. Biol. Chem. *261*, 238–243.

130. Joachimiak, A., Kelley, R.L., Gunsalus, R.P., Yanofsky, C., and Sigler, P.B. (1983) Proc. Natl. Acad. Sci. USA *80*, 668–672.

131. Carey, J. (1988) Proc. Natl. Acad. Sci. USA *85*, 975–979.

132. Rose, J.K., and Yanofsky, C. (1974) Proc. Natl. Acad. Sci. USA *71*, 3134–3138.

133. Kelley, R.L., and Yanofsky, C. (1985) Proc. Natl. Acad. Sci. USA *82*, 483–487.

134. Klig, L.S., Crawford, I.P., and Yanofsky, C. (1987) Nucleic Acids Res. *15*, 5339–5351.

135. Kumamoto, A.A., Miller, W.G., and Gunsalus, R.P. (1987) Genes Dev. *1*, 556–564.

136. Bass, S., Sugiono, P., Arvidson, D.N., Gunsalus, R.P., and Youderian, P. (1987) Genes Dev. *1*, 565–572.

137. Flavin, M. (1975) In Metabolic Pathways, Vol. 7, 457–503, Greenberg, D.M. (ed.). Academic Press, New York.

138. Urbanowski, M.L., and Stauffer, G.V. (1985) Nucleic Acids Res. *13*, 673–685.

139. Greene, R.C., Hunter, J.S.V., and Coch, E.H. (1973) J. Bacteriol. *115*, 56–67.

140. Shoeman, R., Radfield, B., Coleman, T., Greene, R.C., Brot, N., and Weissbach, H. (1985) Proc. Natl. Acad. Sci. USA *82*, 3601–3605.

141. Saint-Girons, I., Belfaiza, J., Guillou, Y., Perrin, D., Guiso, N., Bârzu, O., and Cohen, G.N. (1986) J. Biol. Chem. *261*, 10936–10940.

142. Duchange, N., Zakin, M.M., Ferrara, P., Saint-Girons, I., Park, I., Tran, S.V., Py, M.-C., and Cohen, G.N. (1983) J. Biol. Chem. *258*, 14868–14871.

143. Saint-Girons, I., Duchange, N., Zakin, M.M., Park, I., Margarita, D., Ferrarra, P., and Cohen, G.N. (1983) Nucleic Acids Res. *11*, 6723–6732.

144. Belfaiza, J., Parsot, C., Martel, A., Bouthier de la Tour, C., Margarita, D., Cohen, G.N., and Saint-Girons, I. (1986) Proc. Natl. Acad. Sci. USA *83*, 867–872.

145. Belfaiza, J., Guillou, Y., Margarita, D., Perrin, D., and Saint-Girons, I. (1987) J. Bacteriol. *169*, 670–674.

146. Rafferty, J.B., Phillips, S.E.V., Rojas, C., Boulot, G., Saint-Girons, I., Guillou, Y., and Cohen, G.N. (1988) J. Mol. Biol. *200*, 217–219.

147. Cunin, R., Glansdorff, N., Piérard, A., and Stalon, V. (1986) Microbiol. Rev. *50*, 314–352.

148. Cunin, R., Eckhardt, T., Piette, J., Boyen, A., Piérard, A., and Glansdorff, N. (1983) Nucleic Acids Res. *11*, 5007–5019.

149. Boyen, A., Charlier, D., Crabeel, M., Cunin, R., Palchaudhuri, S., and Glansdorff, N. (1978) Mol. Gen. Genet. *161*, 185–196.

150. Lim, D., Oppenheim, J.D., Eckhardt, T., and Maas, W.K. (1987) Proc. Natl. Acad. Sci. USA *84*, 6697–6701.

151. Hu, M.C.-T., and Davidson, N. (1988) Gene *62*, 301–313.

152. Brown, M., Figge, J., Hansen, U., Wright, C., Jeang, K.-T., Khoury, G., Livingston, D.M., and Roberts, T.M. (1987) Cell *49*, 603–612.

153. Figge, J., Wright, C., Collins, C.J., Roberts, T.M., and Livingston, D.M. (1988) Cell *52*, 713–722.

154. Mukherjee, S., Erickson, H., and Bastia, D. (1988) Cell *52*, 375–383.

155. Chao, M.V., Gralla, J.D., and Martinson, H.G. (1980) Biochemistry *19*, 3254–3260.

Chapter 7
Eukaryotic RNA Polymerases

Ekkehard K.F. Bautz and Gabriele Petersen

The year this volume is published marks the 30th anniversary of the first paper on DNA-dependent RNA polymerase [1]. Considering the development of the field since then, it comes as a surprise that this first paper describes an enzymatic activity of mammalian rather than bacterial origin. In the decade to follow, however, the RNA polymerase of *E. coli* became the prime object of investigation, and no significant advances were made with RNA polymerases from eukaryotes. The reasons for this were twofold: first, there is only one molecular species of RNA polymerase in bacteria, and this could be isolated together with its major initiation factor [2], allowing in vitro transcription assays to be performed with native DNA of defined origin; second, three different enzymes are present in eukaryotes, all of which are purified in the form of a core enzyme whose activities can be measured only under "unnatural" conditions, usually using denatured DNA as templates [for review see 3]. Eukaryotic in vitro transcription systems allowing correct initiation at bona fide promoters have been developed only in the past several years, and only through recombinant DNA techniques has it become possible to start tackling the formidable task of dissecting the structure of the RNA polymerases and to see at last some crude attempts to attribute certain functions like DNA or nucleotide binding to regions in the primary structure of enzyme subunits. The question of how genes are regulated has, over the past decade, been approached almost to the exclusion of the transcribing enzyme in that over 90% of the literature is concerned with the study of promoter and enhancer elements by in vitro mutagenesis and in vivo assays. During the past few years, in vitro transcription systems, which allow the identification of transcription factors and their effects on initiation and elongation of RNA chains mediated by either of the RNA polymerases, have been developed to the point where we begin to perceive a coherent picture of the common features of gene function.

In this review we have decided to restrict ourselves to a description of what is known about the structural composition of the enzymes and about their genes as far as they have been cloned and sequenced. In addition we will discuss briefly some experimental systems that are likely to yield further insights into the important question of how DNA sequence elements exert their regulatory role through interaction with binding proteins, which directly or indirectly regulate the activity of RNA polymerases.

Nomenclature

Twenty years after the discovery of separate forms of RNA polymerases in eukaryotes [4,5], the nomenclature is still split between RNA polymerase I, II, and III (in the United States) and A, B, and C (in most of continental Europe and particularly in France). Subunits are usually given by their molecular weight in SDS gels, and some attempts have been made to give them letters (A, B, C, etc.) in the I, II, III nomenclature and numbers (1, 2, 3, etc.) in the A, B, C nomenclature.

Although we would prefer to see the European nomenclature win eventually, if only for the reason that DNA polymerases are called alpha, beta, and gamma, we feel that we should, at this time, yield to the majority vote and use in this review the I, II, III system (RNA polymerase I = pol I, RNA polymerase II = pol II, and RNA polymerase III = pol III). The subunits we name after their SDS gel molecular weights for the reason that it is still a matter of conjecture as to how many polypeptides are bona fide subunits of the eukaryotic RNA polymerases; i.e., the largest subunit of RNA polymerase II of *Drosophila melanogaster* will be called pol II 215, and the gene will be called RPII215.

Structure of Eukaryotic RNA Polymerases

The three RNA polymerases are multisubunit enzymes. Their subunit compositions are quite similar from yeast to man; therefore, a description of the three yeast enzymes, which are the most thoroughly studied, should be representative for all eukaryotes. As the most detailed descriptions of the yeast enzyme come from the laboratory of A. Sentenac, we confine ourselves to a short summary of the major structural features as he and his co-workers have published on the yeast enzyme, and refer to his 1985 review [6] and others [7–9] for more detailed information.

As illustrated in Table 7.1, all three enzymes consist of two large subunits of greater than 100 kD as well as several small ones. As we shall see later, these large subunits correspond structurally, and probably also functionally, to the β and β' subunits of bacterial RNA polymerase. Very little is known about the structure and function of the smaller-size subunits except that they copurify with enzymatic activity and that they are present in most enzyme preparations in stoichiometric amounts.

Table 7.1 Subunit structure of RNA polymerases from *Saccharomyces cerevisiae*.

Pol I	Pol II	Pol III
<u>190</u>	<u>220</u>	<u>160</u>
	(185)	
<u>135</u>	<u>150</u>	<u>128</u>
		<u>82</u>
49		
43		<u>53</u>
<u>40</u>	44.5	<u>40</u>
		37
<u>34.5</u>	32	<u>34</u>
		31
<u>27</u>	<u>27</u>	<u>27</u>
<u>23</u>	<u>23</u>	<u>23</u>
<u>19</u>	16	<u>19</u>
<u>14.5</u>	<u>14.5</u>	<u>14.5</u>
14		
	<u>12.6</u>	
12.2		
		11
10	10	10

The molecular weights in kD are determined by SDS gel electrophoresis [from 50]. Underlined are the polypeptides that are thought to qualify as bona fide subunits.

Furthermore, at least in the case of *Drosophila* RPII215, they have been shown to each consist of different tryptic polypeptides, excluding the possibility that some of the small subunits are proteolytic products of the large ones [10]. The 27-, 23-, and 14.5-kD subunits are shared by the three isoforms of the enzyme [11,12], and the 40- and 19-kD polypeptides are common subunits of pol I and pol III [12–14]. The structural similarities between the three enzymes are reflected by cross-reactions, albeit weak, found with antibodies raised against individual large subunits of one enzyme with the corresponding subunits of the other two [13]. With polyclonal antibodies, no significant cross-reaction was detected between the largest and second-largest subunit of either of the three enzymes. On the other hand, a monoclonal antibody directed against the 215-kD subunit of *Drosophila* pol II also blotted with the 140-kD subunit [15]. This reaction, however, could be eliminated by raising the salt concentration to 1 M KCl (S. Bialojan, master's thesis, Heidelberg). Some of the small subunits have been shown to be required for enzyme activity, and these represent bona fide subunits of their respective enzymes [for further details see 6].

Primary Structure of Subunit Genes

The cloning of genes coding for RNA polymerase subunits has turned out to be more difficult than expected. But two strategies for cloning have been successful.

The first and more elegant, but quite time-consuming, strategy was the isolation of α-amanitin-resistant mutants of *Drosophila*, mapping of the gene to position 10C on the X chromosome and cloning the gene by transposon tagging [16–18]. This strategy has yielded the gene coding for RPII215. Cross-hybridization with RPII215 gene probes has allowed cloning of this gene from other organisms from yeast to mammals [19–25].

The second approach consisted of screening expression cDNA libraries with subunit-specific antibodies. This approach has yielded clones coding for the large subunits of all three yeast polymerases as well as for a number of the small subunits [26,27]. By now the primary structures of yeast RPI190 [28], RPII220 [19], RPII150 [29], and RPIII160 [19], as well as the common 40-kD subunit of pol I and pol III, have been deduced [30]. In addition, RPII215 [31,32] and RPII140 [33,34], RPIII135 [35] of *Drosophila,* and RPII215 of mouse [24] have been sequenced. Thus, up to now the primary structures of all but one of the six large subunits of the three eukaryotic RNA polymerases are known. A comparison of these sequences suggests a strong phylogenetic relationship between the three enzymes [19,35; A. Sentenac, personal communication].

To deduce functional features from the primary structure of the large subunits is very difficult and probably based on assumptions that are too hypothetical. The only useful approach may derive from a comparison of sequences from enzymes of similar functions and from organisms of distant phylogenetic relationship. The largest subunits of pol II of yeast, *Drosophila,* and mouse share not only extended regions of homology among each other but also, apart from a C-terminal heptapeptide repeat (CTD), exclusively present in the largest subunit of pol II of eukaryotes, with the β' subunit of *E. coli* RNA polymerase. In addition, these same structural domains are conserved between the largest subunits of eukaryotic pols I, II, and III.

This evolutionary conservation suggests a preservation of function. In this context it is of interest to note that two of the regions of homology even display weak homology to *E. coli* DNA pol I and to bacteriophage T7 DNA polymerase. One of these homology regions resides in a two-helix motif which may position itself in the major groove of the duplex product of DNA synthesis [36,37]. Analogously, the corresponding sequences of the RNA polymerase largest subunits could make contact with RNA or the heteroduplex of template DNA and newly synthesized RNA. In addition, a Zn-finger motif of the sequence $C-X_2-C-X_n-H-X_2-H$ [38–40] was observed near the N termini of all largest eukaryotic RNA polymerase subunits sequenced so far. Except for one histidine, an analogous sequence arrangement was found in *E. coli* β' and an almost correct Zn finger in the sequence coded by vaccinia DNA [41].

These structures are in accord with biochemical data suggesting that RNA polymerases require zinc for function [42] and that *E. coli* β' and RPII215 bind DNA in vitro [43–45], as well as with the finding that nascent RNA transcripts can be cross-linked to RPII215 [44,46]. In addition, earlier studies with *E. coli* RNA polymerase mutated in β' led to the conclusion that the largest subunit of the enzyme plays the major role in interacting with the DNA template [47].

The second-largest subunits of yeast and *Drosophila* pol II and of *Drosophila* pol III also possess a Zn finger structure near their C termini. This sequence is entirely lacking in *E. coli* β, suggesting a functional divergence for this part of the molecule. This is again in accord with published data: while no evidence for a significant affinity of *E. coli* β to DNA exists, Gundelfinger reported binding of *Drosophila* pol II 140 to DNA [44]. Another line of evidence that the second-largest subunit interacts with the DNA template comes from studies with antibodies directed against yeast pol I 135 [48] and pol III 128 [49]: preincubation with DNA protects the enzymes against inhibition by mono- and polyclonal antibodies.

The second-largest subunit also harbors the binding sites for the first and second nucleoside triphosphates. Sequence motifs homologous to other nucleotide-binding proteins are present in all pol II second-largest subunits sequenced so far [50]. The most direct evidence for substrate-binding sites comes from a method developed by Grachev and co-workers [51,52] which allows highly selective affinity labeling of RNA polymerase. This procedure makes use of the catalytic activity of the enzymes such that only those sites become affinity labeled at which a second nucleotide becomes covalently linked to the first one. The reaction goes as follows: RNA polymerase is reacted with a benzaldehyde-derived ATP, $NaBH_4$ is added to reduce the Schiff bases, and then DNA template and ^{32}P-UTP are added to allow formation of the first phosphodiester bond. Thus, the radioactive nucleotide becomes covalently linked to the subunit that harbors the binding site for the first nucleotide.

The most prominent feature in the structure of the largest subunits is the already-mentioned heptamer repeat structure YSPTSPS, the CTD. This structure appears to be present in all eukaryotes but only in pol II, and not in pol I or III or in RNA polymerases of eubacteria and archaebacteria. The heptamer unit is tandemly repeated 26 times in yeast [19], 42 times in *Drosophila* [53,54], and 52 times in mammalian polymerases [22,23]. The sequences within the repeats can deviate somewhat, the degree of deviation being largest in *Drosophila*. The amino acid composition of the repeat domain suggests that it is in an extended conformation. This is corroborated by the fact that this is the structural domain that is most sensitive to proteolysis. Without special precautions taken to avoid proteolytic cleavage during enzyme purification, the early purification procedures all yielded an enzyme preparation whose largest subunit had the repeat domain completely missing [55,56]. Yet, these enzyme preparations showed as much activity on denatured DNA as template as did preparations with a complete CTD. Also, correct initiation in an enzyme-dependent promoter-driven transcription-reconstruction system requiring purified pol II and transcription factors was found with a *Drosophila* mutant enzyme lacking all but 19 of the heptamer repeats. In fact, proteolytic removal of the entire repeat domain by chymotrypsin did not affect activity in this system [54].

In contrast, in vivo the CTD appears absolutely essential. A *Drosophila* mutation producing a truncated form of the CTD is a homozygous lethal, as are other mutations deleting or disrupting part of the CTD. Nonet et al. [57] have constructed unidirectional deletions of the RPII220 gene at the 3' end of the coding region with exonuclease III digestion, transformed yeast cells with the constructs, and analyzed for viability. Of the 26 heptamer repeats in yeast, 13 were found to be required for viability. An analogous analysis of the CTD of mouse was done by changing the number of repeats by transformation of these variants in *cis* with an ama^r marker [46]. Between 36 and 78 repeats, lower or higher than wild type, did not affect viability, 25 or fewer repeats were lethal, and 29 to 32 repeats impaired the function of pol II in vivo. These studies also provided interesting data on the quality of the repeats: the more they deviated from the consensus sequence YSPTSPS, the higher the number of repeats required for function. This was also observed by substituting the CTD of yeast with either the hamster or the *Drosophila* CTD. Yeast cells were viable with the heterologous CTD from hamster but not the CTD from *Drosophila* [53]. This surprising result can best be explained by the fact that most *Drosophila* repeats deviate from the consensus sequence by two or more amino acids.

A minor form of pol II (IIo) was repeatedly observed with enzyme preparations from calf thymus. These contained an altered 220-kD subunit that was found to be phosphorylated [58]. A monoclonal antibody directed against the 215-kD subunit reacted with both the phosphorylated and unphosphorylated forms, but not with the 170-kD subunit missing the CTD. There is evidence that phosphorylation of the CTD plays an important role in mRNA synthesis [59].

That the strongly hydrophilic and proline-rich CTD is highly preserved through evolution and unique for pol II has stirred notable interest in speculating about its function. The extendedness of the CTD could make it reach over a considerable distance interacting with transcription factors that bind to promoter upstream-located DNA recognition sites. Alternatively, the CTD may dislodge histones from nucleosomes ahead of polymerase movement. Other possibilities, such as anchoring the enzyme to specific sites within the nucleus or protecting other important sites in the enzyme during transport to the nucleus, were also suggested [46]. The first suggestion seems to be the one most favored by workers in the field of transcription; unfortunately, however, we do not know of any published result that directly supports this hypothesis. On the contrary, the data on in vitro transcription in a system requiring transcription factors argue against its role in a factor-mediated initiation process [60]. The minimal structural requirements of "negative noodles" (see below), on the other hand, keeps the CTD as an appealing candidate for interaction with distantly located activator proteins not measured in an in vitro transcription system. Still, the most obvious difference between in vitro and in vivo transcription is the absence and presence of histones, and as pol II appears to be the only one of the three RNA polymerases transcribing DNA packed in nucleosomes and not naked DNA, the dislodging of histones should be taken seriously, at least as long as there is no experimental proof for interaction of the CTD with transcription factors.

Small Subunits

Little is known about the primary structure and function of the small subunits of the three RNA polymerases except for the yeast RPI/III40, a unique gene coding for a subunit shared between yeast pol I and III [30]. The DNA sequence revealed the presence of an open reading frame coding for a protein of 335 amino acids with a predicted molecular weight that is in quite good agreement with the 40 kD estimated by SDS-PAGE. No sequence homologies were found between this and the known *E. coli* polymerase subunit genes, including factors. With the elegant genetic method of gene disruption in yeast, involving a selectable his$^+$ marker, it could be shown that the function of this gene is essential. Pol I and III share another polypeptide of 19 kD; the only thing known about it is that it can become phosphorylated in vivo [11,12].

Three of the small subunits are shared by pols I, II, and III—namely, polypeptides of 27, 23, and 14.5 kD; these three constitute the core of the common subunits and thus are likely to play a general role in transcription. The 23-kD subunit was found to become phosphorylated in vivo and in vitro [11,12]. A small polypeptide of 10 kD is also a candidate for a common subunit in yeast [49].

Mutant RNA Polymerases

The α-amanitin-resistant mutant C4 of *Drosophila* [16,61] not only was the key to the isolation of the largest subunit gene of pol II from several eukaryotes, it also allowed the isolation of a large number of allelic mutants including several ts mutants and some others showing developmental defects, including mutant UBL, whose nature still needs to be elucidated [62]. A mouse ama^r mutation was found to have resulted in a single base-pair change causing an asparagine-to-aspartate substitution at amino acid 793 of the predicted protein sequence [63]. As expected, this mutation is located in a region that is quite conserved between mouse and yeast pol II, but not between pol II and pol I, pol III, or *E. coli* β′. The *Drosophila* ama^r C4 mutation, whose exact location is not known, has not resulted in an amino acid exchange at or near the position homologous to the mouse mutation, suggesting that mutations affecting at least two sequence areas can lead to the ama^r phenotype [32]. To replace a wild-type gene with an in vitro mutated allele can be done most easily in yeast. Therefore, it is not surprising that lethal mutations have been obtained for all subunit genes that were cloned and sequenced thus far [50].

Three ts conditional lethal mutations were described for RPII215 by Himmelfarb et al. [64]. One was completely deficient in pol II activity in vitro, while the other two showed no effect in vitro, but kinetics on the cessation of cell growth suggested that the mutations affect assembly or holoenzyme stability. Sequence analysis showed nonconservative substitutions leading to changes in net charges in highly conserved domains, one mutation occurring in a possible Zn-binding domain. One of the mutations has two amino acid substitutions; the other two mutations represent single amino acid substitutions. Nonet et al. [65] described a mutant in pol II 220 that displays immediate cessation of RNA synthesis in vivo after a temperature shift and ts sensitivity in vitro. Thus, in their case they are dealing with a mutation rendering the enzyme functionally defective rather than assembly defective. Ts mutants of RPIII160 were constructed by in vitro mutagenesis and substitution for the wild-type allele [66]. Interestingly, one of the mutants investigated showed a greatly reduced synthesis of tRNA in vivo but not of 5.8S RNA and, surprisingly, not of 5S RNA (suggesting that the mutation affected promoter and/or factor recognition). After three doubling times at 37°C, the synthesis of 5S rRNA eventually became arrested. It is likely that this mutant enzyme has a specifically altered recognition site for initiation complexes.

A ts mutant of RPI/III40, the subunit shared by pol I and pol III, was found to be defective in the synthesis of pol I and pol III at the restrictive temperature [30]. The failure of these polymerases to assemble correctly in vivo proves that this subunit is an essential part of both enzymes. In addition, the fact that at the nonpermissive temperature pol II is synthesized normally shows that the level of pol I or pol III does not affect the level of pol II.

Functions

In this section, concerning the function of RNA polymerases in transcription, we give only a very short overview. Given the vast literature on transcription, transcription factors, and regulation (especially for genes transcribed by pol II), it is beyond the limits of this review to cover the topic in depth. Recent reviews on pol I [67], pol III [68], pol II [69–74], and references therein provide more detailed information. Here we try to concentrate on a few selected examples from the most recent literature.

Many details are known about promoter and enhancer DNA sequence organization and interacting protein factors required for specific transcription. The finding and establishing of in vitro and in vivo systems for all three RNA polymerases carrying out accurate transcription of exogenous DNA templates has allowed the identification of promoter and regulatory sequences required for transcription initiation and termination. It has begun the search for general as well as specific transcription factors—protein components that are not part of the complex polypeptide composition of RNA polymerases but that are necessary and essential components for the various steps of RNA synthesis: promoter recognition, chain initiation and elongation, termination, and transcript release. However, the nature of the interaction of these factors with each other and especially with the RNA polymerases themselves remains rather obscure, and is generally referred to as protein-protein interaction and binding.

RNA Polymerase I

The nucleolar enzyme pol I is responsible for transcription of the major ribosomal RNAs. It synthesizes long precursor RNAs that are processed into the mature 18S, 5.8S, and 28S RNA later incorporated into the ribosome. In contrast to the other two polymerases, pol I action is sequestered into a specific cellular component, the nucleolus, where rRNA genes (generally clustered and tandemly repeated) are unfolded for transcription as naked DNA without any histones. Thus the transcription machinery, including promoters, enzymes, and required transcription factors, is concentrated in one location.

Through the use of in vitro transcription systems a variety of nucleotide sequences that promote rDNA transcription have been identified [75–82], located in the "nontranscribed" spacer regions (NTS) separating individual transcription units. The NTS is at least partly transcribed in several organisms: *Drosophila, Xenopus,* mouse. Specific transcription factors in mouse and man bind or interact with these regions and probably act cooperatively to facilitate stable complex formation, initiation, and termination. A core sequence in conjunction with upstream control elements (UCE) both bind TIFIB (analogous to factor D or UBF1).

Core and UCE have to be physically linked, and a synergistic action and cooperative binding of two molecules seems to be required. Once this promoter recognition complex is formed, the polymerase binds efficiently to the protein-promoter complex. The binding is independent of the surrounding DNA sequences; i.e., the polymerase seems to be directed to the promoter solely by protein-protein interactions [79]. Footprinting experiments confirm minimal contacts between polymerase and DNA. Once bound to the initiation complex and in the absence of transcription factors, there is no promoter protection by polymerase alone [79,83].

Transcription initiates at a fixed distance downstream of the TIF-binding site and, although polymerase binding is DNA sequence-independent, subsequent transcription initiation steps require specific DNA sequences. Constructs with mutated transcription start sites can still stably bind transcription initiation factors at their respective binding sites and also bind polymerase, but transcription efficiency is decreased or abolished altogether. The effect of these mutations takes place after polymerase binding, and it is still unknown what else is necessary for accurate transcription initiation. Several possibilities are discussed: (1) specific recognition of DNA sequences by pol I is achieved only subsequent to its binding to the protein-promoter complexes; (2) template melting; (3) additional sequence-specific protein factors; (4) additional protein-protein interactions. Of the transcription factors identified so far, TIFIC and SL1 might fit this last category: in vitro experiments showed that SL1 does not recognize specific DNA sequences but binds to UBF1 [84].

Very little sequence homology could be detected between the promoter regions of several eukaryotic species [67,85]. Since rDNA spacer regions evolve rapidly (with very little sequence homology even between very closely related species), it was not surprising to find that transcription by pol I is highly species specific. Mouse and human rRNA genes are not transcribed in extracts of the other species, and not even rRNA genes of *Drosophila virilis* seem to be transcribed by pol I of *Drosophila mela-nogaster* [86,87].

It has been reported that "mouse and frog violate the paradigm of species specific transcription of ribosomal genes" [88]. It could be demonstrated that *Xenopus laevis* rDNA is specifically transcribed in mouse cell extracts. The heterologous system uses the frog rDNA promoter to accurately initiate transcription. The same transcription factors that are required for the homologous system are involved in the heterologous transcription, including factors D and C.

It has been reported that transcription of mouse rDNA is regulated by an "activated" subform of pol I [89]. A factor C is tightly associated with a small fraction of the polymerase preparation. This factor may be necessary for activation and might be recycled in the process of transcription initiation. It is tempting to compare the action of this kind of factor to prokaryotic sigma factors [90,91].

Several reports have provided evidence that pol I initiation is enhanced by transcription termination sites that coincide with specific promoter elements [67,92–95]. Additional spacer promoters have been identified [78] whose function may be to capture free polymerases and drive them to the gene promoter in order to achieve high levels of transcription characteristic of rRNA genes.

It has been shown that additional protein components are required not only for transcription initiation but also for transcription termination of pol I [96]. A DNA sequence element (SalI box) present several times far downstream the coding region for 28S RNA in the NTS [97] serves as a signal sequence for a specific transcription termination factor [98]. In vitro transcription experiments suggest that both site-specific DNA binding and transcription termination functions (i.e., interaction with the transcription apparatus) reside in the same polypeptide [99]. The molecular mechanisms by which this termination factor causes the stop of RNA elongation and the release of the transcript remain unknown. There is evidence that the action of this factor is specific to pol I: preliminary results introducing Sal boxes downstream pol II transcribed genes and the addition of the termination factor seem to be incapable of terminating pol II transcription in vitro [I. Grummt, personal communication].

RNA Polymerase II

This enzyme is thought to transcribe all protein-coding genes. Transcription can start at a great variety of promotor regions of varying length and sequence elements, depending on the type of gene to be read.

Most eukaryotic genes have a sequence element called a TATA box, with a consensus sequence TATAAA located 25 to 30 bp upstream of the transcription initiation site. Mutations within this sequence element produce heterogeneous upstream start sites without necessarily causing reduction in the rate of initiation [100]. Thus the major effect of a TATA box may be to tell the transcriptional machinery where to start. The first 100 to 200 bp upstream of the start site usually harbor additional promoter elements, such as the CAAT and GC boxes of the sequence CCAAT and GGGCGG, respectively. These sequence elements appear to be of a general nature, in contrast to regulatory elements like the metallothioneine regulatory elements [101], which are binding sites for specific proteins regulating groups of genes with specific functions.

Housekeeping genes, which are transcribed at relatively low levels and in all cell types, often have TATA boxes that are barely recognizable as such; they are in the form of short AT-rich sequences flanked by similarly short sequences of high GC content. In exceptional cases promoter sequences can be completely dissimilar to the commonly found sequence motifs [102]. In addition to these upstream promoter elements, many genes feature additional enhancer elements that can function at a considerable distance from the promoter independent of orientation and relative position to the promoter; some enhancer elements are even located downstream of the promoter site (as in the intron in the kappa light chain [103]). The field of *cis*-acting enhancer elements and of *trans*-acting enhancer-binding proteins has mushroomed in recent years, and it would greatly exceed the format of this review to enter a detailed discussion of enhancer elements and binding proteins. Although little is known about precisely how enhancer elements control the rate of transcription of a particular gene, an interesting fact appears to emerge from the recent reviews of Hatzopoulos et al. [104] and Ptashne [70].

DNA-binding proteins that confer transcriptional activity to specific promoters are highly sequence- and cell type–specific in their DNA-binding function, but the interaction with the enzyme requires very little sequence specificity. In fact, it appears that any "acid blob" or "negative noodle" of a broad general structure appears to interact functionally with the transcriptional machinery [105,106]. Since the only promiscuous sequence domain exclusively present in pol II with an extended structure to potentially interact with the "negative noodle" of the activator protein is the CTD of the largest subunit, this sequence domain has become the prime suspect to be the target for transcription activation. Recent experiments from the laboratory of I. Grummt showed that transcription by pol I cannot be activated simply by the presence of the GAL4 "acid blob". In a mouse in vitro transcription extract supplemented with GAL4, a yeast pol II transcription activator protein [106], transcription of pol I is not increased, although GAL4 binds to the chimeric construction containing the recognition sequences upstream of the transcription initiation site [I. Grummt, personal communication]. Since pol I is lacking the CTD, this result could be interpreted in favor of the CDT factor recognition hypothesis, though it cannot be ruled out that "acid blobs" are interacting with other domains not conserved between pol I and pol II.

The relatively easy experimental setups to test in vitro constructs in vivo by transfection into cultured cells or injection of RNA into oocytes have allowed rapid advances in the analysis of cis-acting regulatory elements affecting transcription of a target gene. In contrast to this rapid progress, biochemical analysis of the transcriptional machinery has been rather slow. Only in the past several years have we seen the establishment of reliable transcription systems from various organisms in which accurate initiation could be shown, as well as a rate of elongation approaching the in vitro rate. Most of these in vitro transcription-initiation assays use a linear template DNA molecule of known sequence and defined length with the promotor a few hundred base pairs from the runoff end of the molecule. In this way, the polymerase start site can be precisely located from the size of the RNA product determined by gel electrophoresis. Factors that stimulate the rate of elongation can be determined by allowing initiation to occur at the single-stranded tail of a "C-tailed" double-stranded DNA template [107]. With this assay, "pause sites" for pol II have been observed, as have potential termination sequences [108].

A major pause site was determined about 14 bp into the double-stranded template that could be released by a factor affecting elongation.

The factor copurified with an RNase H activity and was found to cause transcript displacement, allowing renaturation of the template behind the transcribing polymerase [109]. Thus it appears that transcript displacement speeds up the progress of RNA polymerase during chain elongation. A different route used to identify protein factors affecting the transcribing enzyme was taken by Sopta et al. [7,110], using affinity chromatography on columns containing immobilized calf thymus pol II. This way they obtained three phosphoproteins, which they named RAP72, RAP38, and RAP30, according to their molecular weights. RAP38 was found to correspond to a previously identified pol II factor [111] stimulating nonspecific transcription of native DNA. Most interestingly, RAP30 shows a weak sequence homology with the *E. coli* sigma protein that is conserved between the bacterial and phage sigma factors [J. Greenblatt, personal communication]. More recently RAP74 (formerly RAP72) was found to be associated with RAP30, and this complex is required for efficient initiation at the Adeno 2 major late promoter (Ad2MLP). In addition to Ad2MLP, the RAP74/30 complex seems to be required for a number of promoters including one with no discernible TATA box [7]. Whether the RAP30 protein is similar to, or identical with, the general transcription factor described by Zheng et al. [112], which is also required for initiation at the Ad2MLP by pol II, remains to be seen.

With few exceptions, the sequences at which transcription actually stops are ill defined. Based on the observations of a gradual reduction in hybridization intensity between newly synthesized RNA and DNA probes distal to the gene as well as on SI mapping experiments, the current view is that transcription terminates at multiple sites extending over hundreds or even thousands of nucleotides. Thus the correct location of the poly-A tail does not so much depend on accurate termination as on accurate RNA processing. Studies on transcription termination of histone genes showed that both processing of 3' ends and termination signals in the spacer region are required for termination before the histone genes [113]. Dedrick et al. [108] studied termination by purified pol II in vitro on the tailed native DNA templates. They could show that certain sequences can serve as "intrinsic" terminators for pol II; these are structurally similar (oligo-T runs in the noncoding strand) to the rho-independent termination sites for *E. coli* RNA polymerase. With the strong evolutionary conservation (see above), this is not too surprising.

RNA Polymerase III

Genes transcribed by pol III (i.e., 5S and tRNAs; a variety of small RNAs—7SK, 7SL, U6, and 4.5S; repetitive elements; and some rather uncharacterized products) were the first eukaryotic genes whose promoter and transcription control regions were identified [68,114,115]. In great contrast to all prior investigations and results obtained with prokaryotic genes, the control regions for transcription of 5S RNA genes were found to reside within the coding region. Those internal control regions (ICR) were found to be common to all pol III transcripts. After careful mapping of ICRs, the first transcription factors that are required for accurate transcription have been identified: TFIIIA (which became the prototype of a Zn finger DNA-binding protein [38–40]), TFIIIB, and one or more proteins collectively called TFIIIC [116]. TFIIIC has been resolved into two different protein components. TFIIIA binds the 5S ICR and recruits TFIIIB and TFIIIC to form a stable complex. TFIIIB is the last component to be added (and in contrast to the other factors does not directly interact with DNA), can easily be detached, and probably interacts via protein-protein recognition. Once these components are assembled, pol III, usually absent from isolated chromatin, specifically binds and gets aligned with the upstream transcription initiation start site. It has been suggested that the 5'-positioned part of TFIIIA [117] and *Drosophila* TFIIIC [118] directly interact with the polymerase, but nothing is known about the nature of this interaction.

The fact that protein components other than histones are bound within the coding region has attracted much investigation, but it is still unresolved how transcription proceeds through the complex, which is stable between multiple rounds of transcription. The binding of factors is more stable than the binding of RNA pol III, which is relatively loosely associated and can be selectively dislodged and released between rounds of transcription. In vitro experiments with SP6 polymerase and 5S RNA gene transcription complexes have shown that a preexisting complex is stable throughout transcription but is erased by the progression of a eukaryotic replication fork [119,120].

Transcription of 5S rRNA genes of *Xenopus laevis* in oocytes and somatic cells is differentially regulated during development. It has been shown that transcription complexes of the somatic genes are stable, so these genes remain active while transcription factors dissociate from oocyte-type genes, rendering them inaccessible to factors due to chromatin assembly and thus leading to repression [121].

As for pol I, binding of pol III to promoter-protein complexes is independent of flanking DNA sequences, but accurate transcription and transcription efficiency greatly depend on external promoter domains [122,123]. Mutation and deletion analysis of flanking sequences has shown, at least for some pol III transcribed genes [122,124,125], that there are crucial elements external to the coding region required for specific transcription and strongly affecting efficiency or abolishing transcription altogether. The −30 region, in particular, is thought to directly interact with pol III. Further sequence analysis has revealed that upstream sequences generally considered to be typical for pol II transcription are also present in genes transcribed by pol III: mutations and deletions within a TATA-like sequence [126] dramatically reduce pol III transcription. Whether the pol II TATA-binding factor, which has recently been resolved into more than one component [127], actually interacts with the TATA-like sequence of those genes is not yet clear. Besides that, as shown for other genes, small deletions in the vicinity of this TATA-like sequence and other sequences closer to the initiation site have the same diminishing effect [128,129].

A proximal promoter element (PSE) found in U1 and U2, pol II transcribed genes, is also present in the pol III transcribed U6 gene [130–133]. When deleted, U6 is no longer transcribed by pol III [123].

U6 and 7SK genes have far upstream distal sequence elements (DSE) that have stimulatory effects on pol III transcription [122,133]. Contained within these sequences are copies of an octamer-motif binding protein found in many pol II enhancer elements, although it is not known whether the protein(s) actually bind [122,123,134,135]. Recent results in transcription of U6, 7SK, and 7SL suggest that no intragenic control regions are necessary, and transcription by pol III relies solely on upstream sequences [123,126,134,136].

Detailed analysis of the *Xenopus* U6 promoter revealed PSE and DSE of pol II snRNA gene promoters that enable its transcription by pol II and additional sequences that, in combination with the pol II elements, are required for transcription by pol III. The construction of chimeric genes containing promoter sequences of U6 and U2 resulted in switched promoter recognition specificity from pol II to pol III and vice versa [137].

It has been reported [138] that the major transcription initiation site of the c-myc gene can be used by both pol II and III. Pol II generates the entire transcript, and pol III terminates transcription at the first run of T residues, although it has been suggested that this may be artifactual [139]. Four or more T residues in a GC-rich environment are generally thought to be the transcription termination signal for pol III transcripts. Evidence in all cases is based on α-amanitin resistance, the utilization of pol III termination signals, and competition assays with 5S RNA genes. There is no firm proof that internal sequences are unimportant and/or that transcription is actually performed by pol III.

Our own results (see Fig. 7.1) localizing pol III with specific antibodies on polytene chromosomes of *Drosophila* indicate that there are about twice as many loci served by pol III than previously deduced from the known number of tRNA loci and other presumptive pol III–transcribed sequences. Nevertheless, these findings are intriguing and strongly suggest that factors important for pol II transcription are also involved in pol III transcription; i.e., different eukaryotic polymerases may utilize common transcription factors and enhancers thought to be restricted to pol II transcription. The restrictive notion that one class of polymerases can only recognize one class of promoters seems to be not as tight as formerly thought.

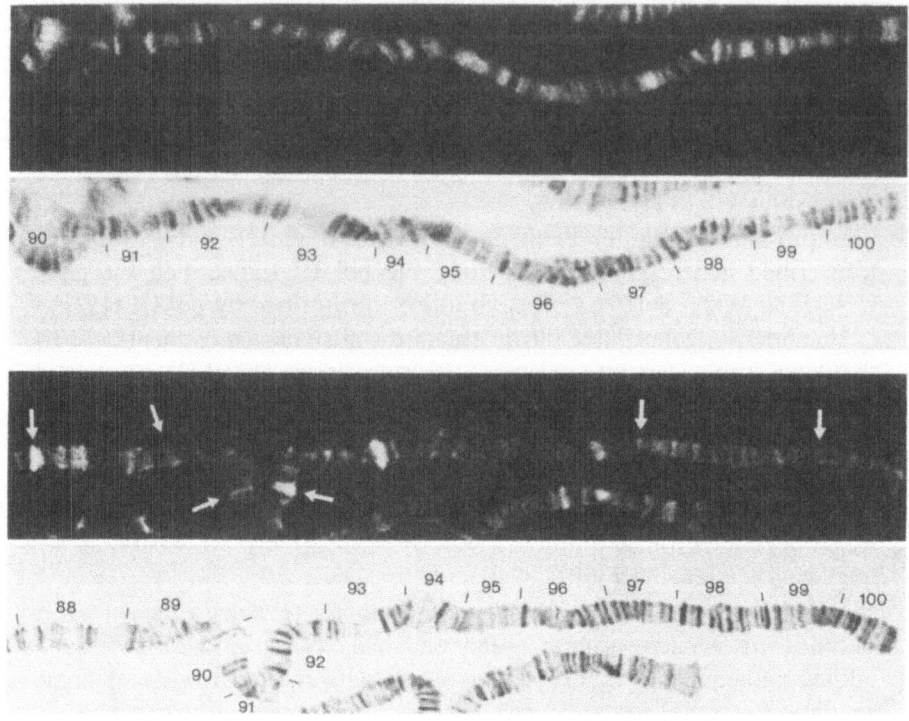

Figure 7.1. Distribution of pol II and pol III on polytene chromosomes (3R) of *Drosophila melanogaster* with specific anti–pol II antibodies (A) and anti–pol III antibodies (B) using FITC-conjugated secondary antibodies. Upper panel shows the immunostaining; lower panel shows the same preparation after orcein staining. Anti–pol II antibodies stain all interbands, whereas anti–pol III generates a more distinct banding pattern. Only some of the fluorescent loci code for tRNAs (99E, 97C, 92A, 90C, 89C, 88A), indicated by arrows. There are additional loci, in some cases brightly stained, where pol III is present, but the gene products are not identified yet.

Results obtained by Sisodia et al. [140] with chimeric genes containing pol III promoters, but with splicing and polyadenylation signals of genes that are transcribed by pol II, show that in vivo splicing and polyadenylation pathways are obligatorily coupled to transcription by pol II. Pol III possesses the inherent capacity to pause transcription (although parameters are unknown)—an obligate prerequisite for transcription termination. Previous studies [141,142] suggested that pol III terminates transcription on its own without any additional factors. There are some hints that termination may exhibit tissue-specific expression and species-specific differences [143]. It has been noted earlier that the La protein is associated with a variety of small RNAs, all of which are thought to be pol III transcripts [144]. Recent experiments from the laboratory of J. Steitz have shown that the La protein is required for accurate transcription termination and transcript release [145]. La interacts with the transcription complex after initiation of synthesis leading to the addition of one or two more U residues to the finally released transcript. There seems to be a feedback regulation in the presence of the La protein in resetting the complex to restart initiation.

Outlook

Over the past dozen years or so, the field of gene expression in eukaryotes has witnessed a two-sided effort. One concerns the identification of *cis*-acting regulatory DNA elements and recognition by DNA binding proteins, some of which could be identified as transcriptional activators. On the other side, a much smaller group of workers busied themselves with the complicated task of continuing to analyze the transcribing enzymes with the so-far limited success of deducing the primary structure of most of the large subunits from DNA clones.

The availability of cloned RNA polymerase subunit genes has led to an interesting approach to determine the sites of interaction of RNA polymerase subunits with individual transcription factors. Rappaport et al. [146] have inserted genomic sequences of the human RPII215 gene, corresponding to parts of the fifth exon, into an expression vector in fusion with the carboxyterminus of β-galactosidase. The resulting fusion protein inhibited specific transcription from the Ad2ML promoter, whereas no effect was seen in a nonspecific transcription assay. It could be shown that the target for the inhibition was the SII transcription elongation factor, which is identical with the RAP38 factor of Sopta et al. [110]. The same approach is used to probe fragments of the large subunits of pol II of *Drosophila* in a specific in vitro transcription assay, and evidence was obtained for a competitive binding of the C-terminal half of the 140-kD subunit with factor 5 [A. Greenleaf, personal communication].

It is clear that this kind of experimental approach will eventually provide a great deal of insight into the interactions between the numerous regulatory proteins and the transcriptional machinery. It may also be worthwhile to switch protein domains between the corresponding subunits of pol II and pol III to see what determines promoter specificity. To obtain a detailed picture of the function of RNA polymerases it will probably be necessary to clone and sequence the small subunit genes and to see whether there are subpopulations of RNA polymerases involved in specific groups of genes.

Acknowledgments. We thank J. Greenblatt, A. Greenleaf, I. Grummt, A. Sentenac, and W. Seifarth for unpublished information.

References

1. Weiss, S.B., and Gladstone, L. (1959) J. Am. Chem. Soc. *81*, 4118.
2. Burgess, R.R., Travers, A.A., Dunn, T.T., and Bautz, E.K.F. (1968) Nature *221*, 43–45.
3. Chambon, P. (1974) The Enzymes, Vol. X, 261–331, Boyer, P.D. (ed.). Academic Press, New York.
4. Roeder, R.G., and Rutter, W.J. (1969) Nature *224*, 234–237.
5. Kedinger, C., Gniazdowski, M., Mandel, T.L., Gissinger, F., and Chambon, P. (1970) Biochem. Biophys. Res. Commun. *38*, 165–171.
6. Sentenac, A. (1985) CRC Crit. Rev. Biochem. *18*, 31–90.
7. Burton, Z.F., Killeen, M., Sopta, M., Ortolan, L.G., and Greenblatt, J. (1988) Mol. Cell. Biol. *8*, 1602–1613.
8. Thuriaux, P., Mann, C., Buhler, J.-M., Treich, I., Gudenus, R., Mariotte, S., Riva, M., and Sentenac, A. (1986) In Extrachromosomal Elements in Lower Eukaryotes, 519–531, Wickner, R.B., et al. (eds.). Plenum Publishing Co., New York.
9. Cornelissen, A.W.C.A., Evers, R., and Köck, J. (1988) In Oxford Surveys on Eukaryotic Genes, Vol. 5, 91–131.
10. Faust, D.M. (1986) PhD Thesis, University of Heidelberg.
11. Bell, G.J., Valenzuela, P., and Rutter, W. (1976) Nature *261*, 429–431.
12. Breant, B., Buhler, J.M., Sentenac, A., and Fromageot, P. (1983) Eur. J. Biochem. *130*, 247–251.
13. Buhler, J.-M., Huet, J., Davis, K.E., Sentenac, A., and Fromageot, P. (1980) J. Biol. Chem. *255*, 9949–9954.
14. Huet, J., Sentenac, A., and Fromageot, P. (1982) J. Biol. Chem. *257*, 2613–2618.
15. Krämer, A., Haars, R., Kabisch, R., Will, H., Bautz, F.A., and Bautz, E.K.F. (1980) Mol. Gen. Genet. *180*, 193–199.
16. Greenleaf, A.L., Borsett, L.M., Jiamachello, P.F., and Coulter, D.E. (1979) Cell *18*, 613–622.
17. Greenleaf, A.L., Weeks, J.R., Voelker, R.A., Ohnishi, S., and Dickson, B. (1980) Cell *21*, 785–792.
18. Searles, L.L., Jokerst, R.S., Bingham, P.M., Voelker, R.A., and Greenleaf, A.L. (1982) Cell *31*, 585–592.
19. Allison, L.A., Moyle, M., Shales, M., and Ingles, C.J. (1986) Cell *42*, 599–610.
20. Ingles, C.J., Biggs, J., Wong, J.K.-C., Weeks, J.R., and Greenleaf, A.L. (1983) Proc. Natl. Acad. Sci. USA *81*, 2157–2161.
21. Ingles, C.J., Himmelfarb, H.J., Shales, M., Greenleaf, A.L., and Friesen, J.D. (1984) Proc. Natl. Acad. Sci. USA *81*, 2157–2161.
22. Ingles, C.J., Moyle, M., Allison, L.A., Wong, J.K.-C., Archambault, J., and Friesen, C.D. (1987) Mol. Cell. Biol. *52*, 383–393.
23. Corden, J.L., Cadena, J.L., Ahearn, J.M., and Dahmus, M.E. (1985) Proc. Natl. Acad. Sci. USA *82*, 7934–7938.
24. Ahearn, J.M., Bartolomei, M.S., West, M.L., Cisek, L.J., and Corden, J.L. (1987) J. Biol. Chem. *262*, 10695–10705.
25. Cho, K.W.Y., Khalili, K., Zandomeni, R., and Weinmann, R. (1985) J. Biol. Chem. *260*, 15204–15210.

26. Riva, M., Mémet, S., Micouin, J.-Y., Huet, J., Treich, I., Dassa, J., Young, R., Buhler, J.-M., Sentenac, A., and Fromageot, P. (1986) Proc. Natl. Acad. Sci. USA *83*, 1554–1558.

27. Young, R.A., and Davis, R.W. (1983) Science *222*, 778–782.

28. Mémet, S.J., Gouy, M., Marck, C., Sentenac, A., and Buhler, J.-M. (1988) J. Biol. Chem. *263*, 2830–2839.

29. Sweetser, B., Nonet, M., and Young, R.A. (1987) Proc. Natl. Acad. Sci. USA *84*, 1192–1196.

30. Mann, C., Buhler, J.-M., Treich, I., and Sentenac, A. (1987) Cell *48*, 627–637.

31. Biggs, J., Searles, L.L., and Greenleaf, A.L. (1985) Cell *42*, 611–621.

32. Jokerst, R.S., Weeks, J.R., Zehring, W.A., and Greenleaf, A.L. (1989) Mol. Gen. Genet. *215*, 266–275.

33. Falkenburg, D., Dworniczak, B., Faust, D.M., and Bautz, E.K.F. (1987) J. Mol. Biol. *195*, 929–937.

34. Faust, D.M., Renkawitz-Pohl, R., Falkenburg, D., Gasch, A., Bialojan, S., Young, R.A., and Bautz, E.K.F. (1986) EMBO J. *5*, 741–746.

35. Kontermann, R., Sitzler, S., Seifarth, W., Petersen, G., and Bautz, E.K.F. (1989) Mol. Gen. Genet. (in press).

36. Ollis, D.L., Brick, P., Hamlin, R., Xuong, N.G., and Steitz, T.A. (1985) Nature *313*, 762–766.

37. Ollis, D.L., Kline, C., and Steitz, T.A. (1985) Nature *313*, 818–819.

38. Vincent, A. (1986) Nucleic Acids Res. *14*, 4385–4391.

39. Berg, J.M. (1986) Science *232*, 485–487.

40. Klug, A., and Rhodes, D. (1987) TIBS *12*, 464–469.

41. Broyles, S.S., and Moss, B. (1986) Proc. Natl. Acad. Sci. USA *83*, 3141–3145.

42. Wu, F.Y.-H., and Wu, C.-W. (1981) In Advances in Inorganic Biochemistry, Vol. 3, 143–166, Eichhorn, G.L., and Marzilli, L.G. (eds.). Elsevier/North-Holland, New York.

43. Fukuda, R., and Ishihama, A. (1974) J. Mol. Biol. *87*, 532–540.

44. Gundelfinger, E.D. (1983) FEBS Lett. *157*, 133–138.

45. Horikoshi, M., Tamura, M., Sekimizu, K., Obimato, M., and Natori, S. (1983) J. Biochem. *94*, 1761–1767.

46. Bartolomei, M.S., Halden, N.F., Cullen, C.R., and Corden, J.L. (1988) J. Mol. Cell. Biol. *8*, 330–339.

47. Gross, G., Fields, D.A., and Bautz, E.K.F. (1976) Mol. Gen. Genet. *147*, 337–341.

48. Huet, J., Phalente, L., Buttin, G., Sentenac, A., and Fromageot, P. (1982) EMBO J. *1*, 1193–1198.

49. Huet, J., Riva, M., Sentenac, A., and Fromageot, P. (1985) J. Biol. Chem. *260*, 15304–15310.

50. Buhler, J.-M., Riva, M., Mann, C.A., Thuriaux, P., Mémet, S., Micouin, J.Y., Treich, I., Mariotte, S., and Sentenac, A. (1987) In RNA Polymerase and the Regulation of Transcription, 25–36, Reznikoff, W.S., et al. (eds.). Elsevier Science Publishing Co., Amsterdam.

51. Grachev, M.A., Hartmann, G.R., Maximova, T.G., Mustaev, A.A., Schäffner, A.R., Sieber, H., and Zaychikov, E.F. (1986) FEBS Lett. *200*, 287–290.

52. Riva, M., Schäffner, A.R., Sentenac, A., Hartmann, G.R., Mustaev, A.A., Zaychikov, E.F., and Grachev, M.A. (1987) J. Biol. Chem. *262*, 14377–14380.

53. Allison, L.A., Wong, J.K.-C., Fitzpatrick, V.D., Moyle, M., and Ingles, C.J. (1988) J. Mol. Cell. Biol. *8*, 321–329.

54. Zehring, W.A., Lee, J.M., Weeks, J.R., Jokerst, R.S., and Greenleaf, A.L. (1988) Proc. Natl. Acad. Sci. USA *85*, 3698–3702.

55. Dezélée, S., Wyers, F., Sentenac, A., and Fromageot, P. (1976) Eur. J. Biochem. *65*, 543–552.

56. Greenleaf, A.L., Haars, R., and Bautz, E.K.F. (1976) FEBS Lett. *71*, 205–208.

57. Nonet, M., Sweetser, D., and Young, R.A. (1987) Cell *50*, 909–915.

58. Christmann, J.L., and Dahmus, M.E. (1981) J. Biol. Chem. *256*, 11798–11803.

59. Cadena, D.L., and Dahmus, M.E. (1987) J. Biol. Chem. *262*, 12468–12474.

60. Price, D.H., Sluder, A.E., and Greenleaf, A.L. (1987) J. Biol. Chem. *262*, 3244–3255.

61. Greenleaf, A.L. (1983) J. Biol. Chem. *258*, 13403–13406.

62. Greenleaf, A.L., Jokerst, R.S., Zehring, W.A., Hamilton, B.J., Weeks, J.R., Sluder, A.E., and Price, D.H. (1987) In RNA Polymerase and the Regulation of Transcription, 459–463, Reznikoff, W.S., et al. (eds.). Elsevier Science Publishing Co., New York.

63. Bartolomei, M.S., and Corden, J.L. (1987) J. Mol. Cell. Biol. *7*, 586–594.

64. Himmelfarb, H.J., Simpson, E.M., and Friesen, J.D. (1987) Mol. Cell. Biol. *7*, 2155–2164.

65. Nonet, M., Scafe, C., Sexton, J.J., and Young, R. (1987) Mol. Cell. Biol. *7*, 1602–1611.

66. Gudenus, R., Mariotte, S., Moenne, A., Ruet, A., Mémet, S., Buhler, J.-M., Sentenac, A., and Thuriaux, P. (1988) Genetics *119*, 517–526.

67. Sollner-Webb, B., and Tower, J. (1986) Annu. Rev. Biochem. *55*, 801–830.

68. Geiduschek, E.P., and Tocchini-Valentini, G. (1988) Annu. Rev. Biochem. *57*, 873–914.

69. Dynan, W.S., and Tjian, R. (1985) Nature *316*, 774–778.

70. Ptashne, M. (1988) Nature *335*, 683–689.

71. Ptashne, M. (1986) Nature *322*, 697–701.

72. Maniatis, T., Goodbourn, S., and Fisher, J. (1987) Science *236*, 1237–1245.

73. Guarente, L. (1988) Annu. Rev. Genet. *21*, 425–452.

74. Struhl, K. (1987) Cell *49*, 295–297.

75. Jones, M.H., Learned, R.M., and Tjian, R. (1988) Proc. Natl. Acad. Sci. USA *85*, 669–673.

76. Haltiner, M.M., Smale, S.T., and Tjian, R.T. (1986) Mol. Cell. Biol. *6*, 227–235.

77. Clos, J., Buttgereit, D., and Grummt, I. (1986) Proc. Natl. Acad. Sci. USA *83*, 604–608.

78. Kuhn, A., and Grummt, I. (1987) EMBO J. *6*, 3487–3492.

79. Kownin, P., Batemann, E., and Paule, M.R. (1987) Cell *50*, 693–699.

80. Windle, J.J., and Sollner-Webb, B. (1986) Mol. Cell. Biol. *6*, 4585–4593.

81. Learned, R.M., Learned, T.K., Haltiner, M.M., and Tjian, R. (1986) Cell *45*, 847–857.

82. Tower, J., Culotta, V., and Sollner-Webb, B. (1986) Mol. Cell. Biol. 6, 3451–3462.
83. Bateman, E., Iida, C.T., Kownin, P., and Paule, M.R. (1985) Proc. Natl. Acad. Sci. USA 82, 8004–8008.
84. Bell, S.P., Learned, R.M., Jantzen, H.-M., and Tjian, R. (1988) Science 241, 1192–1197.
85. Sommerville, J. (1984) Nature 310, 189–190.
86. Grummt, I., Roth, E., and Paule, M.R. (1982) Nature 296, 173–174.
87. Kohorn, B., and Rae, P. (1982) Proc. Natl. Acad. Sci. USA 79, 1501–1505.
88. Culotta, V., Wilkinson, J.K., and Sollner-Webb, B. (1987) Proc. Natl. Acad. Sci. USA 84, 7498–7502.
89. Tower, J., and Sollner-Webb, B. (1987) Cell 50, 873–883.
90. Helman, J.D., and Chamberlin, M.J. (1988) Annu. Rev. Biochem. 57, 839–872.
91. Losick, R., and Pero, J. (1981) Cell 25, 582–584.
92. Henderson, S., and Sollner-Webb, B. (1986) Cell 47, 891–900.
93. Grummt, I., Kuhn, A., Bartsch, I., and Rosenbauer, H. (1986) Cell 47, 901–911.
94. McStay, B., and Reeder, R.H. (1986) Cell 47, 913–920.
95. Baker, S.M., and Platt, T. (1986) Cell 47, 839–840.
96. Platt, T. (1986) Annu. Rev. Biochem. 55, 339–372.
97. Grummt, I., Rosenbauer, H., Niedermeyer, I., Maier, U., and Öhrlein, A. (1986) Cell 45, 837–846.
98. Bartsch, I., Schoneberg, C., and Grummt, I. (1988) Mol. Cell. Biol. 8, 3891–3897.
99. Kuhn, A., Norman, A., Bartsch, I., and Grummt, I. (1988) EMBO J. 7, 1497–1502.
100. Breathnach, R., and Chambon, P. (1981) Annu. Rev. Biochem. 50, 349–383.
101. Seguin, C., and Hamer, D.H. (1987) Science 235, 1383–1387.
102. Valerio, D., Duyvesteyn, M.G.C., Dekker, B.M.M., Weeda, G., Berkvens, T.M., van der Voorn, L., van Ormondt, H., and van der Eb, A.J. (1985) EMBO J. 4, 437–443.
103. Queen, C., and Baltimore, D. (1983) Cell 33, 741–748.
104. Hatzopoulos, A.K., Schlokat, U., and Gruss, P. (1988) In Frontiers in Transcription and Splicing, IRL Press (in press).
105. Sigler, P.B. (1988) Nature 333, 210–212.
106. Ma, J., and Ptashne, M. (1988) Cell 55, 443–446.
107. Kadesch, F.R., and Chamberlin, M.J. (1982) J. Biol. Chem. 257, 5286–5295.
108. Dedrick, R.L., Kane, M.C., and Chamberlin, M.J. (1987) J. Biol. Chem. 262, 9098–9108.
109. Sluder, A.E., Price, D.H., and Greenleaf, A.L. (1988) J. Biol. Chem. 263, 9917–9925.
110. Sopta, M., Carthew, R.W., and Greenblatt, J. (1985) J. Biol. Chem. 260, 10353–10360.
111. Nakanishi, Y., Mitsuhashi, Y., Sekimizu, K., Yokoi, H., Tanaka, Y., Horikishi, M., and Natori, S. (1981) FEBS Lett. 130, 69–72.
112. Zheng, X.-M., Moncollin, V., Egly, J.-M., and Chambon, P. (1987) Cell 50, 361–368.

113. Birnstiel, M.L., Busslinger, M., and Strub, K. (1985) Cell *41*, 349–359.
114. Folk, W.R. (1988) Genes Dev. *2*, 373–375.
115. Sollner-Webb, B. (1988) Cell *52*, 153–154.
116. Yoshinaga, S.K., Boulanger, P.A., and Berk, A.J. (1987) Proc. Natl. Acad. Sci. USA *84*, 3585–3589.
117. Smith, D.R., Jackson, I.J., and Brown, D.D. (1984) Cell *37*, 645–652.
118. Burke, D.J., Schaack, J., Sharp, S., and Söll, D. (1983) J. Biol. Chem. *258*, 15224–15231.
119. Wolffe, A.P., and Brown, D.D. (1988) Science *241*, 1626–1632.
120. Wolffe, A.P., Jordan, E., and Brown, D.D. (1986) Cell *44*, 381–389.
121. Wolffe, A.P., and Brown, D.D. (1987) In RNA Polymerase and the Regulation of Transcription, 465–468, Reznikoff, W.S., et al. (eds.). Elsevier Science Publishing Co., New York.
122. Bark, C., Weller, P., Zabielski, J., Janson, L., and Petterson, U. (1987) Nature *328*, 356–359.
123. Carbon, P., Murgo, S., Ebel, J.-P., Krol, A., Tebb, G., and Mattaj, I. (1987) Cell *51*, 71–79.
124. Sajjadi, F.G., Miller, R.C. Jr., and Spiegelman, G.B. (1987) Mol. Gen. Genet. *206*, 276–284.
125. Garcia, A.D., O'Connell, A.M., and Sharp, S.J. (1987) Mol. Cell. Biol. *7*, 2046–2051.
126. Murphy, S., Di Liegro, C., and Melli, M. (1987) Cell *51*, 81–87.
127. Simson, M.C., Fisch, T.M., Benecke, B.J., Nevins, J.R., and Heintz, N. (1988) Cell *52*, 723–729.
128. Ullu, E., and Weiner, A.M. (1985) Nature *318*, 371–374.
129. Morry, M.J., and Harding, J.D. (1986) Mol. Cell. Biol. *6*, 105–115.
130. Dahlberg, J.E., and Lund, E. (1987) In Structure and Functions of Small Nuclear Ribonucleoprotein Particles, 38–70, Birnstiel, M.L. (ed.). Springer-Verlag, Heidelberg.
131. Kunkel, G.R., Maser, R.L., Calvet, J.P., and Pederson, T. (1986) Proc. Natl. Acad. Sci. USA *83*, 8575–8579.
132. Reddy, R., Henning, D., Das, G., Harles, M., and Wright, D. (1987) J. Biol. Chem. *262*, 75–81.
133. Krol, A., Carbon, P., Ebel, J.-P., and Appel, B. (1987) Nucleic Acids Res. *15*, 2463–2478.
134. Das, G., Henning, D., Wright, D., and Reddy, R. (1988) EMBO J. *7*, 503–512.
135. Kunkel, G.R., and Pederson, T. (1988) Genes Dev. *2*, 196–204.
136. Kleinert, H., and Benecke, B.-J. (1988) Nucleic Acids Res. *4*, 1319–1331.
137. Mattaj, I.W., Dathan, N.A., Parry, H.D., Carbon, P., and Krol. A. (1988) Cell *55*, 432–442.
138. Chung, J., Sussman, D.J., Zeller, R., and Leder, P. (1987) Cell *51*, 1001–1008.
139. Bentley, D., and Groudine, M. (1988) Cell *53*, 245–256.
140. Sisodia, S.S., Sollner-Webb, B., and Cleveland, D. (1987) Mol. Cell. Biol. *7*, 3602–3612.
141. Bogenhagen, B.F., and Brown, D.D. (1981) Cell *24*, 261–270.
142. Cozzarelli, N.R., Gerrard, S.P., Schlissel, M., Brown, D.D., and Bogenhagen, D.F. (1983) Cell *34*, 829–845.

143. Mazabraud, A., Scherly, D., Müller, F., Rungger, D., and Clarkson, S.G. (1987) J. Mol. Biol. *197*, 1–11.
144. Gottlieb, E., and Steitz, J.A. (1987) In RNA Polymerase and the Regulation of Transcription, 465–468, Reznikoff, W.S., et al. (eds.). Elsevier Science Publishing Co., New York.
145. Gottlieb, E., and Steitz, J.A. (1988) EMBO J. (in press).
146. Rappaport, J., Cho, K., Saltzman, A., Prenger, J., Golomb, M., and Weinman, R. (1988) Mol. Cell. Biol. *8*, 3136–3142.

Chapter 8
Applications of Monoclonal Antibodies to the Study of Eukaryotic RNA Polymerases

Nancy E. Thompson and Richard R. Burgess

Transcription is a complex series of carefully regulated events that leads to an RNA product. Central to our understanding of these events is an understanding of how the RNA polymerase (RNAP) interacts with DNA sequences, protein factors, nucleoside triphosphates, and, perhaps, the nuclear suprastructure to faithfully accomplish transcription. Despite the central role that RNAPs play in gene expression, surprisingly little information is available on the mechanisms of these enzymes.

There are three types of nuclear RNA polymerases in eukaryotic cells. They are designated RNA polymerase I, II, and III (or A, B, and C, respectively). Many of the features of these enzymes are contained in other chapters of this volume, and in-depth reviews are available on this subject [1,2]. This chapter will focus on the use of monoclonal antibodies (MAbs) to study RNA polymerases, with particular emphasis on RNA polymerase II. Therefore, the introductory material to this chapter will cover only information that is pertinent to studies that have used antibodies to study nuclear RNA polymerases. Mitochondrial and chloroplast RNA polymerases will not be addressed in this chapter.

Transcriptional Activity of Eukaryotic RNA Polymerases

The most distinguishing feature of each RNA polymerase is the class of genes that each transcribes. The genes for the precursors for the large ribosomal RNA species are transcribed by RNA polymerase I (RNAP I). The genes for transfer RNAs, 5S ribosomal RNA, and U6 small nuclear RNA (snRNA) are transcribed by RNA polymerase III (RNAP III). RNA polymerase II (RNAP II) transcribes the majority of genes that must be transcribed into messenger RNA for translation to a protein product. RNAP II also transcribes the genes for the snRNA species with the exception of U6.

Early attempts to purify RNA polymerase resulted in separation of the three enzymes on DEAE ion exchange chromatography [reviewed in 3]. The enzymatic activity separated in this way showed differential sensitivity to the mycotoxin α-amanitin. RNAP I is essentially insensitive to this inhibitor. RNAP II is sensitive to 1 μg α-amanitin per milliliter, while RNAP III is sensitive to 100 μg/ml. α-Amanitin inhibits transcription by RNAP II and III by inhibiting the formation of phosphodiester bonds at the elongation step [4]. This reagent has been a very useful tool in studies involving transcription.

Assays for RNA Polymerase Activity

Two different in vitro transcription assays are available. The first assay is often regarded as a general or nonspecific assay, because the process that is measured is simple incorporation of radiolabeled UTP into an RNA product (elongation). The procedure described by Jendrisak and Burgess [5] is an example of a nonspecific transcription assay. This assay uses purified or partially purified enzyme, a suitable template (usually denatured calf thymus DNA), nucleoside triphosphates (NTPs) containing a radioactive label, and the proper ionic conditions (8 mM Mg^{+2}, 25 mM $(NH_4)_2SO_4$). Transcription starts at nicks or ends of DNA and is not dependent on the presence of a promoter. It is a useful assay for following purification of RNA polymerase or for examining the effect of a reagent (antibody) on the elongation step of transcription. All three RNA polymerases are active in this assay, and the contribution of each enzyme to the RNA product can be roughly measured by the use of differential amounts of α-amanitin.

The second type of in vitro transcription assay measures transcription from a promoter. It is referred to as specific or promoter-directed transcription. This assay system uses whole-cell extracts [6] or nuclear extracts [7] as the source of RNA polymerase and the necessary transcription factors. Recently, investigators have attempted to fractionate these extracts to arrive at more simplified reaction mixtures [8–10]. In addition to the extract, the investigator supplies a specific template, generally a restriction fragment containing the gene under investigation along with the promoter region for that gene. The template is usually truncated several hundred bp downstream of the initiation site by a restriction enzyme. Thus, the transcript starts at the initiation site and is terminated at the end of the restriction fragment, yielding an RNA product of discrete size that is identifiable by electrophoresis in a denaturing polyacrylamide gel system. Hence the term "runoff" transcription is applied to this type of assay. The RNA product can be detected directly by the use of a radiolabeled NTP in the reaction mixture or by the use of indirect methods such as S1 nuclease mapping or primer extension.

Several in vivo transcription systems are also available. The most commonly used system is the amphibian oocyte, particularly the *Xenopus laevis* oocyte; this system has been reviewed by Gurdon and Wickens [11]. The large size of the oocytes facilitates microinjection of template (usually contained on a bacterial plasmid). The transcript can be detected by standard electrophoretic techniques. Transcripts that are not modified can be detected directly by coinjection of radiolabeled NTP. Modified transcripts can be detected by indirect methods such as S1 nuclease mapping or primer extension. Each oocyte contains sufficient transcript for individual oocyte analysis.

Mammalian cells can also be used to study in vivo transcription. The DNA template is most commonly introduced by transfection [12].

The Transcription Complex

The basic transcription complex can be thought of as consisting of three components: (1) the promoter region and specific DNA sequences (*cis*-acting elements) located upstream of the promoter region that increase the efficiency of the promoter elements; (2) the nuclear proteins that interact with these DNA elements (*trans*-acting elements); and (3) the RNAP and RNAP-associated transcription factors. Recent success in isolating transcription complexes [13,14] should help to define more clearly the involvement of these components in the formation of a transcription complex.

Promoter regions for eukaryotic genes have been intensely studied in recent years. Promoter structure is beyond the scope of this review. However, it is necessary to consider promoter structure in discussions of transcription. The rapid accumulation of data in this field makes timely reviews impossible to find. However, the reader is referred to Dynan and Tjian [15], Maniatis et al. [16], and Jones et al. [17] for general reviews. The promoter region consists of proximal and distal promoter elements. The proximal promoter elements are located within approximatey 100 bp of the initiation site, and distal elements are located within several hundred bp of the initiation site.

Many promoters contain an A-T-rich region referred to as a TATA box. This proximal promoter element seems to function as a point of reference for the transcriptional start site; transcription starts approximately 25 bp downstream from this box. At least one specific protein has been identified that binds to this site [18,19] and helps to position the RNAP with respect to the transcriptional start site. Promoters that do not have a discrete TATA box seem to have a functionally equivalent site.

Another proximal promoter element present in some promoters is a distinctive motif called a CCAAT box which can bind many different proteins [20].

Distal promoter elements are a variety of specific DNA sequences that seem to increase the efficiency of the promoter. Presumably one or more specific proteins interact with these elements to help to position the RNAP at the promoter region.

Proteins, other than RNAP, that are necessary for efficient promoter-directed transcription are referred to collectively as transcription factors. These include the proteins that bind to specific DNA sequences, the proteins that bind to these proteins, and the more general transcription factors that bind to RNAP itself.

The interactions of the *trans*-acting transcription factors with the *cis*-acting DNA sequences are generally considered to give specificity to the process of transcription. However, the basic mechanism of transcription can only be elucidated by understanding the mechanics of RNAP and how it coordinates with the *cis*- and *trans*-acting elements.

Subunit Composition of RNA Polymerases

All RNAPs are complex, multimeric enzymes. The subunit structure consists of two large subunits (140 to 240 kD) and a variety of smaller subunits (less than 100 kD). An SDS-polyacrylamide gel (SDS PAGE) containing the *E. coli* holoenzyme and RNAP II isolated from wheat germ and calf thymus is shown in Figure 8.1. The subunits of prokaryotic RNAP are referred to by Greek letters, while the subunits for the eukaryotic enzymes are generally referred to by subunit size in kilodaltons. Sentenac [2] discusses the criteria that are used to judge the legitimacy of a subunit.

The early literature on eukaryotic RNAP was complicated by the apparent presence of subforms of RNAP II that differ in the size of the largest subunit (approximately 220 vs. 180 kD). Recent evidence has established that the smaller form of this subunit arises during purification of the enzyme owing to proteolytic activity present in the crude steps of purification [21,22]. Proteolysis during the purification seems to be more of a problem with enzymes isolated from mammalian tissue than with enzymes isolated from dormant plant tissue. The calf thymus RNAP II in Figure 8.1 contains predominantly the proteolyzed form, while the wheat germ RNAP II contains predominantly the unproteolyzed form of the largest subunit. A third form of RNAP II whose largest subunit appears larger than the 220-kD subunit on SDS PAGE has been reported in some mammalian systems [23]. This subunit has been shown to be phosphorylated [24,25], which probably accounts for the slower migration. It has been demonstrated that the form containing the unproteolyzed largest subunit is the predominant form in mammalian cells [26], and the form

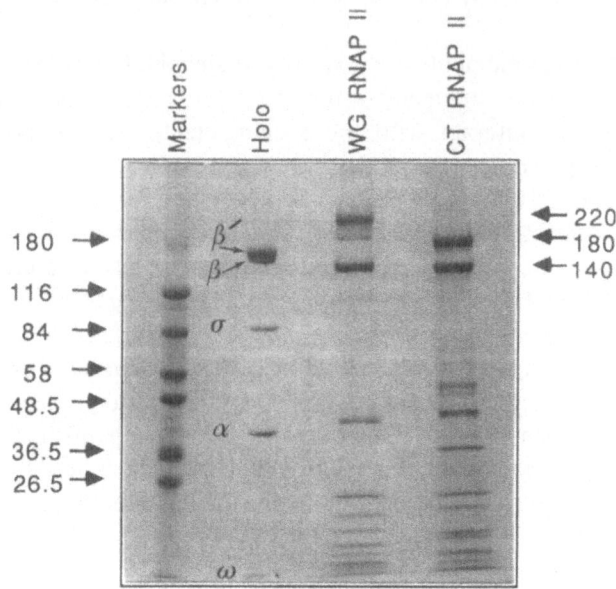

Figure 8.1. Comparison of subunit composition of prokaryotic RNAP and RNAP II isolated from wheat germ and calf thymus. Denatured enzymes were subjected to electrophoresis in a 4% to 20% polyacrylamide gel in the presence of SDS by the method of Laemmli [74]. The holoenzyme from *E. coli* (Holo) was purified by the procedure of Burgess and Jendrisak [75] and consists of subunits β' (155 kD), β (151 kD), σ (70 kD but runs as if it were 90 kD), two copies of α (36 kD), and ω (11 kD). Wheat germ RNAP II (WG RNAP II) was purified by a modification of the procedure of Jendrisak and Burgess [5] and consists of subunits that are 220 kD (and the proteolyzed form that is 180 kD) and 140, 40, 27, 25, 21, 20, 17.8, 17, 16.5, 16, and 14 kD. Calf thymus RNAP II (CT RNAP II) was purified by a modification of the procedure described by Hodo and Blatti [76] and consists of subunits that are 214 kD (and the proteolyzed form, most predominant in this preparation, that is 180 kD) and 140, 34, 25, 20.5, 18, 17.5, and 16.5 kD. Polypeptides of approximately 50 and 40 kD in the calf thymus preparation are contaminants in this particular material.

that can be cross-linked to nascent mRNA is in the phosphorylated state [24]. It is likely that this is the case in all eukaryotic cells.

The cloning and sequencing of the genes for the largest subunit of RNAP II from yeast [27], mouse [28], hamster [29], and *Drosophila* [29,30] have revealed the presence of an unusual heptapeptide repeat on the C terminus of the subunit. This heptapeptide repeat has the consensus sequence Pro–Thr–Ser–Pro–Ser–Tyr–Ser; the number of repeats varies from 26 in yeast to 52 in the mouse. The repeat is less well conserved in *Drosophila* [29]. Conversion from the unproteolyzed to the proteolyzed form of this subunit results in the loss of a large peptide rich in Pro, Ser,

Table 8.1. Subunit genes for eukaryotic and viral RNA polymerases that have been cloned and sequenced.

RNA polymerase	Species	Subunit	Homology	Reference
RNAP I	Yeast	190 kD	β'^a	34
RNAP I[b]	Yeast	40 kD	None	78
RNAP II	Yeast	220 kD	β' RNAP III'(160)[c]	27
RNAP II	Mouse	220 kD	β'	28;79
RNAP II	Hamster	220 kD	β'	29
RNAP II	*Drosophila*	215 kD	β'	29,31
RNAP II	Yeast	140 kD	β^a	32
RNAP II	*Drosophila*	140 kD	β	33
RNAP III	Yeast	160 kD	β' RNAP II(220)	27
RNAP III[b]	Yeast	40 kD	None	78
Viral[d]	Vaccinia	147 kD	β' RNAP II(220) RNAP III(160)	35

[a] β' and β refer to the largest (155-kD) and second largest (151-kD) subunit, respectively, of *E. coli* RNA polymerase.

[b] The 40-kD subunit is common to yeast RNAP I and RNAP III.

[c] Number in parentheses refers to the size of the homologous subunit in kilodaltons.

[d] This subunit shows areas of extensive homology with the 220-kD subunit of yeast RNAP II, the 160-kD subunit of yeast RNAP III, and the 215-kD subunit of *Drosophila* RNAP II.

Thr, and Tyr residues [27,28], which is likely the C-terminal heptapeptide repeat. This domain is likely the major site of phosphorylation [24].

Table 8.1 is a compilation of eukaryotic RNAP subunit genes that have been cloned at this writing. When the C-terminal heptapeptide repeat present on RNAP II is disregarded, computer analysis has demonstrated that the largest subunit of RNAP II shows considerable homology to the largest subunit of RNAP III and to the largest subunit (β') of *E. coli* RNAP [27–29,31]. Likewise, the amino acid sequence of the second-largest subunit of yeast RNAP II [32] and *Drosophilia* [33] shows considerable homology to the second-largest subunit (β) of *E. coli* RNAP. Similar homologies have been reported for the sequences of the largest subunit of yeast RNAP I [34] and the largest subunit of vaccinia virus RNAP [35]. Thus, evolution has conserved some structures that are probably necessary for the enzymatic properties of all RNAPs. Some theoretical considerations of subunit conservation in RNAPs have been addressed by Armaleo [36].

Little is known about the role that each subunit plays in either the catalytic or regulatory activity of each eukaryotic RNAP type. In the prokaryotic enzyme, the β' subunit has been shown to be involved with the DNA-binding function, and the β subunit has been shown to be involved with nucleotide binding. By analogy to the prokaryotic enzyme it would seem likely that these functions would be associated with the largest and second-largest subunits, respectively. Only recently have studies addressed these questions. By probing subunits blotted from an SDS PAGE gel with radioactively labeled DNA, Chuang and Chuang [37] identified the 220-kD subunit (as well as the proteolyzed subunit) of chicken RNAP II as the DNA-binding subunit. These data seem to support the assumption that the largest subunit of the eukaryotic enzyme should have a function homologous to β'. However, by photoaffinity labeling experiments, Freund and McQuire [38] identified only the 37-kD subunit of calf thymus RNAP II as having nucleotide-binding sites. This result seems to be discordant with the homologous function model.

Immunological Studies

Prior to the availability of cloned genes, immunological evidence indicated that considerable common structure existed among the eukaryotic RNAP types. Polyclonal antiserum directed against either the entire RNAP molecule or against isolated subunits often showed cross-reactivity with subunits from the other RNAP types from that species [39,40] or the same RNAP type from a different species [40–44]. Cross-reactions between the same RNAP type from different species were most apparent with the two largest subunits, although some reactivity was apparent with the smaller subunits.

Immunological analyses along with other data established that some of the smaller subunits are common to two or to all three of the RNAP types from a given organism. For example, in yeast, the 27-, 23-, and 14.5-kD subunits are shared by all three RNAPs, whereas the 40- and 19-kD subunits are shared by only RNAP I and RNAP III [40,41].

A systematic study has been conducted using polyclonal antibodies directed against isolated subunits of yeast RNAPs [41,45]. These authors found that antiserum prepared against some, but not all, subunits of RNAP inhibited RNAP activity in nonspecific transcription assays. However, these studies did not establish that the inhibition was due to simply removing the RNAP from solution by immunoprecipitation, and they did not establish that the antiserum to all the subunits was capable of reacting with native RNAP in solution.

Immunological Reagents

The use of polyclonal antiserum has several limitations. First, polyclonal antiserum is dependent on the purity of the immunogen used to prepare it. Although classical serological measures to increase specificity (such as absorption) can be taken, the final antiserum product is dependent on many unknown parameters. Second, because of the processing of the immunogen and changes in the immunological response of the animal, polyclonal antiserum shows considerable variation from animal to animal and even from one bleeding to another of the same animal. Thus, polyclonal antiserum can be difficult to standardize. Finally, once a useful reagent is established, the reagent is available only in limited quantity.

The development of hybridoma technology and the use of this technology to produce MAbs have allowed some of these problems to be addressed. However, MAbs are hardly the panacea of immunological reagents. The greatest attributes of MAb technology are that large amounts of highly uniform reagents can be produced, the antibodies can be well characterized, and, in some cases, the epitope for the antibody can be characterized. In addition, because of the monovalent nature of most epitopes in an antigen, MAbs are generally not precipitating reagents. Thus, the effect of an antibody on enzyme activity can be assessed without removing the enzyme from solution.

Not surprisingly, several groups of investigators have turned to producing MAbs that react with various subunits of RNAP in an attempt to study structure-function relationships within these polypeptides. A survey of these studies is presented in Table 8.2, which also gives a brief synopsis of the application of the particular MAb to the study of RNAP. Some of these experiments are described in detail in the text that follows.

MAbs Produced Against Wheat Germ RNAP II

Our laboratory has studied the structure of wheat germ RNAP II for several years. The advantages of the wheat germ enzyme over the mammalian enzyme are that the starting material is very inexpensive, it is convenient to store, and the yield of enzyme per kilogram of starting material is quite high. The biggest disadvantage is that, until recently [46], a wheat extract that could accurately transcribe genes in a promoter-directed transcription assay was not available. In addition, the wheat germ enzyme could not substitute for the mammalian enzyme in a mammalian cell extract [47]. Despite these problems, many similarities between the wheat germ enzyme and the mammalian enzyme led us to pursue the isolation of MAbs that react with wheat germ RNAP II. We have concentrated on the isolation of MAbs that cross-react with mammalian enzymes and have been able to use these MAbs as tools to investigate basic properties of mammalian RNAP II.

Some of the antibodies that we have isolated are listed in Table 8.3. Most of the methods we have used to isolate these MAbs have been described [48]. MAbs are selected as cross-reacting on the basis of the reaction of the antibody with calf thymus RNAP II on Western blots. To date, all MAbs selected as cross-reactive on this basis have been found to cross-react with the corresponding subunits from HeLa cell and *Xenopus*

Table 8.2. Literature survey of monoclonal antibodies that react with eukaryotic polymerase subunits.

RNAP	Source [Reference]	Functional characterization
RNAP I	Yeast [55]	Primary MAb described reacted with the 135-kD (second-largest) subunit; this MAb inhibited binding of RNAP I to DNA. Reported having isolated MAbs that react with the 190-, 49-, 43-, and 14.5-kD subunits. The 14.5-kD-reactive MAb cross-reacted with the 14.5-kD subunit of RNAP II and RNAP III.
RNAP I	Rat liver [54]	MAb reactive with the 190-kD (largest) subunit; inhibited elongation by both RNAP I and RNAP III, perhaps by masking the NTP-binding site.
RNAP I	Silkworm [56]	MAb reacted with the 132-kD subunit (second-largest), showed weak cross-reactivity with RNAP II and RNAP III from silkworm. MAb inhibited elongation. MAb was used to enrich a sample for RNAP I by immunoaffinity chromatography. MAb stained nuclei of *Chironomus* salivary gland when used for immunohistochemical localization.
RNAP I	Mouse [80]	MAb reacted with the 120-kD (second-largest) subunit. The animal was not immunized with RNAP I; this MAb was isolated from a fusion that used the spleen of a mouse from a strain with an autoimmune disease. The MAb immunoprecipitated RNAP I activity but not RNAP II activity from rat whole-cell extracts; depleted extracts were used in promoter-directed transcription assays. Antigen was localized in the nucleus of cells.
RNAP II	*Drosophila* [43,63]	MAb reacted with both the 215-kD (largest) and 140-kD (second-largest) subunits. This IgG_1 MAb was used to localize RNAP II on polytene chromosomes. This MAb was shown to cross-react with the corresponding subunits of calf thymus RNAP II.
RNAP II	Calf thymus [49,50]	MAb reacted with the 240- and 214-kD (phosphorylated and nonphosphorylated largest) subunits, did not react with the proteolyzed subunit (180 kD). This IgM MAb cross-reacted with HeLa cell RNAP II. It did not inhibit nonspecific transcription but did inhibit promoter-directed transcription.
RNAP II	Calf thymus [57,58]	Primary MAb reacted with the 240- and 214-kD (phosphorylated and nonphosphorylated largest) subunits; this antibody also reacted with the proteolyzed subunit (180 kD). This IgG antibody cross-reacted with human, chicken, *Drosophila,* wheat germ, and yeast RNAP II. It inhibited nonspecific transcription by inhibiting the binding of the enzyme to DNA. A second MAb, an IgM antibody, that did not inhibit nonspecific transcription was isolated but not characterized.
RNAP II	Calf thymus [62]	Primary MAb reacted with both the 220-kD (unproteolyzed and proteolyzed largest) subunit and

Table 8.2. Continued

RNAP	Source [Reference]	Functional characterization
		the 140-kD (second-largest) subunit. It cross-reacted with these two subunits of RNAP II from mouse and wheat germ. This IgM MAb inhibited transcription in promoter-directed transcription assays.
RNAP II	Hen oviduct [81]	Primary MAb described reacted only with native RNAP II. Enzyme containing both the unproteolyzed and the proteolyzed largest subunit could be recovered using this MAb. This antibody cross-reacted with calf thymus RNAP II. It did not inhibit promoter-directed transcription in HeLa cell extracts. The antibody was used to isolate RNAP II–associated transcription factors from HeLa cell extracts. Other MAbs isolated but not characterized in detail include some that react with the 140-, 35-, and 27-kD subunits and one that reacted with the 214- and 140-kD subunits.
RNAP II	*Podospora comata* [59,82]	MAbs were described that react with the 180-kD (proteolyzed largest) and the 145-, 39-, 23.5-, and 11-kD subunits. One of these antibodies reacted with both the 180- and 145-kD subunits. Some of these antibodies showed cross-reactivity with yeast, wheat germ, and calf thymus RNAP II. In the latter study, two MAbs that react with the 11-kD subunit (and cross-react with the 12.5-kD subunit of yeast RNAP II) inhibited nonspecific transcription, perhaps at the initiation step.
RNAP II	*Agaricus bisporus* [83]	Two MAbs were described that react with the 89-kD subunit of RNAP II. One of these antibodies cross-reacted with a number of subunits of *Achlya ambisexualis* RNAP II and the 220-kD subunit of wheat germ RNAP II. A third, apparently conformational antibody was described.

RNAP II. Cross-reacting MAbs have been isolated that react with the 220-, 140-, and 17.8-kD subunits of wheat germ RNAP II and the corresponding subunits from the mammalian enzymes.

Several injection procedures have been tried in an attempt to select for MAbs that react with certain subunits. These injection procedures are described in the footnotes to Table 8.3. When the entire molecule of wheat germ RNAP II is used as an immunogen, the predominant type of MAb that the hybridomas produce reacts with the largest, unproteolyzed subunit but not to any appreciable extent with the proteolyzed subunit. These antibodies cross-react with the unproteolyzed largest subunit of calf thymus RNAP II. A Western blot showing the reaction of antibodies 3WG2, 1CWG1, and 8WG16 is in Figure 8.2. Many MAbs that show this reaction pattern have been isolated, and, with one exception (antibody 8WG16), all are of the IgM isotype. The reaction pattern for these MAbs

Table 8.3. Characterization of some MAbs that have been isolated using wheat germ RNAP II as an immunogen.

MAb designation	Immunization protocol[a]	Isotype	Subunit specificity	CT RNAP II cross-reactivity[b]
3WG3	1	IgG$_{2a}$	Conformational	No
3WG2	1	IgM	220	Yes
1CWG1	1	IgM	220	Yes
1CWG2	1	IgM	220	Yes
1CWG3	1	IgM	220	Yes
8WG8	1	IgG$_1$	220	No
8WG16	1	IgG$_{2a}$	220	Yes
1CWG4	1	IgG$_1$	140	Yes
5WG24	2	IgG$_1$	40	No
5WG26	2	IgG$_1$	40	No
5WG36	2	IgG$_{2b}$	40	No
8WG12	1	IgG$_1$	40	No
8WG21	1	NT[c]	20	No
4WG1	3	IgG$_{2b}$	17.8	Yes
4WG2	3	IgG$_{2a}$	17.8	Yes
4WG3	3	IgG$_{2b}$	17.8	Yes
8WG49	1	NT	17.8	Yes

[a] Protocol 1: Entire wheat germ RNAP II used as an immunogen. Protocol 2: Subunits transferred from SDS-PAGE to nitrocellulose; nitrocellulose dissolved with DMSO [84]. Protocol 3: Subunits <40 kD extracted from gels and renatured [85].
[b] Reacts with the homologous subunit of calf thymus RNAP II in Western blot assays.
[c] Has not been tested.

was similar to the pattern for the MAb (also an IgM) described by Christmann and Dahmus [49] that inhibited promoter-directed transcription but not nonspecific transcription [50]. However, given the information currently available regarding the unproteolyzed largest subunit and the unique C-terminal heptapeptide repeat, we felt that we could establish that our antibodies that gave this type of reaction pattern on Western blots did indeed react with the heptapeptide repeat. Such a highly repetitive domain would be likely to elicit a predominantly IgM response [51].

A peptide containing three repeats of the heptapeptide was synthesized at the University of Wisconsin Biotechnology Center. To facilitate polymerization of the subunits or conjugation of the peptide to a carrier molecule, cysteine residues were placed on each end of the peptide. Thus, the peptide had the following sequence: Cys–(Pro–Thr–Ser–Pro–Ser–Tyr–Ser)$_3$–Cys. Figure 8.3 contains a dot blot showing the reaction of several MAbs with wheat germ RNAP II and the peptide. Antibodies 3WG2, 1CWG1, and 8WG16 react with the unproteolyzed largest subunit (Fig. 8.2). Antibody 4WG2 reacts with the 17.8-kD subunit of wheat germ RNAP II. The control IgM (28.13.3S) reacts with the mouse H2-k^b

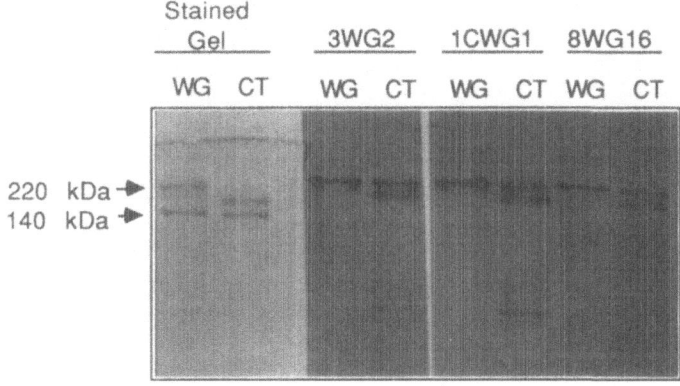

Figure 8.2. Western blots of wheat germ (WG) and calf thymus (CT) RNAP II reacted with MAbs 3WG2, 1CWG1, and 8WG16. RNAP II subunits were separated by SDS PAGE (10%) and transferred to nitrocellulose. Western blots were performed according to Blake et al. (77) using ascites fluid diluted 1:500. Major reaction of each MAb is to the 220-kD band of both polymerases. Some minor reactions to bands of lower molecular weight are seen in the calf thymus RNAP II lanes and are probably due to the presence of some degradation products in the preparation.

antigen [52]. The control IgG (2F8) reacts with the σ subunit of *E. coli* RNAP [53]. Clearly, MAbs 3WG2, 1CWG1, and 8WG16 react with the peptide, whereas the control MAbs do not. We have not yet examined the effect of phosphorylation of the peptide on reactivity with these MAbs.

Antibody 8WG8 is an unusual MAb in that it reacts with the unproteolyzed largest subunit of wheat germ RNAP II and does not react with the proteolyzed subunit (data not shown). However, it does not cross-react with calf thymus RNAP II, nor does it react with the synthetic peptide (Fig. 8.3). We believe that the epitope for this MAb might be contained in a deviation from the consensus sequence contained in the wheat germ RNAP II C-terminal domain.

When the smaller subunits (less than 40 kD) are isolated and used as immunogen (protocol 3 in Table 8.3), the predominant antibody that is isolated reacts with the 17.8-kD subunit. Some MAbs that react with this subunit have also been isolated when the entire wheat germ RNAP II molecule is used as an immunogen. Thus, of the small subunits, the 17.8-kD subunit seems to be immunodominant.

Effect of MAbs on Nonspecific Transcription

Judging from the studies listed in Table 8.2, antibodies to several different subunits of RNAP I and RNAP II have been found to inhibit nonspecific transcription.

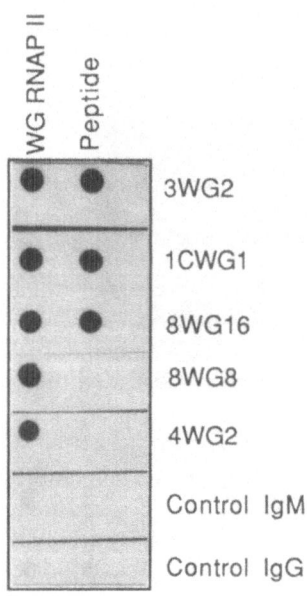

Figure 8.3. Dot blot analysis of the reaction of MAbs with wheat germ RNAP II (WG RNAP II) and the synthetic peptide. Reactions contained 100 ng wheat germ RNAP II and 1 μg of the synthetic peptide. Antigen was applied to the nitrocelluose with the use of a Bio-Dot micro-filtration apparatus (Bio-Rad Laboratories, Richmond, Calif.). Each ascites fluid was diluted 1 : 500. The control IgM (28.13.3S) reacts with the mouse H2-k^b [52] antigen, and the control IgG (2F8) reacts with the σ subunit of E. coli RNAP [53].

MAbs that react with the largest subunit (190 kD) of RNAP I from rat inhibit elongation [54]. This inhibition was affected by the NTP concentration but not the DNA concentration, indicating that the antibody might mask the NTP-binding site. MAbs that react with the second-largest subunit (132 to 135 kD) of RNAP I from yeast [55] and silkworm [56] also inhibit nonspecific transcription. Inhibition due to the yeast MAb could be partially overcome by preincubation of the enzyme with DNA before exposure to the MAb, indicating that this MAb might mask the DNA-binding site. These results seem to contradict the prokaryotic-eukaryotic

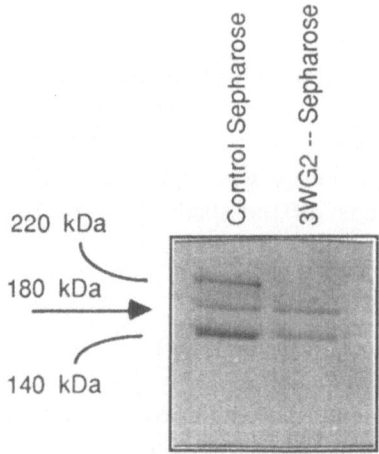

Figure 8.4. SDS PAGE analysis of the large subunits of wheat germ RNAP II that was depleted of the form containing the 220-kD subunit. RNAP II (200 μg/ml) was treated with either 3WG2-Sepharose or the control Sepharose. Sepharose was removed by centrifugation and 15 μl of the supernatant fluid was loaded onto a 10% polyacrylamide gel.

model that the largest subunit is involved with DNA binding whereas the second-largest subunit is involved with substrate binding.

The MAb described by Carroll and Stollar [57,58] that reacts with both the unproteolyzed and proteolyzed largest subunit of calf thymus RNAP II inhibited nonspecific transcription. The binding of this MAb to RNAP II was greatly reduced by preincubation of the enzyme with DNA, indicating that this MAb might mask the DNA-binding site. These results seem to support the prokaryotic-eukaryotic model that the largest subunit is involved in DNA binding.

A MAb that reacts with the smallest subunit (11 kD) of *Podospora* RNAP II [59] inhibited nonspecific transcription. It did not seem to inhibit binding to DNA, and it did not seem to affect elongation. However, it did seem to affect some step in the initiation process. Because the low-molecular-weight subunits are unique to eukaryotic RNAPs, these data are not useful in drawing analogies to the prokaryotic system.

Finally, the MAb described by Christmann and Dahmus [49] that reacted with the unproteolyzed largest subunit but not the proteolyzed largest subunit did not inhibit transcription in the nonspecific assay [50].

We have tested many of the MAbs prepared against wheat germ RNAP II (Table 8.3) for their effect on nonspecific transcription. To date, we have not been able to identify an antibody that inhibits wheat germ RNAP II in the nonspecific transcription assay. Most significantly, none of the MAbs that react with the synthetic peptide containing the hep-tapeptide repeat show any appreciable effect on nonspecific transcription.

Because most preparations of RNAP II contain either predominantly the proteolyzed form or a mixture of forms, it has not been possible to determine unequivocally that the form containing the unproteolyzed largest subunit is active in the nonspecific assay. Previous studies with yeast RNAP II [60] and calf thymus RNAP II [61] have examined fractions from DEAE chromatography that were enriched for one form over the other and, on the basis of specific activity, concluded that both forms are active in the nonspecific transcription assay. The two forms of wheat germ RNAP II are present in many preparations of purified enzyme in approximately equal proportions. We were able to use our MAbs that react with the heptapeptide repeat to deplete a preparation of the form containing the unproteolyzed subunit and assess the contribution of each form to the activity of the enzyme in nonspecific transcription assays.

Antibody 3WG2 was precipitated from ascites fluid with 45% saturated ammonium sulfate, and the precipitate was conjugated to CNBr-activated Sepharose. Control Sepharose was also prepared that was processed identically to the MAb-conjugated Sepharose except that no protein was used in the coupling buffer. Slurries of the 3WG2-Sepharose and control Sepharose were mixed with a preparation of purified wheat germ RNAP II for 2 hours at 4°C and then centrifuged to remove the Sepharose. Supernatant fluids were examined by SDS PAGE and for enzyme activity using denatured calf thymus DNA as the template. The polyacrylamide gel in Figure 8.4 shows that treatment of the enzyme with 3WG2-Sepharose removed virtually all of the form containing the 220-kD subunit. Ten microliters of the material that had been treated with the control Sepharose incorporated 33,000 CPM of ^{3}H-UTP, while 10 μl of the material that had been treated with the 3WG2-Sepharose incorporated 14,000 CPM of ^{3}H-UTP. Thus, the activity of the 3WG2-Sepharose-treated enzyme was approximately one half the activity of the control Sepharose-treated enzyme, and we conclude from this experiment that the two forms of the wheat germ RNAP II are equally active in this assay.

Therefore, if the MAbs that react with the heptapeptide repeat did affect nonspecific transcription (elongation), then a reduction in activity of approximately 50% should be noted. Because no significant inhibition has been observed with the antibodies that react with the heptapeptide repeat, we conclude that reaction of the antibodies with this domain does not affect nonspecific transcription.

Effect of MAbs on Promoter-Directed Transcription

Most of the studies listed in Table 8.2 did not test the effect of the MAbs in a promoter-directed transcription assay. The MAb examined by Dahmus and Kedinger [50] that reacted with the unproteolyzed, but not the proteolyzed, largest subunit was tested in this assay. It was found to inhibit promoter-directed transcription from several promoters that contain a TATA box including the adenovirus 2 major late promoter (Ad2 MLP); this MAb did not inhibit nonspecific transcription. The MAb described by Shastry [62] was also tested in a promoter-directed transcription assay. This MAb, like the MAb described by Kramer et al. [43,63], reacted with both the largest (unproteolyzed and proteolyzed form) and second-largest subunits. The MAb examined by Shastry [62] inhibited promoter-directed transcription from the Ad2 MLP; this MAb was not tested in nonspecific transcription assays.

Some of our MAbs that reacted with the C-terminal heptapeptide repeat were tested in this assay. Antibodies 3WG2 and 1CWG1 were purified from ascites fluid by precipitation with 45% saturated ammonium sulfate and subsequent gel filtration chromatography on Sephacryl S-300. Antibody 8WG16 was purified directly from the ascites fluid by affinity chromatography on protein A–Sepharose. Nuclear extracts were prepared from HeLa cells essentially by the method of Shapiro et al. [64]. A restriction fragment containing the Ad2 MLP was used as a template in a runoff transcription assay. Nuclear extracts were preincubated with varying concentrations of MAbs for 30 minutes at 30°C before the transcription was initiated by the addition of template and NTPs. Figure

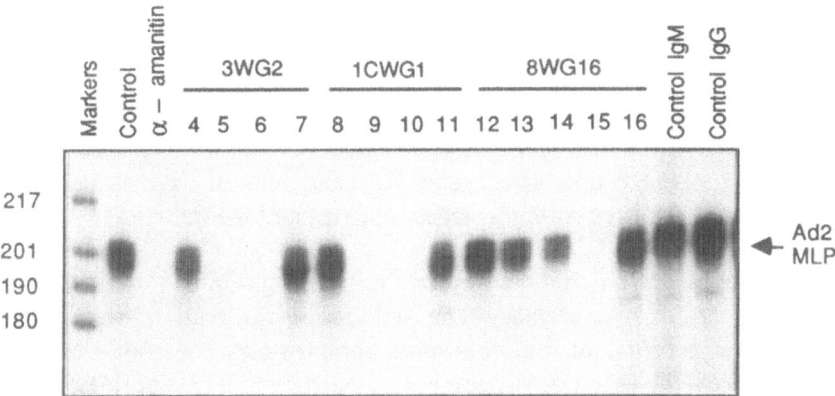

Figure 8.5. Effect of MAbs that react with the heptapeptide repeat on promoter-directed transcription from the adenovirus 2 major late promoter (Ad2 MLP). The specific transcript is 196 nucleotides. Markers (lane 1) are DNA fragments generated by the digestion of pBR322 with Msp I. Reaction in the control lane received no treatment. The reaction in the α-amanitin lane received 1 μg of the mycotoxin per milliliter. Reactions in lanes 4 through 6 received 0.1, 0.5, and 1 μg of antibody 3WG2, respectively. Lanes 8 through 10 received 0.1, 0.5, and 1.0 μg antibody 1CWG1, respectively. Reactions in lanes 12 through 15 received 0.1, 0.5, 1.0, and 2.0 μg of antibody 8WG16, respectively. Reactions in lanes 7, 11, and 16 were terminated with α-amanitin and then postincubated for 30 minutes with the highest concentration of each MAb to test for RNase activity. The control IgM and IgG are described in the legend to Figure 8.3 and used at 1 and 2 μg, respectively.

8.5 contains an autoradiogram prepared from the RNA isolated from these reactions. MAbs 3WG2 and 1CWG1 inhibited transcription from this promoter at 0.5 μg of antibody per reaction. Antibody 8WG16 (the IgG antibody) did not inhibit at this concentration, but it did inhibit at 2 μg antibody per reaction. Neither the control IgM (28.13.3S) nor the control IgG (2F8) showed any effect on transcription. The MAbs did not affect the amount of transcript present if the reaction was terminated with α-amanitin and then postincubated with the highest concentration of MAb, indicating that the absence of transcript in the inhibited reaction was not due to RNase activity in the antibody preparation.

It was of interest to determine if these MAbs could inhibit promoter-directed transcription from a promoter that did not contain a TATA box. For this analysis, we used the promoter for the mouse *dhfr* gene [65]. Antibody 3WG2 was able to inhibit runoff transcription from this promoter at approximately the same concentration that could inhibit transcription from the Ad2 MLP [48].

Effect of MAbs on Promoter-Directed Transcription in Oocytes

Because the MAbs that react with the heptapeptide repeat inhibit runoff transcription in HeLa cell nuclear extracts, we were interested to determine if the same effect can be examined in vivo. For these experiments, we used *Xenopus laevis* oocytes and the human U1 snRNA gene.

Our laboratory has been studying the transcription of the human U1 snRNA gene for several years. This gene is constitutively transcribed by all cells. The promoter region is quite complex and does not contain a TATA box [66]. When a restriction fragment containing the U1 gene is used as the template in runoff transcription assays using HeLa cell extracts, the major in vitro start is 183 bp upstream of the natural +1 start [67]. No accurate start can be detected. However, when a plasmid containing the gene is injected into *X. laevis* oocytes, the gene is transcribed accurately. Because there is no processing of this transcript and because it is stable, the transcript can be detected directly by electrophoresis if a ^{32}P-NTP is coinjected.

We have injected antibody 3WG2 into oocytes with a plasmid containing the human U1 snRNA gene. Inhibition was evident at 30 ng of 3WG2 per nucleus but not with 12.5 ng per nucleus. In addition, endogenous RNAP I and RNAP III were not inhibited by the antibody.

A plasmid containing the gene for human histone H2B has also been tested in this system. The promoter for this gene contains a typical TATA box motif. Transcription was inhibited by 30 ng of antibody 3WG2 per nucleus.

The use of oocyte injections to test the effect of a MAb has not been greatly exploited. Bona et al. [68] examined the effect of the MAb described by Krämer et al. [43,63] on the structure of lampbrush chromosomes. This MAb seemed to disrupt the integrity of these structures. However, this MAb reacted with both the largest (unproteolyzed and proteolyzed form) and the second largest subunit of *Drosophila* RNAP II. Thus, the epitope specificity of this MAb has not been established.

Role of the Heptapeptide Repeat

Several possible functions have been proposed for the unusual heptapeptide repeat on the C terminus of the largest subunit of RNAP II [27–29,69,70]. This domain might provide sites for (1) phosphorylation of the enzyme, (2) binding of transcription factors to the enzyme, or (3) association of the enzyme with chromatin or the nuclear suprastructure. All of these roles could help to regulate transcription. Deletion analysis within the C-terminal domain and subsequent in vivo analysis of the mutants have established that viability requires 10 to 13 repeats of the 26 heptapeptide repeats in the yeast molecule [70], 25 to 36 repeats of the 52 heptapeptide repeats in the mouse molecule [69], and approximately 20 repeats of the 42 repeats in the *Drosophila* molecule [30]. Thus, approximately one half of the repeats are necessary for complete enzyme function in vivo.

The MAb examined by Dahmus and Kedinger [50] and the MAbs described here inhibit promoter-directed transcription but do not inhibit the elongation function of the enzyme (nonspecific transcription). Typically, MAbs are not effective precipitating reagents, because most epitopes are present only in one copy per molecule. However, the heptapeptide repeat presents multiple epitopes. Despite this concentration of epitopes, it seems unlikely that the MAbs are simply removing the enzyme from solution by a precipitation event. The form that contains the heptapeptide repeat is active in the nonspecific transcription assay; the wheat germ preparations that we have used in nonspecific assays contain approximately 50% of this form, and the MAbs have no significant effect on nonspecific transcription.

Thus, it seems that these MAbs inhibit some step in initiation, perhaps by blocking the binding of transcription factors to the heptapeptide repeat. The MAbs inhibit promoter-directed transcription from promoters that contain a TATA box and from those that do not contain this motif. This inhibition can be demonstrated in vitro using HeLa cell extracts and in vivo using *X. laevis* oocytes.

Recently, Zehring et al. [30] were able to combine transcription factors isolated from *Drosophila* nuclear extracts with *Drosophila* RNAP II that lacked the C-terminal domain. Using the *Drosophila* actin 5C gene, they observed accurate in vitro transcription with the reconstituted system from this TATA box–containing gene. These authors concluded that the C-terminal domain was necessary for in vivo transcription but not for in vitro transcription. We have tested this conclusion by inhibiting the endogenous HeLa cell RNAP II with antibody 3WG2 and then supplementing the transcription reaction with purified calf thymus RNAP II that lacks the C-terminal domain. Preliminary results indicate that the truncated enzyme cannot substitute for the C-terminal domain–containing enzyme for accurate transcription from the mouse *dhfr* gene promoter. Transcription from the Ad2 MLP yields a discrete RNA species that seems to initiate correctly. Because the truncated form cannot transcribe from the mouse *dhfr* gene promoter, the ability of the form that lacks the domain to initiate transcription might be promoter dependent. However, as discussed by Zehring et al. [30], the presence of the domain might increase the efficiency, or regulate the usage, of all promoters.

We have examined the effect of the synthetic peptide on transcription. When this peptide is conjugated to bovine serum albumin (BSA) by glutaraldehyde cross-linking and preincubated with HeLa cell nuclear extracts, promoter-directed transcription from the Ad2 MLP and the mouse *dhfr* gene is inhibited. Glutaraldehyde-treated BSA has no effect on transcription. We have not been able to demonstrate any effect of the BSA-peptide conjugate on nonspecific transcription, using wheat germ RNAP II. The BSA-peptide conjugate thus appears to compete with the RNAP II in HeLa cell extracts.

Use of MAbs for Immunoaffinity Chromatography

RNAPs have been difficult to purify from almost any source. Most purification procedures require multiple chromatographic and concentration steps. These procedures result in low yields and reduced specific activity of the enzyme. Several of the studies referenced in Table 8.2 alluded to the possibility of applying the MAbs to immunoaffinity chromatography procedures. In fact, Gowda and Sridhara [56] were able to enrich a fraction for RNAP I by the use of their MAb.

Immunoaffinity chromatography using MAbs has two advantages over conventional chromatographic procedures. First, the extreme specificity of the appropriate antibody would greatly reduce the number of chromatographic steps required for purification. Second, the use of the appropriate cross-reacting antibody should enable one antibody column to be used to isolate RNAP of a given type from virtually any source.

We have examined some of our MAbs for use in immunoaffinity chromatography to purify RNAP II. Our approach has been to identify a cross-reacting antibody that will remove the enzyme from dilute solutions but will also release the enzyme under mild, nondenaturing conditions.

The ideal MAb for this purpose is one that reacts with the C-terminal heptapeptide repeat. This would ensure that the form of RNAP II present in the final product contains the unproteolyzed largest subunit. However, most of our MAbs that react with the heptapeptide are of the IgM isotype. Normally, IgM antibodies do not perform well in immunoaffinity chromatography because of their low intrinsic affinity for the epitope. However, in this case, the highly repetitive nature of the epitope in the antigen results in a high antigen valence and a correspondingly high avidity of the antibody for the antigen. Although we have been unable to show precipitation of RNAP II by these IgM molecules, we have been able to deplete preparations of wheat germ RNAP II prepared by the conventional method of the form containing the unproteolyzed largest subunit by reaction with antibody that had been conjugated to Sepharose (Fig. 8.4).

One of the inherent problems with immunoaffinity chromatography is recovery of the protein from the column with a minimum loss of activity.

The multimeric nature of RNAP II makes it extremely sensitive to inactivation by the reagents and conditions commonly employed for eluting antigen from immunoaffinity columns (i.e., high concentrations of chaotropic agents and extremes of pH value). The advantage of using an MAb over polyclonal antibodies for this purpose is that the homogeneous MAb population will respond uniformly to a given eluting condition. Thus, all antibodies should release the antigen, provided the appropriate release condition can be identified.

To help identify MAbs that will release the antigen under conditions that do not seem to affect enzyme activity, we have employed an enzyme-linked immunosorbent assay (ELISA) procedure to screen antibodies under various conditions. This ELISA procedure is simply a modification of a solid-phase ELISA system used to screen fusions for antigen-specific MAbs. Briefly, the antigen is coated to the wells of a polystyrene plate, and the unreactive sites are blocked. The MAb is allowed to react with the antigen. At this point, various eluting reagents are added to the wells and allowed to react for 20 minutes at room temperature. After washing extensively, the MAb that is remaining is detected with an enzyme-conjugated antibody followed by appropriate washings and substrate addition. The histogram in Figure 8.6 represents the ELISA values obtained from a typical assay. Remember, the low OD values represent the removal of MAb, which means the antigen-MAb complex has been disrupted by the eluting reagent.

We have found that this procedure is not completely representative of the behavior of a MAb when it is used in an authentic immunoaffinity chromatography procedure. For example, it seems that treatment with 50% ethylene glycol (number 3 in Fig. 8.6) disrupts the antigen-antibody reaction. In a chromatographic situation, 50% ethylene glycol will not elute the RNAP II from the column. However, 50% ethylene glycol plus 1 M NaCl (number 4 in Fig. 8.6) and 50% ethylene glycol plus 1 M ammonium sulfate (number 6 in Fig. 8.6) do elute the enzyme from the column. Therefore, we have found this system to be useful in limiting the number of possible eluting reagents and combinations of eluting reagents and conditions that need to be tested in authentic chromatographic procedures.

We have been able to apply immunoaffinity chromatography to the purification of RNAP II from wheat germ. The most useful MAb has been 1CWG3. This MAb reacts with the heptapeptide repeat. Basically, we follow the procedure of Jendrisak and Burgess [5] through the Polymin P precipitation. The enzyme is extracted from the Polymin P pellet with 0.2 M ammonium sulfate and then precipitated with 20 g of ammonium sulfate per 100 ml of material. The pellet is dissolved in 50 mM TRIS-HCl (pH 7.9) containing 0.1 mM EDTA. The material is clarified by centrifugation and applied to the 1CWG3-Sepharose. After extensive washing, the material is eluted with 1M ammonium sulfate plus 50% ethylene

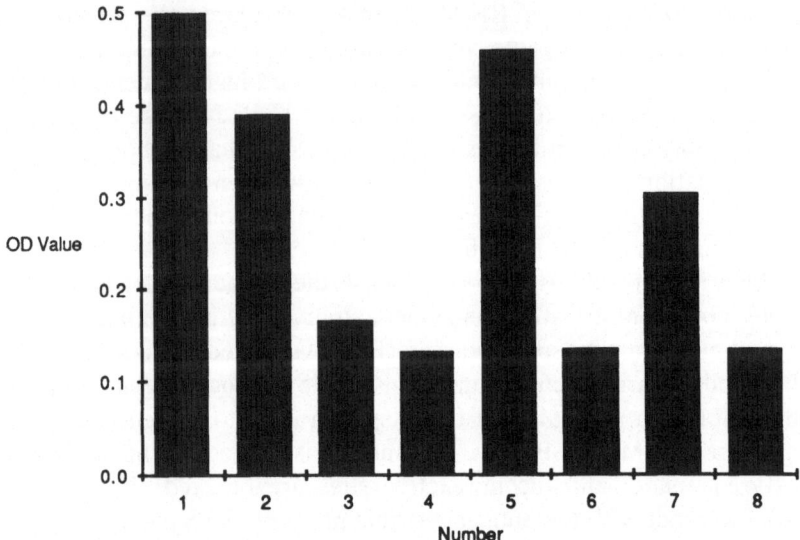

Figure 8.6. Response of the antigen-antibody reaction to various eluting reagents as determined by an enzyme-linked immunosorbent assay (ELISA). Antibody 1CWG3 was reacted with wheat germ RNAP II that was immobilized on a polystyrene plate. The antigen-antibody reaction was treated with 100 μl of (1) 50 mM TRIS-HCl, pH 7.9; (2) 50 mM TRIS-HCl + 1 M NaCl; (3) 50 mM TRIS-HCl + 50% ethylene glycol; (4) 50 mM TRIS-HCl + 1 M NaCl and 50% ethylene glycol; (5) 50 mM TRIS-HCl + 1 M ammonium sulfate; (6) 50 mM TRIS-HCl + 1 M ammonium sulfate and 50% ethylene glycol; (7) 50 mM TRIS-HCl + 2 M NaCl; and (8) 50 mM TRIS-HCl + 2M NaCl and 25% ethylene glycol. The antibody remaining after the above treatment was measured by reaction with an enzyme-labeled secondary antibody and subsequent reaction with the substrate.

glycol. An SDS PAGE gel containing wheat germ RNAP II purified in this way is shown in Figure 8.7. This enzyme preparation if active in the nonspecific transcription assay.

Because of the cross-reactivity of the MAbs that react with the heptapeptide repeat, we should be able to adapt this immunoaffinity chromatography procedure to the purification of RNAP II from virtually any source. We are currently trying to apply it to the isolation of calf thymus RNAP II. The calf thymus material has been more difficult than wheat germ, because the largest subunit is proteolyzed quickly, and most of the enzyme that can react with the immunoaffinity column is lost before the material can be reduced to a manageable volume.

Localization of RNAP II in Nuclei

Several studies described in Table 8.2 used the MAb to localize the antigen on polytene chromosomes or within the nuclei of cultured cells.

Figure 8.7. SDS PAGE of wheat germ RNAP II isolateᵈ ᵇy the conventional procedure (conventional) and by the immunoaffinity chromatography procedure using antibody 1CWG3 (1CWG3-Sepharose). The starting material for the immunoaffinity chromatography column was processed as described in the text. The immunoaffinity-purified material seems to have a contaminating subunit of approximately 20 kD. The identity of this polypeptide is not known, but it is also present when other MAbs are used in the procedure.

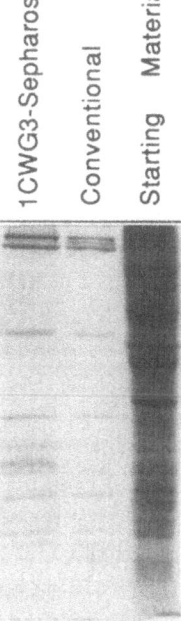

We have just initiated studies to examine the location of RNAP II within the nucleus using the MAbs that react with the heptapeptide repeat. Specifically, we are using antibody 8WG16, because this MAb is an IgG molecule, and it is easier to permeabilize the cells to entry by an IgG than it is to entry by an IgM. Our test system is the plasmodium of *Physarum polycephalum*. The advantage of this system is that the plasmodium has a closed mitosis, and the nuclei go through mitosis synchronously. Thus, we have been able to examine the specific immunofluorescent staining of this MAb through the mitotic cycle. The disadvantage of this system is that the nuclei are very small, and observation necessitates the use of oil immersion. Although the studies are preliminary, we have been able to detect specific "punctate" staining of the nuclear membranes of interphase nuclei. As the nuclei progress into mitosis, no specific staining is observed. During the S phase of the cell cycle, "punctate" staining is again evident.

Use of MAbs for Screening cDNA Libraries

A final study to which we have applied our MAbs is immunological detection of cloned subunit genes. Other investigators [71,72] have used subunit-specific polyclonal serum to screen libraries for yeast subunit genes. We are particularly interested in using our MAbs that cross-react with mammalian RNAP II to screen for human subunit genes. Cho et al. [73] reported the cloning of the largest subunit of human RNAP II using the *Drosophila* gene as a probe. However, the sequence for this gene has not been published.

Initially, we are interested in cloning the exon of the gene for the largest subunit that encodes the heptapeptide repeat. Preliminary screening of a human λ-gt11 cDNA library with antibody 8WG16 yielded a plaque that reacted strongly with this MAb. Lysogens have been prepared with this bacteriophage; after induction with IPTG, a fusion protein of approximately 150 kD is present that reacts strongly with the MAb on Western blots (compared to 116 kD for β-galactosidase). We have subcloned this insert and are currently sequencing the DNA.

Summary

This chapter reviews the literature on the use of monoclonal antibodies (MAbs) to investigate the structure and function of eukaryotic RNA polymerases and describes some of the MAbs that we have isolated using wheat germ RNAP II as the immunogen. Several conclusions can be drawn from the material presented in this chapter.

1. Several investigators have isolated MAbs that react with various subunits of RNA polymerases. Some of these show cross-reactivity with other RNA polymerases. Several investigators have examined their MAbs to determine if nonspecific transcription is inhibited in the presence of the MAb. However, no definitive conclusions can be drawn from this data regarding the role of a particular subunit in this type of transcription.
2. Using wheat germ RNAP II as an immunogen, we have been able to isolate hybridomas that produce MAbs that cross-react with RNAP II from other species. To date, MAbs that react with the 220-, 140-, and 17.8-kD subunits of wheat germ RNAP II and that cross-react with the corresponding subunits of calf thymus RNAP II have been isolated.
3. When wheat germ RNAP II is used as an immunogen, the predominant MAb-producing hybridoma that is isolated produces an IgM antibody that reacts with the heptapeptide repeat on the largest subunit of RNAP II. We have also been able to isolate one IgG molecule that reacts with this domain. Because of the apparent conservation of this domain among RNAP II, these MAbs cross-react with RNAP II from a variety of species. This also establishes that the wheat germ enzyme conforms to the consensus sequence reported for yeast and a variety of mammalian species.
4. The MAbs that react with the heptapeptide repeat have been tested in both nonspecific and promoter-directed transcription assays. These

MAbs do not inhibit nonspecific transcription from denatured calf thymus DNA. However, in extracts prepared from HeLa cell nuclei, these MAbs do inhibit transcription from promoters that contain a TATA box and those that do not contain a TATA box. These experiments confirm and extend the observations made by Dahmus and Kedinger [50] using a MAb that reacts with the unproteolyzed, but not the proteolyzed, largest subunit of calf thymus RNAP II.

5. The MAbs that react with the heptapeptide repeat have been injected into *Xenopus laevis* oocytes. In this in vivo assay, the MAbs inhibit transcription due to RNAP II, but not RNAP I or RNAP III. This seems to be a feasible system for examining the effect of well-characterized antibodies prepared against various components of the transcription machinery.

6. Using the MAbs that react with the heptapeptide repeat, we have developed an immunoaffinity chromatography procedure that can be used to isolate active enzyme from wheat germ extracts. The purified product contains only the unproteolyzed largest subunit. Because of the cross-reactivity of these MAbs, we should be able to adapt this immunoaffinity procedure to the isolation of RNAP II from virtually any source.

7. We are also using the cross-reacting MAbs to screen cDNA libraries and to visualize concentrations of RNAP II within the nucleus.

Acknowledgments. We thank Thomas Steinberg for allowing us to use the results of some unpublished experiments and for help with the promoter-directed transcription experiments. We thank Dayle Hager and Dallas Aronson for assistance with some of the experiments. This research was funded by grants 5-P30-CA07175-24 and 5-P0l-CA23076-10 from the National Cancer Institute.

References

1. Lewis, M.K., and Burgess, R.R. (1982) In The Enzymes, vol. 15B, 109–153, Boyer, P.D. (ed.). Academic Press, New York.
2. Sentenac, A. (1985) CRC Crit. Rev. Biochem. *18*, 31–90.
3. Chambon, P. (1975) Annu. Rev. Biochem. *44*, 613–638.
4. Cochet-Meilhac, M., and Chambon, P. (1974) Biochim. Biophys. Acta *353*, 160–184.
5. Jendrisak, J.J., and Burgess, R.R. (1975) Biochemistry *14*, 4639–4645.
6. Manley, J.L., Fire, A., Samuels, M., and Sharp, P.A. (1983) Methods Enzymol. *101*, 568–582.
7. Dignam, J.D., Lebovitz, R.M., and Roeder, R.G. (1983) Nucleic Acids Res. *11*, 1475–1489.
8. Reinberg, D., Horikoshi, M., and Roeder, R.G. (1987) J. Biol. Chem. *262*, 3322–3330.

9. Reinberg, D., and Roeder, R.G. (1987) J. Biol. Chem. *262*, 3310–3321.
10. Reinberg, D., and Roeder, R.G. (1987) J. Biol. Chem. *262*, 3331–3337.
11. Gurdon, J.B., and Wickens, M.P. (1983) Methods Enzymol. *101*, 370–386.
12. Banerji, J., Rusconi, S., and Schaffer, W. (1981) Cell *27*, 299–308.
13. Broyles, S.S., and Moss, B. (1987) Mol. Cell. Biol. *7*, 7–14.
14. Culotta, V.C., Wides, R.J., and Sollner-Webb, B. (1985) Mol. Cell. Biol. *5*, 1582–1590.
15. Dynan, W.S., and Tjian, R. (1985) Nature *316*, 774–778.
16. Maniatis, T., Goodbourn, S., and Fischer, J.A. (1987) Science *236*, 1237–1245.
17. Jones, N.C., Rigby, P.W.J., and Ziff, E.B. (1988) Genes Dev. *2*, 267–281.
18. Parker, C.S., and Topol, J. (1984) Cell *36*, 357–369.
19. Sawadogo, M., and Roeder, R.G. (1985) Cell *43*, 165–175.
20. Dorn, A., Bollekens, J., Staub, A., Benoist, C., and Mathis, D. (1987) Cell *50*, 863–872.
21. Dahmus, M.E. (1983) J. Biol. Chem. *258*, 3956–3960.
22. Guilfoyle, T.J., Hagen, G., and Malcolm, S. (1984) J. Biol. Chem. *259*, 649–653.
23. Schwartz, L.B., and Roeder, R.G. (1975) J. Biol. Chem. *250*, 3221–3228.
24. Cadena, D.L., and Dahmus, M.E. (1987) J. Biol. Chem. *262*, 12468–12474.
25. Dahmus, M.E. (1981) J. Biol. Chem. *256*, 3332–3339.
26. Kim, W.-Y., and Dahmus, M.E. (1986) J. Biol. Chem. *261*, 14219–14225.
27. Allison, L.A., Moyle, M., Shales, M., and Ingles, C.J. (1985) Cell *42*, 599–610.
28. Corden, J.L., Cadena, D.L., Ahearn, J.M. Jr., and Dahmus, M.E. (1985) Proc. Natl. Acad. Sci. USA *82*, 7934–7938.
29. Allison, L.A., Wong, J.K.-C., Fitzpatrick, V.D., Moyle, M., and Ingles, C.J. (1988) Mol. Cell. Biol. *8*, 321–329.
30. Zehring, W.A., Lee, J.M., Weeks, J.R., Jokerst, R.S., and Greenleaf, A.L. (1988) Proc. Natl. Acad. Sci. USA *85*, 3698–3702.
31. Biggs, J., Searles, L.L., and Greenleaf, A.L. (1985) Cell *42*, 611–621.
32. Sweetser, D., Nonet, M., and Young, R.A. (1987) Proc. Natl. Acad. Sci. USA *84*, 1192–1196.
33. Falkenburg, D., Dworniczak, B., Faust, D.M., and Bautz, E.K.F. (1987) J. Mol. Biol. *195*, 929–937.
34. Mémet, S., Gouy, M., Marck, C., Sentenac, A., and Buhler, J.-M. (1988) J. Biol. Chem. *263*, 2830–2839.
35. Broyles, S.S., and Moss, B. (1986) Proc. Natl. Acad. Sci. USA *83*, 3141–3145.
36. Armaleo, D. (1987) J. Theor. Biol. *127*, 301–314.
37. Chuang, R.Y., and Chuang, L.F. (1987) Biochem. Biophys. Res. Commun. *145*, 73–80.
38. Freund, E., and McQuire, P.M. (1986) Biochemistry *25*, 276–284.
39. Bréant, B., Huet, J., Sentenac, A., and Fromageot, P. (1983) J. Biol. Chem. *258*, 11968–11973.
40. Huet, J., Sentenac, A., and Fromageot, P. (1982) J. Biol. Chem. *257*, 2613–2618.
41. Buhler, J.-M., Huet, J., Davies, K.E., Sentenac, A., and Fromageot, P. (1980) J. Biol. Chem. *255*, 9949–9954.

42. Guilfoyle, T.J., Hagen, G., and Malcolm, S. (1984) J. Biol. Chem. *259*, 640–648.
43. Krämer, A., and Bautz, E.K.F. (1981) Eur. J. Biochem. *117*, 449–455.
44. Weeks, J.R., Coulter, D.E., and Greenleaf, A.L. (1982) J. Biol. Chem. *257*, 5884–5891.
45. Huet, J., Riva, M., Sentenac, A., and Fromageot, P. (1985) J. Biol. Chem. *260*, 15304–15310.
46. Yamazaki, K., and Imamoto, F. (1987) Mol. Gen. Genet. *209*, 445–452.
47. Weil, P.A., Luse, D.S., Segall, J., and Roeder, R.G. (1979) Cell *18*, 469–484.
48. Thompson, N.E., Steinberg, T.H., Aronson, D.B., and Burgess, R.R. (1989) J. Biol. Chem. (in press).
49. Christmann, J.L., and Dahmus, M.E. (1981) J. Biol. Chem. *256*, 11798–11803.
50. Dahmus, M.E., and Kedinger, C. (1983) J. Biol. Chem. *258*, 2303–2307.
51. Basten, A., and Howard, J.G. (1973) In Contemporary Topics in Immunology, Vol. 2, 265–291, Davis, A.J.S., and Carter, R.L. (eds.). Plenum Press, New York.
52. Ozato, K., and Sachs, D.H. (1981) J. Immunol. *126*, 317–321.
53. Strickland, M.S., Thompson, N.E., and Burgess, R.R. (1988) Biochemistry *27*, 5755–5762.
54. Rose, K.M., Maguire, K.A., Wurpel, J.N.D., Stetler, D.A., and Márquez, E.D. (1983) J. Biol. Chem. *258*, 12976–12981.
55. Huet, J., Phalente, L., Buttin, G., Sentenac, A., and Fromageot, P. (1982) EMBO J. *1*, 1193–1198.
56. Gowda, S., and Sridhara, S. (1983) J. Biol. Chem. *258*, 14532–14538.
57. Carroll, S.B., and Stollar, B.D. (1982) Proc. Natl. Acad. Sci. USA *79*, 7233–7237.
58. Carroll, S.B., and Stollar, B.D. (1983) J. Mol. Biol. *170*, 777–790.
59. Vilamitjana, J., and Barreau, C. (1987) Eur. J. Biochem. *162*, 317–323.
60. Dezélée, S., Wyers, F., Sentenac, A., and Fromageot, P. (1976) Eur. J. Biochem. *65*, 543–552.
61. Kedinger, C., and Chambon, P. (1972) Eur. J. Biochem. *28*, 283–290.
62. Shastry, B.S. (1986) Biochem. Biophys. Res. Commun. *136*, 281–287.
63. Krämer, A., Haars, R., Kabisch, R., Will, H., Bautz, F.A., and Bautz, E.K.F. (1980) Mol. Gen. Genet. *180*, 193–199.
64. Shapiro, D.J., Sharp, P.A., Wahli, W.W., and Keller, M.J. (1988) DNA *7*, 47–55.
65. Farnham, P.J., and Schimke, R.T. (1986) Mol. Cell. Biol. *6*, 2392–2401.
66. Murphy, J.T., Skuzeski, J.T., Lund, E., Steinberg, T.H., Burgess, R. R., and Dahlberg, J.E. (1987) J. Biol. Chem. *262*, 1795–1803.
67. Murphy, J.T. Burgess, R.R., Dahlberg, J.E., and Lund, E. (1982) Cell *29*, 265–274.
68. Bona, M., Scheer, U., and Bautz, E.K.F. (1981) J. Mol. Biol. *151*, 81–99.
69. Bartolomei, M.S., Halden, N.F., Cullen, C.R., and Corden, J.L. (1988) Mol. Cell. Biol. *8*, 330–339.
70. Nonet, M., Sweetser, D., and Young, R.A. (1987) Cell *50*, 909–915.
71. Riva, M., Memet, S., Micouin, J.-Y., Huet, J., Treich, I., Dassa, J., Young, R., Buhler, J.-M., Sentenac, A., and Fromageot, P. (1986) Proc. Natl. Acad. Sci. USA *83*, 1554–1558.

72. Young, R.A., and Davis, R.W. (1983) Science 222, 778–782.
73. Cho, K.W.Y., Khalili, K., Zandomeni, R., and Weinmann, R. (1985) J. Biol. Chem. 260, 15204–15210.
74. Laemmli, U.K. (1970) Nature 227, 680–685.
75. Burgess, R.R., and Jendrisak, J.J. (1975) Biochemistry 14, 4634–4638.
76. Hodo, H.G. III, and Blatti, S.P. (1977) Biochemistry 16, 2334–2343.
77. Blake, M.S., Johnson, K.H., Russell-Jones, G.J., and Gotschlich, E.C. (1984) Anal. Biochem. 136, 175–179.
78. Mann, C., Buhler, J.-M., Treich, I., and Sentenac, A. (1987) Cell 48, 627–637.
79. Ahearn, J.M. Jr., Bartolomei, M.S., West, M.L., Cisek, L.J., and Corden, J.L. (1987) J. Biol. Chem. 262, 10695–10705.
80. Barsoum, A.L., Webb, M.L., Balaban, C.D., and Jacob, S.T. (1987) J. Biol. Chem. 262, 12759–12763.
81. Tsai, S.Y., Dicker, P., Fang, P., Tsai, M.-J., and O'Malley, B.W. (1984) J. Biol. Chem. 259, 11587–11593.
82. Vilamitjana, J., Baltz, T., Baltz, D., and Barreau, C. (1983) FEBS Lett. 158, 343–348.
83. Bettiol, M.F., Irvin, R.T., and Horgen, P.A. (1985) Can. J. Biochem. Cell Biol. 63, 1217–1230.
84. Knudsen, K.A. (1985) Anal. Biochem. 147, 285–288.
85. Hager, D.A., and Burgess, R.R. (1980) Anal. Biochem. 109, 76–86.

Chapter 9
RNA Polymerase III and Transcription of 5S Ribosomal DNA

John J. Furth

Background

In 1959 when I went to work in Jerard Hurwitz's laboratory, bacterial enzymes that synthesized DNA and RNA had been discovered. There was good reason to believe that the bacterial DNA polymerase was responsible for the replication of DNA. However, it appeared unlikely that the bacterial enzyme that synthesized RNA, polynucleotide phosphorylase, was responsible for the in vivo synthesis of RNA. A template was not required, and the concentrations of ribonucleoside diphosphates required for synthesis were much higher than present in the cell. (Although no eukaryotic counterpart of the bacterial enzyme had been found, this in itself was not a serious objection).

The thrust of research in Hurwitz's laboratory was to seek, in *E. coli*, the "real" RNA polymerase. It was hypothesized that there were four enzymes, one for each of the ribonucleoside triphosphates. It was further hypothesized that an RNA template was used (mRNA was not known when these studies were initiated). As the least experienced of the four individuals in the laboratory, I was assigned ATP, the one triphosphate that was commercially available (labeled with ^{14}C). After a year of much travail I found an *E. coli* enzyme that converted ATP into RNA and required RNA for the reaction. However, to my dismay only tRNA was effective, and it was not a template but a primer. All the enzyme did was add ATP and CTP to the RNA [1].

Meanwhile the postdoctoral fellow assigned UTP was performing a parallel search. After trying many extracts and many RNA preparations, she obtained an extract that incorporated (into acid-insoluble form) all four ribonucleoside triphosphates in the presence of an "RNA." It was not long before she observed that the reaction was sensitive to DNase as well as RNase; it was DNA "contaminating" her "RNA" preparation that was the template [2]. This was the "breakthrough"; the whole lab switched to the study of this activity, and in short order it was observed that (1) with poly-dT as template only poly-rA was synthesized; (2) with the alternating copolymer of A and T as template, the product was an alternating copolymer of A and U; and (3) with DNA as template the ratio of bases in the synthesized RNA corresponded to the ratio of bases in the DNA [3–5]. Nearest-neighbor analysis, analogous to that carried out in Kornberg's laboratory [6], indicated that synthesized RNA consisted of two strands of opposite polarity and that each strand was of a polarity opposite to the polarity of the DNA template [5]. However, closer analysis suggested that not all the DNA was copied; the RNA was not an exact complement of the DNA (unpublished observations).

In science important discoveries rarely come as a single ripple but as waves. And so it was with RNA polymerase. Observations similar to those of Hurwitz's group were made in the laboratories of Audrey Stevens and Sam Weiss [7,8]. In point of fact, Sam Weiss had been the first to identify an RNA-synthesizing activity requiring triphosphates [9]. His initial observations were made in rat liver extracts, but because of difficulties (at the time) in solubilizing mammalian polymerases he was not able to characterize the activity and switched to *Micrococcus lysodeikticus*.*

Solubilization of eukaryotic RNA polymerase was soon achieved. Enzyme was obtained from a variety of sources including various rat tissues, chicken embryos, bovine lymphosarcoma, and calf thymus [12–15].

It was some time before a second set of discoveries (waves) ensued, including the demonstration that multiple eukaryotic RNA polymerases could be separated on DEAE Sephadex and that the enzymes could be distinguished by sensitivity to α-amanitin [16,17]. RNA polymerase II is inhibited at 1 μg/ml; RNA polymerase III is inhibited at 200 μg/ml; RNA polymerase I is not inhibited. (Earlier studies [18,19] had suggested that these were at least two eukaryotic RNA polymerases.)

Once the fact of multiple RNA polymerases had been established, an appreciation of their functions soon followed. RNA polymerase I (A) synthesized 18S and 28S ribosomal RNA; RNA polymerase II (B) synthesized mRNA; and RNA polymerase III (C) synthesized tRNA and 5S ribosomal RNA.

Pol III also synthesizes a number of other small RNAs, including the adenovirus-associated RNAs VA1 and VA2, analogous Epstein-Barr virus–associated RNAs EBER 1 and EBER 2, U6 RNA, and 7SK and 7SL RNA [20–26]. The VA and EBER RNAs are small and of the same

---

* Other investigators had also been in pursuit of the real RNA polymerase. One was Sol Spiegelman, who had previously reported results that, in retrospect, indicated that he also had found the enzyme [10]. Soon after our report, a publication on the purification and characterization of *E. coli* RNA polymerase appeared [11]; this procedure became the standard protocol for the isolation of the enzyme.

size. EBER 1 is 162 nucleotides (n), and EBER 2 is 172 n. VA1 RNA has been shown to promote viral growth [27 and literature cited]. U6 RNA, in a U4–U6 small ribonucleoprotein, is involved in pre-mRNA splicing [28]. The functions of 7SL and 7SK RNAs, both about 300 n, have not been established. 7SL RNA is found entirely in the cytoplasm, while 7SK RNA is both nuclear and cytoplasmic [29]. 7SK RNA is highly conserved with six base sequence differences between rat and human DNA.

Transcription by RNA Polymerase III (C)

A distinguishing feature of RNA polymerase III is that it leaches out of the nucleus and can be isolated by low salt extraction. Since it was (and still is) technically difficult to separate enzymically active enzyme from the chromatin template with which it is associated, the comparative ease with which RNA polymerase III could be isolated was the reason it was the first polymerase to be reasonably well characterized. The RNA it synthesized was small, had an intermediate sensitivity to α-amanitin, and only a subset of DNA was transcribed [30,31]. (There is a small irony in the fact that the first eukaryotic RNA polymerase to be discovered was to be named III or C.) However, although there was preferential synthesis of 5S RNA, the purified enzyme did not appear to correctly initiate transcription [32,33]. The greater selective transcription observed by Ackerman and Furth [32] was probably due to their preparation containing requisite transcription factors copurified with the enzyme (to be discussed below).

Rapid advances in our understanding of transcription came with the development of recombinant DNA technology and the demonstration by Guang-Jer Wu that a crude cell-free extract accurately initiated transcription from class III genes (wave set 3) [34].

Cell-free transcription systems were soon developed from such diverse sources as yeast, *Drosophila,* and *Xenopus* as well as human cells, and the crude extract was resolved into multiple components in addition to RNA polymerase III [35]. These components form stable transcription complexes with class III genes. Transcription of VA and tRNA genes requires factors present in 0.35 M KCl and 0.6 M KCl step eluates from phosphocellulose chromatography of S100 protein extracts (TFIII B and TFIII C fractions, respectively.) Transcription of 5S RNA requires these two factors plus an additional factor termed TFIIIA (contained in the 0.1 M KCl eluate from phosphocellulose) [reviewed in 36]. A fourth factor, required for transcription of both tRNA and 5S RNA, has recently been reported [37,38].

In addition to RNAs synthesized in vivo, Alu-type repeat DNA is transcribed in vitro. Alu repeat DNA, present in 300,000 to 500,000 copies, is interspersed throughout human DNA with an average spacing of ~4 kb [39]. A typical repeat consists of two sequences of ~300 base pairs (bp) connected by an A-rich linker. Both halves appear related to 7SL RNA [40]. Analysis of 50 Alu sequences reveals no common target or common flanking sequences [41], consistent with the suggestion of Jelinek and Schmid [42] that the Alu family arose as a result of random insertion of the pseudogene. The function, if any, of Alu repeat DNA is not known. While RNA polymerase III is likely to be involved in the evolution of Alu repeat DNA, the "purposeful" transcription in vivo of Alu repeat DNA has not been established. It has been suggested that transcription of repeat elements may influence transcription of nearby genes by Pol II [43,44].

The basis for repression of Alu repeat DNA transcription in vivo has not been elucidated. Human RNA polymerase III more efficiently transcribes the heterologous bovine Alu-type repeat DNA, either cloned or in genomic bovine DNA, than the homologous repeat [45,46]. Also, no transcription is observed of human Alu repeat DNA present in the cloned human 5S gene cluster (Fig. 9.1). These results are consistent with a

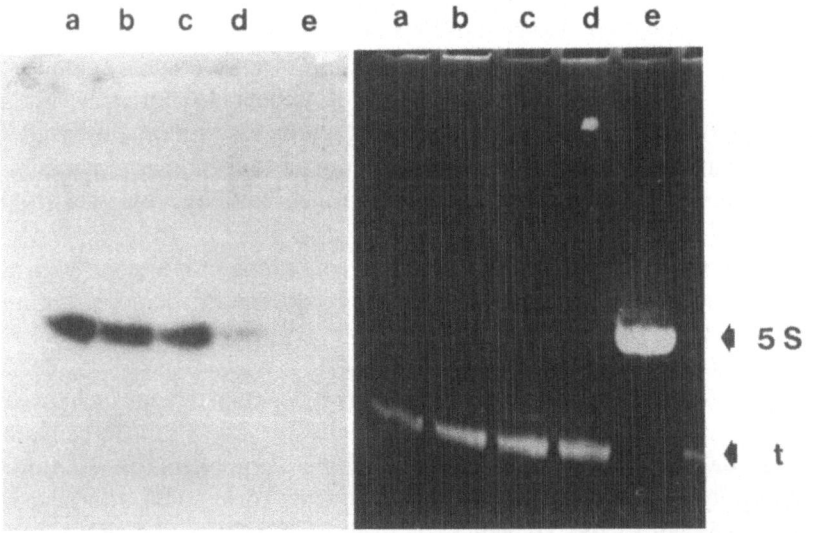

Figure 9.1. Polyacrylamide gel analysis of RNA transcribed from cloned DNA containing the human 5S ribosomal gene. Left panel—radioautograph: (a and b) cloned human 5S DNA, 0.2 μg; (c) as in lanes a and b plus 1 μg/ml α-amanitin; (d) as in lanes a and b plus 200 μg/ml α-amanitin; (e) rat 5S RNA [109], 2 μg. Right panel-same gel, stained with ethidium bromide. Cloned human 5S DNA, a gift of David Schlessinger, contains about six 5S genes and Alu repeat DNA. The 5S gene cluster has been sequenced, and it contains the complement of 5S RNA with no base substitutions [110]. Reactions were carried out as described previously with HeLa cell cytosol extract [46]. tRNA (t) was added as carrier in the isolation of the RNA.

factor, or factors, in the cell-free extract that represses transcription of homologous repeat DNA.

Based on earlier results with prokaryotic RNA polymerases, it was believed that the eukaryotic promoter would be upstream of the initiation site of transcription. However, first for the *Xenopus* 5S gene [47,48] and subsequently for transcription of tRNA genes and the adenovirus-encoded VAl gene, it appeared that promoter sequences were solely within the coding region. Upstream sequences appeared to only be involved in directing the polymer as to the precise initiating nucleotide [47,48]. For tRNA and adenovirus VA RNA genes, the internal control region is composed of two segments of DNA, usually separated by 35 bp, termed the A and B blocks [reviewed in 36,49]. Early results on transcription of the *Bombyx* 5S RNA suggesting essential 5′ regulatory elements [50] were considered a curious anomaly of the silkworm. The anomaly has become of considerable relevance, particularly in regard to 5S RNA transcription (see below).

The 5S Ribosomal Gene Family

5S ribosomal RNA, 120 nucleotides, is a constituent of the large (60S) subunit of the ribosome. A striking feature is its sequence conservation during evolution. All mammalian 5S RNAs are identical, and only about seven nucleotide differences exist between human and frog 5S RNAs. The evolutionary conservation of 5S RNA and the complete absence of viable mutants indicate that it plays a vital role in ribosome structure or function, although this role is not known.

Unlike prokaryotic 5S RNA, which is cotranscribed with 16 and 23S RNA as a large precursor molecule, eukaryotic 5S RNA is independently transcribed by RNA polymerase III as a mature ribosomal component with little or no processing. In contrast to extensive information on the organization of 5S genes in yeast, *Drosophila,* and especially *Xenopus,* comparatively little information is available on the 5S gene family in higher eukaryotes. *Xenopus laevis* contains two kinds of multigene families, oocyte *(Xlo)* and somatic *(Xls),* that encode 5S rRNA. There are over 20,000 oocyte 5S genes per haploid genome but only 400 somatic 5S rRNA genes. There are six nucleotide differences between *Xlo* and *Xls* 5S rRNA genes; the spacers are completely different except for short, conserved sequences near the 5′ and 3′ ends of the genes. Both classes are organized in clusters of tandem repeats. The somatic 5S rRNA genes are always active, while the oocyte 5S genes are not expressed after fertilization and the development of the embryo [51 and literature cited].

There are about 2000 copies of the human 5S gene, and most of these are clustered on chromosome 1 [52,53]. In Chinese hamster cells there are at least two major clusters of 5S RNA genes [55]. One cluster has been localized to 2.2-kb BamHI fragment [56]; these 5S genes are tandemly repeated. Gene variants of pseudogenes have been isolated from mouse, rat, and human DNAs [57–60]. (We distinguish 5S variant genes from 5S pseudogenes in that pseudogenes either are not transcribed in vitro or are transcribed but the RNA differs significantly in size from 5S RNA; variant genes are transcribed and the product is the same size as 5S RNA.)

A 5S gene cluster is found in a BamHl fragment of human DNA as well as in the hamster (Fig. 9.2). In *Bos,* however, Bam Hl does not demarcate a 5S gene-containing cluster. In *Bos,* a 5S gene cluster is present in a Hind III fragment, while Hind III digestion of human DNA does not result in a 5S DNA-containing fragment [46] (Fig. 9.2).

Transcription of the 5S Gene

The best-understood eukaryotic transcription system is that of the *Xenopus laevis* 5S genes. When injected into oocyte nuclei or incubated in extracts of oocyte nuclei, these genes are accurately and efficiently transcribed. While two intragenic control regions (ICRs) are required for tRNA gene transcription, a 34-bp ICR is essential and sufficient for the initiation of 5S RNA transcription [47,48]. Point mutation analysis indicates that the 5S ICR consists of three discrete elements: the 5' element (the A box of tDNA), a spacer element, and a 3' element (C box) [61,62]. In *Drosophila* the ICR contains four elements [63].

The ICR is the binding site for a 38-kD protein (TFIIIA) [64,65]. Two other proteins (TFIIIB and TFIIIC), and possibly other proteins such as TFIIID [37,38], form a transcription complex on the ICR that directs the polymerase to initiate transcription [66–69]. Polymerase III transcribes the 5S gene with the transcription complex remaining intact [70] and terminates transcription at a consensus sequence [71]. Although other factors in the complex may be involved in facilitating termination [72], purified RNA polymerase III, while not recognizing the ICR without TFIIIA, appears to recognize the termination signal [73,74].

Xenopus 5S DNA, both in genomic DNA and cloned, is transcribed by human RNA polymerase III–containing extracts. This indicates that HeLa cell TFIIIA recognizes the *Xenopus* 5S gene as well as the hamster and, probably, the bovine 5S RNA gene [75,76]. The *Xenopus* 5S RNA pseudogene is also transcribed by HeLa cell Pol III [46].

A 37-kD protein related to *Xenopus* TFIIIA has been reported in HeLa cell extracts as a 5S RNA-containing complex [76]. Recently, Seifart and colleagues have isolated and purified human TFIIIA, which they estimated to be 35 kD and have observed functional and structural differences between this protein and *Xenopus laevis* TFIIIA [77]. Human polymerase III in vitro does recognize and transcribe variant human 5S RNA genes [60] but at low efficiency. In addition to the ICR, upstream

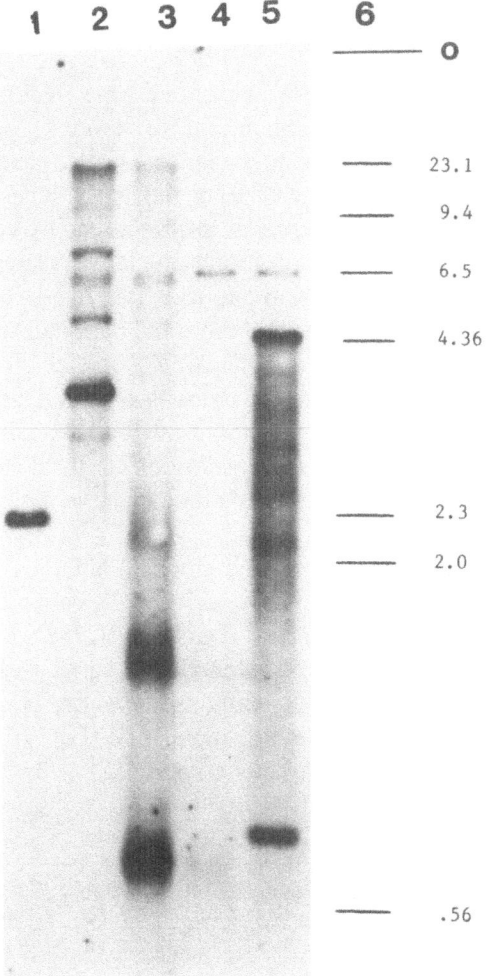

Figure 9.2. Hybridization of 5S DNA to *Xenopus laevis* and bovine DNAs digested with Hind III and to human DNA digested with Bam HI. (1) Human (HeLa cell) DNA, 4.6 μg, digested with Bam HI; (2) Bovine (calf thymus) DNA, 4.5 μg, digested with Hind III; (3) *X. laevis* DNA, 4.0 μg, Hind III; (4) *X. laevis* DNA, 0.4 μg, Hind III; (5) (control) pHU1075 DNA, 0.004 μg, Hind III; (6) (marker) bacteriophage lambda DNA, Hind III. Digested DNAs were electrophoresed on a 1.0% agarose gel, and the DNA was transferred to Gene Screen (DuPont—NEN). The probe, 0.1 μg, was a derivative of pTIII DNA (the hamster 5S gene [56]), a gift of W.R. Folk. The 5S DNA-containing fragment was introduced into bacteriophage M13mp9. Bacteriophage DNA was isolated and labeled with [32P]dCTP.

control regions first identified in *Bombyx* [50] are found in *Drosophila melanogaster* [78,79]. 5' regulatory sequences, both modulators and essential elements, are found 5' to the tRNA, 7SL and 7SK RNA genes. In the case of one polymerase III transcribed gene (7SK), the promoter can be entirely 5' to the coding region [80] [reviewed by Sollner-Webb, 81].

In transcription by RNA polymerase II, two classes of *cis*-acting elements are involved in modulating the basal level transcription observed when all but a small region of promoter DNA has been deleted. The first class are upstream promoter elements located a short distance from the transcription initiation site (the TATA box); the second class, enhancer elements, can be several thousand base pairs from the transcription initiation site. Within the 5'-flanking sequence of *Drosophila, Bombyx,* and *Neurospora* 5S genes, there is a TATA sequence the same distance from the start site. It would not be surprising, in view of the results on 7SL and 7SK RNA transcription, if distal elements control 5S RNA synthesis. These elements could be inactive in vitro or present only in genes transcribed in vivo.

5S RNA is transcribed from the cloned Bam H1 fragment of human HeLa cell DNA (Fig. 9.3). However, this 5S gene could be a variant, similar to the variant isolated by Arnold et al. [60]. Independent of our studies, a bona fide human 5S gene cluster, also present in a Bam H1 fragment, has been identified, isolated, and cloned. This 5S gene cluster is an efficient template for RNA polymerase III (Fig. 9.1).

---▷

Figure 9.3. Polyacrylamide gel analysis of RNA transcribed from plasmid DNA (pF6-28) containing the 5S RNA hybridizing fragment. (a) pF6-28 DNA, 0.2 μg, cytosol extract; (b) pF6-28 DNA, 0.2 μg, nuclear extract; (c) pF6-28 DNA, 0.2 μg, both extracts; (d) adenovirus-2-major late promoter and ε-globin gene DNAs, 0.4 μg, nuclear extract only; (e) marker: ΦX174 DNA, replicative form, digested with Hae III and labeled with [32] P. The ~2.2-kb fragment containing 5S DNA was isolated by electroelution, ligated into pBR322, Bam HI cleaved and dephosphorylated, and transformed into HB101. Transformants were screened using the isolated *Xenopus laevis* oocyte 5S gene cluster [46] labeled with random primer, Klenow fragment, and [32 P]dCTP. Although the probe did not contain vector sequences, hybridization to vector DNA was variably observed, and screening was carried out by growing up transformants, obtaining DNA by "minipreps," digesting the DNA with BamH1, electrophoresing the DNA on 1% agarose gels, and transforming the DNA to Gene Screen (Dupont—NEN). About 600 transformants were screened. Nuclear extract was obtained as described by Dignam et al. [108]. Cytosol extract was obtained as described by Weil et al. [75]. Reaction mixtures contained 110 μM ATP, 110 μM GTP, 110 μM CTP, 6.7 μM UTP (4 μCi), 5.6 mM creatine phosphate, 65 mM KCl, 8 mM MgCl$_2$, DNA, and extract(s). Analysis was on 4% acrylamide gels. This experiment was carried out in collaboration with Roberto Weinmann and Kathy Maguire, Wistar Institute.

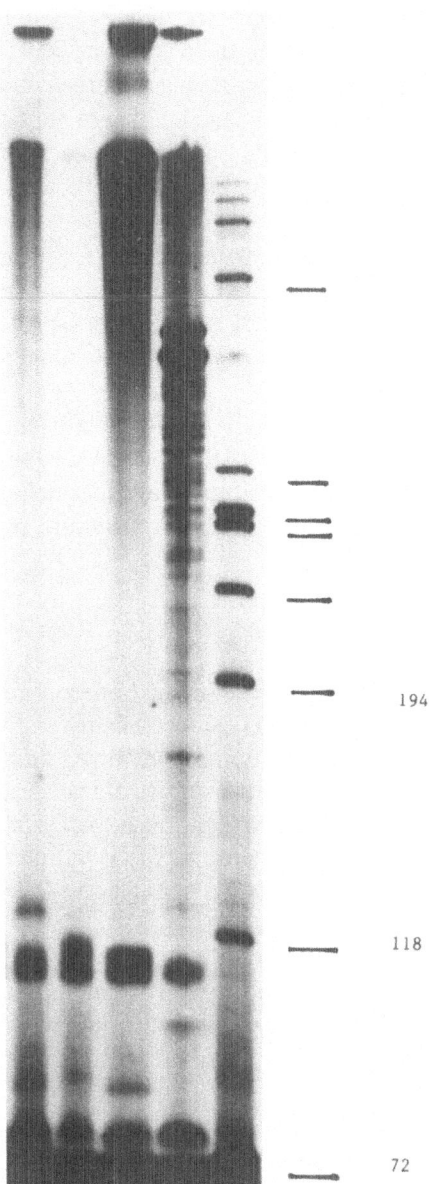

The Cloudy Crystal Ball

The 5S Gene Family

Clearly, the 5S gene family contains pseudogenes and variant genes as well as bona fide genes. Of the approximately 2,000 human 5S genes [52], many, possibly most, will be found to be gene variants or pseudogenes. It has been calculated that a rapidly growing mammalian cell would require 50 active rRNA genes per haploid genome. There may not be many more functional 5S genes than functional genes for the larger ribosomal RNAs, 300 to 400 [82–84], and it is not known whether all these genes are constitutively active.

There is no evidence that mammalian 5S DNA is composed of classes differentially transcribed during development (as in *Xenopus*). In mammalian cells TFIIIA may be stored as a 5S RNA-protein complex [76], suggesting some parallels in the synthesis of ribosomal RNAs in *Xenopus* and mammals.

Repression of transcription of variant and pseudogenes could be due to protein factors in the homologous extract, as suggested above, but could also be due to a positive transcription factor that opens up the chromatin structure analogous to the role chromatin structure plays in permitting transcription of 5S genes [70].

Maintenance of the Differentiated State

In terminally differentiated cells a set of genes remains active while another set remains repressed. The mechanism for this has been suggested by the studies on transcription of the oocyte and somatic 5S genes in *Xenopus*. Based on their extensive studies of transcription of these two sets of genes, Wolffe and Brown propose that stable repression is imposed by regularly spaced nucleosomes that interact with histone HI, preventing the repressed gene from becoming accessible to free transcription factors [51].

Several studies suggest that the potential for gene expression may be correlated with the time of replication. All constitutively active genes examined in several cell types are early replicating, while genes expressed only in specialized tissues—e.g., fibroblasts—appear to be late replicating [85].

Replication disrupts both the active and the repressed state. Wolffe and Brown [51] suggest that in somatic cells, a somatic 5S RNA gene forms a stable complex even at low levels of transcription factors; the oocyte 5S RNA gene forms unstable complexes requiring a high level of transcription factors to maintain activity. Formation of an active somatic 5S gene complex and formation of an inactive oocyte 5S gene complex could result from early replication of active genes in S phase and late replication of inactive genes. Consistent with this theory, somatic 5S RNA genes appear to be replicated early in S phase, while (inactive) oocyte 5S RNA genes are replicated late [86]. Early-replicating somatic genes could deplete a limiting transcription factor.

Coordinate Regulation

Coordinate induction, repression, and regulation of genes is a common phenomenon. For example, extracellular matrix components by fibroblasts are usually, but not invariably, synthesized pari passu [87,88]. This coordination is almost certainly due to both *(cis)* regulatory DNA sequences and gene-specific *trans*-acting protein factors. With linked genes, coordinate regulation can be achieved by the sharing of a common promoter, as is the case for the α-1(IV) and α-2(IV) collagen genes [89–91]. Coordinate regulation of unlinked genes such as the genes for α-1(I) and α-2(I) collagen (which are located on chromosomes 7 and 17, respectively [92]) almost certainly involve *trans*-acting regulatory proteins. Another example is regulation of expression of histone genes [93].

Coordinate regulation of histone and collagen gene expression, as for most protein-coding genes, involves only one RNA polymerase (Pol II). Coordinate regulation of the synthesis of the components of the ribosomes would involve interaction with all three RNA polymerases—Pol I for the transcription of the large ribosomal RNAs, Pol II for the transcription of the ribosomal proteins (rps), and Pol III for the transcription of 5S RNA. The four RNAs and over 70 different proteins that make up the ribosome appear to be synthesized in approximately equal amounts over a wide range of physiologic conditions. In bacteria and lower eukaryotes, this coordinate regulation is controlled by a combination of transcriptional, translational, and posttranscriptional mechanisms [94,95]. Just as the synthesis of extracellular matrix components is not invariably coordinately regulated, the synthesis of the components of the ribosome may be uncoordinated. Uncoordinated synthesis occurs in growth arrest; cyclosporin A–treated mouse lymphosarcoma cells lack the ability to initiate transcription of the large ribosomal RNAs in vitro, while transcription of 5S genes is not affected [96]. We take these results to reflect disturbances of cellular homeostasis akin to a disease state.

Coordinate regulation does not mean that the RNAs are synthesized in equal amounts. In HeLa cells, 5S RNA has been reported to be synthesized in amounts four times greater than required for synthesis of ribosomal subunits [97]. In *Xenopus* development there is a vast disparity in the synthesis of 5S and the larger ribosomal RNAs [51, and literature cited]. The relevant comparison would be the synthesis of 5S RNA and the larger ribosomal RNAs in somatic cells.

Multiple and clustered ribosomal genes are found on the secondary constrictions of chromosomes 13, 14, 15, 21, and 22. The amount of rDNA on different chromosomes varies within and among individuals. The 44-kb rDNA repeat unit contains a 13-kb transcribed portion and 31-kb spacer, 20 kb of which has been sequenced. The 13-kb RNA transcript is processed to yield the 18S, 5.8S, and 28S mature RNA molecules. The transcribed portion of human rDNA is conserved. There is strong homology between mouse and human ribosomal RNA, although there is almost no homology in the spacer regions. The spacer serves multiple functions in directing rRNA transcription and may be partially transcribed [82–84]. Sequences in the spacer may be involved in coordinate regulation. Studies in which protein synthesis is inhibited concomitant with a great decrease in rRNA synthesis suggest that ancillary short-lived proteins are required for the synthesis of both 5S RNA and ribosomal RNA [98]. These proteins may be involved in coordinate regulation. In growth-arrested cells, transcription by both Pol III and Pol I is substantially decreased. Decrease in Pol III transcription results from a reduction in TFIIIB activity; decrease in Pol I activity results from a decrease in the activated subform of Pol I [99]. A common regulatory factor would have to have these two disparate effects.

All three eukaryotic RNA polymerases contain two large subunits that are unique and 4 to 12 smaller subunits (<85 kD), some of which may be shared [100]. In yeast, Pol I and Pol III have been shown to share a common 40-kb subunit [101]. Common subunits may also be involved in coordinate regulation.

Rearrangement of nucleolus organizer regions (NORs) can occur in malignancy, and the existence of additional, presumably inactive, NOR genes has been proposed in centromeric heterochomatin bands, particularly of chromosome 1 [102]. While evidence for additional NORs in the distal part of lq in humans is lacking, it is interesting that a major locus of 5S genes is lq41→q42, corresponding to the site of ectopic Ag-NORs in two of the tumors studied. rDNA is transcribed, the rRNA is processed, and much of the ribosome assembly takes place in nucleoli, and it is particularly significant that 5S rRNA genes are located at the nucleoli during interphase [103]. Taken together, these studies are consistent with coordinated regulation of 5S genes and the larger ribosomal genes.

The chromosomal location of the genes for ribosomal proteins has not been well established, in part because of multiplicity of rp pseudogenes. They appear to be on several chromosomes [104]. Coordinate regulation of the synthesis of ribosomal proteins with the synthesis of the RNAs could involve additional mechanisms, since (1) RNA polymerase II appears to have less in common than polymerases I and III (in terms of subunits), (2) protein stability could be involved, and (3) ribosome assembly takes place in the nucleolus and rp mRNAs are translated in the cytoplasm and the proteins return to the nucleolus.

The Opaque Crystal Ball

We conclude that only a small portion of the human 5S gene family consists of authentic genes. These are probably located on chromosome 1, although it is not inconceivable that bona fide 5S genes are on another chromosome. If, as is likely, the authentic 5S genes are on a different chromosome from the transcribed ribosomal genes (or on the same chromosome but not closely linked), coordinate regulation would involve *trans*-acting protein factors.

The mechanism for coordinate regulation could be analogous to that postulated for the histone genes. The human histone genes are found on multiple chromosomes—1, 6, and 12 [107]. Induction of the H2B gene is mediated by a subtype-specific element containing a core octamer sequence not common to the various histone genes [93]. Coordinate induction of the histone genes could result from a common factor binding to factors specific for the individual histone genes. The common factor could affect binding of the histone-specific factor to its specific DNA element. There need not be common sequences in the histone genes, and control could be positive or negative.

Coordinate regulation of ribosomal RNA synthesis could be by a similar mechanism. There need not be sequences common to the 5S genes and the genes for the larger ribosomal RNAs (and, possibly the ribosomal protein genes). A common factor could bind to factors specific for 5S transcription and to factors specific for the genes for transcription of the larger RNAs.

Transcriptionally active ribosomal DNA is present in a nucleosomelike array, although when visualized in the electron microscope, nucleosomelike beads are not observed, and the nucleosomelike array is less compacted than normal nucleosomal DNA [105]. These two forms of nucleosomes have been termed static and dynamic, respectively [106]. Binding of factors specific for the two classes of RNA genes would convert that area of chromatin from static to dynamic, opening up the chromatin structure and allowing transcription to occur. Opening up chromatin structure could be due to factors binding at a considerable distance from the 5S genes, as a considerable distance from genes in DNA is not necessarily a big distance in the cell because of the compaction of DNA in chromatin and its further compaction in nuclei.

Summary

In this chapter we have rather eclectically reviewed transcription by RNA polymerase III, noting that at least four protein factors in addition to the many subunits of the enzyme are required for correct initiation of transcription. The genes coding for RNAs synthesized by RNA polymerase III have been conserved in evolution; there are many pseudogenes. RNAs synthesized by RNA polymerase III appear to be involved in protein synthesis, transport, or splicing of mRNA; most are not translated.

The 5S gene family consists of many variant and pseudogenes and only a few bona fide genes. Transcription of bona fide 5S genes involves at least one, probably more, protein factors in addition to factors required for transcription of all class III genes. Transcription of 5S RNA is normally coordinated with transcription of other components of the ribosome. We suggest this is due to a common protein factor binding to gene-specific transcription factors which in turn bind to (cis) regulatory DNA sequences in the respective genes. This binding domain could be far outside (5' or 3') the RNA coding region.

Acknowledgments. I thank those investigators who sent me reprints and/or preprints, Steve Ackerman for reading an early version of this chapter, and Patti Roomet for word processing. The experimental results reported were supported by a grant from the Research Foundation of the University of Pennsylvania and were carried out in collaboration with Roy Schmickel and Jim Sylvester, Department of Human Genetics, University of Pennsylvania School of Medicine. Deven Parekh, a University Scholar of the University of Pennsylvania, participated in the isolation of the human 5S DNA containing plasmid.

References

1. Furth, J.J., Hurwitz, J., Krug, R., and Alexander, M. (1961) J. Biol. Chem. *236*, 3317–3322.
2. Hurwitz, J., Bresler A., and Diringer, R. (1960) Biochem. Biophys. Res. Commun. *3*, 15–19.
3. Furth, J.J., Hurwitz, J., and Goldmann, M. (1961) Biochem. Biophys. Res. Commun. *4*, 362–367.
4. Furth, J.J., Hurwitz, J., and Anders, M. (1962) J. Biol. Chem. *237*, 2611–2619.
5. Hurwitz, J., Furth, J.J., Anders, M., and Evans, A. (1962) J. Biol. Chem. *237*, 3752–3759.
6. Josse, J., Kaiser, A.D., and Kornberg, A. (1961) J. Biol. Chem. *236*, 864–875.
7. Stevens, A. (1960) Biochem. Biophys. Res. Commun. *3*, 92–96.
8. Weiss, S.B., and Nakamoto, T. (1961) Proc. Natl. Acad. Sci. USA *47*, 1400–1465.
9. Weiss, S.B., and Gladstone, L. (1959). J. Am. Chem. Soc. *81*, 4118–4119.
10. Spiegelman, S. (1958) In Recent Progress in Microbiology, 81–103, Tunevall, T.G. (ed.). Thomas, Springfield, Illinois.
11. Chamberlin, M., and Berg, P. (1962) Proc. Natl. Acad. Sci. USA *48*, 81–94.
12. Furth, J.J., and Loh, P. (1963) Biochem. Biophys. Res. Commun. *13*, 100–105.

13. Furth, J.J., and Ho, P. (1965) J. Biol. Chem. *240*, 2602–2606.
14. Ramuz, M., Doly, J., Mandel, P., and Chambon, P. (1965) Biochem. Biophys. Res. Commun. *19*, 114–120.
15. Ballard, P., and Williams-Ashman, H.G. (1966) J. Biol. Chem. *241*, 1602–1615.
16. Roeder, R.G., and Rutter, N.J. (1970) Proc. Natl. Acad. Sci. USA *65*, 675–682.
17. Wieland, T.H. (1968) Science *159*, 946–952.
18. Widnell, C.C., and Tata, J.R. (1966) Biochim. Biophys. Acta *123*, 478–492.
19. Chambon, P., Ramuz, M., Mandel, P., and Doly, J. (1968) Biochim. Biophys. Acta *157*, 504–519.
20. Zylber, E.A., and Penman, S. (1971) Proc. Natl. Acad. Sci. USA *68*, 2861–2865.
21. Suzuki, Y., and Giza, P.E. (1976) J. Mol. Biol. *107*, 183–206.
22. Weinmann, R., and Roeder, R.G. (1974) Proc. Natl. Acad. Sci. USA *71*, 1790–1794.
23. Weil, P.A., and Blatti, S.P. (1976) Biochemistry *15*, 1500–1509.
24. Reddy, R., Henning, D., Das, G., Harless, M., and Wright, D. (1987) J. Biol. Chem. *262*, 75–81.
25. Zieve, G., Benecke, B.J., and Penman, S. (1977) Biochemistry *16*, 4520–4525.
26. Reichel, R., and Benecke, B.-J. (1980) Nucleic Acids Res. *8*, 225–234.
27. Glickman, J.N., Howe, J.G., and Steitz, J.A. (1988) J. Virol. *62*, 902–911.
28. Black, D.L., and Steitz, J.A. (1986) Cell *46*, 697–704.
29. Krüger, W., and Benecke, B.-J. (1987) J. Mol. Biol. *195*, 31–41.
30. Furth, J.J., and Austin, G.E. (1970) Cold Spring Harbor Symp. Quant. Biol. *35*, 641–648.
31. Atikkan, E.E., and Furth, J.J. (1977) Cell Differ. *6*, 253–262.
32. Ackerman, S., and Furth, J.J. (1979) Biochemistry *18*, 3243–3248.
33. Parker, C.S., and Roeder, R.G. (1977) Proc. Natl. Acad. Sci. USA *74*, 44–48.
34. Wu, G.-J. (1978) Proc. Natl. Acad. Sci. USA *75*, 2175–2179.
35. Segall, J., Matsui, T., and Roeder, R.G. (1980) J. Biol. Chem. *255*, 11986–11991.
36. Lasser, A.B., Martin, P.L., and Roeder, R.G. (1983) Science *222*, 740–748.
37. Ottonello, S., Rivier, D.H., Doolittle, G.M., Young, L.S., and Sprague, K.U. (1987) EMBO J. *6*, 1921–1927.
38. Yoshinaga, S.K., Boulanger, P.A., and Berk, A.J. (1987) Proc. Natl. Acad. Sci. USA *84*, 3585–3589.
39. Hwu, H.R., Roberts, J.W., Davidson, E.H., and Britten, R.J. (1986) Proc. Natl. Acad. Sci. USA *83*, 3875–3879.
40. Ullu, E., and Tschudi, C. (1984) Nature *312*, 171–172.
41. Kariya, Y., Kato, K., Hayashizaki, Y., Himeno, S., Tarui, S., and Matsubara, K. (1987) Gene *53*, 1–10.
42. Jelinek, W.R., and Schmid, C.W. (1982) Annu. Rev. Biochem. *51*, 813–844.
43. Carlson, D.P., and Ross, J. (1986) Mol. Cell Biol. *6*, 3278–3282.
44. McKinnon, R.D., Shinnick, T.M., and Sutcliffe, J.G. (1986) Proc. Natl. Acad. Sci. USA *83*, 3751–3755.
45. Furth, J.J. (1985) Biochem. Biophys. Res. Commun. *131*, 551–556.
46. Furth, J.J., and Su, C.-Y. (1986) Biochem. J. *237*, 827–835.

47. Sakonju, S., Bogenhagen, D.F., and Brown, D.D. (1980) Cell *19*, 13–25.
48. Bogenhagen, D.F., Sakonju, S., and Brown, D.D. (1980) Cell *19*, 27–35.
49. Ciliberto, G., Castagnoli, L., and Cortese, R. (1983) Curr. Topics Dev. Biol. *18*, 59–88.
50. Morton, D.G., and Sprague, K.U. (1984) Proc. Natl. Acad Sci. USA *81*, 5519–5522.
51. Wolffe, A.P., and Brown, D.D. (1988) Science *241*, 1625–1632.
52. Hatlen, L., and Attardi, G. (1971) J. Mol. Biol. *56*, 535–553.
53. Steffensen, D.M., Duffey, P., and Prensky, W. (1974) Nature *252*, 741–743.
54. Atwood, K.C., Yu, M.T., Johnson, L.D., and Henderson, A.S. (1975) Cytogenet. Cell Genet. *15*, 50–54.
55. Stambrook, P.J. (1976) Nature *259*, 639–641.
56. Hart, R.P., and Folk, W.R. (1982) J. Biol. Chem. *257*, 11706–11711.
57. Emerson, B.M., and Roeder, R.G. (1984) J. Biol. Chem. *259*, 7916–7925.
58. Doran, J.L., Bingle, W.H., and Roy, K.L. (1987) Nucleic Acids Res. *15*, 6297.
59. Reddy, R., Henning, D., Rothblum, L., and Busch, H. (1986) J. Biol. Chem. *261*, 10618–10623.
60. Arnold, G.J., Kahnt, B., Herrenknecht, K., and Gross, H.J. (1987) Gene *60*, 137–144.
61. McConkey, G.A., and Bogenhagen, D.F. (1987) Mol. Cell. Biol. *7*, 486–494.
62. Peiler, T., Hamm, J., and Roeder, R.G. (1987) Cell *48*, 91–100.
63. Sharp, S.J., and Garcia, A.D. (1988) Mol. Cell. Biol. *8*, 1266–1274.
64. Engelke, D.R., Ng, S.-Y., Shastry, B.S., and Roeder, R.G. (1980) Cell *19*, 717–728.
65. Sakonju, S., Brown, D.D., Engelke, D., Ng, S.-Y., Shastry, B.S., and Roeder, R.G. (1981) Cell *23*, 665–669.
66. Shastry, B.S., Ng, S.-Y., and Roeder, R.G. (1982) J. Biol. Chem. *257*, 12979–12986.
67. Bieker, J.J., Martin, P.L., and Roeder, R.G. (1985) Cell *40*, 119–127.
68. Carey, M.F., Gerrard, S.P., and Cozzarelli, N.R. (1986) J. Biol. Chem. *261*, 4309–4317.
69. Setzer, D.R., and Brown, D.D. (1985) J. Biol. Chem. *260*, 2483–2492.
70. Bogenhagen, D.F., Wormington, W.M., and Brown, D.D. (1982) Cell *28*, 413–421.
71. Bogenhagen, D.F., and Brown, D.D. (1981) Cell *24*, 261–270.
72. Rinke, J., and Steitz, J.A. (1982) Cell *29*, 149–159.
73. Cozzarelli, N.R., Gerrard, S.P., Schlissel, M., Brown, D.D., and Bogenhagen, D.F. (1983) Cell *34*, 829–835.
74. Watson, J.B., Chandler, D.W., and Gralla, J.D. (1984) Nucleic Acids Res. *12*, 5369–5384.
75. Weil, P.A., Segall, J., Harris, B., Ng, S.-Y., and Roeder, R.G. (1979) J. Biol. Chem. *284*, 6163–6173.
76. Lagaye, S., Barque, J.-P., LeMaire, M., Denis, H., and Larsen, C.-J. (1988) Nucleic Acids Res. *16*, 2473–2487.
77. Seifart, K.H., Wang, L., Waldschmidt, R., Jahn, D., and Wingender, E. (1989) J. Biol. Chem. *264*, 1702–1709.
78. Selker, E.U., Morzycka-Wroblewska, L., Stevens, J.N., and Metzenberg, R.L. (1986) Mol. Gen. Genet. *205*, 189–192.

79. Garcia, A.D., O'Connell, A.M., and Sharp, S.J. (1987) Mol. Cell. Biol. 7, 2046–2051.
80. Murphy, S., Di Liegro, C., and Melli, M. (1987) Cell 51, 81–87.
81. Sollner-Webb, B. (1988) Cell 52, 153–154.
82. Wilson, G.N. (1982) In The Cell Nucleus, 287–318, Vol. X, Busch, H. (ed.). Academic Press, New York.
83. Baker, S.M., and Platt, T. (1986) Cell 47, 839–840.
84. Worton, R.G., Sutherland, J., Sylvester, J.E., Willard, H.F., Bodrug, S., Dube, I., Duff, C., Kean, V., Ray, P.N., and Schmickel, R.D. (1988) Science 239, 63–68.
85. Taylor, J.H. (1987) Results Probl. Cell Differ. 14, 173–196.
86. Guinta, D.R., and Korn, L.J. (1986) Mol. Cell. Biol. 6, 2536–2542.
87. Cutroneo, K.R., Sterling, K.M., and Shull, S. (1986) In Regulation of Matrix Accumulation, 119–176, Mecham, R.P. (ed.). Academic Press, New York.
88. Myers, J.C., Howard, P.S., Jelen, A.M., Dion, A.S., and Macarak, E.J. (1987) J. Biol. Chem. 262, 9231–9238.
89. Pöschl, E., Pollner, R., and Kühn, K. (1988) EMBO J. 7, 2687–2695.
90. Soininen, R., Huotari, M., Hostikka, S.L., Prockop, D.J., and Tryggvason, K. (1988) J. Biol. Chem. 263, 17217–17220.
91. Burbelo, P.D., Martin, G.R., and Yamada, Y. (1988) Proc. Natl. Acad. Sci. USA 85, 9679–9682.
92. Myers, J.C., and Emanuel, B.S. (1987) Collagen Rel. Res. 7, 149–159.
93. Fletcher, C., Heintz, N., and Roeder, R.G. (1987) Cell 51, 773–781.
94. Meyuhas, O., Thompson, E.A., and Perry, R.P. (1987) Mol. Cell. Biol. 7, 2691–2699.
95. Rhoads, D.D., and Roufa, D.J. (1987) Mol. Cell. Biol. 7, 3767–3774.
96. Mahajan, P.B., and Thompson, E.A. (1987) J. Biol. Chem. 262, 16150–16156.
97. Leibowitz, R.D., Weinberg, R.A., and Penman, S. (1973) J. Mol. Biol. 73, 139–144.
98. Gokal, P.K., Cavanaugh, A.H., and Thompson, E.A. (1986) J. Biol. Chem. 261, 2536–2541.
99. Tower, J., and Sollner-Webb, B. (1988) Mol. Cell. Biol. 8, 1001–1005.
100. Allison, L.A., Moyle, M., Shales, M., and Ingles, C. (1985) Cell 42, 599–610.
101. Mann, C., Buhler, J.-M., Treich, I., and Sentenac, A. (1987) Cell 49, 627–637.
102. DeLozier-Blanchet, C.D., Walt, H., and Engel, E. (1986) Cytogenet. Cell Genet. 41, 107–113.
103. Amaldi, F., and Buongiorno-Nardelli, M. (1971) Exp. Cell Res. 65, 329–334.
104. Human Gene Mapping 9 (1987) Cytogenet. Cell Genet. 46, 1–762.
105. Culotta, V., and Sollner-Webb, B. (1988) Cell 52, 585–597.
106. Kmiec, E.B., Ryoji, M., and Worcel, A. (1986) Proc. Natl. Acad. Sci. USA 83, 1305–1309.
107. Tripputi, P., Emanuel, B.S., Croce, C.M., Green, L.G., Stein, G.S., and Stein, J. L. (1986) Proc. Natl. Acad. Sci. USA 83, 3185–3188.
108. Dignam, J.D., Lebovitz, R.M., and Roeder, R.G. (1983) Nucleic Acids Res. 11, 1475–1489.
109. Ackerman, S., Keshgegian, A.A., Henner, D., and Furth J.J. (1979) Biochemistry 18, 3232–3242.
110. Little, R.D., and Braaten, D.C. (1989) Genomics 4, 376–383.

Chapter 10
Structure and Function of Rho Factor and Its Role in Transcription Termination

Alicia J. Dombroski and Terry Platt

One step in the regulation of gene expression that can be controlled by a variety of mechanisms is the transcription of messenger RNA from the DNA template. It is becoming increasingly apparent that an important aspect of such regulation involves transcription termination and its control. Termination sites at the ends of genes or operons provide a signal for RNA polymerase to cease transcribing, thus defining the 3' end of the primary transcript. Termination signals that reside at other positions in the transcript can also regulate gene expression through the effects of attenuation, retroregulation, and polarity [for reviews see 1,2]. One class of transcription termination events utilized by *Escherichia coli* and various bacteriophages is dependent upon the action of a protein factor called rho, which participates in the recognition of and response to certain termination sites. Though little is understood about the RNA sequence or structural elements required for rho recognition, a great deal has been learned recently about the structure of rho protein itself. This review will focus on the molecular architecture of rho protein with an emphasis on its domain structure, the interaction of each domain with its substrate, and the interactions between domains that are required for catalysis of transcription termination.

Termination Signals

Signals for transcription termination in *E. coli* can be categorized as "simple," or factor-independent, and "complex," or factor-dependent. The simple terminators are characterized by a GC-rich region of dyad symmetry followed by several uridines near the 3' end of the transcript. The stem-loop structure that forms in the RNA causes RNA polymerase to pause, and the transcript is subsequently released owing to the exceptional instability of rU-dA base pairing [for reviews see 3–5].

The best-characterized factor-dependent terminators rely on the action of rho protein, first isolated and characterized by Roberts [6]. A number of articles and reviews have discussed the signals required for rho-dependent termination in detail [1,7–11], so we will only summarize this area. In brief, the signal for rho-dependent termination appears to be complex and consists of a region lacking in secondary structure but rich in cytosine residues that is located in an untranslated region of RNA, 60 to 125 nucleotides upstream from the 3′ end points. Attempts to further define recognition signals in the nascent transcripts of rho-dependent terminators have not revealed any consensus sequences or obvious secondary structures common to all rho-dependent sites [2,8]. Among the few rho-dependent terminators that have been identified, lambda t_{R1} and *E. coli trp t′* have been examined in the most detail. Extensive deletion analysis of the *trp t′* terminator has not revealed a specific sequence element required for termination [12,13]. Similar examination of the lambda t_{R1} terminator [9,14] found its effectiveness to be dramatically reduced by mutations in two pyrimidine-rich regions [14]. A synthetic terminator composed of the sequence $(TCTTC)_{76}$ acts as a very efficient terminator in vitro and suggests that rho recognizes and binds to a polypyrimidine tract [11]. Cytosine residues in particular appear to play an important role [15,16], but beyond the generalizations outlined here, the specific requirements for rho recognition remain elusive.

Rho Protein and Its Domain Structure

History and Genetics

Rho protein was first discovered and characterized as a termination and release factor in an in vitro transcription reaction using bacteriophage lambda DNA [6]. The *rho* gene, located at 84′ on the *E. coli* chromosome [17], has been cloned [18] and sequenced [19]. The unusually long (250-bp) leader region of the gene has been implicated in autoregulation of *rho* expression through an attenuation mechanism [18,20–22]. Numerous mutations in *rho* have been identified, using suppression of mutational polarity as a selectable phenotype. Polarity can be defined as the reduction of gene expression in an operon distal to a site where premature termination of translation has occurred, leading to reduced expression of distal genes [23,24]. Mutations in rho protein have been found to suppress this polarity, which is presumably caused by transcription termination at latent rho-dependent sites that have been exposed by a lack of translation within normally translated genes [for review, see 2]. The mutant *rho* alleles *rho115, rho201, rho221,* and *rho1*(SuA) were all isolated as such polarity suppressors [25–28].

The protein encoded by *rho115* is temperature sensitive for RNA binding in the absence of ATP [29,30] and demonstrates reduced termination efficiency in vitro at the rho-dependent site *trp t′* [30]. *Rho201* drastically reduces termination at the end of the *trp* operon in a *trp-lac* fusion strain [28], but little is known of its characteristics in vitro. *Rho1* (SuA) is defective in interactions with RNA that are coupled to ATP hydrolysis [31]. This mutation is partially dominant and can interfere with the function of wild-type rho by exchange of subunits [32]. *Rho221* behaves similarly to wild-type rho for transcription termination in vitro and was found to have an isoleucine to serine substitution at position 382 [Grant, R.A., Sullivan, K., Ackerson, J., and Platt, T., unpublished results].

Other mutations in rho have been identified based on a variety of pleiotropic effects [33]. *Rho026,* which was selected as defective for coliphage T4 growth [34], has also been shown to exhibit hyperdegradation of abnormal proteins [35] and to inhibit the N protein activity of phage lambda [36]. *Rhots15* carries an IS1 insertion in the carboxyterminal segment of the gene that truncates the polypeptide by 2 kD in size and confers a temperature-sensitive phenotype [37]. *Rhots702* renders the protein extremely labile and severely defective in RNA binding and transcription termination [38,39]. Some revertants of this strain that are hyperactive for termination have been characterized; these have been designated *rho^s-77, rho^s-81,* and *rho^s-82* [40,41]. The diversity of pleiotropic effects observed with some *rho* mutants suggest that rho may have other functions in addition to or auxiliary to termination of mRNA synthesis. Figure 10.1 summarizes the location and relative function of a number of rho mutations that have now been identified; their detailed characterization will be discussed below.

Despite the volume of genetic information, very little is known regarding the mechanism of rho-dependent transcription termination in vivo, and only a few of the genetically identified rho mutant proteins have been purified and characterized in vitro. Based on studies in vitro with the wild-type and genetically engineered mutants, however, a good deal more has been learned about rho factor's activities, and this chapter will be devoted to reviewing the structural and functional aspects of rho's interactions.

Structure of Rho

Rho encodes an essential cellular protein of *E. coli* consisting of 419 amino acids (molecular weight approximately 46.1 kD) with a preponderance of basic residues [19]. Chou and Fasman analysis of the primary amino acid sequence shows no notable clustering of hydrophobic, acidic, or basic residues but predicts a rather disordered amino-terminal third and a tightly ordered structure for the remainder of the polypeptide [19]. A significant amount of sequence homology exists between the ATP-binding region of rho and other ATP-binding proteins where tertiary structures have been analyzed using X-ray crystallography and NMR.

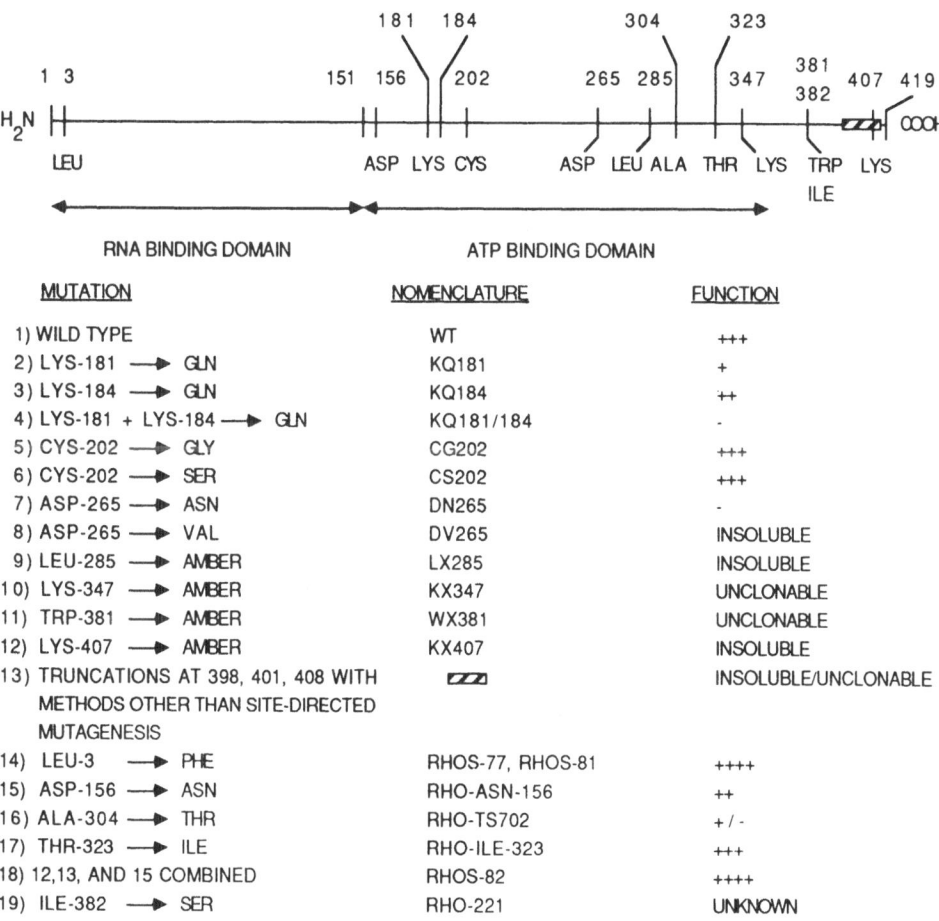

Figure 10.1. Map of identified mutational alterations in rho protein. The locations of altered residues are shown on the linear protein diagram within the RNA- or ATP-binding domains. Below the linear diagram the substitution generated at each site and nomenclature and relative function are indicated. Further details may be found in the published references for mutations 2 to 13 [42,67,81] and 14 to 18 [38,41].

Computer-generated sequence comparisons show rho to be 41% similar to the α and β subunits of the *E. coli* F_1-ATPase and slightly less homologous to adenylate kinase [42]. A similar full statistical analysis done by Doolittle [43] found that sequence similarities between rho and either the α or β subunits of the *E. coli* ATPase extended throughout the full length of the proteins and "left no doubt that these proteins have all descended from a common ancestor." The known structures of adenylate kinase, elongation factor EF-Tu, and phosphofructokinase were used to predict a tertiary structure for the ATP-binding region of the β subunit of the F_1-ATPase based on extensive sequence similarities [44]. Five parallel β

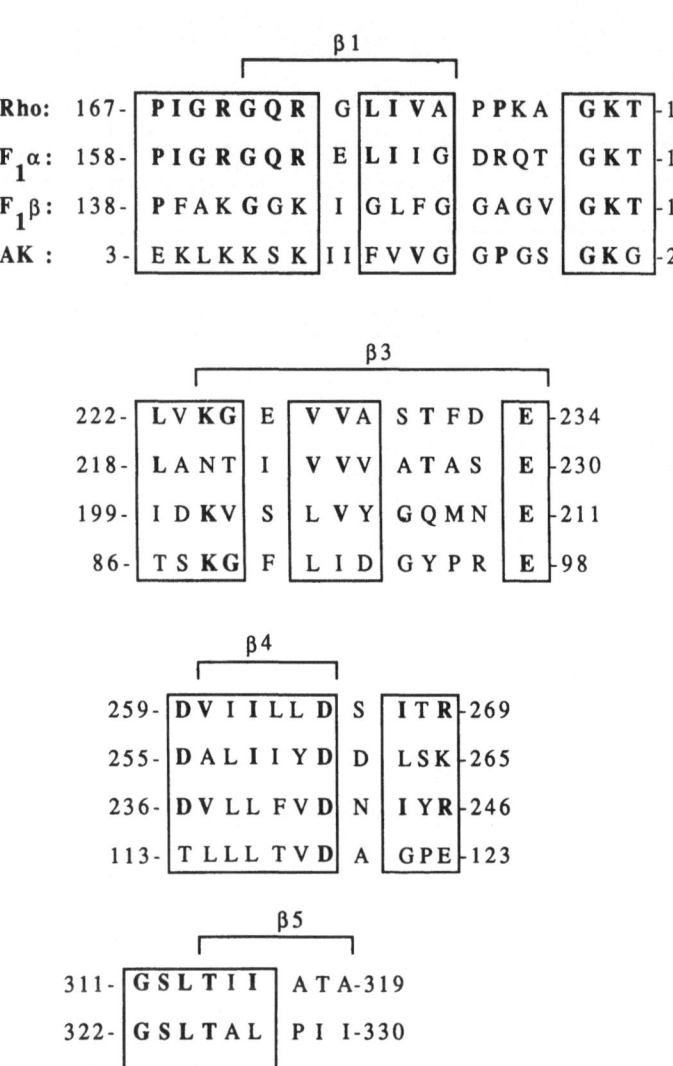

Figure 10.2. Sequence and β structure alignment of rho with three similar ATP-binding proteins. Parallel β-stranded regions 1, 3, 4, and 5 are adjacent and form part of the hydrophobic pocket for ATP in the three-dimensional structure of adenylate kinase and that proposed for $F_1\beta$ [44,45]. The alignment between the α and β subunits of the bacterial F_1-ATPase (for β1, β3, β4, and β5) and adenylate kinase (for β1 and β4) are essentially as determined by Walker et al. [63] with matrix matching programs; adenylate kinase alignment with β3 and β5 is based on its tertiary structure [44,45]. We differ only in β4 of adenylate kinase, which is shifted by one residue to maintain alignment of its critical aspartate-119 with those of the other three proteins. Exact homology is indicated by boldface type; boxes include conservative substitutions except for a few deviations. [From 42.]

strands in adenylate kinase and F_1-β are predicted to form a hydrophobic pocket for the ATP molecule [44,45]. Comparisons of rho to these other ATP-binding molecules can also be used to predict the tertiary structure of rho factor's ATP-binding domain. Figure 10.2 displays the sequence alignment of rho with the α and β subunits of the F_1-ATPase and with adenylate kinase. The most highly conserved sequences in four of the five β strands are shown. The strands of β structure can be folded such that an ATP-binding pocket is formed that closely resembles those for adenylate kinase and F_1-β (see Fig. 10.3). The involvement of specific residues in ATP binding and hydrolysis will be discussed in the section on ATP binding and hydrolysis.

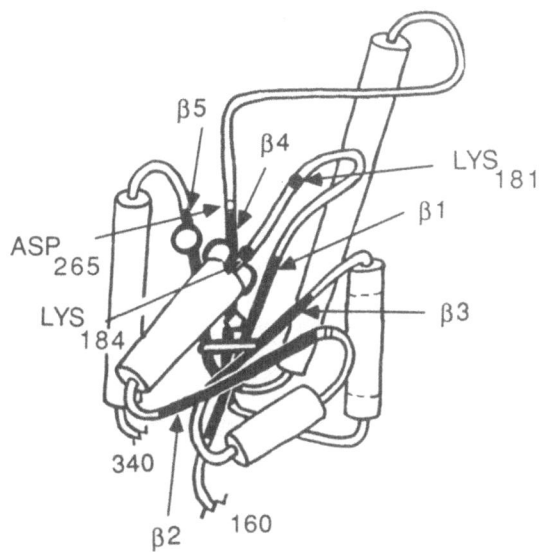

Figure 10.3. A model for the tertiary structure of the ATP-binding domain of rho protein based on homology with the *E. coli* F_1-ATPase [44,88] and adenylate kinase [45]. In the adenylate kinase molecule there are two additional helices deleted in the rho backbone at the site marked by bars following the $\beta2$ strand; rho also has eight additional amino acids in the α-helical region bounded by dashed lines. An ATP molecule is positioned within the hydrophobic pocket formed by the five parallel β strands (represented by black bars). The purine ring is perpendicular to the plane of the page, and the phosphate groups are represented by large open circles. The locations of lys-181, lys-184, and asp-265 are indicated. Strands $\beta1$, $\beta4$, and $\beta5$ are in complete agreement with the Chou and Fasman [89] structure predictions for β sheet applied to rho [19], while our $\beta3$ contradicts their prediction of α helix. We noted [19] that "Chou and Fasman analysis predicts a rather disordered amino-terminal third and a tightly ordered structure for the remainder of the protein," consistent with our independent prediction of separable RNA-binding and ATP-binding domains [42]. [From 67.]

The quaternary structure of rho is a hexamer in solution, as shown by subunit cross-linking studies [46] and by electron microscopy of uranyl acetate-stained rho solutions where subunit assembly in an annular shape suggests a sixfold axis of symmetry [47]. More recent studies employing electron microscopy, however, reveal a cleft in most of the hexameric molecules when RNA is bound to rho, suggesting that rho may not oligomerize into a structure with true sixfold symmetry or that it undergoes conformational changes to an alternative structure [48].

RNA Binding

Rho factor can bind to cytosine-containing synthetic ribopolynucleotides, which serve to activate its ATPase activity [49]. Subsequent studies have shown that rho binds to RNA as a hexamer [46] and has the highest affinity for polyribocytidylate (poly(C)) [15,50]. RNA-dependent ATPase activity can be observed in vitro "uncoupled" from the termination event, and activation of rho is best achieved using unstructured RNA that contains cytosine residues [15]. Rho will also bind to single-stranded DNA containing deoxycytosine but requires the presence of RNA, in the form of oligoribonucleotides, to activate the ATPase activity [50,51]. By fluorescence methods, an RNA-binding site on rho measures 13 nucleotides (nt) of poly(C) per monomer (78 nt per hexamer) [48,52]. Electron microscopy and nuclease protection experiments suggest that RNA is wrapped around or condensed within the rho hexamer and that there is no preference for binding to ends versus internal sites [48].

Richardson [51] proposed that rho actually has two sites for interaction with nucleic acids and that activation of its ATPase activity requires RNA binding at both sites. In this model the higher-affinity primary site can interact with up to 60 nucleotides of either single-stranded RNA or DNA that contains cytosine bases. The secondary, lower-affinity site is specific for RNA, prefers cytosine bases, interacts with only eight nucleotides, and is dependent on saturation of the substrate site with ATP. Mutations selected as hyperactive in termination, rho^s-77 and rho^s-81, carry a leucine to phenylalanine substitution at position 3, resulting in altered selectivity in the primary polynucleotide binding site. Three simultaneous mutations changing Leu-3 to Phe, Asp-156 to Asn, and Thr-33 to Ile in rho^s-82 caused altered interactions at both the primary and secondary polynucleotide-binding sites [40]. The requirement for both sites in transcription termination has not been demonstrated, although analysis of the rho^s-81 and rho^s-82 mutations suggests that the primary polynucleotide-binding site is important in determining termination efficiency, and the secondary site is important in the selection of specific endpoints [41].

The binding of rho to RNA induces conformational changes in the protein, rendering it more susceptible to degradation by trypsin [53,54]. Delineation of the RNA-binding region within the rho polypeptide originally took advantage of the observation that limited tryptic digestion of rho in the presence of RNA and ATP yields predominantly two fragments, F1 (283 amino acids) and F2 (136 amino acids). F1 (the amino-terminal fragment) was isolated, renatured, and found to bind poly(C) similarly to intact rho [54]. Neither F1 nor F2 regained ATPase activity. Within the F1 fragment is the sole cysteine residue (202) of each rho subunit, and one hypothesis to account for rho's preferential binding to poly(C) suggested that Cys-202 might "nucleate" rho binding by formation of a transient covalent Michael adduct at cytosine bases in the RNA [1].

This hypothesis was consistent with observations that rho could be inactivated by the sulfhydryl-modifying reagents N-ethylmaleimide (NEM) and p-hydroxymercuribenzoate (pHMB) [55] and that the spacing of cytosine residues in several rho-dependent terminators was sufficient to allow interaction at the unique Cys on each of the six rho subunits. Surprisingly, conversion of Cys-202 to Ser and Gly by oligonucleotide site-directed mutagenesis (see Fig. 10.1) did not affect any of the protein's in vitro activities, demonstrating that Cys-202 was in fact dispensable [42]. The RNA-binding domain was further defined by hydroxylamine cleavage of the rho polypeptide to yield two fragments, N1 and N2, which correspond to the amino terminal 151 amino acids and the carboxy-terminal 268 amino acids. N1 retains full RNA-binding activity and specificity, so the first 151 amino acids or less of the protein have sufficient information for RNA recognition [42].

Consensus sequences for RNA binding have been reported in other systems. Several proteins involved in snRNP assembly contain the "RNP consensus" of Arg/Lys–Gly–Phe/Tyr–Gly/Ala–Phe/Tyr–Val–X–Phe/Tyr, which is also seen in poly(A) binding protein; hnRNP proteins A1, A2, and C; and the pre-rRNA binding protein C23 [56]. The A1 hnRNP protein was photochemically cross-linked to oligodeoxynucleotides at four phenylalanine residues which are highly conserved among eukaryotic RNA-binding proteins [57]. The binding of the gene 5 protein of bacteriophage fd and the gene 32 protein of bacteriophage T4 to single-stranded DNA is facilitated by the stacking of aromatic side chains with nucleic acid bases [58,59]. Similar hydrophobic interactions may contribute to interactions between rho and RNA. Since the amino acid sequence of rho protein contains seven phenylalanine residues within the RNA-binding domain (N1 fragment) and Phe-62 and Phe-64 are located in a region that partially conforms to the RNP consensus, photochemical cross-linking of rho to RNA may reveal a similar specificity of binding. It is notable that a mutation of Leu-3 to phenylalanine in rho results in significantly tighter RNA binding at the primary site [40,41].

ATP Binding and Hydrolysis

Early investigations of rho function led to the discovery that the RNA-dependent ATPase activity observed alone is maximally stimulated by poly(C) [49]. It was subsequently shown that ATP hydrolysis is required for termination of transcription in vitro at the lambda rho-dependent termination site, t_{R1} [60]. The reaction is dependent on the presence of Mg^{2+} and RNA containing cytidylate residues with little ordered structure [15,55]. All four ribonucleoside triphosphates are hydrolyzed, with ATP having the lowest K_m (9 μM) [55]. A kinetic analysis stimulated by poly(C) suggested that ATP binds to rho first followed by poly(C) and that ADP is the first product released followed by P_i, then poly(C) [61].

The region of rho protein where ATP binds is now known. Computer-generated sequence comparisons brought to light the extensive similarity between rho, the α and β subunits of E. coli F_1-ATPase, and other nucleotide-binding proteins [42,43,62]. These homologies led to the prediction of an ATP-binding domain similar to those for adenylate kinase and F_1-ATPase (see "Structure of Rho," above). As shown in Figure 10.2 (β-strand 1), many nucleotide-binding proteins display the consensus Gly–X–X–X–X–Gly–Lys–Thr [63]. Rho contains the sequence Ala–Pro–Pro–Lys_{181}–Ala–Gly–Lys_{184}–Thr at the homologous position. Of particular note is the presence of an additional Lys at 181 not normally part of the consensus. The appearance of Ala rather than Gly at the first position in the consensus is not uncommon, representing a conservative substitution [64], and several ATP-dependent helicases specifically share this feature [65]. This region forms a flexible loop that may interact with bound nucleotides, and the lysine following the loop is believed to contact the α phosphate of ATP [45]. The observation that a 50-residue synthetic peptide encompassing this region, corresponding to amino acids 135 to 184 of the $F_1\beta$ subunit, can alone bind ATP lends support to this hypothesis [66]. A second region of strong homology is the B consensus sequence [63] that comprises β4; a very hydrophobic region that forms a β-sheet at the back of the nucleotide pocket is followed by a highly conserved aspartate residue that has been implicated in interactions with Mg^{2+} of Mg-ATP [45].

Oligonucleotide site-directed mutagenesis has been used to examine the importance of several conserved amino acids in the nucleotide binding site of rho [67]. Figure 10.1 lists these mutations, including changes at Asp-265, Lys-181, and Lys-184, with their effect on function relative to wild-type rho. Each of the mutant proteins was evaluated in vitro for RNA binding, ATPase activity, and helicase and termination ability. None of the mutations affected RNA binding ability of the proteins, but the activities that are dependent on ATP hydrolysis were affected to varying extents. Substitution of Lys-184 by Gln improves the ATPase and related activities, while the same substitution at Lys-181 reduces but does not eliminate activity. The double mutation changing both Lys-181 and Lys-184 to Gln abolishes ATPase activity, as does a mutation changing Asp-265 to Asn. Each mutant protein appears to be binding ATP but is defective in interactions required for hydrolysis. These residues may therefore be involved directly in the catalytic step or provide an environment with the proper charge distribution to promote catalysis. Whatever the mechanism of involvement, the results support the proposed position for ATP within the ATP-binding domain (see Fig. 10.3) [67].

ATP-binding sites can often be more precisely defined using purine nucleotide affinity analogs [68]. Bear et al. [54] attempted to label the ATP site of rho with the photoaffinity analog 8-azidoadenosine 5'-triphosphate and the chemical affinity label adenosine 2', 3'-dialdehyde 5'-triphosphate, but the results were not consistent, and both analogs bound nonspecifically in more than one tryptic fragment of rho. A different affinity analog, pyridoxal 5'-diphospho-5'-adenosine (PLP-AMP), structurally resembles ATP and covalently modifies lysyl residues in adenine nucleotide-binding sites [69] and was successfully used to label the ATP-binding site of rho [70]. PLP-AMP competes directly with ATP for binding and binds at a stoichiometry of approximately three molecules per rho hexamer, in agreement with an analysis of ATP hydrolytic sites showing that the rho hexamer can only bind three molecules of ATP in the presence or absence of poly(C) [71]. Subunit mixing experiments demonstrate that the incorporation of two to three mutant subunits in a hybrid hexamer inactivates ATPase activity [32].

PLP-AMP binds very specifically in the N2 ATP-binding domain as long as poly(C) is present. The affinity of rho for PLP-AMP is improved in the presence of the RNA substrate poly(C), supporting the notion that poly(C) induces a conformational change that enhances the binding of ligand. Proteolytic digestion of [3H]PLP-AMP-labeled rho yields a single labeled peptide that has been purified and sequenced. Labeling occurs uniquely at Lys-181 [70], which is not the highly conserved Lys in the nucleotide-binding consensus sequence, but other proteins including adenylate kinase and $F_1\alpha$ appear to have more than one lysyl residue present at the nucleotide-binding site. The residue that reacts with PLP-AMP may be the one that is best positioned to form a Schiff-base complex. Site-directed mutagenesis supports the participation of both Lys-181 and Lys-184 at the nucleotide site of rho [67]. Based on our model for the tertiary structure of rho's nucleotide-binding site, we predict that Lys-181 resides at a position of potential interaction with the γ-phosphoryl of ATP leaving Lys-184 in a position to interact near the α-phosphate as seen for the homologous lysine in adenylate kinase [45].

Additional sequence homology to the F_1-type ATPases occurs between positions 345 and 359 in rho (see Fig. 10.4). Phe-355 of rho is homologous to Tyr-345 of the beef heart mitochondrial ATPase (MF$_1$) that is affinity

A.

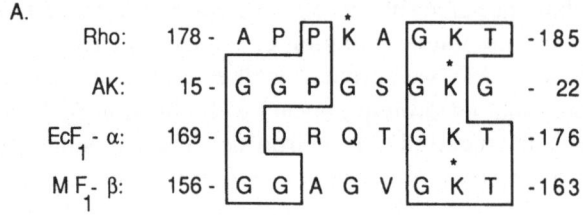

B.

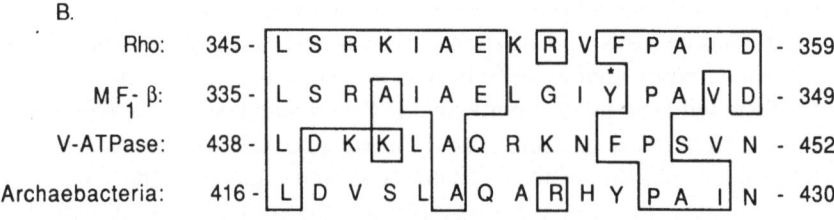

Figure 10.4. Homologous sequences contributing to the ATP-binding domain on rho and other nucleotide-binding proteins. Amino acid sequence alignment is as presented in Figure 10.2 using standard single-letter codes. A) Corresponds to the region immediately following the $\beta1$ strand. B) Corresponds to a region approximately 26 amino acids beyond the $\beta5$ strand. Adenylate kinase, AK; α subunit of *E. coli* F_1-ATPase, EcF_1-α; β subunit of beef heart mitochondrial F_1-ATPase, MF_1-β; carrot vacuolar ATPase, V-ATPase; and archaebacteria ATPase. Identical sequences are boxed. Positions of labeling with adenine nucleotide affinity analogs are indicated with asterisks. [From 70].

labeled by both 2-azido-ATP [72,73] and FSBI [74] and is present at the catalytic site on the β subunit of MF_1 [73]. The carrot vacuolar ATPase [75] and the archaebacteria ATPase also have a phenylalanine present in this position. Either tyrosine or phenylalanine could interact with the adenine ring of bound ATP. It is fairly clear that rho protein is structured very similarly to the F_1-type ATPases in the region of ATP binding. Because rho factor's ATPase activity is RNA dependent, it is obvious that some type of interaction must take place between the RNA-binding domain and the ATP-binding domain to facilitate hydrolysis of ATP. Domain interactions will be discussed later in this chapter.

Coupling of Rho Activity to Termination

Helicase, Release, and Termination Activities

The role of ATP hydrolysis in transcription termination was unknown for many years. It was suggested at one point that the energy provided by the ATPase activity may allow rho to translocate along the RNA chain toward the RNA polymerase paused at the termination site [50; for review see 76]. The possibility that the ATPase could also be involved in reactions that modify one of the transcriptional components was also considered. NMR studies of the stereochemistry of rho-mediated ATP hydrolysis, however, indicate that the phospho group of ATP is transferred directly to water without the existence of a phosphoenzyme or phospho-RNA intermediate [77].

Hydrolysis of ATP was essential for rho-dependent termination and release of RNA from ternary transcription complexes using bacteriophage T7 DNA templates [60,78,79]. The release of RNA from the DNA template and from RNA polymerase, when characterized with a T7 template, indicated that the release of DNA accompanies the release of RNA [80]. In an assay independent of transcription, Brennen et al. [16] then discovered that rho had a helicase activity that could release RNA from an RNA-DNA hybrid in the absence of any other components of a normal transcription complex. The helicase substrate consists of a hybrid formed between a polylinker sequence at the 3′ end of RNA derived from the rho-dependent terminator *trp t′* and the complementary sequence in a single-stranded DNA molecule. The ability to release the RNA from this duplex substrate is dependent on NTP hydrolysis, proceeds in a 5′ to 3′ direction along the RNA, and requires a 5′ single-stranded region of RNA that contains the signal(s) needed for rho-dependent termination.

We are currently examining the NTP requirement of the reaction, effects of sequence deletion on RNA recognition, effects of distance between the rho recognition site and the RNA-DNA duplex, and the effects of sequences of varying structural potential between the rho site and the RNA-DNA hybrid. While all four rNTPs are hydrolyzed by rho to the same extent under helicase reaction conditions, only ATP and GTP activate RNA-DNA unwinding. At low concentrations CTP and UTP are poor activators of release, and at high concentrations they can inhibit the reaction even with ATP present [Brennan and Platt, unpublished]. Thus, NTP hydrolysis appears necessary but not sufficient for RNA release. Tests of trypsin sensitivity indicate that rho is much less stable with CTP or UTP present than with ATP [Spear and Platt, unpublished], so we suspect that the "fit" in the NTP site must help couple hydrolysis to helicase activity. Deletions of the *trp t′* terminator region were used to determine minimum sequence requirements for rho recognition and binding. A minimum length of 100 nucleotides of unhybridized RNA is apparently required for efficient unwinding of the RNA-DNA duplex, and more than one region of *trp t′* can be recognized by rho. Differences have been observed between regions that function as helicase substrates and those that function in in vitro transcription termination assays. This may be due to the effects of flanking sequences on termination in these assays or the rate at which polymerase transcribes through the region. Preliminary experiments to test the distance between the rho recognition site and the hybrid have indicated that (1) as many as 450 nucleotides could be inserted between *trp t'* and the duplex, (2) the nature of the sequence inserted has little effect (even a stable hairpin structure does not interfere), and (3) the *trp t'* region must be upstream of the inserted sequences [Brennan, Steinmetz, and Platt, unpublished]. The results of the rho helicase studies suggest that the RNA-dependent ATPase activity is used to unwind the RNA-DNA hybrid in the transcription bubble, thus facilitating release of the transcript from the transcription complex.

Structure/Function Relationships and Models for Rho Action

Some important questions remain regarding the details of rho participation in transcription termination and RNA release. Basic to the RNA-dependent nature of the ATPase activity is rho recognition of and activation by RNA. The binding of RNA to rho induces a conformational change in the protein, as do the binding and hydrolysis of ATP [53,54]. It seems likely that although single-stranded DNA can also bind to rho, it is unable to produce the correct conformational alterations that then allow ATP hydrolysis. The RNA dependence of rho's ATPase activity also necessitates some interaction between the RNA-binding and ATP-binding domains either directly through domain-domain contacts or indirectly through a chain of conformational changes. It is clear that each domain can be affected separately through mutational and chemical modifications [67,81].

Oligonucleotide site-directed mutagenesis of Cys-202 [42,81] and attachment of sulfhydryl-specific labels at Cys-202 [82] do not significantly change the properties of rho protein. Yet inactivation can be induced by p-hydroxymercuribenzoate (pHMB), which is also specific for Cys-202. Despite its inability to hydrolyze ATP, pHMB-modified rho maintains full RNA-binding ability as well as ability to bind the ATP analog PLP-AMP [81]. It is thus possible that merely the introduction of a negatively charged group at Cys-202 prevents proper domain interactions required for hydrolysis. N-ethylmaleimide (NEM) also abolishes ATPase activity [55] but must do so by alkylation of sites other than or in addition to Cys-202. Inactivation of rho is reduced considerably when ATP is bound, but once NEM has inactivated the protein, it can no longer bind to RNA. Thus, the conformational state of rho affects its sensitivity to NEM inactivation [81], as is also the case for trypsin digestion of rho [53,54].

A simple model to illustrate domain interactions for rho is shown in Figure 10.5. In this scheme, binding of RNA to rho alters the protein conformation from a "closed" state to an "open" one, where it is very accessible to reagents such as trypsin and NEM. This conformation may allow greater access of ATP to the ATP-binding domain, as suggested by nucleotide analog studies [70]. Upon binding ATP, the protein assumes a much tighter "active" conformation preventing access by NEM and trypsin. ATP hydrolysis is enabled when the two domains interact

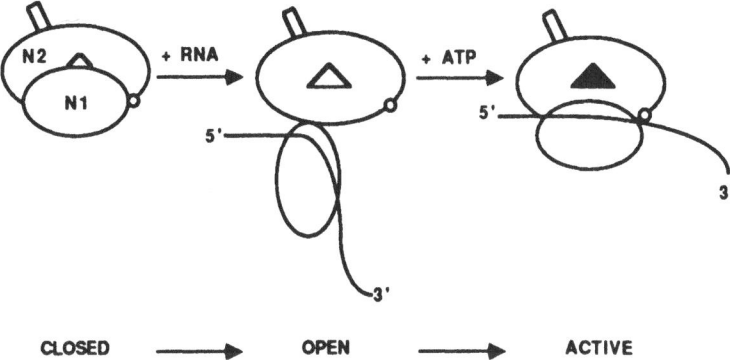

Figure 10.5. Cartoon model for conformational changes of a rho subunit. The RNA-binding and ATP-binding domains are N1 and N2, respectively. The open triangle represents an empty ATP-binding site, and the closed triangle an occupied one. The small circle is Cys-202, and the rectangle represents the α-helical C-terminal region, which may be involved in subunit association. See text for discussion. [From 81.]

correctly, while helicase and termination additionally require that the ATP site is correctly filled (with ATP rather than CTP or UTP). At first glance this model appears to contradict the kinetic findings that ATP binds rho first followed by poly(C) [61], and it also does not take into account the proposed two-site model for RNA binding [51]. However, it seems plausible that the primary RNA-binding site, described in the two-site model, is filled first (this site can also bind DNA, but activation of ATPase requires RNA binding at the secondary site) followed by ATP binding and then RNA binding at the secondary site. This modification then would account for the order of substrate binding defined by kinetics as well as the two-site RNA-binding hypothesis.

The observation that rho productively binds only three ATP molecules per hexamer [71] remains unexplained in mechanistic terms since each of the six rho subunits contains the same type of ATP-binding site. It is possible that the domains of adjacent subunits in the hexamer must also interact in some manner or that the oligomeric arrangement of monomers allows accessibility of only three ATP-binding sites per hexamer. It is also still unknown how rho associates into its hexameric form, but determinations of the sites of subunit-subunit contacts and symmetry of oligomerization are being addressed. Mutations introduced in the carboxy-terminal region of rho in vitro (see Fig. 10.1) typically result in an insoluble protein product, and some evidence suggests that this region could be involved in intersubunit interactions. Several carboxy-terminal alterations have been unclonable, implying that a complete carboxy-terminus is necessary for rho function and cell viability [81].

The function of rho in terminating transcripts is still not entirely understood. Studies by Galluppi and Richardson [50] resulted in the proposal that rho binds to RNA and wraps the polynucleotide chain around its hexameric structure to eventually reach the paused RNA polymerase molecule and the end of the transcript. The RNA-DNA helicase activity of rho is consistent with a translocation model where rho enters at the rho binding site and, once bound, can move 5' to 3' along the RNA, driven by ATP hydrolysis, to reach the RNA-DNA duplex. The processivity of 5'–3' translocation in the helicase reaction is substantial, as judged by the lack of effect of inserting spacer sequences of various lengths (up to 450 nucleotides) and of different sequence content between the 220 nucleotide *trpt'* region and the duplex [Brennan, Steinmetz, and Platt, unpublished]. The simplest model is that rho uses the energy of ATP hydrolysis to move along the RNA until it reaches a duplex region, and that duplex disruption is the inevitable consequence of continued translocation. No evidence yet bears on whether rho can recognize an RNA-DNA duplex region per se, and the processivity of rho on RNA-DNA duplex molecules is unknown.

A comparison of the structure and function of rho to other proteins reveals that F_1-type ATPases, myosin, and the dnaB helicase protein all exist as hexamers and perform some type of mechanical work in the cell. Extensive sequence homologies are only apparent between rho and the α and β subunits of F_1-ATPases, but structural similarities may be more important when comparing the types of activities these proteins are performing. Both rho and dnaB act as helicases to mechanically remove one polynucleotide chain from another , while myosin is involved in the mechanics of muscle contraction. A rotary mechanism has been proposed for the F_0F_1 ATPase where the three β subunits undergo ATP-dependent conformational changes as they are forced to move relative to some asymmetric structural component [83].

A final and important question is how rho works in vivo when the entire transcription complex is present. Rho is certainly capable of releasing transcripts in vitro (as already discussed) using the helicase assay or release from isolated paused complexes, without any other components of a transcription complex. Participation of the nontemplate strand may be important if some of the free energy for release comes from rewinding of the DNA helix. The presence of other transcriptional components may also be necessary for maximum helicase efficiency. One hypothesis has an obligatory template-encoded pause by RNA polymerase at the termination region independent of rho, with termination occurring when rho encounters the paused complex. Pausing by RNA polymerase is likely to play a role in termination, although no polymerase pause sites can be detected in the *trp t'* terminator [Brennan and Platt, unpublished]. In any case, some evidence implicates an interaction between rho and RNA polymerase. Several studies have demonstrated a restoration of normal termination activities in strains carrying a defective rho allele by the presence of an RNA polymerase mutation [84–86]. Rho can be retarded on an RNA polymerase-Sepharose column if the nusA protein is present, but it is not clear whether rho binds a particular conformational state of polymerase or if nusA couples rho to polymerase [87]. Thus far, direct physical contact of rho with polymerase remains to be observed, and the interactions between rho and nusA are not well understood.

Concluding Statement

We have learned a great deal about the structure and function of rho protein in the past few years. Rho is active as a hexamer that binds three molecules of ATP and 78 nucleotides of RNA at a time. Distinct binding domains exist for ATP and RNA which must interact in some fashion to activate hydrolysis of ATP. The energy of ATP hydrolysis is probably utilized for 5' to 3' translocation of rho along the transcript and for removing the nascent RNA transcript from the DNA template via the helicase activity. Future studies should further illuminate the nature of domain interactions, subunit interactions, mechanisms for binding and release of the transcript, and interactions with other transcriptional components. An improved understanding of gene regulation through transcription termination will undoubtedly result as well as the establishment of some common structural and mechanistic characteristics between evolutionarily related enzymes.

References

1. Platt, T. (1986) Annu. Rev. Biochem. *55*, 339–372.
2. Platt, T., and Bear, D.G. (1983) In *Gene Function in Prokaryotes*, 123–126, Beckwith, J., Davies, J., and Gallant, J.A. (eds.). Cold Spring Harbor Laboratory, Cold Spring Harbor, NY.
3. Adhya, S., and Gottesman, M. (1978) Annu. Rev. Biochem. *47*, 967–996.
4. Rosenberg, M., and Court, D. (1979) Annu. Rev. Genet. *13*, 319–353.
5. Platt, T. (1981) Cell *24*, 10–23.
6. Roberts, J. (1969) Nature *224*, 1168–1174.
7. Sharp, J.A., and Platt, T. (1984) J. Biol. Chem. *259*, 2268–2273.
8. Morgan, W.D., Bear, D.G., Litchman, B.L., and Von Hippel, P.H. (1985) Nucleic Acids Res. *13*, 3739–3754.
9. Lau, L.F., and Roberts, J.W. (1985) J. Biol. Chem. *260*, 574–584.
10. Yager, T.D., and Von Hippel, P.H. (1987) In *E. coli and S. typhimurium: Cellular and Molecular Biology*, 1241–1275, Neidhart, F. (ed.). American Society of Microbiologists, Washington, D.C.
11. Bear, D.G., and Peabody, D.S. (1988) Trends Biochem. Sci. *13*, 343–347.
12. Platt, T., Mott, J.E., Galloway, J.L., and Grant, R.A. (1985) In *Sequence Specificity in Transcription and Translation*, 151–160, UCLA Symp. Mol. Cell. Biol., New Ser. *30*, Calendar, R., and Gold, L. (eds.). Alan R. Liss, New York.

13. Galloway, J.L., and Platt, T. (1988) J. Biol. Chem. *263*, 1761–1767.

14. Chen, C.-Y., and Richardson, J.P. (1987) J. Biol. Chem. *262*, 11292–11299.

15. Lowery, C., and Richardson, J.P. (1977) J. Biol. Chem. *252*, 1381–1385.

16. Brennan, C.A., Dombroski, A.J., and Platt, T. (1987) Cell *48*, 945–952.

17. Calhoun, D.H., Traub, L., Wallen, J.W., Gray, J.E., and Guterman, S.K. (1984) Mol. Gen. Genet. *193*, 205–209.

18. Brown, S., Albrechtsen, B., Pederson, S., and Klemm, P. (1982) J. Mol. Biol. *162*, 283–298.

19. Pinkham, J.L., and Platt, T. (1983) Nucleic Acids Res. *11*, 3531–3545.

20. Tsurushita, N., Hirano, M., Shigesada, K., and Imai, M. (1984) Mol. Gen. Genet. *196*, 458–464.

21. Barik, S., Bhattacharya, P., and Das, A. (1985) J. Mol. Biol. *182*, 495–508.

22. Matsumoto, Y., Shigesada, K., Hirano, M., and Imai, M. (1986) J. Bacteriol. *166*, 945–958.

23. Franklin, N., and Yanofsky, C. (1976) In *RNA Polymerase*, 693–706, Losick, R., and Chamberlin, M.J. (eds.). Cold Spring Harbor Laboratory, Cold Spring Harbor, New York.

24. Oppenheim, D.S., and Yanofsky, C. (1980) Genetics *95*, 785–795.

25. Malamy, M. (1966) Cold Spring Harbor Symp. Quant. Biol. *31*, 189–201.

26. Fiandt, M., Szybalski, W., and Malamy, M.H. (1972) Mol. Gen. Genet. *119*, 223–231.

27. Beckwith, J. (1963) Biochim. Biophys. Acta *76*, 162–164.

28. Guarente, L.P., Mitchell, D.H., and Beckwith, J. (1977) J. Mol. Biol. *112*, 423–436.

29. Kent, R.B., and Guterman, S.K. (1982) Mol. Gen. Genet. *187*, 330–334.

30. Sharp, J.A., Guterman, S.K., and Platt, T. (1986) J. Biol. Chem. *261*, 2524–2528.

31. Richardson, J.P., and Carey, J.L. III (1982) J. Biol. Chem. *257*, 5767–5771.

32. Richardson, J.P., and Ruteshouser, E.C. (1986) J. Mol. Biol. *189*, 413–419.

33. Fassler, J.S., Arnold, G.F., and Tessman, I. (1986) Mol. Gen. Genet. *204*, 424–429.

34. Stitt, B.L., Revel, H.R., Lielausis, I., and Wood, W.B. (1980) J. Virol. *35*, 775–789.

35. Simon, L.D., Gottesman, M., Tomczak, K., and Gottesman, S. (1979) Proc. Natl. Acad. Sci. USA *76*, 1623–1627.

36. Das, A., Gottesmann, M.E., Wardwell, J., Trisler, P., and Gottesman, S. (1983) Proc. Natl. Acad. Sci. USA *80*, 5530–5534.

37. Gulletta, E., Das, A., and Adhya, S. (1983) Genetics *105*, 265–280.

38. Imai, M., and Shigesada, K. (1978) J. Mol. Biol. *120*, 451–466.

39. Shigesada, K., and Imai, M. (1978) J. Mol. Biol. *120*, 467–486.

40. Mori, H., Imai, M., and Shigesada, K. (submitted).

41. Tsurushita, N., Shigesada, K., and Imai, M. (submitted).

42. Dombroski, A.J., and Platt, T. (1988) Proc. Natl. Acad. Sci. USA *85*, 2538–2542.

43. Doolittle, R.F. (1986) In *Protein Engineering: Applications in Science, Medicine, and Industry*, 15–27, Inouye, M., and Sarma, R. (eds.). Academic Press, New York.

44. Duncan, T.M., Parsonage, D., and Senior, A.E. (1986) FEBS Lett. *208*, 1–6.

45. Fry, D.C., Kuby, S.A., and Mildvan, A.S. (1986) Proc. Natl. Acad. Sci. USA *83*, 907–911.

46. Finger, L.R., and Richardson, J.P. (1982) J. Mol. Biol. *156*, 203–219.

47. Oda, T., and Takanami, M. (1972) J. Mol. Biol. *71*, 799–802.

48. Bear, D.G., Hicks, P.S., Escudero, K.W., Andrews, C.L., McSwiggen, J.A., and Von Hippel, P.H. (1988) J. Mol. Biol. *199*, 623–635.

49. Lowery-Goldhammer, C., and Richardson, J.P. (1974) Proc. Natl. Acad. Sci. USA *71*, 2003–2007.

50. Galluppi, G.R., and Richardson, J.P. (1980) J. Mol. Biol. *138*, 513–539.

51. Richardson, J.P. (1982) J. Biol. Chem. *257*, 5760–5766.

52. McSwiggen, J.A., Bear, D.G., and Von Hippel, P.H. (1988) J. Mol. Biol. *199*, 609–622.

53. Engel, D., and Richardson, J.P. (1984) Nucleic Acids Res. *12*, 7389–7400.

54. Bear, D.G., Andrews, C.L., Singer, J.D., Morgan, W.D., Grant, R.A., Von Hippel, P.H., and Platt, T. (1985) Proc. Natl. Acad. Sci. USA *82*, 1911–1915.

55. Lowery, C., and Richardson, J.P. (1977) J. Biol. Chem. *252*, 1375–1380.

56. Wickens, M.P., and Dahlberg, J.E. (1987) Cell *51*, 339–342.

57. Merrill, B.M., Stone, K.L., Cobianchi, F., Wilson, S.H., and Williams, K.R. (1988) J. Biol. Chem. *263*, 3307–3313.

58. Brayer, G.D., and McPherson, A. (1984) Biochemistry *23*, 340–349.

59. Prigodich, R.V., Casa-Finet, J., Williams, K.R., Konigsberg, W., and Coleman, J.E. (1984) Biochemistry *23*, 522–529.

60. Howard, B.H., and DeCrommbrugghe, B. (1976) J. Biol. Chem. *251*, 2520–2524.

61. Sharp, J.A., Galloway, J.L., and Platt, T. (1983) J. Biol. Chem. *258*, 3482–3486.

62. Doggett, P.E., and Blattner, F.R. (1986) Nucleic Acids Res. *14*, 611–619.

63. Walker, J.E., Saraste, M., Runswick, M.J., and Gay, N.J. (1982) EMBO J. *1*, 945–951.

64. Higgins, C.F., Hiles, I.D., Salmond, G.P.C., Gill, D.R., Downie, J.A., Evans, I.J., Holland, I.B., Gray, I., Buckel, S.D., Bell, A.W., and Hermodson, M.A. (1986) Nature *323*, 448–450.

65. Ford, M.J., Anton, I.A., and Lane, D.P. (1988) Nature *332*, 736–738.

66. Garboczi, D.N., Shenbagamurthi, P., Kirk, W., Hullihen, J., and Pederson, P. (1988) J. Biol. Chem. *263*, 812–816.

67. Dombroski, A.J., Brennan, C.A., Spear, P., and Platt, T. (1988) J. Biol. Chem. *263*, 18802–18809.

68. Colman, R.F. (1983) Annu. Rev. Biochem. *52*, 67–91.

69. Tamura, J.K., Rakov, R.D., and Cross, R.L. (1986) J. Biol. Chem. *261*, 4126–4133.

70. Dombroski, A.J., LaDine, J.R., Cross, R.L., and Platt, T. (1988) J. Biol. Chem. *263*, 18810–18815.

71. Stitt, B.L. (1988) J. Biol. Chem. *263*, 1130–1137.

72. Garin, J., Boulay, F., Iffartel, J.P., Lunardi, J., and Vignais, P.V. (1986) Biochemistry *25*, 4431–4437.

73. Cross, R.L., Cunningham, D., Miller, C.G., Xue, Z., Zhou, J.-M., and Boyer, P.D. (1987) Proc. Natl. Acad. Sci. USA *84*, 5715–5719.

74. Bullough, D.A., Verburg, J.G., Yoshida, M., and Allison, W.S. (1987) J. Biol. Chem. *262*, 11675–11683.

75. Zimniak, L., Dittrich, P., Gogarten, J.P., Kibak, H., and Taiz, L. (1988) J. Biol. Chem. *263*, 9102–9112.
76. Von Hippel, P.H., Bear, D.G., Morgan, W.D., and McSwiggen, J.A. (1984) Annu. Rev. Biochem. *53*, 389–446.
77. Stitt, B.L., and Webb, M.R. (1986) J. Biol. Chem. *261*, 15906–15909.
78. Richardson, J.P., and Conaway, R. (1980) Biochemistry *19*, 4293–4299.
79. Shigesada, K., and Wu, C.-W. (1980) Nucleic Acids Res. *8*, 3355–3369.
80. Andrews, C., and Richardson, J.P. (1985) J. Biol. Chem. *260*, 5826–5831.
81. Platt, T., Brennan, C.A., Dombroski, A.J., and Spear, P. (1989) In *Molecular Biology of RNA*, 325–334, UCLA Symp. Mol. Cell. Biol., New Ser., Vol. 94, Ceck, T. (ed.). Alan R. Liss, New York.
82. Seifried, S.E., Wang, Y., and Von Hippel, P.H. (1988) J. Biol. Chem. *263*, 13511–13514.
83. Kironde, F.A.S., and Cross, R.L. (1987) J. Biol. Chem. *262*, 3488–3495.
84. Guarente, L.P., and Beckwith, J. (1978) Proc. Natl. Acad. Sci. USA *75*, 294–297.
85. Das, A., Merril, C., and Adhya, S. (1978) Proc. Natl. Acad. Sci. USA *75*, 4828–4832.
86. Guarente, L. (1979) J. Mol Biol. *129*, 295–304.
87. Schmidt, M.C., and Chamberlin, M.J. (1984) J. Biol. Chem. *259*, 15000–15002.
88. Parsonage, D., Duncan, T.M., Wilke-Mounts, S., Kironde, F.A., Hatch, L., and Senior, A.E. (1987) J. Biol. Chem. *262*, 6301–6307.
89. Chou, P.Y., and Fasman, G.D. (1978) Adv. Enzymol. *47*, 45–148.

Chapter 11
The Mechanism and Control of Pre-mRNA Splicing

Jonathan C.S. Noble and James L. Manley

Gene expression in eukaryotes requires that primary transcripts of structural genes be modified in a variety of ways before translation as mature mRNAs. One of the most demanding, in terms of mechanism, of these required modifications is the process of splicing, by which noncontiguous stretches of sequences (exons) are brought into juxtaposition with the removal of intervening sequences (introns). Since splicing was discovered several years ago [1,2], enormous strides have been made in the progress of our understanding of this exacting process. This progress has encompassed the basic mechanism or pathway of the reactions involved, the sequences within the pre-mRNA that direct splicing, the factors required and their interactions with the primary transcript, and, recently, the mechanism and control of alternative splicing. References 3 and 4 present related reviews.

The Pathway of Pre mRNA Splicing

It is only over the last 3 or 4 years that the broad outline of the "enzymatic" steps involved in pre-mRNA splicing has become apparent. This outline owes a great deal to the advent of cell-free systems capable of splicing pre-mRNAs in vitro, which has made the process far more accessible to study [5–8]. The results of these studies, which include kinetic analyses in vitro [9,10] and the characterization of splicing intermediates and products in vitro [8–11] and in vivo [12–14], have described a two-step model for the splicing reaction (Fig. 11.1). In the first step, cleavage of the pre-mRNA at the 5' splice junction together with formation of a lariat-structured intermediate, in which the 5' end of the intron is joined by a 2'–5' phosphodiester linkage, or branch, to a nucleotide upstream of the 3' splice junction. In the second step, cleavage of the pre-mRNA at the 3' splice junction occurs together with ligation of the 5' and 3' exons, resulting in the release of a lariat intron and generation of the spliced product. Although both of the steps in this

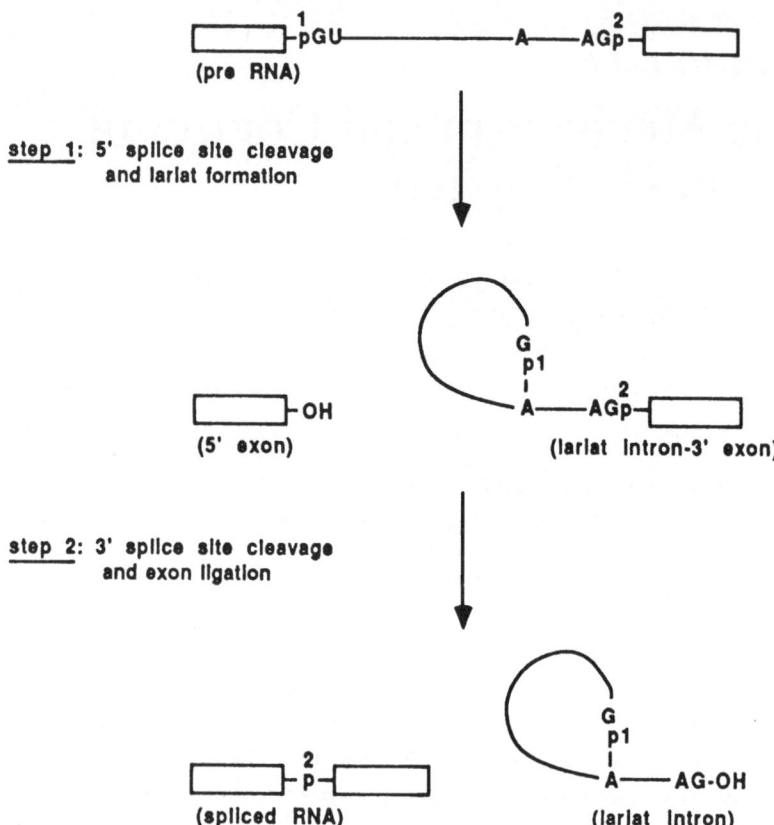

Figure 11.1. Schematic outline of the pre-mRNA splicing pathway. Boxes indicate exons, and the line indicates an intron. Splicing products and intermediates are identified in parentheses. The reactions involved in steps 1 and 2 are described in summary form. The A that serves as the branch nucleotide is indicated, as are the invariant GU and AG dinucleotides at the 5' and 3' splice junctions, respectively. The fate of the phosphates at both splice junctions is traced throughout the pathway.

process involve what would appear to be two-component reactions, these are believed to occur in a concerted fashion.

The Nucleotide Sequences Involved in Pre-mRNA Splicing

One of the principal characteristics of the splicing process is its remarkable accuracy. From the very first it was apparent that this accuracy must be determined by sequences within the pre-mRNA that direct the precise cleavage and ligation reactions. The sequences that define the 5' and 3' splice sites within a given pre-mRNA have been shown to comprise a number of elements.

5' and 3' Splice Junctions

An extensive comparison of sequences surrounding 5' and 3' splice junctions in higher eukaryotes has identified conserved sequences that conform at the 5' splice site to the consensus CAG/*GU*RAGU and at the 3' splice site to the consensus $(Y)_{n>11}NCAG$/G [15]. Within these sequences, the dinucleotides 5' *GU* and 3' *AG,* which lie immediately adjacent to the 5' and 3' splice junctions on the intron side, are essentially invariant [16], while the remaining positions demonstrate considerable sequence variation [15]. The role of these conserved sequences in defining 5' and 3' splice junctions has been confirmed by mutation. This analysis has shown that single base substitutions within the 5' splice site consensus sequence, particularly those involving the invariant GU dinucleotide, in most cases abolished the use of the mutated 5' splice site in vivo [reviewed in 17,18]. These mutations also resulted in the activation of cryptic 5' splice sites in vivo [17] and in vitro [7], which were used in place of the authentic site. Unexpectedly, some mutants containing substitutions within the invariant GU were found to undergo the first step of the splicing reaction in vitro but were halted at the second step [18]. Similar analysis of mutants involving the 3' splice site consensus, including those containing deletions of the polypyrimidine stretch [19–21], has shown that these sequences are essential for efficient use of the authentic site in vivo. Interestingly, deletions of the polypyrimidine stretch and substitutions in the invariant AG were also shown to inhibit the first step in the splicing reaction in vitro [18,22,23], although this step does not involve cleavage at the 3' splice junction.

The analysis of sequences located at the 5' and 3' splice junctions of yeast introns has demonstrated that they differ in several respects from those found in higher eukaryotes. First, 5' splice site sequences in yeast conform strictly to the consensus /GUAYGU [24,25], with little or no variation permitted at any position. Deletions or point mutations within these sequences, which in most instances either eliminated or greatly reduced the efficiency of splicing using the authentic site or else caused the accumulation of intermediates [26–31], did not, in contrast to higher eukaryotes, result in activation of cryptic 5' splice sites. Interestingly, a mutant containing a substitution at the fifth position of the consensus was found to undergo the first reaction of splicing but was blocked for the second step [27], implying, as for higher eukaryotes, that recognition of 5' splice site sequences is also important in the second step of the reaction. Second, 3' splice site sequences in yeast do not include a polypyrimidine stretch but instead contain YAG immediately adjacent to the 3' splice junction as the only conserved element [32]. While the analysis of mutations has shown that this AG is essential for the use of the authentic site in vivo [30,33], it has been noted that, in contrast to higher eukaryotes, it is not required for the first step in the splicing reaction in vivo [30,33] or in vitro [34] in yeast.

Branch Site

Only a single authentic branch site has been found to be used in any one intron in most instances thus far examined, both in higher eukaryotes [10,11,14,22; with one exception: 35] and in yeast [8,12,13]. Although most branch sites have been mapped in vitro, in the cases where it has been investigated the same branch site has been found to be used in vivo and in vitro [8,13,14,35]. A comparison of branch site sequences and locations has indicated that branch site selection may be conditioned by two factors in higher eukaryotes. First, the sequences surrounding the branch site conform to a rather liberal consensus YNCURAY [14,22,36], of which the penultimate nucleotide is the actual site of the 2'–5' phosphodiester linkage, or branch nucleotide. Within this sequence only the branch nucleotide itself appears to be invariant (see Table 11.1). Second, all branch sites so far analyzed lie within 18 to 37 nucleotides of the 3' splice junction (Table 11.1). This strict distance constraint, together with the demonstrated requirement for 3' splice site sequences in the first step of the splicing reaction in vitro, has been interpreted as suggestive of a role for 3' splice site sequences in determining the location of the branch site in higher eukaryotes.

The relative importance of these features for branch site selection has been explored by mutation. The results of this analysis were several. First, while invariant, the A comprising the branch nucleotide could be exchanged with any of the three alternative bases without drastically affecting the efficiency of splicing in vitro [37]. In each case, branching was observed to the replacing nucleotide, and to a varying degree to the nucleotide preceding the normal branch position (also an A). Of the three bases, U appears to serve least efficiently as a branch "acceptor." Second, when the authentic branch site sequence is deleted, cryptic branch sites within the prescribed distance of the 3' splice junction are used instead [38,39]. These cryptic branch sites, whose use correlates with a reduced efficiency of splicing in vitro, demonstrate considerably greater variation from the branch site consensus.

Analysis of the sequences and locations of branch sites in yeast introns has suggested that branch site selection in yeast, in contrast to branch site selection in higher eukaryotes, is determined principally by the sequences surrounding the branch site and not by spatial considerations. The relative unimportance of distance constraints for branch site selection in yeast is supported by two observations. First, the spacing between the branch site and 3' splice junction varies over a considerable range in

Table 11.1. Catalogue of mammalian branch site sequences.

Intron	Branch site	Position	Reference
Adenovirus E1A (13S)	*UUUAAAG*	30	22
Adenovirus MLP (L1-L2)	*UUCUUAU*	24	11
Human α-globin (1VS1)	*CCCUCAC*	19	22
Human α-globin (1VS2)	*CACUGAC*	18	22
Human ε-globin (1VS1)	*CUCUAAU*	31	22
Human γ-globin (1VS1)	*UUCUGAC*	30	22
Human β-globin (1VS1)	*CACUGAC*	37	22
Mouse β-globin (1VS1)	*CACUAAC*	36	22
Rabbit β-globin (1VS1)	*UGCUGAC*	34	22
			14
Rabbit β-globin (1VS2)	*UGCUAAC*	32	14
Rat insulin II (1VS1)	*CCUCAAC*	18	22
SV40 T/t	*UUCUAAU*	18	35
	UACUGAU	24	35
	AACUACU	27	35
	UGUUAAA	32	35
Higher eukaryotes consensus	YNCURAY	18–37	22
			14
Yeast consensus	*UACUAAC*	6–53	24
			42

For each of the introns listed on the left, the sequence surrounding the branch nucleotide(s) is given. Within each branch site sequence, the homology to the consensus is indicated in italics. The distance of the branch nucleotide from the 3' splice junction is recorded on the right.

yeast—6 to 53 nt [25,32]. Second, and more convincingly, 3' splice site sequences have been shown to be dispensable for the first step of the splicing reaction in yeast [30,31,34]. In yeast, the sequences around the branch site conform strictly to the consensus UACUA AC, which shares homology with the branch site consensus in higher eukaryotes (Table 11.1). The importance of this sequence for branch formation has been investigated by mutational analysis, which has shown that point mutations involving the branch nucleotide dramatically influence the efficiency of splicing in vivo [24,28–31,40].

Interestingly, when the effects of different substitutions were analyzed, it was found that an A → C inhibited both the first and second steps in the reaction, whereas A → G inhibited only the second step. Regardless of the efficiency of lariat formation, the nucleotide at the "branch position" in the consensus was invariably used as the acceptor [31]. Single base substitutions involving nucleotides other than the branch acceptor have been found to have varied effects on the efficiency of splicing in vivo, depending on the position mutated. In contrast to higher eukaryotes, deletion of the entire branch site consensus in yeast was shown not to result in the activation of cryptic branch sites [with one exception; 41] but instead resulted in the complete inhibition of splicing in vivo [32,42].

Interestingly, although unimportant for the first step in the reaction, the distance between the branch site and the 3' splice junction in yeast introns has been shown to be significant for the second step [33,43].

Overall, in terms of the sequence requirements of splicing in higher eukaryotes compared to yeast, it is apparent that the former is characterized by a considerably greater variation in the sequences that define splice junctions and the branch site than the latter. This greater sequence variation may provide the degree of flexibility to the process of splice site selection in higher eukaryotes that is necessary to allow for such refinements as alternative splicing. In yeast, among the 20 or so intron-containing pre-mRNAs analyzed so far, none have been found to undergo alternative splicing, while this feature is common in higher eukaryotes [reviewed in 44]. This flexibility would also appear to be responsible for the phenomenon of cryptic splice site utilization in higher eukaryotes, which is not observed in yeast.

Factors Interacting with the Pre-mRNA During Splicing

The advent of cell-free systems capable of splicing pre-mRNA in vitro has permitted considerable progress in defining not only the factors involved but also the nature of their interactions with each other and with the pre-mRNA. Chief among the factors identified are several small nuclear ribonucleoprotein particles, or snRNPs. These particles, which are found in the nuclei of all eukaryotic cells, each contain a single RNA component (U1, U2, U5) or, in one case, two RNAs (U4/U6) complexed with approximately half a dozen proteins [see 45 for review].

U1 snRNP

The proposal that U1 snRNP might be involved in pre-mRNA splicing in higher eukaryotes was made several years ago on the basis of an observed complementarity between the highly conserved 5' end of U1 snRNA and the 5' splice site consensus sequence [46,47]. Since then, a wealth of evidence has accumulated to support the involvement of U1 in this role. First, U1 snRNP has been shown to bind a 5' splice site both in purified preparations [48] and in crude extracts [49–51], and binding occurs very early in the splicing reaction [50]. Second, oligonucleotide-targeted degradation of the 5' end of U1 RNA [49,52] and immunoprecipitation of U1 snRNP have been shown to inhibit splicing both in vivo [53] and in vitro [54]. Third, and perhaps most convincingly, direct genetic evidence for a base-pairing interaction between U1 RNA and the 5' splice site consensus sequence has been provided by the demonstration that a mutant U1 snRNA, containing a single base change in its 5' end, can compensate for an otherwise splicing-deficient complementary mutation in a 5' splice site in vivo [55].

The analysis of the interactions of U1 snRNP with the 5' splice site using the techniques of RNase protection and immunoprecipitation has indicated the existence of two distinct phases of interaction. The initial binding, which protects primarily the 5' splice site consensus sequence, occurs very early in the reaction and depends critically on the integrity of the 5' end of U1 snRNA [50,51]. This interaction appears to be rather tenuous [56] and occurs essentially independently of 3' splice site sequences or intact U2 snRNP [50]. The second phase of interaction with the 5' splice site, which does not appear to require the integrity of the 5' end of U1 RNA, is characterized by the protection of sequences extended in both 5' and 3' directions from the initially protected core [50]. This second phase, which appears more stable, depends on interaction with U2 snRNP and 3' splice site sequences [50,57]. The interaction of U1 snRNP with cryptic 5' splice sites activated by mutation was found to be distinct from that characteristic of authentic 5' splice sites, to the extent that only the second phase of interaction was detectable in vitro [51].

U2 snRNP and Associated Factors

The role of U2 snRNP in recognition of the branch site in higher eukaryotes was also initially suggested on the basis of an observed complementarity between an internal conserved sequence in U2 snRNA and the branch site consensus [36]. Since then, a considerable body of evidence has accumulated to support the involvement of U2 in higher eukaryotes, and its equivalent in yeast (snR20) [58], in this role. However, the nature of the interaction between U2 and the branch site in higher eukaryotes appears quite different from that between U1 and the 5' splice site. While U2 snRNP was shown to bind to the branch site in crude extracts [49,50], this interaction could not be reproduced using purified preparations of U2 [49,50]. Furthermore, binding of U2 was found to occur with a lag. The explanation for this difference was suggested by the analysis of mutant pre-mRNAs, which demonstrated that 3' splice site sequences are required for U2 association with the branch site [50]. It may be recalled that in higher eukaryotes, sequences at the 3' splice junction are essential for the first step in the splicing reaction in vitro, consistent with a role for these sequences in the association of factors with the branch site. Two factors have been described that appear to bind specifically to the 3' splice site: a 70K protein, which also interacts with an snRNP, probably U5, under selected conditions [60,61], and another, less well defined protein, designated U2AF [59]. Using RNA substrate exclusion and competition assays, the second of these factors, U2AF, has been shown to bind to 3' splice site sequences prior to the interaction of U2 snRNP with the branch site and to facilitate this interaction.

The precise boundaries of the interaction of U2 with the branch site

A

                          ★
                          A
                5'— UACUA CA — 3'   branch site
    YEAST ||||| ||
                3'— AUGAU GU — 5'   snR20 (nt33-39)

                          ★
                          A
    GROUP II   5'— UACCU UCAC — 3'
    (self-splicing | ||• •|||
    intron b1-1) 3'— AAGGG GGUG — 5'

                          ★
                          A
                5'— YNCUR Y — 3'   branch site
    MAMMALIAN | ||| |
                3'— AUGAU G — 5'   U2 snRNA (nt33-39)

B yeast branch   ★
       site consensus   U A C U A A C

                        A   1 6 0 1 8 14 1

                        G   0 3 0 0 5 0 1

                        C   6 2 12 1 1 1 8

                        U   8 4 3 13 1 0 5

       mammalian branch   ★
       site consensus   Y N C U R A Y

       match   93% 80% 87% 87%
                                87% 93%

C mammalian branch   ★
       site consensus   5'— Y N C U R A Y — 3'
       U2 snRNA | ||| | |
                          3'— A U G A U   G — 5'

       position base-paired   53% 100% 87% 87%
       (including G.U)   60% 87%

have been defined using the techniques of RNase protection and immunoprecipitation. This analysis has indicated that the region protected by this interaction extends from 20 nt upstream of the branch nucleotide to 10 nt downstream [49,50]. Interestingly, the region protected by this and other factor interactions with the branch site/3' splice site region appears to change through the course of the splicing reaction, so that at later stages both branch site and the 3' splice site are protected in what appears to be a continuous fashion. However, the significance of this change in protection remains unknown.

In yeast, the association of the equivalent of U2 with the branch site has been shown to involve a base-pairing interaction between snR20 and the conserved branch site consensus sequence, UACUA AC. The basis for this conclusion is the demonstration that a mutation in an internal conserved sequence in snR20 can suppress otherwise splicing-deficient complementary mutations in the branch site in vivo [62]. Interestingly, alignment of the snR20 and branch site sequences indicated that the branch nucleotide itself might be bulged from the RNA:RNA duplex generated by this base pairing (Fig. 11.2A). A bulged A-helix motif has also been noted to be conserved in self-splicing group II introns, which involve lariat-structured intermediates [63]. Although the RNA duplex in this case is generated by intramolecular base pairing between sequences within the intron (see Fig. 11.2A), the bulged A has been shown to be the site of the branch in these self-splicing introns also [63] and to be essential for lariat formation [64,65]. The similarity extends to sequences surrounding the branch site in group II introns, which exhibit a high degree of homology to the yeast branch site consensus (Fig. 11.2A). In yeast, the analysis of mutants containing substitutions of the branch nucleotide has demonstrated, consistent with the model for the proposed interaction between snR20 and the branch site, that these mutations are not compensated for by complementary mutations in snR20. Instead, they appear to

◁―――――――――――――――――――――――――――――――

Figure 11.2. Comparison of mammalian, yeast, and group II branch site sequences [drawn in part from 62]. (A) The potential for base-pairing between the yeast branch site and a region of snR20 is presented schematically [62]. A similar base-pairing potential involving the mammalian branch site consensus and a region of U2 snRNA is described. The bulged adenosine that serves as the branch nucleotide in each case is indicated. These features are compared to the proposed structure of sequences surrounding the branch site in self-splicing group II introns (specifically of intron b1-1 [91]). (B) The nucleotide frequencies for each position in the mammalian branch site consensus are tabulated. A consensus is derived, and the fraction of mammalian branch site sequences conforming to this consensus at each position is expressed as a percentage. This consensus is compared to that described for yeast. (C) The fraction of mammalian branch sites able to base-pair with U2 snRNA at each position in the consensus is described as a percentage (G–U pairings included).

be suppressed by mutations in another locus [66]. On the basis of this suppression it has been suggested that a factor distinct from snR20 might be involved in specific recognition of the branch nucleotide. If so, the ability of mutant pre-mRNAs containing an A → C substitution at the branch position to interact stably with the yeast equivalent of U2 snRNP in vitro [67] suggests that this recognition of the branch nucleotide probably occurs subsequent to formation of the base-pairing interaction between snR20 and the branch site contextual consensus.

In spite of the difference in requirements for 3' splice site sequences for the first step of the splicing reaction in yeast and higher eukaryotes, it would appear likely that the interactions between U2 snRNP (and its equivalent in yeast) and the branch site share a number of important features. The possibility that a base-pairing interaction between U2 snRNA and the branch site consensus may occur in higher eukaryotes, while not addressed directly by experiment, is supported by the demonstrated conservation of the sequences involved, both in U2 RNA and the branch site (see Fig. 11.2B and C). As with yeast, base pairing between U2 snRNA and the branch site consensus in higher eukaryotes would also generate a bulged A at the branch position (Fig. 11.2A) However, it remains to be established whether recognition of this "bulged A" also involves a factor analogous to that proposed in yeast.

U4/U6 and U5 snRNPs

In addition to U1 and U2, several other snRNPs have been shown to be involved in pre-mRNA splicing, both in higher eukaryotes and in yeast. Indirect evidence exists to suggest that one of these snRNPs, U5, may bind specifically to the 3' splice site. This evidence is essentially twofold. First, it has been demonstrated that a particle that is resistant to micrococcal nuclease digestion in the same measure as U5 snRNP and that contains an RNA with a cap structure characteristic of snRNAs binds stably to a 3' splice site in vitro [68]. Second, a 70K protein that binds specifically to 3' splice site sequences in vitro has been shown to be associated with an snRNP, probably U5, under selected conditions [60,61]. To reconcile these observations with the finding that U2AF also binds to the 3' splice site, it has been suggested that U2AF and this "U5 complex" bind sequentially to these sequences [59]. This explanation is supported by studies of the order of assembly of snRNPs into the splicing complex, which indicate that the binding of U2 snRNP to the pre-mRNA, and therefore presumably U2AF, precedes the association of U5 (see below).

An additional snRNP, U4/U6 in higher eukaryotes, has also been shown to be required for pre-mRNA splicing in vitro [69,70]. However, it is believed not to be involved in direct contact with the pre-mRNA [50], and its function remains essentially unknown.

Although far less studied, in most cases, in terms of their function, each of the snRNPs described above for higher eukaryotes has been shown to possess a functional equivalent in yeast (U1: snR19 [71]; U2: snR20 [58]; U4/U6: snR14/snR6 [72]; U5: snR7 [73]. This conservation of components is yet another illustration of the similarity between yeast and higher eukaryotes in regard to the overall splicing mechanism.

Splicing Complexes—The Spliceosome

During the course of the splicing reaction in vitro, rapidly sedimenting complexes containing both the pre-mRNA and snRNPs have been shown to form both in higher eukaryotes [74–76] and in yeast [77]. The formation of these multicomponent complexes, termed spliceosomes, has been shown to precede the earliest enzymatic event of splicing and to provide the framework within which these reactions occur. The order of assembly of snRNPs into the spliceosome has been studied intensively for both yeast and higher eukaryotes. The results of these analyses have agreed, with one important qualification, in suggesting the assembly pathway illustrated in Figure 11.3 [78–81]. The essential qualification of this proposed pathway concerns the involvement of U1 snRNP in the complex throughout assembly. A number of studies of spliceosome assembly have failed to detect association of U1 (or its yeast equivalent) with early and late-stage intermediate spliceosomal complexes [81–84]. The consensus interpretation of these observations is that the interaction between U1 snRNP and the pre-mRNA during assembly is considerably less stable than that involving other snRNPs [76]. The alternative interpretation, that U1 is involved only transiently in assembly of the complex, is contradicted by a large body of evidence [50,57,74,76,78].

Overall, the assembly pathways in yeast and higher eukaryotes appear remarkably similar. As shown in Figure 11.3, the interaction of U1 and U2 snRNPs (or their yeast equivalents) with the pre-mRNA is the earliest event in assembly. This interaction has been shown for higher eukaryotes to be dependent on both 5' and 3' splice site sequences [57,76,78,79, 83,84]. In yeast, formation of this prespliceosomal complex requires only the 5' splice site and branch site sequences [30,34,67,81]. The possibility that U1 and U2 snRNPs might interact with each other at this stage has been suggested by the finding that mutations involving the 5' splice site delay the formation of a complex containing U2 snRNP bound at the branch site. In the next step in assembly, a preformed poly-snRNP containing U4, U5, and U6 snRNAs, or their yeast analogs, associates with this complex. This association has been shown in yeast to be blocked by mutations involving the branch nucleotide. The nature of the interactions between each snRNP and other components of this intermediate complex is essentially a mystery. Finally, U4 snRNA (or its equivalent in

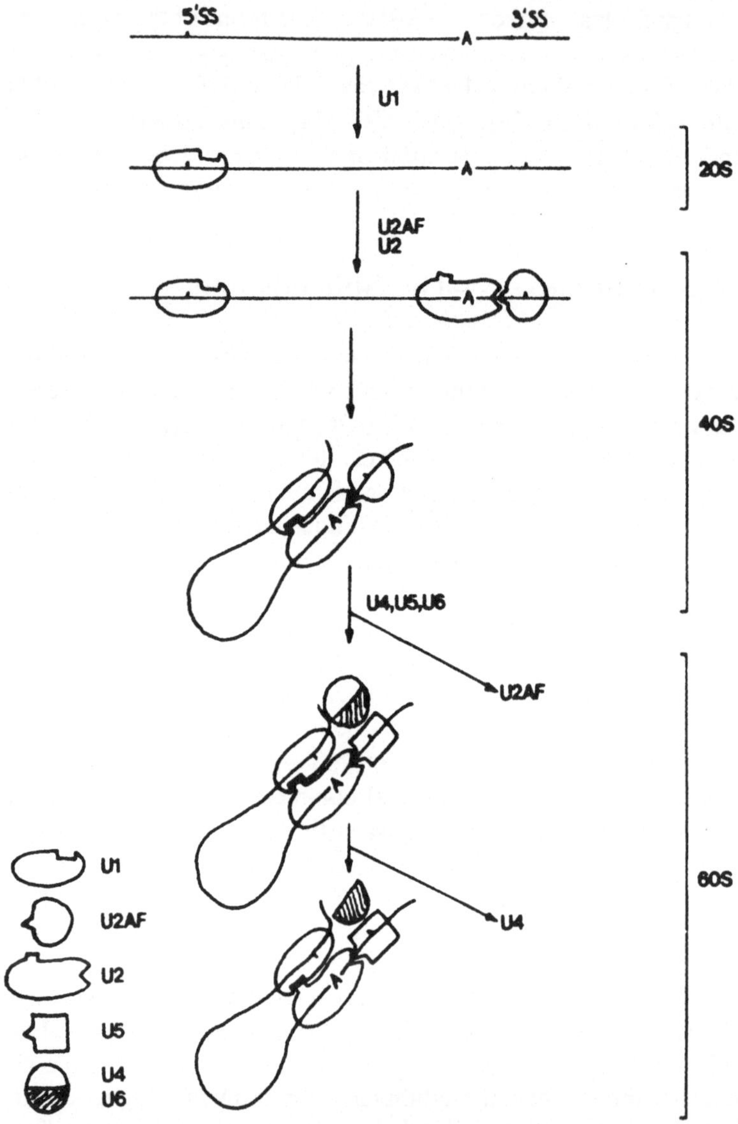

Figure 11.3. Schematic outline of the assembly of the mammalian spliceosome. The sizes of the intermediate and "mature" splicing complexes are indicated on the right. The possibility that U1 and U2 snRNA might interact at the stage of the 40S complex is suggested, but not proved, by the apparent interaction between 5' splice site and branch site regions which characterizes formation of this complex [46,52].

yeast) dissociates from the complex, with the resultant formation of the "mature" spliceosome. At this stage, immediately preceding the first enzymatic reaction of splicing, U2 snRNP has been shown to be in contact with the 5' splice site, while U1 snRNP is in contact with the branch site region [50].

Alternative Splicing

Alternative splicing of pre-mRNAs is known to be of widespread importance in higher eukaryotes in the generation of protein diversity and, possibly, in the control of gene expression [reviewed in 85]. In spite of the considerable progress that has been made in defining the mechanism of splicing of simple pre mRNAs—i.e., precursors that are not able to be spliced by alternative pathways (reviewed above)—little is known of the detailed mechanism and regulation of alternative splicing. The basic pathway by which alternatively spliced RNAs are processed does, however, appear to be the same as simple precursors [e.g., 35,86,87].

One factor that has been shown to be an important determinant of alternative splice site selection in regard to both alternative splicing and the phenomenon of cryptic splice site activation, is the degree of homology of a given splice site to its consensus sequence. The activation of cryptic 5' splice sites in vivo has been shown to result from mutations which increase their homology to the consensus [reviewed in 17]. For SV40 early pre-mRNA, the adenovirus E1A precursor and other, artificially constructed substrates that are spliced alternatively in vivo using multiple 5' splice sites, mutations that improve the match of a particular 5' splice site to the consensus (thus improving its complementarity with the 5' end of U1 snRNA), have been shown to enhance the use of the mutated site [55,88,89]. Together, these observations have indicated that, as for authentic vs. cryptic 5' splice sites, a competition exists between alternative 5' splice sites for factor binding (presumably of U1 snRNP) [48,50,51], the outcome of which is determined, at least in part, by their homology to the consensus sequence.

The homology of a given 3' splice site to the consensus sequence is similarly thought to be an important determinant of 3' splice site selection. Mutations involving the CAG/G at the 3' splice junction have been shown, in some instances, to result in the activation of cryptic 3' splice sites in vivo [63], as have deletions involving the polypyrimidine tract [19,20]. These studies have indicated that a competition exists between authentic and cryptic 3' splice sites for factors, the outcome of which is determined largely by their homology to the 3' splice site consensus sequence. A role for the polypyrimidine stretch specifically in alternative splicing has been suggested in the case of the SV40 early pre-mRNA. The pattern of splicing demonstrated by this pre-RNA in vivo, which involves the use of two alternative 5' splice sites either of which may be joined to a single 3' splice site, has been shown to be strongly influenced by the pyrimidine/purine content of the "polypyrimidine" tract [21] as well as the length of this sequence [90]. A correspondence has also been noted between the existence of purine-rich, as opposed to pyrimidine-rich, stretches and the ability of several pre-mRNAs to undergo alternative splicing using multiple 3' splice sites. This correlation has led to the proposal that such purine-rich regions might act as alternative splicing signals, recognized by as yet unidentified *trans*-acting factors [91–93].

Another factor that has been shown to influence splice site selection is the size of the intron removed. The efficiency of splicing has been found to be dramatically reduced in vivo and in vitro for introns truncated beyond a length of approximately 80 nt [19,39,94], resulting at the same time in the activation of 5' or 3' splice sites upstream or downstream, respectively, of the authentic sites. A likely explanation of this minimum length requirement is the necessity for simultaneous binding of factors, among them U1 and U2 snRNPs, to the 5' splice site and branch site, respectively, which is sterically constrained for introns smaller than the minimum. The possibility that intron size might be a determinant of alternative splicing has been explored for the SV40 early pre-mRNA, which involves the alternative removal of introns of 346 nt (large T) and 66 nt (small t). The pattern of splicing described by this pre-RNA in vivo [90,95] and in vitro [96] was found to be profoundly affected by the size of the small t intron. Expansion mutants containing sequences inserted between the small t 5' splice site and branch site were found to splice small t RNA far more efficiently, relative to large T, than the wild type, and deletions as small as 2 nt greatly reduced small t splicing in vivo.

Another possible determinant of alternative splicing in cases involving multiple 5' splice sites joined to a shared 3' splice site, which has been suggested by the example of the SV40 early pre-mRNA, is a competition among alternative pathways for factors, primarily U2 snRNP, bound to the branch site region. Splicing of SV40 early pre-RNA to large T mRNA has been shown to use multiple alternative branch sites, the most downstream of which are used also by the small t splice [35]. Mutational analysis of the T/t branch site region has indicated that the relative efficiencies of large T vs. small t splicing are determined in part by a competition between these alternative splicing pathways for factor interactions with the available branch sites [96,a,b]. Interestingly, the predominant branch site used in vitro in extracts prepared from two different human cell types (HeLa and 293) and in vivo in microinjected *X. laevis* oocytes was found to vary [35] in a manner correlated with previously observed cell-specific differences in the ratio of alternatively spliced large T and small t mRNAs [53,95]. Thus an attractive model is that differential interactions of factors with the lariat branch site region dictate the choice of alternative 5' splice sites.

The multiple branching pattern observed for large T may not be a universal feature of alternative splicing involving multiple 5' splice sites; splicing of the adenovirus E1A pre-mRNA appears to use only a single shared branch site for all three alternative splicing pathways demonstrated by this precursor [87]. But it is possible that a similar competition for factors bound to the branch site region is a determinant of alternative splicing pattern even in instances in which a single branch site is used. A variation on this theme appears to occur in the case of the polyoma early pre-mRNA. Alternative splicing of this pre-RNA involves two 5' splice sites that are joined in three of four possible combinations to two alternative 3' splice sites [97]. In this instance, the close positioning of 3' splice junctions places the branch site for the downstream 3' splice site, used by middle t splicing, within the polypyrimidine stretch of the upstream 3' splice site, shared by large T and small t splices [Noble et al., manuscript submitted]. As a result, it appears that a competition between these overlapping splicing signals for different factors determines at least in part the relative efficiencies of the large T and small t vs. the middle t splice.

In addition to primary sequence effects, it has been suggested that the secondary, or higher-order, structure of pre-mRNA might also influence alternative splice site selection. This suggestion is supported both by circumstantial evidence, the correlation between differences in 5' or 3' ends and a given splicing pattern in vivo, and by the observed effects of distal mutations in alternative splice site selection in vivo [22,98–101]. A more direct test of the role of secondary structure as a determinant of alternative splicing has been made using mutant pre-RNAs that contain inverted repeats of either intron or splice site sequences. Analysis of such precursors has demonstrated that splice site selection in vitro can be dramatically influenced by these mutations [102,103]; however, similar studies in vivo produced conflicting results [88,95,102,103]. Interestingly, for pre-mRNAs that have the capacity to be spliced alternatively, including SV40 early pre-mRNA, these mutations appeared to influence the efficiency of splicing in vivo using a particular splice site in the anticipated way [88,95], whereas precursors that could normally be spliced in only one way demonstrated little or no effect of mutation.

Another factor that appears to play a role in determining the efficiency with which a given splice site is selected is its position within the precursor in relation to other splice sites, as observed in mutant globin genes containing duplicated splice sites [104,105]. This influence has been suggested to reflect a role for exon sequences in splice selection, with longer lengths of exon sequence adjacent to duplicated splice sites enhancing their use [106]. However, again, in a pre-RNA capable of alternative splicing, this phenomenon was not observed [95].

Finally, it has been suggested that *trans*-acting factors, in some instances at least, play a role in determining alternative splice site selection. Although the evidence for this is somewhat circumstantial and derives principally from observations that certain pre-mRNAs may be spliced alternatively in distinct ways at different stages of differentiation or in different tissues or cell types [reviewed in 85], it would appear that the role of such factors in determining splice site selection is important for the regulation of alternative splicing [e.g., 44,95]. One possible way in which *trans*-acting factors might influence splice site selection is as components of the spliceosome. Indeed, some evidence exists to suggest that distinct subpopulations of some snRNAs, including U1, U4, and U6, are expressed according to developmental stage [107,108]. However, no evidence has been obtained to link these changes with differences in alternative splicing.

In summary, a number of factors have been defined that influence alternative splice site selection. Future efforts at elucidating the mechanism of alternative splicing for specific pre-mRNAs must focus on the detailed interactions involved in the assembly of "alternative spliceosomes." A principal aim of these studies must be to establish the precise stage, or stages, at which assembly becomes committed to a particular alternative splicing pathway.

References

1. Berget, S.M., Moore, C., and Sharp, P.A. (1977) Proc. Natl. Acad. Sci. USA *74*, 3171–3175.
2. Chow, L.T., Gelinas, R.E., Broker, T.R., and Roberts, R.J. (1977) Cell *12*, 1–8.
3. Green, M.R. (1986) Annu. Rev. Genet. *20*, 671–708.
4. Padgett, R.A., Grabowski, P.J., Konarksa, M.M., Sellers, S., and Sharp, P.A. (1986) Annu. Rev. Biochem. *55*, 898–903.
5. Padgett, R.A., Hardy, S.F., and Sharp, P.A. (1983) Proc. Natl. Acad. Sci. USA *80*, 5230–5234.
6. Hernandez, N., and Keller, W. (1983) Cell *35*, 89–99.
7. Krainer, A.R., Maniatis, T., Ruskin, B., and Green, M.R. (1984) Cell *36*, 993–1005.
8. Lin, R.-J., Newman, A.J., Cheng, S.-C., and Abelson, J. (1985) J. Biol. Chem. *260*, 14780–14792.
9. Grabowski, P., Padgett, R.A., and Sharp, P.A. (1984) Cell *37*, 415–427.
10. Ruskin, B., Krainer, A.R., Maniatis, T., and Green, M.A. (1984) Cell *38*, 317–331.
11. Konarska, M.M., Grabowski, P.J., Padgett, R.A., and Sharp, P.A. (1985) Nature *313*, 552–557.
12. Rodriguez, J.R., Pikielny, C.W., and Rosbash, M. (1984) Cell *39*, 603–610.
13. Domdey, H., Apostol, B., Lin, R.-J., Newman, A., Brody, E., and Abelson, J. (1984) Cell *39*, 611–621.
14. Zeitlin, S., and Efstratiadis, A. (1984) Cell *39*, 589–602.
15. Mount, S. (1982) Nucleic Acids Res. *10*, 459–472.
16. Breathnach, R., Benoist, C., O'Hare, K., Gannon, F., and Chambon, P. (1978) Proc. Natl. Acad. Sci. USA *75*, 4853–4857.

17. Mount, S., and Steitz, J. (1983) Nature *303*, 380–381.
18. Aebi, M., Hornig, H., Padgett, R.A., Reiser, J., and Weissmann, C. (1986) Cell *47*, 555–565.
19. Wieringa, B., Hofer, E., and Weissman, C. (1984) Cell *37*, 915–925.
20. Van Santen, V.L., and Spritz, R.A. (1985) Proc. Natl. Acad. Sci. USA *82*, 2885–2889.
21. Fu, X.-Y., Ge, H., and Manley, J.L. (1988) EMBO J. *7*, 809–817.
22. Reed, R., and Maniatis, T. (1985) Cell *41*, 95–105.
23. Ruskin, B., and Green, M.R. (1985) Nature *317*, 732–734.
24. Langford, C.J., Klinz, F.-J., Donath, C., and Gallwitz, D. (1984) Cell *36*, 645–653.
25. Teem, J.L., Abovich, N., Kaufer, N.F., Schwindinger, W.F., Warner, J.R., Levy, A., Woolford, J., Leer, R.J., Van Raamsdonk-Duin, N.N.C., Nager, W.H., Planta, R.J., Schultz, L., Friesen, J.D., Fried, H., and Roshbash, N. Nucleic Acids Res. *12*, 8295–8312.
26. Gallwitz, D. (1982) Proc. Natl. Acad. Sci. USA *79*, 3493–3497.
27. Parker, R., and Guthrie, C. (1985) Cell *41*, 107–118.
28. Newman, A.J., Lin, R.-J., Cheng, S.-C., and Abelson, J. (1985) Cell *42*, 335–344.
29. Jacquier, A., and Roshbash, M. (1986) Proc. Natl. Acad. Sci. USA *83*, 5835–5839.
30. Vijayraghavan, V., Parker, R., Tamm, J., Iimura, Y., Rossi, J., Abelson, J., and Guthrie, C. (1986) EMBO J. *5*, 1683–1695.
31. Fouser, L.A., and Friesen, J.D. (1986) Cell *45*, 81–93.
32. Langford, C.J., and Gallwitz, D. (1983) Cell *33*, 519–527.
33. Fouser, L.A., and Friesen, J.D. (1987) Mol. Cell. Biol. *7*, 225–230.
34. Rymond, B.C., and Roshbash, M. (1985) Nature *317*, 735–737.
35. Noble, J.C.S., Pan, Z.-Q., Prives, C., and Manley, J.L. (1987) Cell *50*, 227–236.
36. Keller, E.B., and Noon, W.A. (1984) Proc. Natl. Acad. Sci. USA *81*, 7417–7420.
37. Hornig, H., Aebi, M., and Weissmann, C. (1986) Nature *324*, 589–591.
38. Padgett, R.A., Konarska, M.M., Aebi, M., Hronig, H., Weissmann, C., and Sharp, P.A. (1985) Proc. Natl. Acad. Sci. USA *82*, 8349–8353.
39. Ruskin, B., Green, J.M., and Green M.R. (1985) Cell *41*, 833–834.
40. Jacquier, A., Rodriguez, J.R., and Roshbash, M. (1985) Cell *43*, 423–430.
41. Cellini, A., Parker, R., McMahon, J., Guthrie, C., and Rossi, J. (1986) Mol. Cell. Biol. *6*, 1571–1578.
42. Pikielny, C.W., Teem, J.L., and Roshbash, M. (1983) Cell *34*, 395–402.
43. Cellini, A., Felder, E., and Rossi, J. (1986) EMBO J. *5*, 1023–1030.
44. Brietbart, R.E., Andreadis, A., and Nadal-Ginard, B. (1987) Annu. Rev. Biochem. *56*, 467–495.
45. Birnstiel, M.L., ed. (1988). Structure and Function of Major and Minor Small Nuclear Ribonucleoprotein Particles. Springer-Verlag, Berlin, 216 pp.
46. Lerner, M.R., Boyle, J.A., Mount, S., Wolin, S.L., and Steitz, J.A. (1980) Nature *283*, 220–224.
47. Rogers, J., and Wall, R. (1980) Proc. Natl. Acad. Sci. USA *77*, 1877–1879.
48. Mount, S.M., Petterson, I., Hinterberger, M., Karmas, A., and Steitz, J.A. (1983) Cell *33*, 509–518.

49. Black, D.L., Chabot, B., and Steitz, J.A. (1985) Cell *42*, 737–750.
50. Chabot, B., and Steitz, J.A. (1987) Mol. Cell. Biol. *7*, 281–293.
51. Chabot, B., and Steitz, J.A. (1987) Mol. Cell. Biol. *7*, 698–707.
52. Kramer, A., Keller, W., Appel, B., and Luhrmann, R. (1984) Cell *38*, 299–307.
53. Fradin, A., Jove, R., Hemenway, C., Keiser, H.D., Manley, J.L., and Prives, C. (1984) Cell *37*, 927–936.
54. Padgett, R.A., Mount, S.M., Steitz, J.A., and Sharp, P.A. (1983) Cell *35*, 101–107.
55. Zhuang, Y., and Weiner, A.M. (1986) Cell *46*, 827–835.
56. Ruskin, B., and Green, M.R. (1985) Cell *43*, 131–142.
57. Zillman, M., Rose, S.D., and Berget, S.M. (1987) Mol. Cell. Biol. *7*, 2877–2883.
58. Ares, M. (1986) Cell *47*, 49–59.
59. Ruskin, B., Zamore, P.D., and Green, M.R. (1988) Cell *52*, 207–219.
60. Tazi, J., Alibert, C., Temsamani, J., Reveillaud, I., Cathala, G., Burnel, C., and Jeanteur, P. (1986) Cell *47*, 755–766.
61. Gerke, V., and Steitz, J.A. (1986) Cell *47*, 973–984.
62. Parker, R., Siliciano, P.G., and Guthrie, C. (1987) Cell *49*, 229–239.
63. Van der Veen, R., Arnberg, A.C., Van der Horst, G., Bonen, L., Tabak, H.F., and Grivell, L.A. (1986) Cell *44*, 225–234.
64. Schmelzer, C., and Muller, M.W. (1987) Cell *51*, 753–762.
65. Van der Veen, R., Kwakman, J.H., and Grivell, L.A. (1987) EMBO J. *6*, 3827–3831.
66. Couto, J.R., Tamm, J., Parker, R., and Guthrie, C. (1987) Genes Devel. *1*, 445–455.
67. Pikielny, C.W., Rymond, B.C., and Roshbash, M. (1986) Nature *324*, 341–345.
68. Chabot, B., Black, D.L., LeMaster, D.M., and Steitz, J.A. (1985) Science *230*, 1344–1349.
69. Berget, S.M., and Robberson, B.L. (1986) Cell *46*, 691–696.
70. Black, D.L., and Steitz, J.A. (1986) Cell *46*, 697–704.
71. Kretzner, L., Rymond, B.C., and Roshbash, M. (1987) Cell *50*, 593–602.
72. Siliciano, P.G., Brow, D.A., Roiba, H., and Guthrie, C. (1987) Cell *50*, 585–592.
73. Patterson, B., and Guthrie, C. (1987) Cell *49*, 613–624.
74. Grabowski, P.J., Seiler, S.R., and Sharp, P.A. (1985) Cell *42*, 345–353.
75. Frendeway, D., and Keller, W. (1985) Cell *42*, 355–367.
76. Bindereif, A., and Green, M.R. (1986) Mol. Cell. Biol. *6*, 2582–2592.
77. Brody, E., and Abelson, J. (1985) Science *228*, 963–967.
78. Bindereif, A., and Green, M.R. (1987) EMBO J. *6*, 2415–2424.
79. Konarska, M.M., and Sharp, P.A. (1987) Cell *49*, 763–774.
80. Lamond, A.I., Konaraka, M.M., Grabowski, P.J., and Sharp, P.A. (1988) Proc. Natl. Acad. Sci. USA *85*, 411–415.
81. Cheng, S.-C., and Abelson, J. (1987) Genes Devel. *1*, 1014–1027.
82. Grabowski, P.J., and Sharp, P.A. (1985) Science *223*, 1294–1299.
83. Konarska, M.M., and Sharp, P.A. (1985) Cell *48*, 845–855.
84. Lamond, A.I., Konarska, M.M., and Sharp, P.A. (1987) Genes Devel. *1*, 532–543.

85. Breitbart, R.E., and Nadal-Ginard, B. (1987) Cell *49*, 793–803.
86. Noble, J.C.S., Prives, C., and Manley, J.L. (1986) Nucleic Acids. Res. *14*, 1219–1235.
87. Schmitt, P., Gattoni, R., Keohavong, P., and Stevenin, J. (1987) Cell *50*, 31–39.
88. Eperon, L.P., Estibeivo, J.P., and Eperon, I.C. (1986) Nature *324*, 280–282.
89. Zhuang, Y., Leung, H., and Weiner, A.M. (1987) Mol. Cell. Biol. *7*, 3018–3020.
90. Fu, X.-Y., Colgan, J., and Manley, J.L. (1988) Mol. Cell. Biol. *8*, 3582–3590.
91. Falkendahl, S., Parker, V.P., and Davidson, N. (1985) Proc. Natl. Acad. Sci. USA *82*, 449–453.
92. Rozek, C.E., and Davidson, N. (1986) Proc. Natl. Acad. Sci. USA *82*, 2128–2132.
93. Bernstein, S.I., Hansen, C.J., Becker, K.D., Wassenberg, D.R., Roche, E.S., Donady, J.J., and Emerson, C.P. (1986) Mol. Cell. Biol. *6*, 2511–2519.
94. Ulfendahl, P.J., Pettersson, U., and Akusjarvi, G. (1985) Nucleic Acids Res. *13*, 6299–6315.
95. Fu, X.-Y., and Manley, J.L. (1987) Mol. Cell. Biol. *7*, 738–748.
96. Manley, J.L., Fu, X.-Y., Noble, J.C.S., and Ge, H. (1987) In Molecular Biology of RNA: New Perspectives. Inouye, M., and Dudock, B.S. (eds.). Academic Press, New York, pp. 97–112.
96a. Noble, J.C.S., Prives, C., and Manley, J.L. (1988) Genes Devel. *2*, 1460–1475.
96b. Noble, J.C.S., Ge, H., Chaudhuri, M., and Manley, J.L. (1989) Mol. Cell. Biol. *9*, in press.
97. Treisman, R., Cowie, A., Favaloro, J., Jat, P., and Kamen, R. (1981) J. Mol. Appl. Genet. *1*, 83–92.
98. Khoury, G., Gruss, P., Dhar, R., and Lai, C.-J. (1979) Cell *18*, 85–92.
99. Piatak, M., Subramanian, K.N., Roy, P., and Weissmann, S.M. (1981) J. Mol. Biol. *153*, 589–618.
100. Somasekhar, M.B., and Mertz, J.E. (1985) Nucleic Acids. Res. *13*, 5591–5609.
101. Plotch, S.J., and Krug, R.M. (1986) Proc. Natl. Acad. Sci. USA *83*, 5444–5448.
102. Solnick, D. (1985) Cell *43*, 667–676.
103. Solnick, D., and Lee, S.I. (1987) Mol. Cell. Biol. *7*, 3194–3198.
104. Kuhne, T., Wieringa, B., Reiser, J., and Weissman, C. (1983) EMBO J. *2*, 727–733.
105. Lang, K.M., and Spritz, R.A. (1983) Science *220*, 1351–1355.
106. Reed, R., and Maniatis, T. (1986) Cell *46*, 681–690.
107. Lund, E., and Dahlberg, J.E. (1987) Genes Devel. *1*, 39–46.
108. Lund, E., Kahan, B., and Dahlberg, J.E. (1985) Science *229*, 1271–1274.

Chapter 12
Ribonucleoproteins and the Structure of Nascent Transcripts

Stanley F. Barnett, Stephanie J. Northington, and Wallace M. LeStourgeon

Many ultrastructural and biochemical findings suggest that pre-mRNA is packaged during transcription into a repeating array of regular ribonucleoprotein particles by the six "core" proteins of 40S nuclear ribonucleoprotein particles (40S hnRNP particles). If true, then pre-mRNA is packaged in a fashion not totally dissimilar to the nucleosomal packaging of DNA. While the packaging of nascent transcripts into a repeating "ribonucleosome" structure [1,2] seems to complicate our present view of RNA splicing (i.e., splice site recognition, spliceosome assembly, lariat formation, excision, and ligation), there is considerable evidence that all these events do occur while RNA exists in a highly packaged state. On this point it is worth noting that biological systems have worked through the structural and sequence recognition problems created by the packaging of DNA into nucleosomes. More important, it now seems likely that the events of RNA processing can be understood in mechanistic detail because much progress has recently been made in the ability to isolate and study the fundamental repeating element of nascent transcripts and to monitor the biochemical events of RNA packaging in vitro. A better knowledge of monoparticle structure is a prerequisite to a detailed understanding of the events of RNA processing and will help to clarify present attempts to "characterize" the spliceosome complex, a structure quite different from the 40S hnRNP particles that function to package the bulk of nascent pre-mRNA transcripts and prevent the formation of heteroduplex RNA.

In this article we will review the evidence which argues (1) that the fundamental repeating structural element of nascent transcripts is synonymous with isolated 40S nuclear ribonucleoprotein particles (40S hnRNP particles), (2) that these structures are mostly composed of three unique tetrameric groupings of the six major hnRNP core proteins, and (3) that during transcription three or four copies of each tetramer assemble through an RNA-activated mechanism to package 700-nt increments of RNA into regular repeating structures possessing a defined mass and protein stoichiometry.

Ultrastructural Evidence

The ultrastructural evidence that pre-mRNA is packaged into a repeating array of regular ribonucleoprotein particles has come from studies involving two different experimental approaches. In the more traditional procedures, fixatives have been used to immobilize and stabilize the contents of isolated nuclei or chromosomes [3–8] in situ. Both thin and thick sections of these materials have been examined with low-voltage [3–6] and high-voltage electron microscopy [7–9]. In these studies nascent transcripts are identified as template-associated particulate complexes that form lateral elements analogous to the branches of the Christmas tree structures seen in spreads of active genes. This experimental approach has consistently demonstrated that newly synthesized RNA exists as an array of more or less contiguous particles of uniform size which, depending on the methodology used, vary in diameter from 20 to 30 nm. While fixation in situ circumvents to some extent the concern of artifactual generation of particles via protein rearrangement during sample preparation, this visualization of particles in sections does not allow one to unambiguously distinguish between particles which represent a fundamental repeating structural element and larger spliceosome complexes [10].

In the second ultrastructural approach, several techniques have been used to spread transcriptionally active chromatin on grids coated with various supports [1,10–21]. When aggressive procedures are used to spread active genes, for example the s36-1 and s38-1 chorion genes of *Drosophila,* the transcripts appear as protein-coated fibrils with a diameter near 5 nm [10]. The only particulate complexes observed are at the 5' and 3' splice sites flanking introns; these are too large to be 40S monoparticles and are probably remnants of larger spliceosome complexes. However, as discussed in Osheim et al. [10], when gentle procedures are used to spread transcriptive units, one observes either large nondispersed granules closely associated with the chromatin strand or, in more dispersed areas, a contiguous array of uniform particles with the same dimensions as negatively stained 40S monoparticles (near 22 nm).

Among the most convincing studies to date are the ultrastructural and biochemical analyses conducted in the laboratory of John Sommerville on isolated transcription complexes from *Triturus* ovaries [15–17,21]. In this system nascent transcripts are organized into a repeating array of uniform 20 nm RNP "beads" which are converted into monoparticles upon mild RNase treatment. Malcolm and Sommerville [21] went on to show that the structure of the hnRNP particles is dependent on the RNA substrate (i.e., that the RNA-associated proteins alone will not form particles), that the particles are homogeneous in protein composition regardless of their location on the transcript, that the particles dissociate in high salt but are not affected by EDTA, that the particles appear to package RNA in a sequence-independent manner, that in a very low salt environment at pH 8 to 9 protein rearrangement leads to the formation of a fibrillar structure, and that the RNA substrate assumes a highly complex secondary structure when stripped of its associated proteins. It must be noted here that all of these properties are true of isolated hnRNP from exponentially growing HeLa cells (discussed below).

While the observations briefly summarized above have mostly been obtained through studies on amphibian or insect cells, the basic RNA packaging system appears to be well conserved in mammalian cells. For example, in the study of Tsanev and Djondurov [18] nascent transcripts were gently released from the chromatin of mouse erythroleukemia cells on EM grids and were found to exist as contiguous arrays of hnRNP-size particles. Moreover, not only were the individual particles the size of isolated 40S monoparticles, but in many of the oligomeric complexes monoparticles with a centrally located granule, often seen in EMs of isolated 40S monoparticles, were observed. It should also be remembered that the first ultrastructural studies demonstrating the relationship between isolated monomer hnRNP particles and oligomeric hnRNP complexes were conducted in the laboratory of Georgiev [22–24] on hnRNA released from isolated rat liver nuclei.

Biochemical Evidence

The biochemical evidence that nascent transcripts are packaged into a contiguous array of regular particles by multiple copies of a few abundant nuclear proteins is equally convincing. For example, if nuclei isolated under conditions that minimize nuclease and protease activity are dispersed by ultrasound or gently extracted with low-salt solutions, almost all of the hnRNA is recovered in the supernatant after the chromatin is removed by centrifugation [22,25–32]. If the supernatant is then subjected to sedimentation through 15% to 30% sucrose density gradients, the hnRNA is recovered in a broad zone between 30S and 300S in association with the six major "core" proteins of oligomeric hnRNP complexes [25,33,34]. Numerous minor high-molecular-weight polypeptides can also be seen in gels of gradient fractions containing the hnRNP complexes. Electron micrographs of this material reveal mostly monomer particles in the 30S to 50S gradient fractions and polyparticles in the faster-sedimenting regions [22–24]. Moreover, if HeLa nuclear sonicates that contain mostly polyparticles are incubated at 37°C for 15 minutes to facilitate endogenous nuclease activity or if RNase-A is added at low levels, then essentially all of the hnRNA is recovered in uniform 22 to 24-nm monoparticles which sediment as a single peak at 40S in gradients [27,35–37]. These 40S monomers possess the same core protein composition and stoichiometry as the polyparticle complexes sedimenting at 30S to 300S prior to nuclease treatment [27]. When polyparticle complexes are converted to monoparticles, most of the minor high-molecular-weight proteins remain in a dissociated form at the top of gradients. The six major hnRNP proteins migrate in SDS PAGE as three groups of doublet bands (the A1–A2, B1–B2, and C1–C2 doublets) [25,38] with apparent molecular weights between 32 and 44 kD [25,27,28,33–36,39–42]. As discussed in the following sections, several experimental findings now argue that multiple copies of these proteins assemble to form monoparticles in the molar ratio $n(3A1{:}3A2{:}1B1{:}1B2{:}3C1{:}1C2)$ where $n = 3$ or 4.

Several lines of evidence demonstrate that gradient-purified 40S hnRNP particles, composed mostly of the A, B, and C group doublets, correspond to the 20 to 22-nm particles associated with gently spread or glutaraldehyde-fixed transcripts. First, the RNA packaged in isolated polyparticles is quantitatively converted to 40S monoparticles upon mild nuclease activity, and the RNA recovered from the latter has the same sequence complexity as total nuclear hnRNA [22,23,30,32,43,44]. Second, gradient-purified 40S hnRNP particles possess the same ultrastructural morphology as those seen in spread transcriptive units [18] and in isolated polyparticle complexes [22,24]. Third, monoclonal and polyclonal antibodies against the "core" proteins of 40S hnRNP specifically localize over the ribonucleoprotein granules containing the nascent transcripts of lampbrush and polytene chromosomes [45–48]. Fourth, the "core" proteins of 40S hnRNP are present in transcriptionally active nuclei at very high concentrations. In HeLa cells, for example, there is 30% to 40% as much of the core hnRNP as the core histones [38]. This simple observation alone argues strongly that the function of the six core proteins is primarily to package RNA. Finally (as discussed in the following section), we have shown through reconstitution studies that the core proteins assemble in vitro on appropriate lengths of exogenous RNA to form 40S monoparticles, dimers, and oligomeric complexes [26]. The correct assembly of hnRNP particles is highly dependent on RNA length: monoparticles assemble on 700 ± 20 nucleotides of RNA, and integral multiples of this length support the assembly of dimers and oligomeric complexes [26]. This fundamental packaging activity occurs in a sequence-independent manner and does not require nucleotide triphosphates as an energy source.

Taken together, the ultrastructural and biochemical observations summarized above demonstrate that during transcription 700-nt increments of pre-mRNA are packaged by multiple copies of a few abundant nuclear proteins to form a repeating array of regular particles. It is, however, well established that the six core proteins of 40S particles are not the only proteins associated with packaged RNA in vivo. While the core proteins are easily seen to be the major components of gradient-isolated hnRNP and although these proteins by themselves possess the intrinsic ability to package RNA into a repeating array of regular particles like that observed in vivo (discussed below), numerous other proteins are associated with packaged RNA in vivo. For example, as shown by the Dreyfuss group, if antibodies against the core proteins are used to isolate hnRNP by immunoadsorption, numerous higher-molecular-weight proteins are observed [41,41]. These probably correspond to the minor high-molecular-weight proteins present in isolated polyparticle preparations prior to dissociation into monoparticles by nuclease treatment. At least two of these proteins (68 and 120 kD) are defined as true RNA-binding proteins by cross-linking to RNA upon UV irradiation and by the coisolation of the core hnRNP proteins by immunoadsorption using antibodies against these proteins [41,49]. More recently, it has been argued that at least 24 polypeptides are true components of hnRNP complexes in vivo because they are not dissociated by heparin from hnRNP complexes isolated by immunoadsorption [50]. While resistance to heparin dissociation is not likely to define hnRNPs, and one should not conclude that proteins dissociated by heparin are not truly associated with hnRNP in vivo, it is likely that, given adequate sensitivity, one can probably detect far more than 24 hnRNP-associated polypeptides. On this contention we mention the recent approach of Joseph Gall's laboratory [51] where a family of monoclonal antibodies against germinal vesicle proteins was screened for the ability to bind nascent transcripts in lampbrush chromosome preparations. Two minor proteins of 90 and 120 kD were found to possess the same distribution on nascent transcripts as the core proteins of hnRNP.

The minor high-molecular-weight proteins, which are more apparent in polyparticle preparations or in hnRNP complexes isolated by immunoadsorption, are not likely to be structural elements of the core particle. They easily dissociate without affecting core hnRNP composition or stability [26,35], most are present at less than one copy per monoparticle, and they are not consistently recovered in a fixed stoichiometry [26,27,35]. However, these components may be dynamically involved in the events of RNA packaging. Some may function to modulate the assembly of core particles at specific sites, while others may be enzymes involved in particle dissociation and RNA processing. Various enzymatic activities have been detected in preparations of isolated hnRNP [52–56], but it is not clear which activities are relevant and which are artifactually associated with the macromolecular complex during isolation.

Packaging of 40S hnRNP Particles

Information on the evolutionary conservation, physical chemical properties, primary structure, immunological properties, RNA-binding domains, and synthesis of the six major core proteins of hnRNP particles can be found in various review articles [2,57–59]. Consistent with the major arguments of this article, emphasis will focus here on the evidence that multiple copies of the six core proteins possess the intrinsic ability to package, through a sequence-independent mechanism, 700-nt increments of RNA into an array of regular particles with the protein composition $n(A1_3, A2_3, B1, B2, C1_3, C2)$ where $n = 3$ or 4. Much of the evidence for this activity of the six major core particle proteins has come from a series of in vitro assembly studies using RNA substrates of defined length and sequence [26].

Until the development of a reliable procedure for the in vitro assembly of 40S hnRNP particles, three properties of isolated 40S hnRNP particles greatly slowed progress in the field. First, the pre-mRNA moiety packaged in monoparticles is only a fragment of the nascent transcript. This is consistent with the facts that most transcripts are packaged into an array of several to many particles and that the RNA fragments in typical monoparticle preparations reflect the sequence complexity of total nuclear pre-mRNA [30,32,43,60,61]. Second, unlike nucleosomes, hnRNP particles do not provide a high degree of nuclease protection for the packaged RNA moiety. Because of this it has not been possible to define with adequate precision the length of RNA actually packaged in each particle. In our early efforts to determine the length of RNA packaged in isolated monoparticles, we observed that even in the best preparations the isolated RNA migrates in gels as a heterodisperse population of fragments between 600 and 800 nt in length [25,28]. Third, although isolated monoparticles sediment in gradients as a more or less homogeneous population of particles, electron micrographs typically reveal size heterogeneity in negatively stained preparations, with diameters between 22 and 24 nm. It now appears that this is caused by the association of accessory proteins with some of the core particles because particles reconstituted on defined-length RNA are highly homogeneous 22-nm structures, and gels of these reveal only trace levels of the minor higher-molecular-weight components [26]. Although whole particles also migrate in electrophoresis as a more or less homogeneous population [62], it has not been possible to determine a precise mass for 40S monoparticles using traditional procedures. As a result, it has not been possible to determine precisely the protein copy number, the ratio of protein:RNA, or, of course, the topology of protein and RNA in 40S hnRNP particles.

We have identified the parameters required for the assembly of highly homogeneous monomers, dimers, and polyparticle complexes on pre-mRNA molecules of defined length and sequence synthesized in vitro[26]. In brief, gradient-isolated 40S hnRNP particles from exponentially growing HeLa cells are digested aggressively with protease-free micrococcal nuclease to achieve complete protein dissociation. The nuclease is then inactivated with EDTA, and RNA is added at a protein:RNA ratio of 10:1. This is higher than the ratio of protein and RNA in native hnRNP (6 to 8:1) but has proved to yield optimal results. Fortunately, one must exceed a

20:1 ratio in the reaction mix before particles with incorrect protein composition are formed. The particles that assemble spontaneously may be purified by gel filtration chromatography or by sedimentation through sucrose density gradients. Numerous observations argue that the particles that assemble in vitro are structurally identical to native hnRNP. They sediment in gradients exactly like native monoparticles, they have the same protein composition and stoichiometry, they show the same pattern of protein dissociation upon increasing salt concentration, and they have the same sensitivity to nuclease as native 40S hnRNP particles. Also, as in hnRNP in living cells and isolated 40S monoparticles, the C proteins (C1 and C2) of reconstituted particles selectively cross-link to the RNA substrate upon UV irradiation. Finally, native and reconstituted particles have the same appearance in electron micrographs, including the centrally located electron-dense granule. The most important new information derived from the reconstitution studies is listed below:

1. The length of RNA required for the correct assembly of HeLa 40S monoparticles is 700 ± 20 nt. This corresponds well with the heterodisperse 600 to 800-nt peak previously seen when the RNA of native monoparticles is analyzed in denaturing gels [28] or in gradients [25]. Furthermore, integral multiples of this length support the assembly of dimers and oligomeric complexes possessing the same protein composition as monoparticles. In previous reconstitution studies on various homoribopolymers, conducted in the laboratories of Martin [40] and Shafer [35], the length of RNA in monomer particles was estimated from electron micrographs of polyparticle complexes to be 1,000 to 1,200 nt. The RNA length per particle we have recently reported (700 ± 20) was determined by identifying the minimum length required for the stoichiometric assembly of 40S monoparticles [26]. This was then confirmed by showing that dimers assemble on twice this length and by dividing the length of the ϕX-174 genome by 7.5 (the average number of particles that form on this ssDNA of 5,386 nt).
2. If RNA lengths other than integral multiples of 700 are used as the assembly substrate, then aggregate complexes form that sediment in a heterodisperse manner in gradients and that do contain the stoichiometric mix of native particle protein (i.e., $A1_3$, $A2_3$, B1, B2, $C1_3$, C2).
3. If the molar ratio of the core proteins in the reaction mix is intentionally skewed (made to contain less A1, C1, and C2 relative to A2, B1, and B2) by overdigesting the initial nuclear sonicate, the protein stoichiometry in the reconstituted particles is nevertheless correct (i.e., $A1_3$, $A2_3$, B1, B2, $C1_3$, C2) if RNA in the reaction mix is limiting. In these experiments the excess A2, B1, and B2 remain at the top of gradients in a low-molecular weight form.
4. Observations 1, 2, and 3 argue that the core proteins possess the necessary information to assemble into a regular structure when bound to RNA such that an RNA length of 700 just accommodates all

necessary protein-protein interactions. The absolute requirement for RNA in this protein assembly mechanism argues that RNA "activates" the protein-protein binding domains and suggests that as the proteins associate with RNA they become competent to bind other proteins during some kind of stepwise event.

5. The core particle proteins do not themselves possess the ability to bind RNA in a sequence-specific manner, nor do they possess the ability to distinguish between RNA and ssDNA. This is seen in the correct assembly of monoparticles on isolated exons and introns, on phage T4 transcripts that contain no eukaryotic processing signals, on normal mouse and human β-globin pre-mRNA transcripts, on normal and intron-truncated adenovirus transcripts, and on the ϕX-174 ssDNA genome [26]. Based on the ability of particles to assemble on homoribopolymers, Martin's group and Shafer's group also concluded that the core proteins lack sequence binding specificity [35,40]. This was also the basic conclusion of Malcolm and Sommerville, as they always observed long contiguous arrays of particles on newly synthesized RNA [21].

6. Neither a 5' cap structure, a 3' poly(A) moiety, divalent cations, nor nucleotide triphosphates are required for the correct assembly of 40S hnRNP complexes in vitro [26].

The above observations argue that during transcription in HeLa cells, 700-nt lengths of RNA are packaged by multiple copies of the core hnRNP proteins into a repeating array of regular particles. We believe that these findings confirm the ultrastructural and biochemical findings summarized above. The ability of the core proteins to package RNA into a regular repeating structure regardless of sequence suggests that *trans*-acting elements such as snRNP may function in vivo to phase the assembly of some or all particles in a manner consistent with the events of RNA splicing. More importantly, from an experimental perspective, the ability to assemble 40S hnRNP in vitro finally opens the door to a series of rather straightforward studies designed to determine the topology of protein and RNA in monoparticles and to better understand the assembly, disassembly, and possible phasing of monoparticles on nascent transcripts. In the absence of an assay for 40S hnRNP *function*, one might argue that we cannot know if the particles that assemble in vitro are truly identical to native hnRNP. To this we reply that the overwhelming preponderance of evidence argues that the *function of hnRNP is to package RNA*. This is exactly what the core proteins do when confronted with RNA in solution.

Formation of 40S hnRNP Particles

Several experimental observations now demonstrate that in HeLa cells 40S hnRNP particles are formed by the stoichiometric self assembly of the six major core particle proteins in the presence of RNA such that the molar ratio of these proteins in intact particles is $n(A1_3, A2_3, B1, B2, C1_3, C2)$. In this section we will present the findings that demonstrate the basis for this stoichiometry and the evidence that the core proteins exist as three unique tetrameric groupings in intact particles. Unequivocal evidence exists for the $A2_3B1$ [63] and the $C1_3C2$ [37] tetramers, and it is likely that protein A1 exists as a tetramer of $A1_3B2$ (discussed below). The molar ratio of the core particle proteins listed above holds if care is taken to use rapidly growing cells free of mycoplasma and if nuclei are isolated quickly so as to minimize nuclease and protease activity during nuclear isolation and particle preparation. Proteins A1, B2, C1, and C2 are especially labile to protease and also dissociate preferentially upon excessive exposure to nuclease [27]. In addition, particles isolated from slowly growing or quiescent cells are deficient in A1 and frequently reveal less C protein as well [33,64,65]. While densitometry of Coomassie-stained bands in single-dimension SDS PAGE indicated in our early studies that a 1:1:1 relationship existed between A1, A2, and C1, that these proteins were present at about three times the amount of B1 and B2, and that the latter were also present at a 1:1 molar ratio, the experiments summarized below were necessary to identify and confirm the stated stoichiometry and to determine the structural basis for this molar ratio.

Core Particle Proteins

In our first efforts to obtain information on the arrangement of the core particle proteins in intact 40S hnRNP, the cleavable protein cross-linking reagent 3,3'-dimethyl dithiobispropionimidate dihydrochloride (DTBP) was used to cross-link the proteins in sucrose-gradient–isolated monoparticles [27]. This reagent contains two identical reactive groups that react covalently with primary amines spaced about 11 Å apart. It also contains a disulfide bond that can be cleaved with mercaptoethanol. After exposing isolated monoparticles to DTBP, the cross-linked species were identified by the method of Wang and Richards [66,67]. Briefly, the cross-linked particles were solubilized in electrophoresis sample buffer without reducing agent and resolved in cylindrical gels. These gels were then soaked in sample buffer containing 2-mercaptoethanol to cleave the cross-linking reagent and placed horizontally across the second-dimension slab gel. Following second-dimension electrophoresis, three major well-resolved spots were observed. These were identified as homotrimers of proteins A1, A2, and C1. While spots corresponding to heterotypic contacts between the trimers and proteins B1, B2, and C2 were apparent in the gels, it was not possible to identify the minor proteins in contact with each trimer because of inadequate resolution in the first-dimension gel [27]. This problem results from the similarity of molecular weights between these proteins.

The demonstration that the three major proteins A1, A2, and C1 exist as three different homotrimers in intact particles confirmed the 1:1:1 molar ratio previously indicated by gel densitometry. As described below we have found it necessary to isolate and characterize each of the three major protein complexes to identify the proteins in contact with each trimer.

C Proteins in hnRNP and in Solution

In our first characterization of isolated 40S hnRNP from HeLa cells, we demonstrated that as the salt concentration increases from 0.1 to 0.4 M, the A and B group doublets dissociate from the RNA packaged in monoparticles, leaving the C proteins as the only major RNA-bound protein [25]. We have taken advantage of this intrinsic property of the C proteins in the first step of a two-step procedure for the purification of native C protein (C1 and C2) [68]. Briefly, the A and B group proteins are dissociated from gradient-isolated monoparticles by 0.4 M NaCl. The stable C protein–RNA complex is collected in the exclusion volume from gel filtration columns, then aggressively digested with micrococcal nuclease before further purification using a mono-Q anion exchange matrix. The C protein complex purified through this procedure is at least 95% homogeneous (by SDS PAGE); it is stable in high and low salt, in 0.5% deoxycholate, and in EDTA; and it retains its ability to participate in the *in vitro* assembly of 40S monoparticles [37].

During the course of the C protein purification studies we observed that C1 and C2 copurify in a densitometric ratio very near 3:1 [68]. This, of course, is the approximate densitometric ratio seen for these proteins in single-dimension PAGE of intact 40S hnRNP and is consistent with the cross-linking studies demonstrating that a fundamental trimeric association of C1 exists in intact 40S monoparticles [27]. We also observed that the C protein complex elutes from non–agarose gel filtration columns in high and low salt with an apparent mass near 500,000. This finding could only result if the C protein complex existed as a spherical group of 15 or more polypeptides or if fewer polypeptides were present in an asymmetric arrangement. We determined the Stokes radius, sedimentation coefficient ($S_{20,w}$), partial specific volume, and frictional ratio of the purified C protein complex, and from these hydrodynamic properties we calculated the approximate molecular weight to be near 135,000. This molecular weight was most consistent with the presence of four copies of a polypeptide with the molecular weight of C1 (MW = 31,931) [69].

To further prove the existence of a tetramer and to identify its composition we conducted a series of sedimentation and protein cross-linking studies using the reagent DSP. These studies clearly demonstrate that the C protein complex exists as a $C1_3C2$ tetramer [37]. The tetramer is stable in low and high salt and in 0.5% Na^+ deoxycholate. This stability is not dependent on divalent cations or the presence of disulfide bridges between the polypeptides. Several lines of evidence strongly argue that the tetramer exists as such in intact 40S hnRNP. For example, our previous protein-protein cross-linking studies with DTBP and with CuP convincingly demonstrate that C1 (as well as A1 and A2) exists as a homotrimer in native monoparticles. While these cross-linking reagents (unlike DSP) do not quantitatively cross-link the $C1_3C2$ tetramer, a reexamination of our original data [27] clearly shows the presence of the tetramer. In addition, the C protein complex released from freshly isolated monoparticles with nuclease or with high salt possesses the same hydrodynamic properties as highly purified tetramers. Finally, the stability of the tetramer itself argues for its existence in intact monoparticles. The very large apparent mass of the tetramer (from gel filtration analyses) together with its frictional ratio of 2 argues that it resembles either a prolate ellipsoid with an axial ratio of 20 or an oblate ellipsoid with an axial ratio of 30. Knowledge of the tetramer's structure will ultimately be an important objective because the spatial arrangement of the RNA-binding domains (four per tetramer) will determine the amount of RNA in contact with each tetramer. For example, if the tetramer is organized like the four spokes of a wheel that lie in a single plane (oblate) and if the RNA-binding domains are terminal and peripheral, then, given a Stokes radius of 6.0 nm, such complex could bind as much as 115 nt of RNA. The reader is reminded here that there are at least three, and probably four, C protein tetramers in each 40S monoparticle and that each monoparticle packages 700 nucleotides of RNA.

Proteins A2 and B1

We have previously reported that if the hnRNP complexes in nuclear sonicates are digested with ribonuclease beyond the point necessary to convert polyparticles to monomers, then one sees a progressive dissociation of proteins A1, C1, and C2 [27]. This event is followed by the rearrangement of proteins A2 and B1 to form highly homogeneous 43S particles that possess the same 3:1 molar ratio (via densitometry) as in intact monoparticles. We also observed that in the absence of RNA and at temperatures above 10°C, proteins A2 and B1 self-assemble to form extremely long helical filaments that can only be solubilized in hot SDS-BME [27]. The filamentous structures also contain A2 and B1 in the 3:1 ratio seen in monoparticles and in the 43S rearrangement structure [27]. During the course of the reconstitution studies summarized above, we wondered if A2 and B1 might alone package RNA into a repeating array of particles and if all structures composed of these proteins are formed by the assembly of tetramers of $A2_3B1$.

Our recent studies have in fact shown that proteins A2 and B1 exist in solution at salt concentrations below 0.2 M as a tetramer of $A2_3B1$ [63] and that they assemble in vitro in a sequence-independent manner to form 43S complexes that, like intact 40S hnRNP, package 700-nt lengths of RNA into monomer, dimer, and oligomeric complexes [70]. Our previous studies showing that A2 exists as trimers in intact 40S hnRNP argue that $A2_3B1$ tetramers exist as a fundamental structural component of native hnRNP. The packaging of 700 ± 20 nucleotides of RNA into regular repeating structures either by all six of the 40S hnRNP core particle proteins or by two of the proteins alone was unexpected but informative as follows.

The ability of $A2_3B1$ tetramers to assemble into a stable particle that binds the same length of RNA as reconstituted hnRNP suggests that $A2_3B1$ tetramers can replace tetramers of $C1_3C2$ and $A1_3B2$ in the particle structure. The characteristic stoichiometry of native particles may therefore be determined by regulatory interactions among the different tetramers. This possibility is supported by the finding that the interactions among the $A2_3B1$ tetramers show some distinct differences from the protein-protein interactions that occur when all the tetramers are present [70]. For example, during hnRNP assembly in vitro, the assembly of particles of correct stoichiometry and size is dependent on appropriate protein : RNA ratios in the reaction mix. By contrast, assembly of the 43S particle occurs in a cooperative, all-or-none manner regardless of protein : RNA ratio.

Like native hnRNP particles, the 43S rearrangement complex dissociates completely on extensive treatment with nuclease. Moreover, when 43S particles are reconstituted in the presence of limiting RNA, excess protein remains near the top of gradients in a low-molecular-weight form. Both findings indicate that the protein-protein binding domains among tetramers function only when the tetramers are bound to RNA. The dependence of particle assembly on RNA length must result from the formation of a stable structure when the correct number of tetramers have bound to the RNA substrate.

We were surprised to find that although the 40S hnRNP particles and the 43S particles package the same length of RNA in the repeating unit, the ratio of protein to RNA is distinctly larger in the 43S complex (9 : 1 rather than 6 : 1). The only simple explanation of these results taken together is that a complex of $A2_3B1$ tetramers and RNA first forms, and then (perhaps because of the cooperative nature of tetramer association) additional tetramers that are not bound to RNA add to the complex. This interpretation, if correct, suggests that hnRNP forms a structure with open symmetry rather than one with closed symmetry (polyhedral shell). A similar suggestion was presented by Thomas et al. [71] based on their electron microscopic and hydrodynamic studies on HD-40, a single A2-like protein in *Artemia* that has the ability to package RNA into a repeating array of RNP particles.

Protein Stoichiometry

We have shown that the core proteins A2 and B1 exist as tetramers of $A2_3B1$ in solution and in intact 40S particles [63]. We have also shown that proteins C1 and C2 exist as tetramers of $C1_3C2$ in solution and in intact 40S particles [37]. We have not isolated tetramers of $A1_3B2$, but it is likely that such structures exist in monoparticles. For example, cross-linking studies have shown that A1 (like A2 and C1) exists in intact monoparticles as a trimer and that A1 is present in the same molar ratio as A2 and C1 (i.e., $1:1:1$). Gel densitometry reveals that B2 and B1 are present at a $1:1$ molar ratio and at one third the amount of A1. This $3:1$ relationship between the three major core proteins (A1, A2, and C1) and the B group proteins and C2 is confirmed by the existence of the tetramers. In addition, A1 and B2 show similar sensitivity to protease cleavage and to nuclease-induced dissociation [27]. Finally, tryptic finger-prints and preliminary peptide sequence analyses indicate that B2 is homologous to A1 and B1 is homologous to A2 [B.M. Merrell and K.R. Williams, personal communication]. For these reasons we conclude that the core protein stoichiometry in 40S monoparticles is $n(A1_3, A2_3, B1, B2, C1_3, C2)$ and that this stoichiometry results from the assembly of three tetramers composed of $A1_3B2$, $A2_3B1$, and $C1_3C2$.

It is of course important to know the actual copy number of each polypeptide in monoparticles. This information is a prerequisite in elucidating monoparticle structure. If either the precise mass of monopar-ticles were known or the number of any individual tetramer in monopar-ticles could be experimentally determined, then the protein copy number could be calculated. Several lines of reasoning argue at this time that there are either three or four copies of each tetramer in monoparticles. One means of approximating the mass of monoparticles is based on their protein-to-RNA ratio. Since monoparticles package 700 ribonucleotides, the mass contributed by the RNA moiety is near 237,000 daltons. Using a protein-to-RNA ratio of $7:1$ (discussed below), monoparticles should have a mass near 1.9×10^6. If monoparticles contain four copies of each tetramer, then, from the molecular weights of the individual proteins [69, 72], the protein mass alone is near 1.6×10^6. When the RNA mass is added, a value of 1.84×10^6 is obtained for the mass of monoparticles. If, however, three copies of each tetramer are present, then the mass of monoparticles would be 1.44×10^6. Using this approach it appears likely that four copies of each tetramer are present per monoparticle. However, this approximation is dependent on the accuracy of the protein-to-RNA ratio in monoparticles. We have used the buoyant density of glutaral-dehyde-fixed monoparticles in cesium chloride gradients and also colorimetric and spectrophotometric procedures to determine the amount of protein and RNA present in gradient-isolated and reconstituted mono-particles [26, 27]. Depending on the methodology used, our values have ranged from 6 to 8:1. If three copies of each tetramer are present in monoparticles and if the mass is accordingly as low as 1.44×10^6, then the protein-to-RNA ratio must be close to 5:1. While the reasoning pursued here does not unequivocally distinguish between three and four copies of each tetramer per monoparticle, it does argue that the actual number of tetramers is not likely to be less than three or more than four per particle.

Structure of 40S hnRNP Particles

Many of the observations discussed in this article are summarized below. Any attempt to build a model describing the structure of the 40S particle must be consistent with these observations.

1. The particles are built of three similar but different tetramers composed of $A1_3B2$, $A2_3B1$, and $C1_3,C2$ such that the core protein composition and stoichiometry is $n(3A1, 3A2, B1, B2, 3C1, C2)$ where $n = 3$ or 4 [25, 27, 37, 63].
2. Seven hundred plus or minus twenty nucleotides of RNA are packaged in each particle in such a way that most of the RNA is not sterically protected from nuclease [26, 27].
3. Monoparticles have a mass between 1.44×10^6 and 1.9×10^6.
4. The tetramers are arranged in monoparticles such that the $A2_3B1$ tetramers exist in a protease-protected position [27]. Purified $A2_3B1$ tetramers are very sensitive to protease [63].
5. The RNA moiety is obligately required for particle assembly and "activates" the protein-protein binding domains among tetramers [25–27].
6. The six major core particle proteins possess the intrinsic ability to package RNA in vitro through a mechanism that is dependent on RNA length and protein-to-RNA ratio. Monoparticles assemble on 700 ± 20 nucleotides, while dimers and oligomeric complexes assemble on integral multiples of this length [26]. The in vitro assembly of 40S hnRNP particles is not dependent on nucleotide triphosphates, a 3' poly(A) moiety, or a 5' cap structure.
7. All of the proteins possess at least one conserved RNA-binding domain, and each is probably in contact with RNA [58, 73].
8. The C protein tetramer dissociates from RNA at relatively high salt concentrations (0.6 M salt) [25]; the tetramer itself is stable in 2 M salt [37, 68], it can be selectively cross-linked to RNA with UV irradiation [26], and it is highly asymmetric [37].
9. Proteins A1 and A2 dissociate from RNA at relatively low salt concentrations (0.15 M), while the B group proteins dissociate at slightly higher concentrations (0.2 to 0.3 M salt) [25].

10. Unlike the C protein tetramer, purified tetramers of $A2_3B1$ dissociate into soluble polypeptides at low salt (0.15 to 0.2 M) [63].

11. The $A2_3B1$ tetramers alone can package 700 nucleotides of RNA into a highly homogeneous rearrangement structure which sediments as 43S, has a mass near 2.3×10^6, and has a protein : RNA ratio near 9 : 1 [70].

12. The $A2_3B1$ tetramers also self-assemble at temperatures above 10°C in the absence of RNA to form highly insoluble helical filaments of indeterminate length [27].

13. The packaged RNA can be cleaved to fragments about 150 nt in length without particle dissociation [58].

14. The core proteins lack the ability to bind RNA in a sequence-specific manner, and they cannot distinguish between RNA and ssDNA [26, 35, 40].

15. Electron micrographs of negatively stained particles reveal a more or less flattened spherical structure with a diameter of 220 Å. In some preparations there is often a hint of six-sided symmetry, and a centrally located electron dense area can be seen [18, 27]. As revealed by the STEM and neutron scatter facilities at the Brookhaven National Laboratory, the particles have a diameter in solution near 180 Å (unpublished observations). This indicates that the RNA is foreshortened by a factor of about 10 through its association with the core proteins.

16. The particles probably possess the ability to aggregate in vivo to form large RNP granules which are likely to be synonymous with "perichromatin granules" [5, 21, 45, 74].

It is likely that the general topology of protein and RNA in 40S monoparticles can be deduced from the above information if a few additional facts are known. For example, it is important to know the location and linear order of the tetramers along the 700-nt length of packaged RNA and if the RNA wraps around subunits before particle assembly or if the RNA is wrapped on the surface of a protein core following RNA-activated tetramer assembly. For a more detailed structural understanding of monoparticles, it will be important to know the length of RNA bound by each tetramer, each tetramer's shape, and the location of the protein and RNA-binding domains.

References

1. Oda, T., Nakamura, T., and Watanabe, S. (1977) J. Electron. Microsc. (Tokyo) 26, 203–207.
2. Chung, S.Y., and Wooley, J. (1986) Proteins Struct. Func. Genet. 1, 195–210.
3. Derenzini, M., Pession-Brizzi, A., and Novello, F. (1981) J. Ultrastruct. Res. 77, 66–82.
4. Mott, M.R., and Callan, H.G. (1975) J. Cell. Sci. 17, 241–261.

5. Skoglund, U., *et al.* (1983) Cell *34*, 847–855.
6. Lamb, M., and Daneholt, B. (1979) Cell *17*, 835–848.
7. Olins, A., *et al.* (1984) Eur. J. Cell Biol. *35*, 129–142.
8. Olins, D., *et al.*, (1983) Science *220*, 498–500.
9. Skoglund, U., Anderson, K., Strandberg, B., and Daneholt, B. (1986) Nature *319*, 560–564.
10. Osheim, Y.N., Miller, O.L. Jr., and Beyer, A.L. (1985) Cell *43*, 143–151.
11. Kierszenbaum, A.L., and Tres, L.L. (1974) J. Cell Biol. *63*, 923–935.
12. Beyer, A., Miller, O., and McKnight, S. (1980) Cell *20*, 75–84.
13. Beyer, A.L. (1983) Mol. Biol. Rep. *9*, 49–58.
14. Beyer, A.L., Bouton, A.H., and Miller, O.L. Jr. (1981) Cell *26*, 155–163.
15. Malcolm, D.B., and Sommerville, J. (1974) Chromosoma *48*, 137–158.
16. Sommerville, J.J. (1973) J. Mol. Biol. *78*, 487–503.
17. Sommerville, J. (1981) In The Cell Nucleus Vol. 8, 1–57, Busch, H. (ed.). Academic Press, New York.
18. Tsanev, R.G., and Djondurov, L.P. (1982) J. Cell Biol. *94*, 662–666.
19. Fakan, S., Leser, G., and Martin, T.E. (1986) J. Cell Biol. *103*, 1153–1157.
20. N'Da, E., Bonnanfant-Jais, M.L., Penrad-Mobayed, M., and Angelier N. (1986) J. Cell Sci. *81*, 17–27.
21. Malcolm, D., and Sommerville, J (1977) J. Cell. Sci. *24*, 143–165.
22. Samarina, O., and Krichevskaya, A. (1981) In The Cell Nucleus Vol. 9, 1–48, Busch, H. (ed.). Academic Press, New York.
23. Samarina, O.P., Lukanidin, E.M., Molnar, J., and Georgiev, G.P. (1968) J. Mol. Biol. *33*, 251–263.
24. Samarina, O.P., Lukanidin, E.M., and Georgiev, G.P. (1967) Biochim. Biophys. Acta *142*, 561–564.
25. Beyer, A.L., Christensen, M.E., Walker, B.W., and LeStourgeon, W.M. (1977) Cell *11*, 127–138.
26. Conway, G., Wooley, J., Bibring, T., and LeStourgeon, W.M. (1988) Mol. Cell. Biol. *8*, 2884–2895.
27. Lothstein, L., *et al.* (1985) J. Cell Biol. *100*, 1570–1581.
28. Walker, B.W., Lothstein, L., Baker, C.L., and LeStourgeon, W.M. (1980) Nucleic Acids Res. *8*, 3639–3657.
29. Martin, T.E., *et al.* (1978) Cold Spring Harbor Symp. Quant. Biol. *42*, 899–909.
30. Kinneburgh, A.J., Billings, P.B., Quinlan, T.J., and Martin, T.E. (1976) Prog. Nucleic Acid Res. Mol. Biol. *19*, 335–351.
31. Kish, V.M., and Pederson, T. (1978) Methods Cell Biol. *17*, 377–399.
32. Munroe, S.H., and Pederson, T.J. (1981) Mol. Biol. *147*, 437–449.
33. LeStourgeon, W.M., *et al.* (1977) Cold Spring Harbor Symp. Quant. Biol. *42*, 885–898.
34. Economidis, I.V., and Pederson, T. (1983) Proc. Natl. Acad. Sci. USA *80*, 1599–1602.
35. Wilk, H.E., Angeli, G., and Schafer, K.P. (1983) Biochemistry *22*, 4592–4600.
36. Wilk, H.E., *et al.* (1985) Eur. J. Biochem. *146*, 71–81.
37. Barnett, S.F., Friedman, D.L., and LeStourgeon, W.M. (1989) Mol. Cell Biol. *9*, 492–498.
38. LeStourgeon, W.M., Lothstein, L., Walker, B., and Beyer, A. (1981) In The Cell Nucleus, 49–87, Busch, H. (ed.). Vol. 9, Academic Press, New York.

278   S.F. Barnett et al.

39. Karn, J., Vidali, G., Boffa, L., and Allfrey, V. (1977) J. Biol. Chem. *252*, 969–976.
40. Pullman, J.M., and Martin, T.E. (1983) J. Cell Biol. *97*, 99–111.
41. Choi, Y.D., and Dreyfuss, G. (1984) Proc. Natl. Acad. Sci USA *81*, 7471–7475.
42. Dreyfuss, G., Choi, Y.D., and Adam, S.A. (1984) Mol. Cell. Biol. *4*, 1104–1114.
43. Kinneburgh, A., and Martin, T. (1976) Proc. Natl. Acad. Sci. USA *73*, 2725–2729.
44. Pederson, T., and Davis, N.G. (1980) J. Cell Biol. *87*, 47–54.
45. Fakan, S., Leser, G., and Martin, T. (1983) J. Cell Biol. *98*, 358–363.
46. Spirin, A.S., Belitsina, N.V., and Lerman, M.I. (1965) J. Mol. Biol. *14*, 611–615.
47. Jones, R.E., Okamura, C.S., and Martin, T.E. (1980) J. Cell Biol. *86*, 235–243.
48. Christensen, M.E., *et al.* (1981) J. Cell Biol. *90*, 18–24.
49. Choi, Y.D., and Dreyfuss, G. (1984) J. Cell Biol. *99*, 1997–2004.
50. Pinol-Roma, S., Choi, Y.D., Matunis, M.J., and Dreyfuss, G. (1988) Genes Dev. *2*, 215–227.
51. Roth, M., and Gall, J. (1987) J. Cell Biol. *105*, 1047–1054.
52. Holcomb, E.R., and Friedman, D.L. (1984) J. Biol. Chem. *259*, 31–40.
53. Blanchard, J.M., Brunel, C., and Jeanteur, P. (1977) Biochem. Soc. Trans. *5*, 670–671.
54. Periasamy, M., Brunel, C., and Jeanteur, P. (1979) Biochimie *61*, 823–826.
55. Periasamy, M., Brunel, C., Blanchard, J.M., and Jeanteur, P. (1977) Biochem. Biophys. Res. Commun. *79*, 1077–1083.
56. Niessing, J., and Sekeris, C.E. (1970) Biochim. Biophys. Acta *209*, 484–492.
57. Molnar-Kimber, K., Summers, J., Taylor, J., and Mason, W.J. (1983) J. Virology *45*, 165–172.
58. Dreyfuss, G. (1986) Annu. Rev. Cell. Biol. *2*, 457–495.
59. LeStourgeon, W.M., Lothstein, L., Walker, B., and Beyer, A. (1981) In The Cell Nucleus Vol. 9, 49–87, Busch, H. (ed.). Academic Press, New York.
60. Kinneburgh, A.J., and Martin, T.E. (1976) Biochem. Biophys. Res. Commun. *73*, 718–726.
61. Augenlicht, L.H., and Lipkin, M. (1976) J. Biol. Chem. *251*, 2592–2599.
62. Volkova, I.V., and Gauze, L.N. (1982) Mol. Biol. (Mosk.) *16*, 123–128.
63. Barnett, S.F., and LeStourgeon, W.M. (1989) Mol. Cell Biol. *9*, 492–498.
64. Celis, J.E., Bravo, R., Arenstorf, H.P., and LeStourgeon W.M. (1986) FEBS Lett. *194*, 101–109.
65. Loeb, J., Ritz, E., Creuzet, C., and Jami, J. (1976) Exp. Cell. Res. *103*, 4540–4553.
66. Wang, K., and Richard, F. (1974) J. Biol. Chem. *249*, 375–389.
67. Barnett, S.F., LeStourgeon, W.M., and Friedman, D.L. (1988) J. Biochem. Biophys. Methods *16*, 87–98.
68. Swanson, M.S., Nakagawa, T.Y., LeVan, K., and Dreyfuss, G. (1987) Mol. Cell. Biol. *7*, 1731–1739.
69. Thiery, T., Barnett, S.F., Bibring, T., and LeStourgeon, W.M. (1989) (in press).

70. Thomas, J.O., Glowacka, S.K., and Szer, W. (1983) J. Mol. Biol. *171*, 439–455.
71. Cobianchi, F., *et al.* (1988) J. Biol. Chem. *263*, 1063–1071.
72. Wooley, J., Chung, S.Y., Wall, J., and LeStourgeon W.M. (1986) Biophys. J. *49*, 17–19.
73. Daskal, Y., Komaromy, I., and Busch, H. (1980) Exp. Cell. Res. *126*, 39–46.

Section III Chromosome Organization

Chromosome function in replication and transcription is determined by the structure of the genome, the nucleus, and the cell. At the shortest range of structure, understanding the interactions of DNA and RNA polymerases, and accessory proteins, with the DNA double helix is a three-dimensional question. Thus, a complete description of the activities of the genome ultimately means defining the spatial positions of the atoms of the protein and nucleic acid components. The functioning of the genome must also be influenced by the three-dimensional organization of the cell at the longest range of structure. This level of structure concerns the spatial distribution of chromosomes, for higher eukaryotes, in mitotic and interphase cells. As investigations of chromosome function become increasingly refined, a full understanding of these processes will require that the results be directly related to the substructure of the cell. Because of the importance of structure, from the level of the arrangement of proteins, DNA, and smaller molecules to the level of the organization of the intact cell, several chapters in this section are devoted to these topics.

Replication and transcription take place within the living cell. And while it is difficult to determine the composition and concentrations of the cellular components present at the sites of genome activity, they are certainly not identical to the experimental in vitro conditions. In particular, DNA of eukaryotes is complexed with histone proteins to produce the higher-order structures of chromosomes and nuclei. Such structures must be considered along with the enzymes and associated proteins that are immediately involved in replication and transcription. The interaction of histones and DNA creates the nucleosomes and "30-nm" fibers, and folding of the chromatin fibers forms the interphase and metaphase chromosomes. This process places constraints on the DNA and its functional activities and establishes the higher-order chromosome organization of the cell. Furthermore, the chromosomes of interphase cells are arranged within a nucleus, and experiments have indicated that nuclear substructure may have significant roles in both DNA replication and mRNA transcription. Since chromosomes are the functionally active form of the genome, and not naked DNA, chapters are included that deal with aspects of the histones. Protamines, structural chromosomal proteins of sperm, are also discussed.

A connection therefore exists between the molecular biology and cell biology of chromosome function. It is not simply that the larger cellular structures are built up of smaller substructures, and these in turn are built from assemblies of molecules. The connection is more profound in suggesting that large-scale cellular structure and the functioning of the genome are directly and intimately associated.

Chapter 13
Spatial Distribution of Chromosomes in Human Mitotic Cells

Kenneth W. Adolph

The highest level of chromosome organization concerns the three-dimensional in vivo arrangement of chromosomes. The 3-D distribution of chromosomes is most readily investigated with mitotic cells, since individual chromosomes can be distinguished, especially in metaphase. The problem being addressed therefore differs from those involving the lower levels of chromosome structure. These relate to the interaction of histones with DNA to form nucleosomes and the coiling of the nucleosome filament into the chromatin fiber. An intermediate level of structure relates to metaphase chromosome substructure—that is, the mode of folding the chromatin fiber into the characteristic morphology of a metaphase chromosome.

For metaphase cells, describing the spatial arrangement of chromosomes means giving their position, orientation, and configuration. Metaphase cells of higher eukaryotes, including humans, contain a variety of chromosomes with different sizes, shapes, and genetic information. These chromosomes are normally characterized by cytogeneticists through disrupting the cells and preparing chromosome spreads. Such treatments destroy the 3-D distribution of chromosomes that existed in the living cell. Few reports have dealt with chromosome organization in intact cells, and these have been concerned with insect cells and grasses. Investigating human cells is therefore of great importance.

HeLa cells were chosen for the approach described in this chapter because they are the most widely studied human cell line. Results with these cells have appeared in innumerable papers in biochemistry and cell biology journals. More specifically, isolated metaphase chromosomes from HeLa cells have been extremely valuable for electron microscopic and biochemical studies of chromatin fiber packaging [1] and cell cycle changes in chromosome-associated proteins [2]. Since HeLa cells were originally derived from cancer tissue, comparing the spatial distributions of chromosomes in HeLa cells (with about 68 chromosomes per cell) and normal human cells (with 46) should be revealing. Differences in metaphase chromosome packing and chromosome structural alterations through the cell division cycle would be demonstrated.

Understanding the organization of chromosomes in metaphase cells is the central goal of the approach described in this chapter. Chromosomes in metaphase are highly condensed and have characteristic and readily distinguishable morphologies that recur from mitosis to mitosis. Metaphase chromosomes are also of medical significance because they are employed for cytogenetic analysis of abnormal and disease conditions. An additional goal is to determine the arrangement of chromosomes in anaphase, when sister chromatids have separated and are moving to opposite poles of the cell. The association of chromosomes with the spindle is a unique feature of anaphase cells that influences chromosome arrangement. Another cellular component that interacts with chromosomes is the nuclear membrane. In telophase, chromosomes act as a template around which the nuclear envelope assembles, and following the organization of chromosomes into telophase is thus a significant aspect of this research. Metaphase, anaphase, and telophase: these were the cell cycle phases for which 3-D reconstructions were undertaken.

Mitosis is a continuous process, which implies that the arrangement of chromosomes in metaphase is related to chromosome organization in the interphase nucleus. Since structure determines function, knowing the spatial distribution of chromosomes in metaphase should be significant for understanding the functioning of the interphase nucleus. Nuclei are active in gene transcription and DNA replication, whereas metaphase chromosomes are largely inactive, but, as these considerations suggest, their organization is important for genome activity. Three-dimensional reconstructions of metaphase cells should also shed light on the mechanisms and roles of exchanges of genetic information between different chromosomes. Exchanges of this sort may, for example, activate oncogenes in carcinogenesis.

The spatial arrangements of chromosomes in metaphase, anaphase, and telophase HeLa cells were determined by combining electron microscopy and computer graphics. The initial stage of the reconstructions involved electron microscopy of thin sections through intact cells fixed to preserve the in situ chromosomal structure. The images were digitized in the second stage of the procedure, and thresholding and boundary-tracking computer programs were utilized to obtain the chromosome boundary contours. In the final stage, advanced computer graphics software was applied to create solid surfaces around the chromosomes, thus completing the 3-D reconstructions.

Examination of the micrographs gave considerable information about chromosome organization in mitotic cells even before the reconstruction procedure was carried out. It was clear that chromosomes are not randomly distributed throughout the metaphase plate. In the thin sections, the longer chromosome arms were seen to be located around the perimeter of the plate and radially oriented, while the shorter chromosomes were more centrally located. The reconstructions provided definitive evidence that the positions and orientations of metaphase chromosomes depend on their sizes. Reconstructed large chromosomes were found to be spatially arranged near the perimeter of the plate. In addition, unexpected features of the configuration of individual large chromosomes became apparent. The chromosome arms were not extended, as seen in spreads of metaphase cells or for isolated chromosomes. Instead, each chromosome arm was bent at its centromere so that the telomeres were directed outward. Thus the telomeres of both the long and short arms of each chromosome were distal to the centromere relative to the center of the metaphase plate.

The reconstructions further demonstrated that the plate or disk arrangement of chromosomes in metaphase is continued during anaphase and into telophase. The anaphase disk of chromosomes has a more amorphous structure since many chromosomes are in contact and the nuclear membrane is beginning to coat the cluster of chromosomes. The main structural feature that is prominent in the reconstructions is the difference between the two sides of the anaphase disk. One side, facing the direction of movement on the spindle, is relatively smooth and has rounded edges. The opposite side is characterized by protruding chromosome arms which are trailing the direction of movement.

For telophase cells, the boundaries of the newborn nuclei were reconstructed, and not individual chromosomes. The nucleus at this early stage has a striking, disklike morphology. The sides are not perfectly flat but are uneven, with bulges and concave regions. The dimensions of the disk are close to those of the metaphase and anaphase structures. A continuity therefore exists in the spatial distribution of chromosomes through cell division.

Electron Microscopy of Thin Sections

It was important in the reconstruction procedure to preserve the distribution of chromosomes in living, intact cells. Electron micrographs were therefore taken of serial, thin sections through cells treated with chemical fixative to "freeze" the in vivo arrangement of chromosomes. Consecutive thin sections were collected through the entire volume occupied by the chromosomes. This allowed the complete distribution of chromosomes to be reconstructed while also enabling the structures of individual metaphase chromosomes to be resolved.

Mitotic cells were prepared by mechanical shake-off of cells from monolayers. Drugs such as colchicine and nocodazole were avoided, since their use resulted in abnormal arrangements of chromosomes. Cells detached from monolayers were treated with glutaraldehyde in warm growth medium (serum free) [3]. This treatment cross-links proteins and thereby preserves the in situ cellular morphology. The cells were then additionally fixed with osmium tetroxide.

An embedding medium of Quetol 651 [Andre Abad and Rod Kuehn, personal communication] was used. Quetol 651 was chosen because it is a water-soluble, low-viscosity medium that sections well and gives high electron image contrast. Embedding the cells involved dehydration through 50% and 75% Quetol in water, 100% Quetol monomer, and 100% Quetol polymerization mixture, followed by curing the samples as thin layers and mounting small squares for sectioning.

Ribbons of consecutive sections were obtained using a Reichert Ultracut E microtome with a diamond knife [3]. Sections about 100 nm thick were picked up on single-slot grids covered with carbon-coated Formvar films. The grids were stained with uranyl acetate in water and with lead citrate. Micrographs were taken with an Hitachi H600 electron microscope, typically operated at 50 kV and at a magnification of 5,000×. The film used to capture the images was DuPont Cronalar EM-7.

Samples prepared by mitotic shake-off contain cells in all phases of mitosis. Therefore, typical examples in metaphase, anaphase, and telophase were selected from among the cells in the sections, and the complete set of sections containing the chromosomes or nuclei was photographed in the electron microscope. Mitosis is a continuous process, and the designation of the phase of mitosis is somewhat arbitrary. Metaphase cells were selected that showed chromosomes densely aligned in the metaphase plate and with the plate at its minimum thickness. Anaphase cells were defined by the presence of two sets of sister chromosomes. Telophase cells were photographed that showed two progeny cells still connected at the midbody.

Electron micrographs of serial sections are included in Figure 13.1 which show a typical metaphase chromosome. The example is a chromosome with long arms and is located near the perimeter of a metaphase plate that is similar to the plate in Figure 13.2. The arms of these peripheral chromosomes are bent so that the telomeres extend outward. This would appear to be a consequence of the dense packing of chromosomes in forming the plate.

Understanding the precise structures of individual chromosomes and their positions in the plate is very difficult from direct examination of the micrographs. It was for this reason that three-dimensional reconstructions were carried out, based upon the electron micrographs of serial sections. The section thickness of 100 nm was chosen so that the morphology of individual chromosomes could be resolved. Since chromosome arms are about 0.6 μm wide, the minimum number of sections to completely contain a chromosome, favorably oriented, would be approximately six or seven. Thinner sections would give higher resolution of individual chromosomes but would require the collection of a cum-

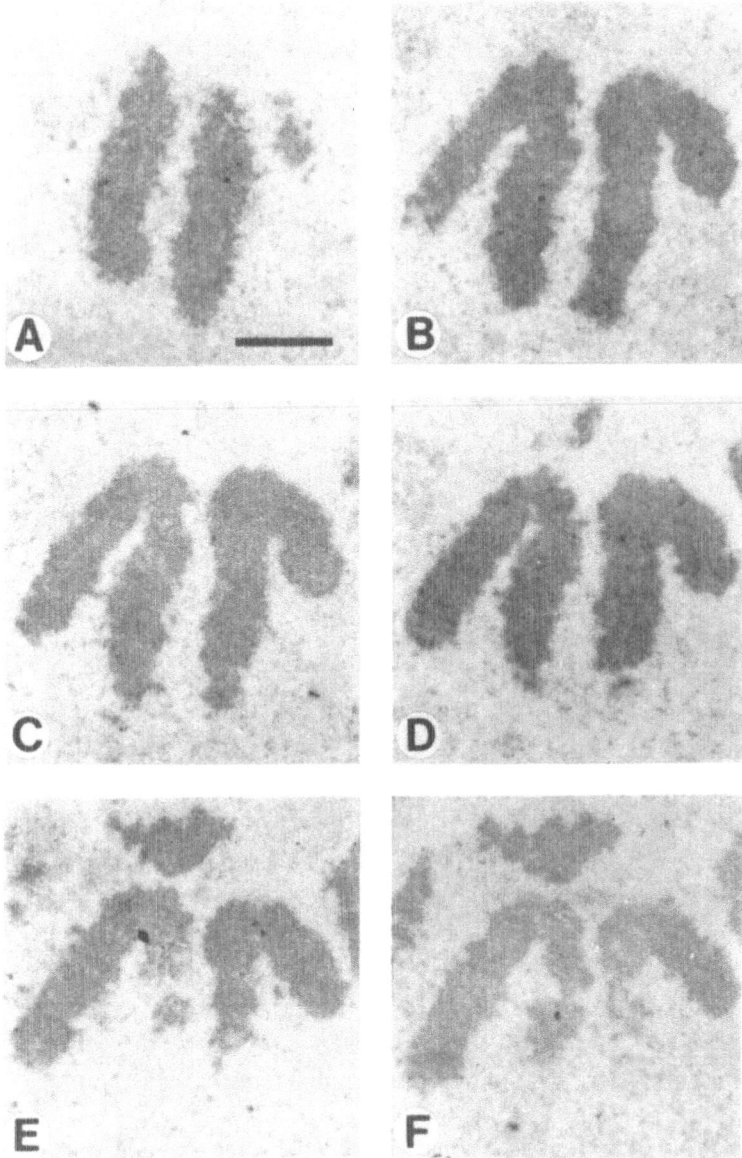

Figure 13.1. Electron micrographs of serial sections of a chromosome in a HeLa metaphase plate. The chromosome is one of the large chromosomes and is found close to the perimeter of the plate. The arms are oriented radially outward with the telomeres distal and the centromeres proximal to the plate's center. Since the sections are 100 nm thick and the chromosome arms are 0.6 μm wide, about 6 or 7 sections are required to pass through the chromosome. Magnification bar, 1.0 μm.

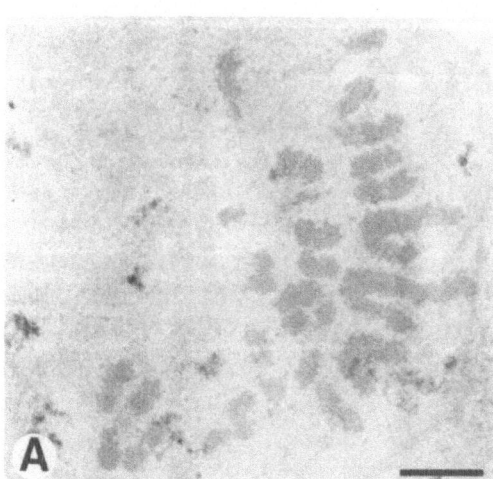

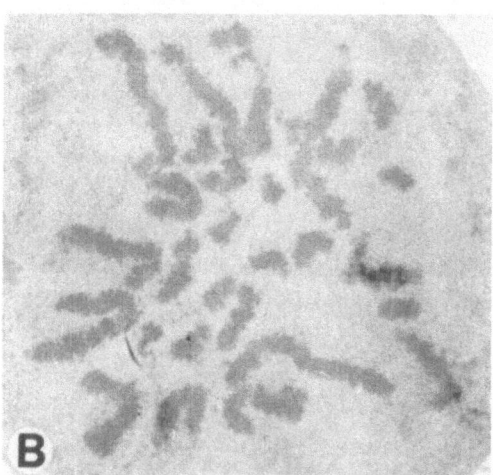

Figure 13.2. Distribution of chromosomes in the HeLa metaphase plate shown by electron micrographs of thin sections. Micrographs from the same consecutive series of sections are included in panels A and B; the example in (A) is just penetrating the plate, while the section in (B) is closer to the center. The magnification bar represents 2.0 μm.

bersome number of sections to pass completely through the metaphase plate.

To give an overview of the distribution of chromosomes, Figures 13.2A and 13.2B contain electron micrographs of sections cut into the face of a metaphase plate. The characteristic feature observed in these micrographs is the peripheral location and radial orientation of the large chromosomes; small chromosomes are near the center. The microtome knife has cut sections that do not intersect the face of the plate directly flat-on. This is seen in Fig. 13.2A which shows a section with chromosomes along the right edge of the plate, cut as the microtome knife first encountered the chromosome cluster. A section near the middle of a plate is in panel B. The chromosomes are distributed evenly throughout a circle. To pass entirely through the metaphase disk required the collection of 82 such sections.

The appearance of the sections could be due to the chromosomes near the center of the metaphase plate being oriented normal to the plane of the plate. But sections cut 90° to those in Figure 13.2 demonstrate that the smaller chromosomes are indeed located near the center of the plate. However, these chromosomes are not radially oriented like the chromosomes at the periphery, and some preferential orientation normal to the plate is observed. The locations of the longer chromosome arms at the perimeter of the plate and their outward orientation are also clearly observed. However, edge-on sections are impractical to work with since about 140 are needed for a complete set.

The central event of cell division occurs with the separation of sister chromatids at the onset of anaphase. This phase of mitosis is also distinguished by movement of the separated chromosomes to opposite poles of the cell by interaction with the spindle. As anaphase progresses, nuclear membrane forms around the chromosomes, which act as nucleating centers for membrane assembly. These major events result in substantially different structures for the anaphase chromosome cluster and for individual anaphase chromosomes. Figure 13.3 shows the typical appearance of thin sections of late anaphase HeLa cells. At this point, the two sister complements of chromosomes are close to the opposite poles, and patches of membrane are visible. The anaphase disk of chromosomes in Figure 13.3 is encountered edge-on. The section in C is just entering the disk, while A and B are near the middle; 2.2 μm (22 sections) separate B and C, and 0.8 μm (8 sections) separate A and B. Individual chromosomes can still be recognized and, in fact, appear sharply defined owing to their coating of nuclear membrane. However, because the chromatids have separated, single chromosome arms are present, but the characteristic X-shaped structures of metaphase chromosomes are not. In addition, many chromosomes are in the process of fusing, so a typical thin section displays a mixture of isolated and fused chromosomes. One side of the anaphase chromosome cluster is flatter than the opposite side and has rounded edges; the other side is marked by protruding chromosome arms. These features result from movement of the mass of chromosomes on the mitotic spindle which flattens the disklike cluster on one side and leaves trailing chromosome arms on the other. The electron micrographs included in Figure 13.3 are just three sections out of a total of 123 that were needed to go through, edge-on, the entire anaphase disk of chromosomes. To completely reconstruct the detailed morphology of the entire anaphase disk from all 123 micrographs is obviously a formidable task.

Telophase cells contain nuclei (Fig. 13.4), and not chromosomes, because a single nuclear membrane now surrounds the cluster of fused and decondensed chromosomes. Dense foci of chromatin are observed within the nuclei, and these may be the remnants of the metaphase and

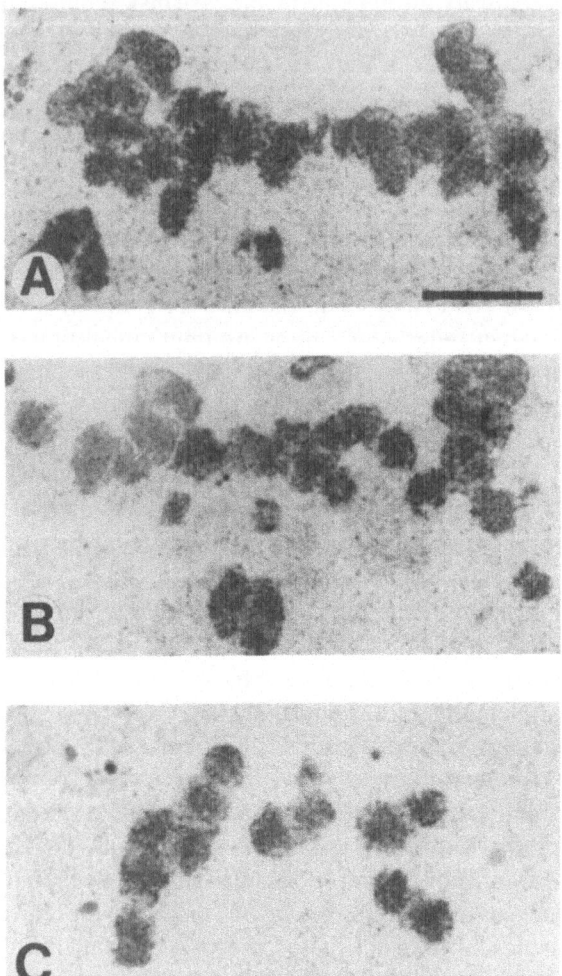

Figure 13.3. Characteristic electron micrographs of the arrangement of chromo-somes in the anaphase cell. The thin sections shown in the three panels are from the same consecutive series. The microtome knife has cut into the anaphase chromosome cluster from the side. Panels A and B are near the middle of the cell, while panel C includes an early section in the series. Bar, 2.0 μm.

anaphase chromosomes. The newborn nuclei are not spherical structures, like mature nuclei, but have a disk morphology with overall dimensions (width 14 μm, thickness 4 μm) similar to the metaphase and anaphase structures. The position in the cell cycle of the nucleus in Figure 13.4, which has the disklike appearance, is precisely known, since the cell is connected to its sister at the midbody. Cell division seems imminent

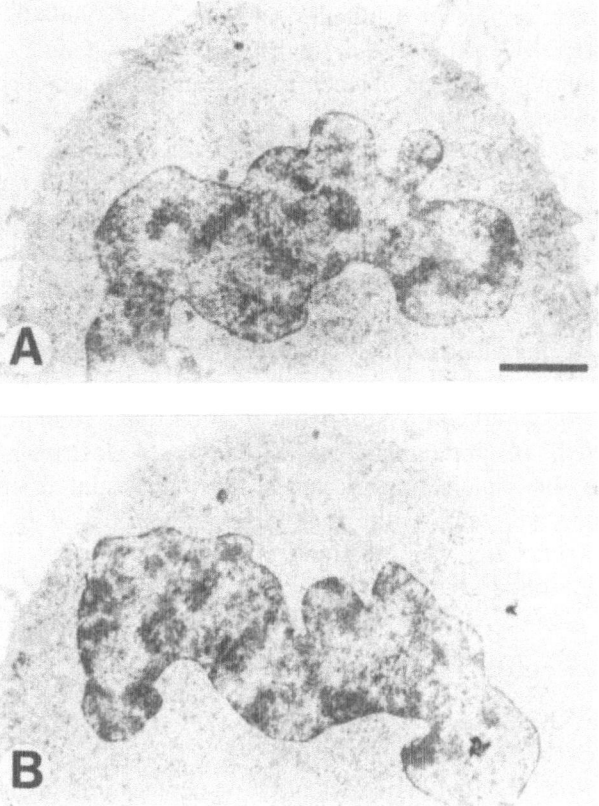

Figure 13.4. Structure of the telophase nucleus. The electron micrographs are part of a series intersecting the side of the same nucleus. The micrographs shown are in the body of the nucleus; they are separated by seven sections (0.7 μm). Bar, 2.0 μm.

because of the narrowness of the midbody connection. The telophase nucleus is not, however, a perfect disk, like a hockey puck, but its sides and perimeter are uneven. (The 3-D reconstructions made these structural features much more readily perceivable.) The two sides show different degrees of unevenness: one side has more prominent concave and bulging regions than the other. Over 130 thin sections were required to pass through the entire nucleus shown in Figure 13.4. Not all of these are needed, however, to provide detailed 3-D reconstructions of nuclei, because of their lesser degree of structural detail relative to chromosomes.

Comparison of the chromosomal and nuclear structures of metaphase, anaphase, and telophase cells demonstrates the continuity that exists through the different cell cycle phases. This continuity is due to the close

and temporally overlapping interactions of the various components—
chromosomes, spindle microtubules, nuclear membrane—that make up
the structures. Its existence also implies that understanding the arrangement of metaphase chromosomes is relevant to an understanding of
nuclear structure and function.

The advantage of using electron micrographs of thin sections as the
basis of 3-D reconstructions should be emphasized: the arrangement of
chromosomes in intact, living cells is preserved. In other techniques such
as whole-mount EM and scanning EM, the chromosomal structures must
be isolated from their native cellular environment using relatively harsh
conditions. And they are subject to flattening, shearing, and other
distortions during electron microscopy. These problems are avoided by
use of the thin-sectioning procedure, which thus gives results that
accurately record the in vivo spatial distributions. High resolution is
another benefit of employing thin sectioning and electron microscopy.
Quantifying the dimensions, volumes, positions, and orientations of
chromosomes will require the high resolution provided by EM. And
accurately measuring the locations of centromeres and the sites of
monoclonal antibody binding and in situ hybridization with nucleic acid
probes will also be possible with this procedure.

Image Processing

Reconstructing the 3-D boundaries of chromosomes first involved reducing each set of electron microscopic images to a 3-D computer database of
chromosome contours. Micrographs were digitized to gray level images of
$512 \times 512 \times 8$ bits per pixel using a Dage MTI68 Newvicon video camera.
The camera was coupled to an International Imaging Systems model 75
image processor and a Masscomp MC535 minicomputer. The registration
capabilities of the image processing system allowed each section to be
visually aligned with the immediately preceding, digitized section. A
radiometric transform was then applied to enhance contrast. Since the
heavily stained chromosomes have a higher density than the background
cellular material, chromosomes were isolated from the background using
a thresholding function. (The chromosomes were distinguished as light
regions since EM negatives, not prints, were digitized.) Single-pixel noise
was removed with a rectangular 3×3 median filter operation [4]. A
boundary tracking or "contouring" program was then applied to compute
the chromosome boundaries in each section. The complete set of contours was stored in a chain-coded format giving the x, y, z coordinates of
each contour [5].

Figure 13.5 shows aligned contours corresponding to an individual
metaphase chromosome (A) and a telophase nucleus (B). The stacked
sections give a 3-D effect, and, in fact, contour representations such as

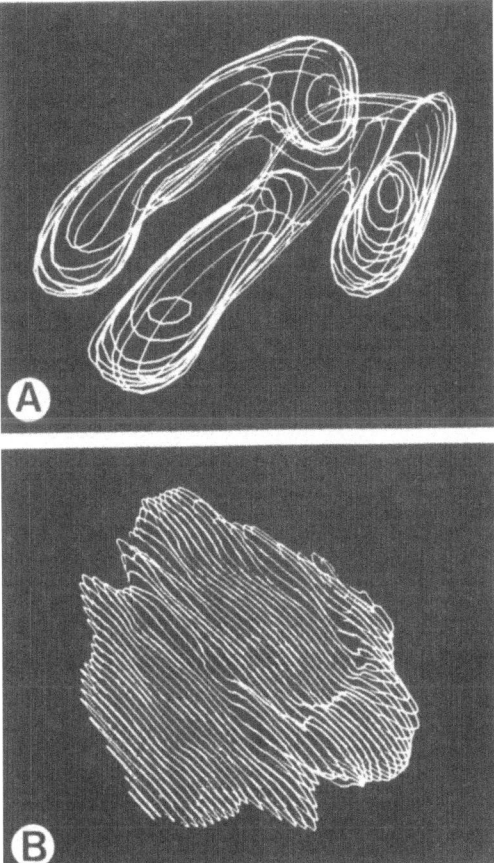

Figure 13.5. Aligned contours of a metaphase chromosome and a telophase nucleus. Electron micrographs such as those in Figures 13.2 and 13.4 were digitized, and chromosome and nuclear boundaries were computed by using a boundary tracking program and other applicable programs. In panel A, stacked sections are displayed that correspond to the metaphase chromosome in Figure 13.6C. Panel B shows the set of contours representing a telophase nucleus as in Figure 13.8. (For clarity, every third section through the nucleus is included.) (Copr. 1989 Kenneth W. Adolph)

these are often considered to be 3-D reconstructions. With the approach described below, however, they represent the unprocessed 3-D data and the basis for the full reconstructions. Figure 13.5 shows the contours in different sections to be well registered and to progress smoothly from section to section. Distortion of the sections was not a significant problem with these examples, and the fact that the images are projections through the 100-nm-thick sections was not a limitation. The aligned sections demonstrate that this procedure is not restricted to simply defining the position and orientation of each chromosome axis. On the contrary, the entire volume of each metaphase chromosome can be reconstructed (volume structuring).

At this point in the analysis, metaphase chromosomes are seen to have characteristic morphologies and spatial locations. The arms of the large chromosome included in Figure 13.5A do not form a simple, extended, X-shaped structure but are folded back on themselves in a more complex configuration. Examination of the aligned sections for the entire metaphase plate shows, even more definitively than did the original micrographs, that the size of chromosomes determines their position and orientation. Chromosomes with long arms are found near the perimeter of the plate; the arms lie close to the plane of the plate and are directed outward. The value of a 3-D perspective is more evident for sections of the telophase nucleus (Fig. 13.5B), since the original micrographs gave a very limited view of the whole. A full view of the undulating morphology of the disk-shaped telophase nucleus begins to emerge as a result of stacking the contours. Although arranging the contoured sections in this way gives a 3-D effect, details of the structures are seen only after proceeding with the full reconstructions, as described below.

The entire image-processing procedure avoided manual tracing of chromosome and nuclear boundaries. Digitized images of the densely stained chromosomes were highlighted with a thresholding function, and a boundary-tracking program was applied to compute the contours. Having to decide exactly where the boundaries were located was thus not required. Another benefit of the image-processing procedure is that distortions due to stretching, shrinkage, etc. could be corrected to give the best alignment of sections.

Three-Dimensional Computer Reconstructions

Chromosome surfaces were rendered by employing computer graphics software to connect the contours between sections with a polygonal mesh. These were subsequently represented as solid surfaces and viewed with artificially chosen colors and reflective properties. Toward this end, the 3-D database derived by image processing was input to an IRIS 2400T graphics workstation (Silicon Graphics, Mountain View, CA) running Wavefront 3-D modeling and rendering software (Wavefront Technologies, Santa Barbara, CA). The complete set of contours could be viewed interactively on the graphics workstation, which allowed small corrections to be made in the alignment of sections. In addition, the contours could be viewed at any chosen angle and at different scales to obtain the most informative perspectives.

In this study, two methods of 3-D representation were used. The detailed structures of individual metaphase chromosomes and telophase nuclei were reconstructed by carefully connecting vertices of adjacent contours with the polygonal mesh. The surface models produced by this complete Wavefront procedure were highly realistic and had smooth 3-D boundaries. Models were generated by the second method through "extruding" the contours in a section to abut the next section. The thick-slice method gave a quicker rendering and was applied to reconstruct the entire metaphase plate or anaphase chromosome cluster. The final images of the reconstructions were photographed with a Dunn Instruments Multicolor 35-mm film recorder.

Figure 13.6 contains 3-D reconstructions of individual metaphase chromosomes (A–C) and also shows the positions of these chromosomes in the metaphase plate (D). The computer renderings are of large chromosomes, since these reveal informative structural details. The 3-D configurations of the chromosome arms are more difficult to distinguish for small chromosomes, which are not much longer than they are wide. A striking feature of the chromosomes in panels A to C is that the chromatids are bent at the centromeres. Chromosomes in situ are therefore not the extended and planar structures seen in chromosome spreads. Instead, large chromosomes have an intricate configuration. The three chromosomes in Figure 13.6 show that chromatid bending is present both when the two arms are similar and when they are very different in length. On placing these chromosomes in their correct locations in the metaphase plate (panel D), they are seen to have similar general positions and orientations. All three are found near the perimeter of the plate and have telomeres distal to the center of the plate relative to the centromeres. Furthermore, the chromatids are all close to the plane of the plate: the chromatid axes, though not exactly in the plane, form only small angles with it. The bending of the chromatids, permitted by their flexibility, would appear to be a consequence of the dense packing of chromosomes that is present in the metaphase plate.

The 3-D reconstructions also indicate that the identical arms of sister chromatids may not, at a particular moment, have the same lengths. The uneven lengths of homologous arms in Figure 13.6 suggest that the contraction and lengthening of sister chromatids are somewhat asynchronous. Another interesting aspect of the reconstructions is that, in B and C, the sister chromatids are slightly separated. This may represent the beginning of anaphase.

Reduction of the diploid to haploid complement of chromosomes is accompanied by major alterations in chromosome interactions and structure. Movement of separated chromosomes on the spindle is brought about by interactions with kinetochore microtubules at the centromere region. Nuclear membrane formation around the surfaces of the chromosomes is initiated even before movement on the spindle is completed. Chromosome structure is also changing as individual chromosomes lose definition by coming into extensive contact and fusing. A more amorphous clustering of chromosomes therefore exists in anaphase, although the platelike arrangement established in metaphase continues to be present. Figure 13.7 includes two views of a reconstruction of the anaphase chromosome cluster. The chromosomes are arranged as a disk with the two sides showing morphological differences. The side trailing the direction of movement (panel A) is characterized by an irregular surface due to protruding chromosome arms. The opposite side (panel B) is somewhat smoother and displays pits instead of protruding chromosomes. The anaphase disk is slightly bowed away from the direction of motion. These asymmetric structural features can be explained as resulting from hydrodynamic resistance of the cell contents to chromosome movement. But it appears that the relative locations of chromosomes in the disk are the same as in metaphase.

The telophase nucleus reconstruction in Figure 13.8 reveals a disklike morphology that continues the structure established in metaphase and anaphase. The dimensions of the structures are similar, and the nucleus has bulges and grooves that seem to correspond to the distribution of chromosomes. The nucleus included in Figure 13.8 is in a cell still attached to its sister at the midbody, and so the position of the nucleus in the cell division cycle is precisely known. The nucleus thus resembles the anaphase and metaphase clusters of chromosomes, except that its external boundary is completely covered by the nuclear envelope. As the cell cycle progresses, further dispersion of chromatin accompanies the rounding up of the nucleus. The two sides of the nucleus in Figure 13.8 are not flat but show significant structural detail. The side in panel A has large grooves, which generally circle the edge of the face, and bulging regions, particularly near the center. The view in panel B is rotated 180° and shows that the reverse side of the nucleus has less pronounced ridges and concave regions. But although the structural details are less prominent, the nuclear envelope is far from smooth and featureless. Structural variability is therefore a characteristic of the telophase nucleus. But the dominant characteristic is the disklike morphology that has been preserved from anaphase.

These are the fundamental observations and conclusions to be derived from 3-D reconstructions of metaphase, anaphase, and telophase cells. A primary aim with the approach, successfully carried out, was to obtain precise and readily perceivable representations of the in vivo configurations of metaphase chromosomes; chromosome structural changes were also followed during anaphase and telophase. The computer reconstruction procedure will, in further investigations, allow accurate determination of the dimensions, volumes, shapes, and positions of chromosomes. The substructure and identity of individual chromosomes could also be studied.

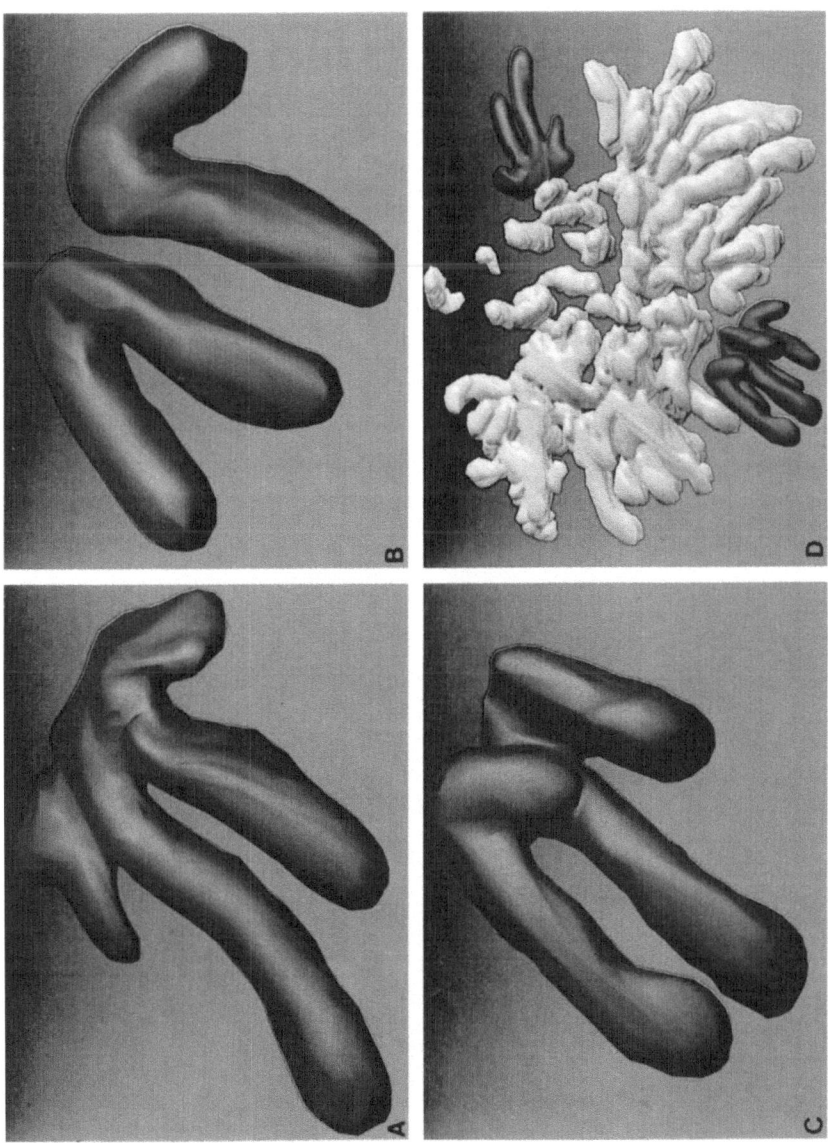

Figure 13.6. Three-dimensional computer reconstructions of metaphase chromosomes (A-C) and the metaphase plate containing the chromosomes (D). In D, the fully rendered chromosomes shown individually in A through C can be recognized by their deep red color. (Copr. 1988 Kenneth W. Adolph)

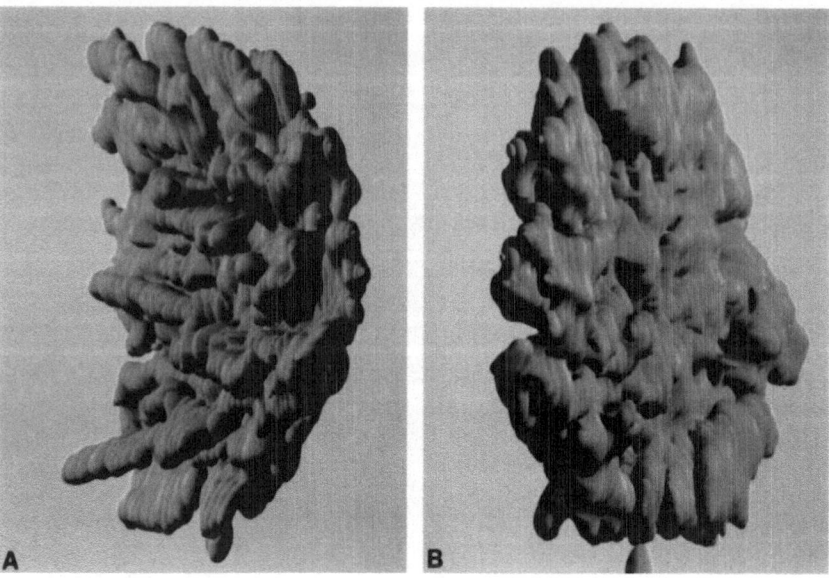

Figure 13.7. Spatial arrangement of chromosomes in the anaphase cell. The two views are rotated, from edge-on, counterclockwise about a vertical axis by 55° (A) or 305° (B). (Copr. 1988 Kenneth W. Adolph)

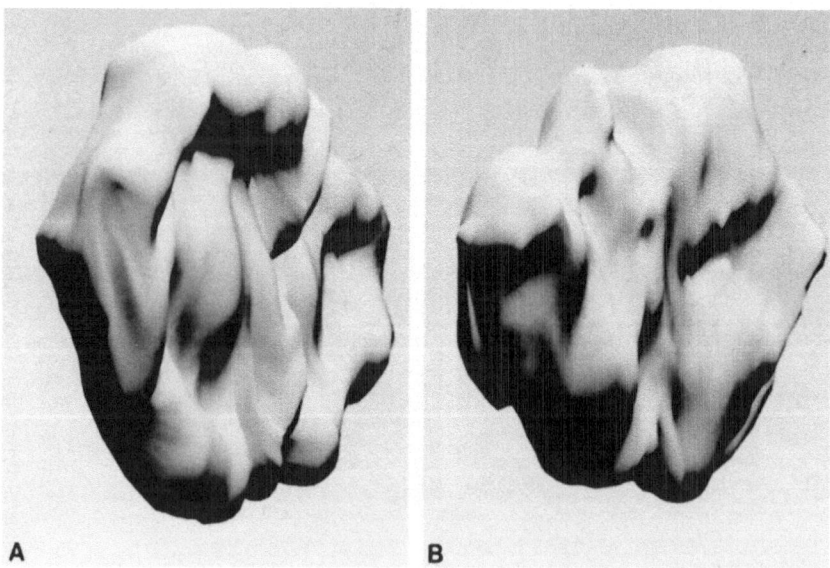

Figure 13.8. A reconstructed telophase nucleus. The two views of the same nucleus differ by a 180° rotation. (The counterclockwise angle of rotation, from edge-on, is 120° in A and 300° in B.) (Copr. 1988 Kenneth W. Adolph)

A major conclusion was that the spatial location of metaphase chromosomes is determined by their size: large chromosomes are located at the periphery of the plate, and their arms are directed radially outward. Since this arrangement is a general consequence of the dense packing of chromosomes in the metaphase plate, and same mode of organization probably exists for other cell types. This should be true even though the cells may have different numbers, sizes, and shapes of chromosomes. But the number would have to be large enough so that the distribution of chromosomes is determined by general packing considerations. For higher eukaryotes including humans, monkeys, rabbits, and rats, this condition would certainly hold. But for cells with few chromosomes (e.g., *Drosophila melanogaster*) or with chromosomes having unusual sizes or shapes, the 3-D arrangement of chromosomes may be unique to that cell type.

The location of the larger chromosomes around the perimeter of the metaphase plate and their radial orientation would promote the maximum density of chromosome packing. The longer chromosome arms are unwieldy, and their distribution around the perimeter would minimize disruption of the dense packing of chromosomes. The smaller chromosomes, however, can arrange themselves efficiently near the center of the plate. These considerations assume that individual chromosomes are free to take any position in the plate to maximize the packing density. But it is possible that DNA connects one chromosome to another and thus creates a specific pattern of attachments between chromosomes. The existence of these connections could be a major influence in establishing the 3-D arrangement of chromosomes. Large chromosomes may necessarily be attached to other large chromosomes, and small to small, to account for the observed spatial distribution. Other factors may contribute to determining the arrangement of chromosomes. For instance, the function of the metaphase plate is to facilitate the anaphase separation of sister chromatids; the 3-D distribution of chromosomes is therefore closely associated with the formation of the mitotic spindle. Thus, attachment of microtubules to the kinetochores is probably also a major influence in fixing the spatial positions of chromosomes.

The 3-D reconstructions do not stand alone. Recent advances have provided information that is relevant to an understanding of eukaryotic chromosome structure and function and that should be integrated with the 3-D results. Information is accumulating regarding the DNA and protein components of chromosomes, in particular DNA sequences [6] and topological properties [7], and the role of DNA-binding proteins [8]. Other active research areas concern the cell biology of mitosis [9], the X and Y chromosomes [10,11], and meiosis [12-14]. Specialized eukaryotic cells, including yeast [15] and polytene chromosomes [16], have proved to be useful for investigations of eukaryotic molecular biology. The 3-D reconstructions of mitotic HeLa cells are especially important because no comparable results have been published for other mammalian cells. A few different systems have provided noteworthy data. Mitotic cells of grasses have been reconstructed from electron micrographs of serial sections [17]. And optical sectioning microscopy was applied to determine the coiling mode of *Drosophila melanogaster* polytene chromosomes [18].

The 3-D computer reconstructions of chromosomes in mitotic HeLa cells should have important implications for understanding human chromosome function. The serial sectioning and reconstruction technique has great potential because of the high resolution provided, with the possibility of investigating molecular aspects of 3-D structure and its relation to function. Thus, further studies will be concerned with extending this approach to chromosome substructure through determining the spatial locations of particular gene sequences and chromosome regions. The distributions of chromosomes in human mitotic cells are complex, as the reconstructions have demonstrated. But investigating this highest level of chromosome organization is now, and should be increasingly, a valuable and informative approach.

Acknowledgments. The image processing and computer graphics results included in this review were carried out with Dr. Charles Knox in the Biomedical Image Processing Laboratory, University of Minnesota. Serial sections were expertly prepared by Andre Abad, and Rod Kuehn is to be thanked for advice regarding electron microscopy. Funds from the National Institutes of Health provided the facilities of the Biomedical Image Processing Laboratory. A grant from the University of Minnesota graduate school supported this study.

References

1. Adolph, K.W. (1988) In Chromosomes and Chromatin, Vol. II, 3–27, Adolph, K.W. (ed.). CRC Press, Boca Raton, Florida.
2. Adolph, K.W. (1988) In Chromosomes and Chromatin, Vol. III, 59–85, Adolph, K.W. (ed.). CRC Press, Boca Raton, Florida.
3. Adolph, K.W. (1981) Eur. J. Cell Biol. *24*, 146–153.
4. Rosenfeld, A., and Avinash, C.K. (1982) Digital Picture Processing, 261–264. Academic Press, New York.
5. Castleman, K.R. (1979) Digital Image Processing, 316–317. Prentice Hall, New York.
6. Schmid, C.W., and Jelinek, W.R. (1982) Science *216*, 1065–1070.
7. White, J.H., Cozzarelli, N.R., and Bauer, W.R. (1988) Science *241*, 323–327.
8. Goodwin, G.H., Nicolas, R.H., Cockerill, P.N., Zavou, S., and Wright, C.A. (1985) Nucleic Acids Res. *13*, 3561–3579.
9. Balczon, R.D., and Brinkley, B.R. (1987). J. Cell Biol. *105*, 855–862.
10. Gartler, S.A., and Riggs, A.D. (1983) Annu. Rev. Genet. *17*, 155–190.
11. Hennig, W. (1985) Adv. Genet. *23*, 179–234.

12. Hotta, Y., and Stern, H. (1984) Chromosoma *89*, 127–137.
13. Busby, S., and Bakken, A.H. (1980) Chromosoma *79*, 84–104.
14. Risley, M.S., Einheber, S., and Bumcrot, D.A. (1986) Chromosoma *94*, 217–227.
15. Zakian, V.A., Blanton, H.M., and Wetzel, L. (1986) In Extrachromosomal Elements in Lower Eukaryotes, 493–498, Wicker, R.B., Hinnebusch, A., Gunsalus, I.C., Lambowitz, A.M., and Hollaender, A. (eds.). Plenum Press, New York.
16. Jamrich, M. (1986) In Chromosome Structure and Function, 221–242, Risley, M.S. (ed.). Van Nostrand Reinhold, New York.
17. Heslop-Harrison, J.S., Smith, J.B., and Bennett, M.D. (1988) Chromosoma *96*, 119–131.
18. Hochstrasser, M., Mathog, D., Gruenbaum, Y., Saumweber, H., and Sedat, J.W. (1986) J. Cell Biol. *102*, 112–123.

Chapter 14

Structure/Function Relationships in the Bacteriophage T4 Single-Stranded DNA-Binding Protein

Yousif Shamoo, Kathleen M. Keating,
Kenneth R. Williams, and William H. Konigsberg

Biological Role of the Bacteriophage T4 ssDNA Binding Protein Encoded by Gene 32

Bacteriophage T4 gene 32 protein (gp32) is a prototype for a class of proteins whose fundamental feature includes a very strong affinity for single-stranded nucleic acids. This binding appears to be largely nonspecific with respect to sequence and favors single-stranded over double-stranded nucleic acids. Tight binding to single-stranded nucleic acids is undoubtedly important for the essential role of gp32 in T4 DNA replication, repair, and recombination [for a review see 1]. When *E. coli,* infected with T4 bacteriophage carrying temperature-sensitive mutations in gene 32, is shifted to a nonpermissive temperature, T4 DNA replication ceases [2], and the intracellular T4 DNA is rapidly degraded to small fragments [3]. The unbound concentration of gp32 in vivo is in the range of 2 to 3 μM [4], and it is thus present in "stoichiometric" amounts relative to the number of replication forks present. This finding is consistent with the presumed role of gp32 in T4 DNA replication which is to remove adventitious secondary structures from the single-stranded DNA template just ahead of the advancing replication fork and to protect the resulting ssDNA from nuclease degradation. In vitro DNA replication assays carried out with T4 DNA polymerase (gp43) and three T4 DNA polymerase accessory proteins (gp44, gp45, and gp62) indicate that the addition of gp32 stimulates the rate of DNA synthesis by 100- to 200-fold [5,6]. Stimulation of the rate of DNA replication by gp32 can be accounted for by a combination of ssDNA stabilization and direct protein : protein contacts between gp32 and the other components of the replication complex [7]. Gene 32 protein participates passively in DNA repair by binding to ssDNA that has been exposed as a result of damage, thus protecting it from nuclease digestion [8]. It has been recently shown that gp32 in conjunction with uvsX catalyzes homologous DNA strand exchange [9], and thus gp32 participates actively in T4 DNA recombination [7]. Studies on amber mutants of gp32 have also shown that gp32 is essential for the formation of "joint" DNA molecules that are required for T4 DNA recombination [10]. Alberts and Frey [11] showed that gp32 was capable of accelerating renaturation of T4 DNA under physiological conditions. In addition, T4 gene 32 mutants were also found to have a decreased frequency of recombination between two closely linked rII mutations [12]. Direct gp32 : protein interactions as well as the ability of gp32 to protect ssDNA from nucleases and to facilitate pairing of homologous strands of DNA, all appear to contribute to the overall role of gp32 in DNA recombination.

Gp32 binds ssDNA at least an order of magnitude more tightly than ssRNA [13]. This behavior underlies an essential feature of the observed autoregulation exhibited by gp32 in T4 infected *E. coli*. Krisch et al. [14] observed that the intracellular amount of gp32 was directly related to the concentration of unbound ssDNA present. Further work carried out in vivo and with in vitro systems has shown that gp32 concentrations are tightly controlled at the level of translation [15,16]. Once gp32 saturates all of the ssDNA present in the cell it then binds to the 5'-untranslated region of its own mRNA and in so doing blocks the initiation of translation of gene 32 mRNA. The apparent specificity that gp32 exhibits in binding to the 5' region of its mRNA can be accounted for by the presence of an unusual stem-loop "pseudoknot" structure about 35 bases upstream from the 5' end of gene 32 mRNA, which has a particularly high affinity for gp32 [17]. It is essential that the in vivo gp32 concentration be tightly coupled to levels of ssDNA, since failure to repress gp32 synthesis would allow the gp32 concentration to rise to a level where it would bind nonspecifically to single-stranded regions on other mRNA molecules, eventually shutting down protein synthesis entirely.

Gp32 : Nucleic Acid Interactions

Gp32 binding significantly alters several of the physicochemical properties of the nucleic acid. Electron microscopy of complexes composed of ssDNA from bacteriophage fd and gp32 indicate that gp32 extends the distance between adjacent DNA bases by approximately 50% and that in the presence of excess ssDNA gp32 binds in long clusters indicative of cooperative gp32 : gp32 interactions [18]. Hydrodynamic studies of gp32 : ssDNA complexes suggest that gp32 binding imposes a rather rigid, extended conformation on the nucleic acid lattice with internucleotide spacings of about 5 Å, in agreement with results obtained from electron microscopy [11,18]. Gp32 binding also results in changes in polynucleotide hyperchromicity [20], in a reduction of the molar ellipticity of the DNA [20], and in an increase in the fluorescence emission of poly(1,N⁶-ethenoadenylic acid) [21]. Although all of these perturbations have been ascribed to a general unstacking of the nucleic acid bases induced by complex formation, circular dichroism (CD) of gp32 : ssDNA complexes has shown that neither solvent nor temperature-induced unstacking alone can account for the observed change in absorbance or CD spectra obtained on complex formation. Hence, instead of being disordered and flexible as the nucleic acid would be at high temperature, the nucleic acid lattice in a gp32 : ssDNA complex may actually be regular and rather rigid in conformation [22]. Scheerhagen et al. [22] have developed a model where the spectroscopic properties associated with gp32 binding are accounted for by extended base-base distance (4.6 Å) and a substantial tilt of the bases ($\leq 10°$).

In addition to altering several of the physicochemical properties of the nucleic acid, formation of a gp32 : ssDNA complex leads to changes in the structure of gp32 that can also be detected by physicochemical probes. The quenching of the intrinsic protein fluorescence of gp32 by the binding of oligo- and polynucleotides is an example of one change that has been used extensively to characterize the binding properties of gp32. Using fluorescence quenching as well as a variety of other techniques, the site size of gp32 has been estimated to range from 5 to 11 [11,13,20,23–26] with an average value of about 7 nucleotides. Kelly and Von Hippel [27] demonstrated that the maximum quenching associated with binding was directly related to oligonucleotide length, such that binding of the dinucleotide $d(pA)_2$ resulted in only 3.3% quenching compared to the 30.5% quenching observed with $d(pA)_8$. These results suggest that ssDNA binding might be polar, with the 5' end of the oligonucleotide being "fixed" at a defined locus and then extending toward a tryptophan residue that is postulated to lie near the 3' end of the oligonucleotide binding groove [27]. As the oligonucleotide becomes long enough for its 3' end to approach the "reporter" tryptophan residue, the fluorescence arising from that tryptophan is increasingly quenched [27]. Near-maximal fluorescence quenching is achieved with oligonucleotides of eight bases. No significant further increase in fluorescence quenching is seen when gp32 is titrated with longer oligonucleotides such as $d(pT)_{16}$ or denatured calf thymus DNA [27].

In addition to permitting an estimate of the binding site size for gp32, fluorescence quenching has also been useful for estimating the affinity of gp32 for a variety of nucleic acid lattices of differing base compositions and lengths. The apparent affinity (K_{app}) of gp32 for an infinite single-stranded lattice has been represented by McGhee and Von Hippel [28] as $K_{app} = K_{int} \omega$, where K_{int} is the intrinsic affinity of gp32 for a single-site lattice and ω is a unitless cooperativity parameter defined as the equilibrium constant for moving a gp32 molecule between an isolated and a contiguous binding site. The cooperativity parameter thus provides a measure of the strength of protein: protein interactions between adjacent gp32 molecules bound to a long ssDNA lattice. Because the overall apparent affinity of gp32 for a long polynucleotide is too large to be able to measure accurately, an alternative approach is to evaluate ω using a lattice such as $d(pT)_{16}$ that is just long enough to accommodate two adjacent gp32 molecules. Under these circumstances the measured K_{app} corresponds to $K_{int} \omega^{1/2}$ rather than to the larger $K_{int} \omega$. An approximate value of K_{int} can be obtained by determining the affinity of gp32 for a single-site lattice such as $d(pT)_8$.

Binding studies carried out with oligonucleotides ranging from two to eight bases (one site size and less) indicate that in all cases, the affinity of gp32 for these oligonucleotides is betweeen 2×10^5 M^{-1} and 6×10^5 M^{-1}. Thus, no significant increase in binding affinity occurs with increasing oligonucleotide length within this range [21,26]. The implication of this finding is that the majority of the overall free energy of oligonucleotide binding is achieved with binding of the first two bases. Additional interactions between gp32 residues and ssDNA bases may therefore be more important for determining specificity, for binding ssDNA as opposed to either ssRNA or double-stranded nucleic acids, rather than for directly contributing to the binding affinity. Additional binding studies on gp32 utilizing oligonucleotides of varying base compositions and different sugar type revealed little effect of these parameters on the binding affinity of gp32 for oligonucleotides. Thus when constrained to bind one site size or less, gp32 binding exhibits very little ribo- or deoxyribonucleic acid base specificity or salt concentration dependence [21,26]. In contrast to the results seen for gp32: oligonucleotide binding, binding experiments carried out with polynucleotides under in vivo salt conditions indicate a significant preference for deoxy- over ribopolynucleotides and a pronounced effect of base composition on binding with the affinity of gp32 for polynucleotides decreasing in the following order: poly(dT) > poly(dG) > poly(dC) > T4 ssDNA > poly(dU) > poly(dA) > poly(rU) > poly(rA) > poly(rC) [4]. Affinity constants derived for these gp32: deoxypolynucleotide complexes range from about 2×10^7 M^{-1} to 2×10^9 M^{-1}, whereas affinity constants derived for the analogous gp32: ribopolynucleotide complexes are, in general, at least 10-fold less [4].

Another characteristic feature of polynucleotide compared to oligonucleotide binding to gp32 is a strong dependence on salt concentration, which has been postulated to arise from three additional electrostatic interactions between positively charged amino acid side chains and the phosphate backbone of the polynucleotide which do not seem to be able to occur when gp32 binds oligonucleotides [21]. Most of the difference in binding affinities between oligonucleotides ($K_{app} = 10^5$ M^{-1}) and polynucleotides ($K_{app} = 10^8$ to 10^9 M^{-1}) has been ascribed to the ability of gp32 to undergo cooperative gp32: gp32 interactions on nucleic acid lattices that are longer than two site sizes. The cooperativity parameter (ω) for gp32 binding to polynucleotides is equal to about 1,000 for both ssDNA and ssRNA and is largely independent of nucleic acid base type and salt concentration [13,21]. The most straightforward interpretation of these observations is that gp32 can be thought of as possessing two distinct but interconvertible conformations, a noncooperative mode observed in oligonucleotide binding, which has a cooperativity parameter of 1, and a cooperative mode seen with longer nucleic acid lattices, which has a cooperativity parameter of about 1,000 [21].

Jensen and co-workers [20] used sedimentation velocity to show that gp32 also has an appreciable affinity for dsDNA. Since the affinity of gp32 for dsDNA (K_{app} = 10^4 M^{-1}) is about 4 orders of magnitude less than for ssDNA, gp32 should be an effective helix-destabilizing protein. Indeed, experiments performed with the synthetic polynucleotide d(A-T) show that in 150 mM NaCl, gp32 is able to reduce the melting temperature of this polynucleotide from 65°C to 25°C. At first glance it is surprising that gp32 is unable to melt native T4 or other naturally occurring dsDNA [11,20]. One rationalization for this apparent discrepancy derives from the fact that to melt dsDNA, gp32 must first nucleate cooperative binding at ssDNA loops that occur normally via dsDNA "breathing." In kinetic terms, Von Hippel's group [20] have suggested that the ssDNA loops that form normally in naturally occurring dsDNA are too transient to allow the initial gp32 nucleation step to occur. At the in vivo concentration of unbound gp32 (2 to 3 μm), it should, however, bind all ssDNA and ssRNA regions as well as melt most adventitious secondary structures in ssDNA, but not those present in ssRNA (keeping in mind the lowered affinity of gp32 in the cooperative binding mode for ssRNA as compared to ssDNA [4]).

Based on knowledge derived from equilibrium fluorescence quenching studies, Lohman has extended our understanding of the mechanism and kinetics of gp32 dissociation from nucleic acids by using stopped-flow fluorescence quenching techniques [29,30]. The results indicate that gp32 dissociation from contiguously bound sites proceeds from the ends, perhaps directly or by first translocating to a noncontiguous site and then dissociating [29,30]. One implication of this work is that gp32 may have a kinetically viable means of translocating along the ssDNA until it reaches another gp32 molecule or cluster. This ability to translocate may help to explain how gp32 binding in vivo can keep pace with movement of the replication complex. That is, under in vivo conditions the approximate association rate of gp32 is only 15 to 20 sec^{-1} [30] compared to the in vivo rate of fork movement of 500 to 1,000 base pairs sec^{-1} [31].

Domains in Gp32 Involved in Protein : Protein and Protein : ssDNA Interactions

Based on the amino acid sequence of gp32 [32,33] and nucleotide sequence of gene 32 [34], gp32 has 301 amino acids and a molecular weight of 33,487. Gp32 has an acidic isoelectric point of 5.0 [35], and at pH 7.0 it has a net charge of -10. Although gp32 has an overall negative charge, examination of the primary structure reveals that this charge is asymmetrically distributed with the NH_2-terminal half of the protein having a net charge of $+10$, while the COOH-terminal portion has a charge of -20. Recently, gp32 has been shown to be a zinc metalloprotein with one zinc atom tetrahedrally coordinated to three cysteines and perhaps one histidine [36]. Secondary structure estimates for gp32 based on circular dichroism are in good agreement with Chou-Fasman predictions [33] and suggest that gp32 contains approximately 20% α-helix, 20% β-sheet, and 60% random coil [23]. Based on a site size of seven bases [20] with an internucleotide distance of 0.46 nm [18,19], the protein : nucleic acid interface of gp32 may span about 3.2 nm. Since the monomeric form of gp32 behaves hydrodynamically like a prolate ellipsoid with an axial ratio of 4 : 1 and an overall length approaching 12 nm [11], it seems possible that adjacent molecules of gp32 that are cooperatively bound to single-stranded nucleic acid might overlap one another to some extent. Carroll et al. [37] demonstrated that gp32 molecules are capable of interacting with each other even in the absence of nucleic acid. The size of these aggregates increases with gp32 concentration; however, at the in vivo concentrations of unbound gp32, dimers or trimers would be expected to predominate.

In the mid-1970s several groups observed [23,38,39] that gp32 could be subjected to partial proteolysis. The resulting fragments were then characterized with the hope of identifying discrete functional domains. With a variety of proteases it has been shown that cleavage occurs primarily at two regions in gp32. The first cleavage occurs at a position within residues 253 to 275. The precise cleavage point depends on the particular protease chosen. The second site is located within the region spanning residues 9 to 21 [32,35,39–41]. Removal of the acidic COOH-terminal "A" domain (consisting of residues 253 to 301 when trypsin is used for proteolysis) results in a fragment referred to as gp32*-A. Cleavage of the NH_2-terminal "B" domain at lysine-21 with trypsin gives gp32*-B. Removal of both the "A" and "B" domains with trypsin results in core gp32*-(A + B) comprising residues 21 to 253 and having a predicted molecular weight of about 26,024 [33].

Characterization of gp32*-B' (produced by Staph. aureus protease cleavage at residue 9) and of gp32*-(A + B) has shown that removal of the NH_2-terminal B region completely abolishes gp32 : gp32 interactions involved in both gp32 indefinite self-aggregation and cooperative gp32 binding to polynucleotides [24,42–44]. Even though intact gp32 and core gp32*-(A + B) have the same intrinsic affinity of about 10^6 M^{-1} for $d(pT)_8$, the affinity of gp32 for poly(dT) (10^9 M^{-1} or above) is approximately 1,000 times higher than that of core gp32*-(A + B) for poly(dT) (about 10^6 M^{-1} [24]). This difference in apparent binding affinities for poly(dT) compared to $d(pT)_8$ results from the fact that the cooperativity parameter (ω) for the gp32:ssDNA complex is about 10^3 compared to only 1 for the core gp32*-(A + B) complex [24,43]. It appears from these results that the NH$_2$-terminal "B" region is essential for gp32:gp32 interactions.

When the COOH-terminal "A" domain is removed from gp32, the resulting fragment (gp32*-A) differs from intact gp32 in several properties, which include a greatly increased ability to melt native dsDNA [35] and a loss of gp32:T4 DNA polymerase (gp43) and gp32:T4 RNA priming protein (gp61) interactions [45]. Although both gp32*-A and gp32 have approximately the same apparent affinity for oligo- and polynucleotides, oligonucleotide binding by gp32*-A is more salt dependent than gp32 [43]. Partial trypsin digestion of gp32 demonstrates that binding of gp32 to polynucleotides is accompanied by a conformational change in the protein that may be correlated with the postulated change from the oligonucleotide to the polynucleotide binding mode [21]. This change causes a decrease in the rate of cleavage of the NH$_2$-terminal "B" region and an increase in the rate of release of the COOH-terminal "A" region [40]. Taken together, these data suggest that the inability of gp32 to melt native dsDNA results from a "kinetic block" that results from the presence of the COOH-terminal "A" domain. The "A" region can be thought of as a hinged domain that occludes part of the gp32:ssDNA interface in a manner that precludes the polynucleotide binding mode. In the cooperative (polynucleotide) binding mode the "A" domain may shift its position, exposing basic residues that can interact with the phosphodiester backbone of ssDNA [21]. If this postulated conformational change is slow relative to the rate of "breathing" of dsDNA, then gp32 would be kinetically blocked from denaturing dsDNA. This model predicts that once the acidic "A" domain has shifted out of the ssDNA binding groove, it would be positioned to interact with other T4 DNA replication and recombination proteins.

The results obtained from these and other studies of gp32*-A and gp32*-B indicate that the core gp32*-(A + B) fragment contains the critical nucleic acid binding region [24,43,44]. Using hydroxylamine as a general mutagenizing agent, Doherty et al. [46] found that 21 out of 22 isolated missense gene 32 mutants are contained within the DNA sequence coding for core gp32*-(A + B). Further support for this assignment of the DNA-binding domain comes from physicochemical experiments including both fluorescence quenching and ^{1}H-NMR. Results of these studies show that both gp32 and core gp32*-(A + B) have site sizes of approximately 7 nucleotides and both possess about the same intrinsic affinity (K_{int}) for oligonucleotides of one site size or less [24,25,43,44]. In addition, Williams and Konigsberg [44] demonstrated that when intact gp32 is cross-linked to ssDNA with UV irradiation, all of the cross-linked residues are within the core gp32*-(A + B) domain.

^{1}H-NMR studies are beginning to define particular gp32 amino acids involved in binding. ^{1}H-NMR spectra of core gp32*-(A + B): oligo-nucleotide complexes indicate that a set of resonances corresponding to a discrete group of aromatic protons shift upfield upon oligonucleotide binding [25]. Since the number of aromatic protons that shift upfield increases with increasing length of the oligonucleotide (from about two to eight bases [25]), it appears likely that binding is polar, as was originally suggested by Kelly and Von Hippel [27]. That is, the 5'-phosphorylated end of the oligonucleotide occupies a unique position on gp32, and as the oligonucleotide is lengthened, it approaches an increasing number of aromatic protons on gp32. ^{1}H-NMR difference spectra generated by subtracting the spectra of core gp32*-(A + B) from that of the core gp32*-(A + B): d(pA)$_8$ complex demonstrate that resonances produced from five tyrosine and two phenylalanine residues give rise to all the observed chemical shifts [25,47].

Chemical modification studies on gp32 with tetranitromethane are in agreement with the ^{1}H-NMR results in that they suggest that five out of eight tyrosines in intact gp32 can be protected from tetranitromethane modification by the prior addition of ssDNA [23]. More recent studies have also shown that phenylalanine-183 of gp32 can be photo-cross-linked to d(pT)$_8$, suggesting that this phenylalanine residue is also situated at the gp32: ssDNA interface [48]. In addition, chemical modification of trypto-phan residues [49–51] suggest that at least one of the five tryptophan residues in gp32 is near the ssDNA binding surface of gp32.

Taken together, these studies demonstrate that the ssDNA binding site of gp32 is contained within core gp32*-(A + B) and that residues 71 to 137, which contains seven of the eight tyrosines and three of the five tryptophans in gp32, must make a contribution to the ssDNA binding surface. Although a model of the gp32: ssDNA complex awaits X-ray diffraction studies, the available data suggest that gp32 binding involves a combination of electrostatic interactions between approximately three basic amino acids and the phosphodiester backbone [21], hydrophobic interactions between five tyrosine and two phenylalanine residues and the nucleic acid bases, and cooperative interactions between adjacent gp32 molecules bound to the ssDNA lattice.

The Role of Tyrosine Residues in gp32: ssDNA Interactions

In vitro mutagenesis can provide information about individual amino acid residues that are involved in gp32 binding to ssDNA. Since five tyrosine residues in gp32 had already been implicated in ssDNA binding, these residues were attractive targets for this approach. Sequential substitution of each tyrosine residue in gp32 by a nonaromatic amino acid provided a family of mutant gp32 structural genes that could then be expressed in

E. coli so that the altered proteins could be characterized. The physico-chemical data acquired from this set of singly substituted gene 32 proteins allowed a rigorous test of current models for the involvement of individual aromatic amino acids in single-stranded nucleic acid binding. In general, substitution of a nonaromatic amino acid for a tyrosine in gp32 may be expected to result in a decreased affinity for nucleic acids either with or without accompanying changes in the overall gp32 structure. Those mutations that result in a decrease in the affinity of gp32 for nucleic acids without a concomitant alteration in the overall secondary and tertiary gp32 structure are most likely to define amino acids that are directly involved in binding. To detect changes in the secondary and higher-order structure of gp32 we have relied on a variety of approaches including differential scanning microcalorimetry, circular dichroism, and limited proteolysis. Finally, each of the mutants has been subjected to atomic absorption to determine if the intrinsic zinc ion that has previously been shown to make a significant contribution to the thermal stability [52] of gp32 is still present.

Differential scanning microcalorimetry provides an array of thermody-namic parameters that reflect the thermal stability of the protein. These include the temperature of denaturation, the denaturation enthalpy, and the degree of stabilization of the protein as a consequence of substrate binding. Large changes in thermal stability have generally been observed in cases where the mutation either occurs at a structurally important residue or at a site that has a low solvent accessibility [53,54]. Small changes (1°C to 3°C for gp32) in thermal stability are mostly observed for positions at the protein surface, where there is more flexibility and greater solvation stabilization [54].

Circular dichroism studies were also carried out on the mutant gp32 proteins in order to detect any large changes in the overall conformation of the polypeptide backbone [23,56]. Limited proteolysis provides an-other useful probe for detecting structural perturbations that have been induced in proteins by amino acid substitutions. With gp32 the rate of formation and disappearance of core gp32*-(A + B) is sensitive to changes that affect the tertiary structure in the vicinity of the protease-sensitive peptide bonds. Removal of the zinc ion from gp32 results in a protein that is hypersensitive to protease digestion [36] and that, based on differential scanning microcalorimetry [52], is considerably less stable than native gp32.

Table 14.1 summarizes the preliminary binding data available on the eight mutant gp32 proteins. Each mutant contains a nonaromatic amino acid in place of one of the eight tyrosines in gp32. Two of the mutant proteins, one of which contains leucine in place of tyrosine 92 and the other of which contains serine in place of tyrosine 73, appear to have undergone major overall structural changes compared to native gp32. The leucine 92 protein probably does not fold properly, as evidenced by the

Table 14.1. Binding affinities of wild-type and mutant gp32 proteins for d(pT)$_{16}$.

	K$_{app}$	ω
gp32	1.5×10^7	410
Ser 73	No measurable affinity	
Ala 84	0.8×10^7	590
Leu 92	?	
Val 99	0.5×10^7	430
Leu 106	0.4×10^7	130
Ser 115	0.9×10^7	460
Ser 137	1.7×10^7	410
Ile 186	0.4×10^7	340

fact that it is the only gp32 mutant that appeared to be degraded in vivo and that could therefore not be isolated by column chromatography [57]. SDS gel electrophoresis, carried out on crude extracts derived from cells in which the leucine 92 protein had been overexpressed, showed a large amount of an approximately 15,000-dalton protein, which direct amino-terminal sequencing demonstrated was actually a fragment containing the amino terminus of gp32 [57]. In addition to the absence of the expected zinc ion, the serine 73 mutant protein had an unusually low calorimetric transition temperature (48°C compared to the value of 55°C for native gp32) and showed characteristic changes in its circular dichroism spectrum that, as in the case of apo-gp32, were consistent with the almost complete loss of β-pleated sheet structure [57]. All of the changes evident in the serine 73 protein, including its inability to be converted by limited proteolysis to a stable core species, suggest major alterations in the overall secondary and tertiary structure that together probably account for the low affinity of this mutant protein for ssDNA (Table 14.1). From the available data, therefore, both tyrosine 73 and 92 seem to play essential roles in maintaining the native gp32 conformation.

The stability of the serine 137 mutant protein to limited proteolysis, its intrinsic zinc ion, and its near-normal circular dichroism spectrum suggest an overall structure at 25°C that is similar to that of native gp32. In addition, the similar binding affinities of the wild-type and the serine 137 mutant gp32 proteins for d(pT)$_{16}$ (Table 14.1) suggest that tyrosine 137 is not directly involved in binding.

In contrast to the three tyrosine mutants that have been discussed, corresponding to serine 73, leucine 92, and serine 137, the five remaining tyrosine mutants listed in Table 14.1 all appear to have significantly decreased apparent affinities for d(pT)$_{16}$ and yet retain overall structures that closely resemble the native conformation. The serine 115 mutant is somewhat unusual in this regard in that, despite its normal circular dichroism spectrum at 25°C, this mutant protein had the lowest calorimetric transition temperature measured (43°C compared to the value of

55°C for native gp32 [57]). Since the ^{1}H-NMR spectrum of the serine 115 mutant is sufficiently similar to that of native gp32 to allow the tyrosine 115 resonances to be assigned (by subtracting the serine 115 from the wild-type spectrum [47]), it is clear that the overall structure of this gp32 protein at 25°C has not been significantly altered by the tyrosine to serine substitution at position 115. As a general rule, therefore, it would seem that while a near normal calorimetric transition temperature provides evidence that the native structure is intact, an unusually low transition temperature does not necessarily indicate a significantly altered overall structure at 25°C. Since all five of the remaining mutant proteins listed in Table 14.1, corresponding to alanine 84, valine 99, leucine 106, serine 115, and isoleucine 186, appear to share a near native conformation as evidenced by limited proteolysis and circular dichroism, the decreased affinities of these proteins for ssDNA may reflect the loss of specific tyrosine : nucleotide base interactions [57]. The leucine 106 protein provides the only instance where substitution of a tyrosine residue has significantly decreased the cooperativity of binding (Table 14.1). It is difficult to determine, however, if the decreased cooperativity parameter for the leucine 106 protein results from the direct loss of an amino acid that is involved in gp32 : gp32, protein : protein interactions or, alternatively, is a secondary effect perhaps resulting from an improper alignment of gp32 molecules along a ssDNA lattice.

Although ^{1}H-NMR spectra are only available for the serine 115 mutant protein, these preliminary results [47] are in good agreement with the conclusions reached from the fluorescence studies. That is, ^{1}H-NMR difference spectra of the wild type and serine 115 mutant protein complexed with oligonucleotides reveal that tyrosine 115 is among the first tyrosines to become involved in ssDNA binding and that resonances corresponding to the 2,6 and 3,5 protons of this tyrosine residue are shifted upfield during complexation [47]. Upfield shifts such as those observed for tyrosine 115 suggest that the aromatic nucleic acid base and tyrosine residue are in close enough proximity (<4.5 Å) to be affected by each other's aromatic ring current shifts [58,59].

The overall conclusions from the in vitro mutagenesis studies concerning the extent of the involvement of aromatic amino acids in the binding of gp32 are also in good agreement with previous ^{1}H-NMR studies on gp32*-(A + B) : oligonucleotide complexes [24,47]. That is, while proton NMR difference spectra suggest that as many as five tyrosine and two phenylalanine residues may be involved in binding, the magnitude of the chemical shifts (0.06 to 0.4 ppm) are smaller than would be expected for complete overlap of the aromatic rings of these amino acid residues with the nucleotide bases at the minimum stacking distance of 3.4 Å [25]. Thus, while the bases of the bound nucleotide appear to approach as many as seven aromatic rings in the base-binding pockets of gp32*-(A + B), the stacking distances are larger than optimum, and/or the ring overlap is less than maximal. From ^{1}H-NMR data, Prigodich et al. [25] thus concluded that "other than providing complementary pockets for a ladder-type orientation of the bases, the contribution of the stacking interactions to the thermodynamic stability of the complexes is probably not great."

As shown by the data in Table 14.1, this hypothesis is supported by our in vitro mutagenesis studies, which suggest that the maximum possible contribution of any single tyrosine residue to the overall free energy of binding of gp32 to $d(pT)_{16}$ ranges from a low of only about 3% of the total for tyrosine 115 to a high of only about 8% of the total binding energy for tyrosine 106 or 186. Assuming that none of the five mutations listed in Table 14.1 that we believe correspond to tyrosine residues involved in binding (i.e., tyrosine residues 84, 99, 106, 115, and 186) result in any significant changes in the secondary or tertiary structure and that each tyrosine that has been mutated represents a completely independent interaction, then, overall, these five tyrosine residues could account for a maximum of 30% of the free energy of binding of gp32 to $d(pT)_{16}$. Since the cooperativity parameter derived for wild-type gp32 (Table 14.1) is sufficiently large to account for about 36% of the overall apparent affinity of gp32 for $d(pT)_{16}$, the proposed hydrophobic interactions between aromatic amino acid side chains of gp32 and the bases of the single-stranded DNA might potentially account for as much as 50% of the *intrinsic* affinity of gp32 for an oligonucleotide such as $d(pT)_{16}$.

Photochemical Cross-Linking of gp32 to $d(pT)_8$

Photochemical cross-linking has previously proved to be extremely useful for identifying amino acid residues involved in the binding of the *E. coli* SSB [60], bacteriophage fd gp5 [61,62], and rat hnRNP A1 protein [63] to single-stranded nucleic acids. In the case of the fd gp5 protein, X-ray crystallographic data have shown that the cross-linked residue, cysteine 33, is indeed at the gp5 : ssDNA interface [64]. When the complex formed between the core gp32*-(A + B) protein and $[^{32}P]d(pT)_8$ is exposed to ultraviolet light, as much as 25% of the core gp32*-(A + B) becomes covalently linked to the $[^{32}P]d(pT)_8$ [48]. As expected, the photo-cross-linking efficiency of the native gp32 is very similar to that of the core gp32*-(A + B) fragment [44,48]. This finding suggests that the topology of the gp32 : oligonucleotide interface, at least with respect to the site of photo-cross-linking, is not significantly altered by removal of both the NH$_2$-terminal "B" and COOH-terminal "A" regions.

Tryptic digestion of the cross-linked core gp32*-(A + B): $[^{32}P]d(pT)_8$ complex followed by anion exchange HPLC and amino acid analysis of the resulting ^{32}P-labeled peptide peak demonstrated that all of the photo-cross-linking occurs within a single tryptic peptide that spans residues 179 to 190 in gp32 [48]. Amino acid sequencing of this ^{32}P-labeled peptide demonstrated that phenylalanine 183 represents the single site of photochemical cross-linking of core gp32*-(A + B) to d(pT)$_8$ [48]. It is interesting that as in the case of the *E. coli* SSB [60] and the A1 hnRNP [63], protein cross-linking occurs at a phenylalanine residue. This finding presumably reflects both the general involvement of aromatic amino acids in the binding of proteins to single-stranded nucleic acids and the ease of cross-linking to phenylalanine. Phenylalanine 183 is also close to tyrosine 186, which is one of the five tyrosine residues that in vitro mutagenesis studies suggest may be involved in binding ssDNA [57]. Assuming that phenylalanine 183 is one of the two phenylalanines whose proton resonances shift upon the additon of an oligonucleotide [47], there should be at least one additional phenylalanine residue involved in binding that has not yet been elucidated.

The Function of Zinc in Gene 32 Protein

Gene 32 protein is a metalloprotein containing a single Zn(II) ion in a tetrahedral coordination liganded by three cysteines and presumably a histidine [36]. The amino acids that are proposed to be ligands for the metal are Cys-77, His-81, Cys-87, and Cys-90. Site-directed mutagenesis has provided further evidence for this set of ligands by showing that removal of the only other cysteine in the protein does not affect the zinc content [Giedroc and Coleman, personal communication].

The resistance of core gp32*-(A + B) to proteolysis is sharply reduced by the removal of the Zn(II) ion. While the intact gp32 is very resistant to any proteolysis beyond the removal of the "A" and "B" domains, the apo-gp32 is rapidly proteolyzed to several different fragments that have lower molecular weight than gp32*-(A + B) [36]. This demonstrates that removal of the zinc(II) alters the structure of gp32 in a way that makes other cleavage sites accessible to trypsin. Removal of the zinc ion from gp32 is also accompanied by changes in the circular dichroism spectrum that are indicative of alterations in the overall secondary structure [Keating, Giedroc, and Coleman, unpublished observations].

The extent to which Zn(II) stabilizes the folded form of gp32 was determined using differential scanning calorimetry (DSC) [52]. The DSC scans of the Zn(II) and apo forms of gp32 in Figure 14.1 reflect the 6° decrease in the thermal denaturation temperature and the 40% decrease in denaturation enthalpy that accompanies Zn(II) removal. Apo-gp32 begins to denature below 45°C. This result explains why limited proteolysis of this protein at 45°C produces only small peptides [36]. The free energy of denaturation of the apo-gp32 is −1.7 kcal mol^{-1} at 55.4°C (the denaturation temperature of Zn(II)-gp32, where its free energy of denaturation is zero). This is a relatively small decrease in denaturational free energy compared to that induced by the removal of Zn(II) from other metallo-

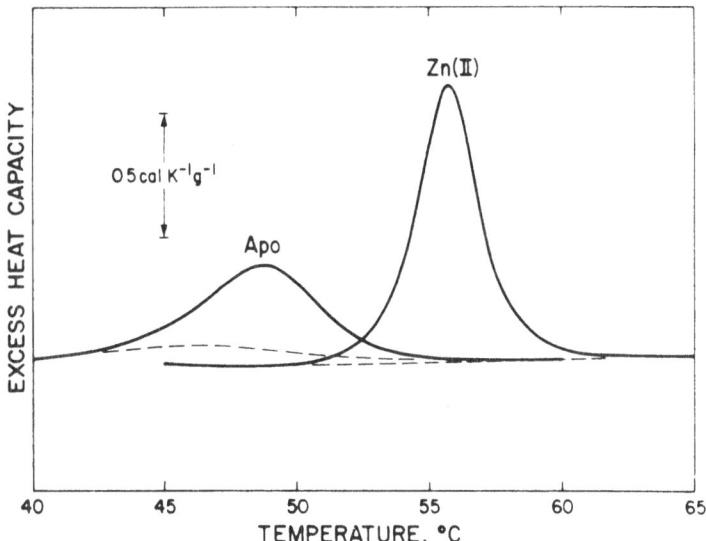

Figure 14.1. Effects of zinc(II) removal on the thermal denaturation of gp32. The curves were obtained on a DASM-4 differential scanning calorimeter at a scan rate of 1°C/min on 2.0 mg/ml solutions of protein [52].

proteins [65–67] where decreases in denaturation temperatures of 20°C to 28°C were observed. The relatively small free-energy change associated with the removal of zinc from gp32 may be the result of an entropic compensation for the enthalpic destabilization. An entropy-associated increase in conformational fluctuation could account for the increased susceptibility of apo as compared to native gp32 to proteolysis by trypsin. It is clear from the differential scanning calorimetry as well as from the previous limited proteolysis and circular dichroism studies that the zinc ion is an important structural determinant of gp32.

Despite the close proximity in the primary sequence of gp32 between the zinc-binding domain (defined here as extending from cysteine 77 to cysteine 90 and specifically including histidine 81 and cysteine 87) and a region of gp32 (spanning tyrosine 84 to 115) that contains four tyrosines that appear to be involved in ssDNA binding, removal of the zinc ion from gp32 has relatively little effect on the intrinsic gp32 : ssDNA interaction. Binding studies with $d(pT)_8$ demonstrate that removal of Zn(II) reduces the affinity of gp32 for an oligonucleotide only by a factor of about 5, from 4×10^5 M^{-1} to 8×10^4 M^{-1} [68]. The decrease corresponds to about 13% of the total free energy of binding of gp32 to $d(pT)_8$ and, in fact, based on the in vitro mutagenesis studies, more than 50% of this decrease might be accounted for by loss of the proposed tyrosine 84 interaction with ssDNA. This assumes that zinc removal is accompanied by changes in gp32 conformation such that the nearby tyrosine 84 is no longer able to interact with the nucleic acid lattice. In contrast to the relatively small effect of the intrinsic Zn(II) on gp32 binding to $d(pT)_8$, the intrinsic Zn(II) appears to make a significant contribution, accounting for about 30% of the overall binding energy [68], to the affinity of gp32 for $d(pT)_{16}$.

The decreased affinity of apo-gp32 for $d(pT)_{16}$ appears to result largely from the decreased cooperativity of apo-gp32 binding rather than from any direct effect on the intrinsic gp32 : $d(pT)_{16}$ interaction. The cooperativity parameter (ω) is about 4 for apo-gp32 as compared to about 4,000 for native gp32 binding to $d(pT)_{16}$ [68]. Surprisingly, in the case of apo-gp32 the cooperativity of binding appears to increase upon going from $d(pT)_{16}$ to an indefinite lattice. Hence, the estimated cooperativity parameter for apo-gp32 binding to polyriboethenoadenylic acid in 0.3 M NaCl is close to 200 [Keating, unpublished observations]. Even though additional studies are clearly needed because of possible problems that may arise in comparing cooperativity parameters for ssDNA versus ssRNA, it appears that zinc(II) is essential for cooperative gp32 : gp32 interactions on a defined lattice that is only long enough to allow for one gp32 : gp32 interaction. Perhaps the decrease in gp32 cooperativity that accompanies Zn(II) removal can be partially compensated for by having an indefinite lattice where each gp32 molecule is involved in two gp32 : gp32 interactions as compared to only one such protein : protein interaction on $d(pT)_{16}$. The additional gp32 : gp32 interaction may thus stabilize a particular conformation or decrease the degree of conformational flux about a particular conformation that Zn(II) removal has weakened and that is required for cooperative binding.

Several other methods of studying protein : DNA interactions reinforce the assertion that the ability of gp32 to bind ssDNA in a highly cooperative manner is reduced by the removal of zinc. Electron microscopy studies of gp32-circular ssDNA complexes show that, with subsaturating protein levels, apo-gp32 forms relatively short oligomers, whereas Zn(II)-gp32 oligomerizes to the point where each ssDNA molecule that binds a gp32 molecule is completely covered [69]. Similarly, the helix-coil transition of poly[d(A-T)] that is induced by apo-gp32 is broad and is qualitatively similar to that seen with a noncooperatively binding gp32 proteolysis fragment, gp32*-B' [68]. Finally, the lack of any significant effect of polynucleotides on the calorimetric transition of apo-gp32 is similar to the inability of oligonucleotides to alter the thermal denaturation profile of Zn(II)-gp32 [52]. In the case of Zn(II)-gp32, the addition of poly(dT) results in several changes with respect to thermal denaturation of the protein (increased T_m, increased free energy of stabilization and cooperativity of the thermal transition [42,52]) that, because they have not been seen with the gp32 : $d(pT)_8$ complex, have been ascribed to effects arising from gp32 : gp32 cooperative protein : protein interactions. Since none of these characteristic changes occur when poly(dT) is added to apo-gp32, the calorimetric data further reinforce the idea that apo-gp32 binding is less cooperative than is Zn(II)-gp32 binding to poly(dT) [68]. According to the results of protein affinity chromatography, the intrinsic Zn(II) is also essential for gp32 : gp32 interactions (perhaps related to the ability of gp32 to undergo indefinite aggregation) that occur in the absence of ssDNA. Hence, whereas Zn(II)-gp32 readily binds to a gp32 affinity column, apo-gp32 fails to bind to a similar column containing covalently linked apo-gp32 [69].

An interesting notion concerning possible functions of the Zn(II) in gp32 has arisen from the discovery of a number of gene 32 mutations that have been isolated as phages unable to grow on the restrictive Tab 32-4 bacterium [70]. These mutations include amino acid substitutions not only in the putative zinc binding domain of gp32 but throughout the core domain of the protein, and it is suggested that because the characteristic phenotype of these mutants is reversed by the addition of exogenous Zn(II), they all reduce the affinity of gp32 for Zn(II). Because two of three mutations that are more deleterious to late transcription than to DNA replication are located in the sequence in between the cysteines that are thought to ligate Zn(II), the authors propose that gp32's "zinc finger" may be required for the activation of late transcription via sequence recognition [70].

Conclusions

The picture that emerges concerning relationships between the structure and functions of gp32 is that of a protein that contains a minimum of three functionally but not necessarily structurally discrete domains. While the complete loss of cooperativity that accompanies removal of the NH$_2$-terminal "B" region from gp32 clearly indicates that the first nine amino acids are essential for gp32 : gp32 interactions whether they occur between gp32 molecules free in solution or between adjacent gp32 molecules bound to ssDNA, there is no evidence that this domain is *directly* involved in gp32 : gp32 interactions. In addition, recent studies on the role of the intrinsic Zn(II) in gp32 strongly suggest that zinc binding leads either to a change in the gp32 conformation or to a reduction in conformational flux that not only stabilizes the overall gp32 structure but is also essential to allow highly cooperative binding of gp32 to ssDNA. It appears therefore that neither the NH$_2$-terminal "B" region nor the zinc binding region is in and of itself sufficient to permit maximum cooperativity of gp32 binding to ssDNA. Similarly, while loss of the COOH-terminal "A" region relieves the kinetic block that normally prevents gp32 from destabilizing native T4 dsDNA and is associated with the loss of several heterologous gp32 : protein interactions, it is not known if the "A" region is a direct participant in these latter interactions. Binding of gp32 to ssDNA appears to involve a minimum of three ionic interactions

[21] and as many as seven hydrophobic interactions between aromatic amino acid side chains and the nucleotide bases [47].

There is evidence from photo-cross-linking studies [48] for the direct involvement of phenylalanine 183 and from ^{1}H-NMR studies for the involvement of tyrosine 115 in binding [47]. In vitro mutagenesis studies further suggest that tyrosines numbered 84, 99, 106, and 186 also contribute directly to binding [57]. While the contribution of any one of these hydrophobic interactions to the overall binding affinity for an oligonucleotide such as $d(pT)_{16}$ is likely to be small, with the largest perhaps corresponding to 8% of the overall free energy of binding in the case of tyrosine 106 or 186, together these hydrophobic interactions may constitute a ssDNA binding groove on the surface of gp32 (Fig. 14.2) that could account for a maximum of 30% of the overall binding energy.

Since the cooperativity parameter for native gp32 is sufficiently large to account for about 36% of the apparent affinity of gp32 for $d(pT)_{16}$, ionic and other kinds of interactions probably account for the remaining 34% of the binding energy. The real significance of the hydrophobic interactions may not be so much in their energetic contribution to binding but rather in the ability of these numerous small interactions to impart a postulated "ladderlike" structure onto the ssDNA [25] that is optimum to then serve as a substrate for T4 DNA polymerase and other enzymes involved in T4 DNA repair and recombination. These hydrophobic interactions may also contribute to the specificity of gp32 in terms of its preferential binding to single-stranded as opposed to double-stranded nucleic acids. Intuitively, these interactions should be more favored for single-stranded nucleic acids, and indeed even very short peptides of the general structure Lys–X–Lys, where X is an aromatic amino acid, show preferential binding for single-stranded nucleic acids [72]. In this regard gp32 seems to serve as an excellent prototype for a large number of prokaryotic and eukaryotic proteins that bind to single-stranded nucleic acids. There is now evidence for the involvement of aromatic amino acids in the binding of the bacteriophage fd gene 5 [73], the E. coli SSB [60], and the rat A1 hnRNP protein [63]. Four phenylalanine residues in the A1 hnRNP protein that have been identified by photo-cross-linking studies as being at the interface of the A1 : nucleic acid complex are among a set of highly conserved residues that are found in the same relative positions in all major hnRNP proteins, the poly(A) binding protein, nucleolin, a U1 snRNP-associated protein, and numerous other eukaryotic proteins that bind single-stranded RNA [see 63]. Clearly, the involvement of aromatic amino acids in the binding of proteins to single-stranded nucleic acids represents a general mechanism that is not restricted just to gp32. While considerable knowledge has been gained concerning the relationship of structure and function in the bacteriophage T4 gene 32 protein, significant further progress will require high-resolution X-ray crystallographic studies on a suitable gp32*-(A + B) : oligonucleotide complex.

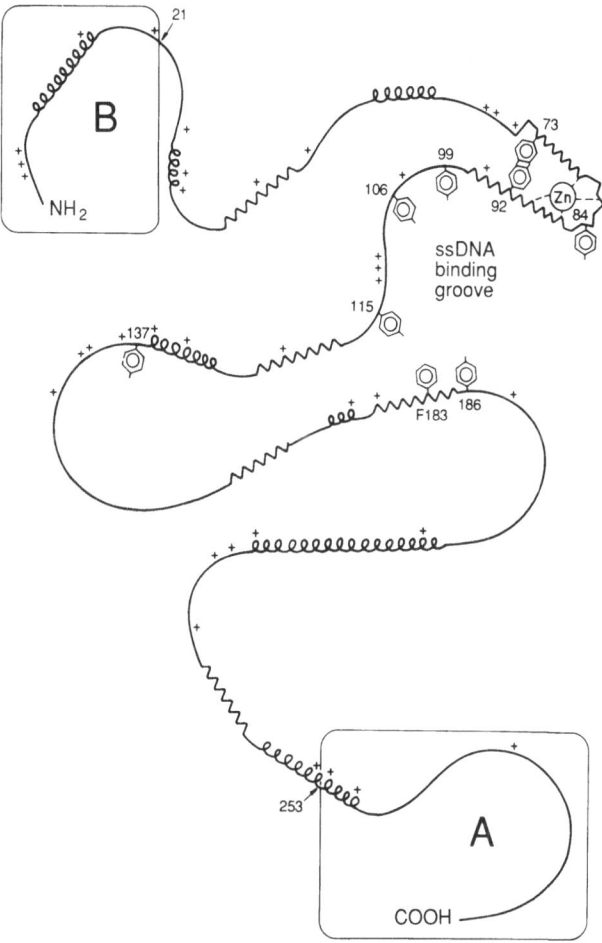

Figure 14.2. Cartoon depiction of functional regions in gp32. The α-helices, β-structure, and turns are positioned where their probabilities are greatest as predicted by the DELPHI program which is based on the method of Garnier et al. [71]. The overall extent of α-helix (20%) and β-sheet (20%) were derived from circular dichroism data. The small arrows near the "A" and "B" domains represent limited proteolysis sites for trypsin. The Zn(II) ion is thought to be coordinated to cysteine residues numbered 77, 87, 90 and histidine 81. Tyrosines 73 and 92 appear to play a structural role while tyrosines 84, 99, 106, 115, and 186 all appear to be involved in binding to ssDNA. Replacement of any of these latter five tyrosine residues by a nonaromatic amino acid results in a two- to four-fold reduction in affinity of gp32 for $d(pT)_{16}$ [57]. Phenylalanine 183 is positioned in the proposed ssDNA binding groove because it is the only site of photo-cross-linking between gp32 and oligo $d(pT)_8$ [47].

References

1. Chase, J.W., and Williams, K.R. (1986) Annu. Rev. Biochem. *55*, 103–136.
2. Riva, S., Cascino, A., and Geiduschek, E.P. (1970) J. Mol. Biol. *54*, 85–102.
3. Curtis, M.J., and Alberts, B. (1976) J. Mol. Biol. *102*, 793–816.
4. Von Hippel, P.H., Kowalczykowski, S.C., Lonberg, N., Newport, J.W., Paul, L.S., Stormo, G.D., and Gold, L. (1982) J. Mol. Biol. *162*, 795–818.
5. Nossal, N.G., and Peterlin, B.M. (1979) J. Biol. Chem. *254*, 6032–6037.
6. Alberts, B.M., Barry, J., Bedinger, P., Burke, R.L., Hibner, U., Liu, C.-C., and Sheridan, R. (1980) In Mechanistic Studies on DNA Replication and Genetic Recombination, Vol. 19, 449–471, Alberts, B.M., and Fox, C.F. (eds.). Academic Press, New York.
7. Formosa, T., Burke, R.L., and Alberts, B.M. (1983) Proc. Natl. Acad. Sci. USA *80*, 2442–2446.
8. Wu, J.-R., and Yeh, Y.-C. (1973) J. Virol. *12*, 758–765.
9. Kodadek, T., Wong, M.L., and Alberts, B.M. (1988) J. Biol. Chem. *263*, 9427–9436.
10. Tomizawa, J.-I., Anraku, N., and Iwama, Y. (1966) J. Mol. Biol. *21*, 247–253.
11. Alberts, B.M., and Frey, L. (1970) Nature *227*, 1313–1318.
12. Berger, H., Warren, A., and Fry, K. (1969) J. Virol. *3*, 171–175.
13. Newport, J.W., Lonberg, N., Kowalczykowski, S.C., and Von Hippel, P. (1981) J. Mol. Biol. *145*, 105–121.
14. Krisch, H.M., Bolle, A., and Epstein, R.H. (1974) J. Mol. Biol. *88*, 89–104.
15. Russel, M., Gold, L., Morrissett, H., and O'Farrell, P. (1976) J. Biol. Chem. *251*, 7263–7270.
16. Lemaire, G., Gold, L., and Yarus, M. (1978) J. Mol. Biol. *126*, 73–90.
17. McPheeters, D.S., Stormo, G.D., and Gold, L. (1988) J. Mol. Biol. *201*, 517–535.
18. Delius, H., Mantell, N.J., and Alberts, B. (1972) J. Mol. Biol. *67*, 341–350.
19. Sheerhagen, M.A., Kuil, M.E., Van Grondelle, R., and Blok, J. (1985) FEBS Lett. *184*, 221–225.
20. Jensen, D.E., Kelly, R.C., and Von Hippel, P.H. (1976) J. Biol. Chem. *251*, 7215–7228.
21. Kowalczykowski, S.C., Lonberg, N., Newport, J.W., and Von Hippel, P.H. (1981) J. Mol. Biol. *145*, 75–104.
22. Scheerhagen, M.A., Bokma, J.T., Vlaanderen, C.A., Blok, J., and Van Grondelle, R. (1986) Biopolymers *25*, 1419–1448.
23. Anderson, R.A., and Coleman, J.E. (1975) Biochemistry *14*, 5485–5491.
24. Spicer, E.K., Williams, K.R., and Konigsberg, W.H. (1979) J. Biol. Chem. *254*, 6433–6436.
25. Prigodich, R.V., Casas-Finet, J., Williams, K.R., Konigsberg, W., and Coleman, J.E. (1984) Biochemistry *23*, 522–529.
26. Kelly, R.C., Jensen, D.E., and Von Hippel, P.H. (1976) J. Biol. Chem. *251*, 7240–7250.
27. Kelly, R.C., and Von Hippel, P.H. (1976) J. Biol. Chem. *251*, 7229–7239.
28. McGhee, J.D., and Von Hippel, P.H. (1974) J. Mol. Biol. *86*, 469–489.
29. Lohman, T.M. (1984) Biochemistry *23*, 4656–4665.
30. Lohman, T.M. (1984) Biochemistry *23*, 4665–4675.

31. Liu, C.C., Burke, R.L., Hibner, U., Barry, J., and Alberts, B. (1978) Cold Spring Harbor Symp. Quant. Biol. *43*, 469–487.
32. Williams, K.R., LoPresti, M.B., Setoguchi, M., and Konigsberg, W. (1980) Proc. Natl. Acad. Sci. USA *77*, 4614–4617.
33. Williams, K.R., LoPresti, M.B., and Setoguchi, M. (1981) J. Biol. Chem. *256*, 1754–1762.
34. Krisch, H.M., and Allet, B. (1982) Proc. Natl. Acad. Sci. USA *79*, 4937–4941.
35. Hosoda, J., and Moise, H. (1978) J. Biol. Chem. *253*, 7547–7555.
36. Giedroc, D., Keating, K., Williams, K., Konigsberg, W., and Coleman, J. (1986) Proc. Natl. Acad. Sci. USA *83*, 8452–8456.
37. Carroll, R.B., Neet, K., and Goldthwait, D.A. (1975) J. Mol. Biol. *91*, 275–291.
38. Hosoda, J., Takacs, B., and Black, C. (1974) FEBS Lett. *47*, 338–342.
39. Moise, H., and Hosoda, J. (1976) Nature *259*, 455–458.
40. Williams, K.R., and Konigsberg, W.H. (1978) J. Biol. Chem. *253*, 2463–2470.
41. Tsugita, A., and Hosoda, J. (1978) J. Mol. Biol. *122*, 255–258.
42. Williams, K.R., Sillerud, L.O., Schafer, D.E., and Konigsberg, W.H. (1979) J. Biol. Chem. *254*, 6426–6432.
43. Lonberg, N., Kowalczykowski, S.C., Paul, L.W., and Von Hippel, P.H. (1981) J. Mol. Biol. *145*, 123–138.
44. Williams, K.R., and Konigsberg, W.H. (1981) In Gene Amplification and Analysis, Vol. 2, 475–508, Chirikjian, J.G., and Papas, T.S. (eds.). Elsevier/North-Holland, Amsterdam.
45. Burke, R.L., Alberts, B.M., and Hosoda, J. (1980) J. Biol. Chem. *255*, 11484–11493.
46. Doherty, D.H., Gauss, P., and Gold, L. (1982) Mol. Gen. Genet. *188*, 77–90.
47. Prigodich, R.V., Shamoo, Y., Williams, K.R., Chase, J.W., Konigsberg, W.H., and Coleman, J.E. (1985) Biochemistry *25*, 3666–3672.
48. Shamoo, Y., Williams, K.R., and Konigsberg, W. (1988) Proteins Struct. Function Genet. *4*, 1–6.
49. Toulme, J.-J., LeDoan, T., and Helene, C. (1984) Biochemistry *23*, 1195–1201.
50. LeDoan, T., Toulme, J.-J., and Helene, C. (1984) Biochemistry *23*, 1202–1207.
51. Casas-Finet, J., Toulme, J.-J., Cazenave, C., and Santus, R. (1984) Biochemistry *23*, 1208–1213.
52. Keating, K.M., Ghosaini, L.R., Giedroc, D.P., Williams, K.R., Coleman, J.E., and Sturtevant, J.M. (1988) Biochemistry *27*, 5240–5245.
53. Hecht, M.H., Sturtevant, J.M., and Sauer, R.T. (1984) Proc. Natl. Acad. Sci. USA *81*, 5685–5689.
54. Alber, T., Sun, D.P., Wilson, K., Wozniak, J.A., Cook, S.P., and Matthews, B.W. (1987) Nature *330*, 41–46.
55. Howell, E.E., Villafranca, J.E., Warren, M.S., Oatley, S.J., and Kraut, J. (1986) Science *231*, 1123–1128.
56. Greenfeld, N., and Fasman, G.D. (1969) Biochemistry *8*, 4108–4116.
57. Shamoo, Y., Ghosaini, L.R., Keating, K.M., Williams, K.R., Sturtevant, J.M., and Konigsberg, W.H. (1989) Biochemistry, in press.
58. Giessner-Prettre, C., and Pullman, B. (1971) J. Theor. Biol. *31*, 287–294.

59. Giessner-Prettre, C., and Pullman, B. (1976) Biochem. Biophys. Res. Commun. *70*, 578–581.
60. Merrill, B.M., Williams, K.R., Chase, J.W., and Konigsberg, W.H. (1984) J. Biol. Chem. *259*, 10850–10856.
61. Paradiso, P.R., Nakashima, Y., and Konigsberg, W.H. (1979) J. Biol. Chem. *254*, 4739–4744.
62. Paradiso, P.R., and Konigsberg, W.H. (1982) J. Biol. Chem. *257*, 1462–1467.
63. Merrill, B.M., Stone, K.L., Cobianchi, F., Wilson, S.H., and Williams, K.R. (1988) J. Biol. Chem. *263*, 3307–3313.
64. Brayer, G.D., and McPherson, A. (1983) J. Mol. Biol. *169*, 565–596.
65. Chlebowski, J.F., and Mabrey, S. (1977) J. Biol. Chem. *252*, 7042–7052.
66. Engeseth, H.R., and McMillin, D.R. (1986) Biochemistry *25*, 2448–2455.
67. Lepock, J.R., Arnold, L.D., Torrie, B.H., Andrews, B., and Kruuv, J. (1985) Arch. Biochem. Biophys. *241*, 243–251.
68. Giedroc, D.P., Keating, K.M., Williams, K.R., and Coleman, J.E. (1987) Biochemistry *26*, 5251–5259.
69. Keating, K.M., Giedroc, D.P., Harris, L.D., Ghosaini, L.R., Williams, K.R., Sturtevant, J.M., and Coleman, J.E. (1987) In Protein Structure, Folding, and Design 2, 35–44, Oxender, D. (ed.). Alan R. Liss, New York.
70. Gauss, P., Krassa, K., McPheeters, D.S., Nelson, M.A., and Gold, L. (1987) Proc. Natl. Acad. Sci. USA *84*, 8515–8519.
71. Garnier, J., Osguthorpe, D., and Robson, B. (1978) J. Mol. Biol. *120*, 97–120.
72. Brun, F., Toulme, J., and Helene, C. (1975) Biochemistry *14*, 558–563.
73. O'Connor, T.P., and Coleman, J.E. (1983) Biochemistry *22*, 3375–3381.

Chapter 15
Crystallographic Studies of Two Proteins That Bind Single-Stranded DNA

Alexander McPherson

Although most protein-DNA recognition phenomena governing the expression of genetic information are exercised at the level of double-stranded DNA, a vast number of salient cellular events utilizing nucleic acids involve primarily or exclusively ssDNA or RNA. Indeed, one of the first events to occur following recognition and binding of activating proteins to native DNA is the unwinding of the duplex to expose the single strands. Events that include the replication of DNA to produce daughter strands, the transcription of DNA to RNA, viral replication and infection, the repair of DNA, genetic recombination, and DNA modification and degradation all involve primarily single-stranded forms of DNA. Viruses frequently possess single-stranded nucleic acid genomes, and virtually all RNA structures such as tRNA, ribosomal RNAs, and messenger RNAs are single stranded. It is further likely that other forms of nucleic acid that we now know little about, such as the viroids, will involve exclusively single-stranded forms. All of these single-stranded nucleic acids, both RNA and DNA, exhibit a host of both intra- and intermolecular interactions in arriving at their active or native state.

To understand, at the atomic level, how the transfer of information encoded in ssDNA and its structures is mediated, we must understand how proteins bind to ssDNA and identify what chemical interactions, of both a sequence-specific and nonspecific nature, are involved in the interactions. Finally, we need to delineate the structural alternatives available to ssDNA under fully hydrated conditions, the range of conformations and dynamic modes that are assumed, and how these are affected by the binding of ions, ligands, and proteins in the environment.

Currently, the only technique that permits the precise determination of macromolecular structures, that is, their direct visualization, is X-ray diffraction analysis of single crystals. This method may be applied to proteins, nucleic acid molecules, or complexes between the two. The only requirement is that the objects of investigation be structurally and conformationally uniform—that is, chemically homogeneous.

Application of the X-ray technique [for detailed descriptions see 1–3] requires the measurement of hundreds of thousands of diffraction intensities from both native and heavy-atom-modified crystals, their correlation, and their incorporation into a Fourier synthesis. The Fourier synthesis, or electron density map, serves as a three-dimensional image of the electron density distribution over the volume in space occupied by the macromolecule. From this distribution and available chemical information, the crystallographer can deduce, and later refine, a detailed atomic model of the protein or nucleic acid.

A unique feature of macromolecular crystals is that they have a very high solvent content. This allows the investigator to diffuse various substrates, inhibitors, coenzymes, or other ligands directly into the crystals. The ligands, in general, freely complex with the macromolecules even as they are constrained to their lattice positions. By application of a second diffraction procedure, known as a difference Fourier analysis, which utilizes the diffraction intensities from the complex crystals, an image of the macromolecule-ligand association can be obtained [4]. From images of such complexes, the means by which specific binding is established, the mechanisms of enzyme catalysis, and the manner by which effectors or drugs perturb macromolecular behavior can be elucidated. In a sense, protein and nucleic acid crystals can be thought of as ligand-binding laboratories that offer the X-ray crystallographer the opportunity to see at the molecular level the atomic interactions that mediate and control the vital processes of living tissues.

Two proteins that bind to ssDNA have been solved and refined to a good level of precision (2.5 Å Bragg spacings or better), one of these while complexed to oligomers of deoxyribonucleic acid. From these two structures we have gotten pictures of how at least these two proteins bind to nucleic acid and the chemical mechanisms used to bring about complex formation and stabilization. The first of these proteins, the gene 5 DNA unwinding protein from bacteriophage fd, is described below.

Biological Properties of the Gene 5 Protein

Bacteriophage fd is a filamentous virus that infects *E. coli*. It is closely related to phages M13 and f1. The genome of this phage provides for the expression of 10 genes encoded in a circular, single DNA strand consisting of 6,408 nucleotides whose sequence is known [5]. The native phage particle is a long, flexible cylinder of length about 9,000 Å and a diameter of about 60 Å. The core of the virus consists of two antiparallel strands derived from the collapsed circular DNA. These strands are complexed on the exterior by the coat protein, the product of gene 8 [6]. In addition, the capsid contains five gene 3 and five gene 6 proteins that cap one end of the particle and three or four gene 7 and gene 9 protein molecules that complete the opposite end [7–9]. Following infection of the *E. coli* cell by absorption of the phage to the bacterial pili and insertion of viral DNA, the replication of fd follows via the rolling circle mechanism described by Denhardt [8].

The replication process occurs in three stages. Initially, the parental DNA is converted into a double-stranded replicative form (RF) composed of the viral DNA strand paired with a complement, and this includes synthesis of an RNA primer at a specific locus on the DNA circle, extension by DNA polymerases III and I, and ligation into a closed circle [10–12]. The DNA is then supercoiled by a cellular gyrase. In the second state, the RF-DNA is cleaved specifically by the product of gene 2, unwound by an *E. coli* enzyme, and replicated by the cellular DNA polymerase.

In the final stage of infection, termination of RF form DNA must be effected and the synthesis of single-stranded daughter virions completed. When the RF-DNA pool reaches a size of 100 to 200 copies per cell, the gene 5 protein is synthesized [13,14]. These protein molecules bind with high affinity to newly synthesized viral DNA and preclude complementary strand formation by polymerase. Combination of gene 5 protein with daughter strands is stoichiometric, and the complex is observed to be a long helical filament that somewhat resembles the mature phage. This gene 5 protein-DNA complex migrates to the host cell membrane where the gene 5 protein is displaced, in a poorly characterized process, by the gene 8 coat protein. The virus is capped and transported across the membrane to the exterior. The gene 5 protein returns to the cytoplasm for reuse in the formation of new nucleoprotein particles [10–12].

A feature of the infection cycle that has provided molecular biologists, physical chemists, and X-ray crystallographers with a very useful tool is the appearance in the individual *E. coli* cells of upwards of 100,000 copies of the gene 5 protein [13]. In practical terms, this means that hundreds of milligrams of a pure DNA-binding protein can be isolated from a rather modest amount of infected cells. Thus, the gene 5 protein is one of the few protein molecules of this kind available in adequate amounts for crystallographic and physical studies.

The amino acid sequence of the gene 5 DNA-binding protein (gene 5 protein) has been determined [15,16]. It contains 87 amino acids in a single polypeptide chain, giving a molecular weight of 9,700. No disulfide bridges are present. This relatively small protein exists predominantly as a dimer in solution, although at elevated concentrations higher-order aggregates are observed [17,18]. Both circular dichroism (CD) experiments [19] and sequence predictions [20] indicated that the gene 5 protein was composed for the most part of β-pleated sheet structure with little or no helical component.

The binding affinity of the gene 5 protein for bacteriophage fd DNA has been estimated to be on the order of 10^{-9} M [9]. Chemical modification studies have shown three of five tyrosines to be involved in DNA binding

[17,20]. At least one phenylalanine is also involved in the binding process [21–23]. In addition, lysine and arginine residues have been shown to participate in liganding DNA [20,24].

The stoichiometry of gene 5 protein-DNA binding has also been investigated. The number of nucleotides bound per protein monomer has been variously reported to be five [14,18,25,26], four [13,18,19,27], and three [20,28]. It is significant that the protein will tightly bind single-stranded DNA of any base sequence or composition. A large cooperative effect is associated with complexation of longer polynucleotides. That is, once one gene 5 protein dimer is bound to a DNA strand, association of the next dimer immediately adjacent to the first is enhanced an estimated 60-fold [13,18]. It is also known that bound DNA is completely unstacked and in an extended conformation regardless of its original state [13,19].

Although the primary physiological role of the gene 5 protein is to switch on synthesis of single-stranded DNA daughter virions and then stabilize as well as protect these strands, it has also been found to be a strong helix-destabilizing protein. That is, the melting temperatures of a variety of double-stranded DNAs are lowered by approximately 40°C in the presence of the gene 5 protein, irrespective of nucleotide composition [13]. This melting or unwinding capacity appears to result from a strong preference for single-stranded DNA over its double-stranded counterpart. In this way, gene 5 protein disturbs the natural equilibrium between native and single-stranded regions of duplex DNA, leading to unwinding

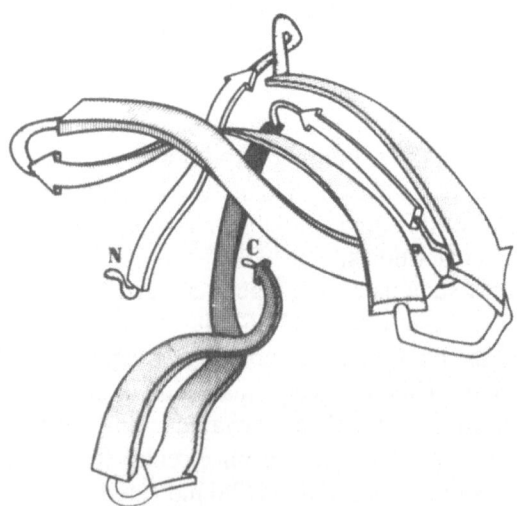

Figure 15.1. Schematic drawing of the polypeptide chain backbone of gene 5 binding protein (G5BP) illustrating the three major β loops present. The three β loops are termed the DNA-binding loop, the complex loop, and the dyad loop in proceeding from the N to the C terminus. The N and C termini are also indicated.

of the nucleic acid. Thus, this protein belongs to a class of helix-destabilizing proteins that includes the *E. coli* DNA unwinding protein SSB [29], calf thymus unwinding protein [30], the gene 32 protein from T_4 [31], and bovine pancreatic ribonuclease [32].

The Crystal Structure of the Gene 5 Protein

A schematic drawing illustrating the overall polypeptide chain folding of the gene 5 protein is shown in Figure 15.1, and a stereo drawing of the gene 5 protein is shown in Figures 15.2A and B. In accordance with the

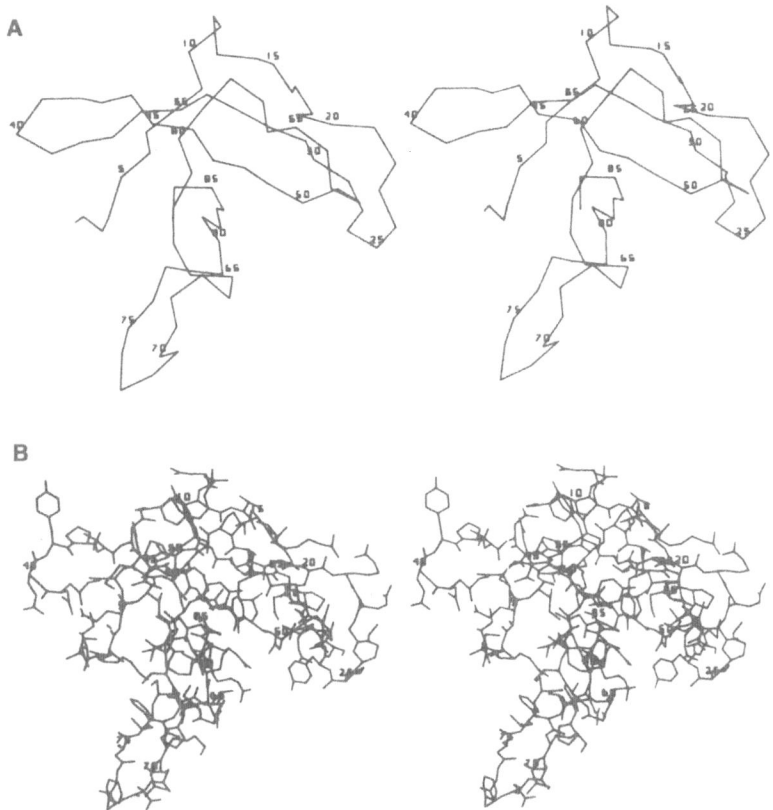

Figure 15.2. (A) Stereo drawing of the α-carbon backbone of G5BP following completion of 2.3 Å resolution refinement. Orientation is that of Figure 15.1, for ease of correlating the structural features illustrated in both. (B) Stereo drawing of all the nonhydrogen atoms of the G5BP structure. The view presented here is the same as that of the α-carbon drawing above. In both drawing the α-carbon of every fifth amino acid is labeled as to its position in the amino acid sequence.

CD-ORD studies [19] and structure-sequence prediction rules [20], the molecule is composed entirely of β structure. The overall dimensions are approximately 35 × 30 × 47 Å. These are rather large for a protein composed of only 87 amino acids, and as one might expect, the gene 5 protein monomer structure is quite open.

Figure 15.1 shows that from the N terminus, the first 10 residues form a short β strand. Then, following a β turn, the polypeptide chain completes the first major loop of the molecule. It is composed of residues 15 through 21, and we have termed it the "DNA binding loop." This loop rises well above the central mass of the protein seen best in Figure 15.2B. The return strand of polypeptide chain then goes on to create a second major β loop composed of residues 33 through 49. We have termed this the "complex loop." The polypeptide chain then descends to the back of the molecule to form a short eight-residue strand (residues 52 to 59) before reemerging to generate a third large β loop. This final β loop is composed

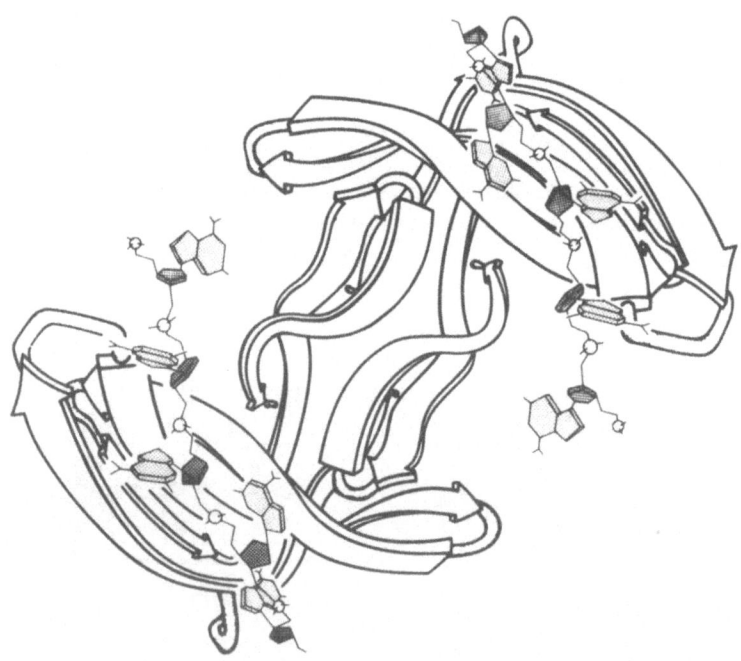

Figure 15.3. Stylized drawing of the G5BP dimer and the course of bound single-stranded DNA. Features of the bound chains are coded as follows: open circles, phosphate groups; dark gray rings, sugar moieties; light gray rings, bases. Differentiation of polypeptide chains and the location of specific amino acids can be made by comparing with the stereo drawing of Figure 15.4, all of which view the G5BP dimer from a similar orientation. A total of 10 nucleotides are illustrated, five in each of the two bound DNA strands.

of residues 61 to 82 and is termed the "dyad loop." In this orientation, the C terminus lies near the N terminus, toward the back of the molecule.

The major species of gene 5 protein found in solution is the dimer unit [17,33], which is seen schematically in Figure 15.3 and in stereo in Figure 15.4. It has been proposed that the forces involved in dimer association are mainly hydrophobic, and it is known this species is stable under conditions of high ionic strength, pH extremes, dilution, and elevated temperatures [18]. Other observations indicate that dimer units do not self-associate into larger complexes unless single-stranded nucleic acid material is present. The major physiological role the gene 5 protein plays in the bacteriophage fd life cycle is the protection of newly formed, covalently closed, circular, single viral strands. This is accomplished by coalescing with the viral strand to form a helical, linear, nucleoprotein rod. No such self-assembling complexes are observed in the absence of viral DNA.

It is also the dimer form of the gene 5 protein, seen in different orientations in Figures 15.5 and 15.6, that is found in the crystalline state. That is, two gene 5 molecules closely associate about a perfect dyad axis, which is coincident with a twofold symmetry element of the crystallographic unit cell. Examination of the dimer unit reveals a striking feature

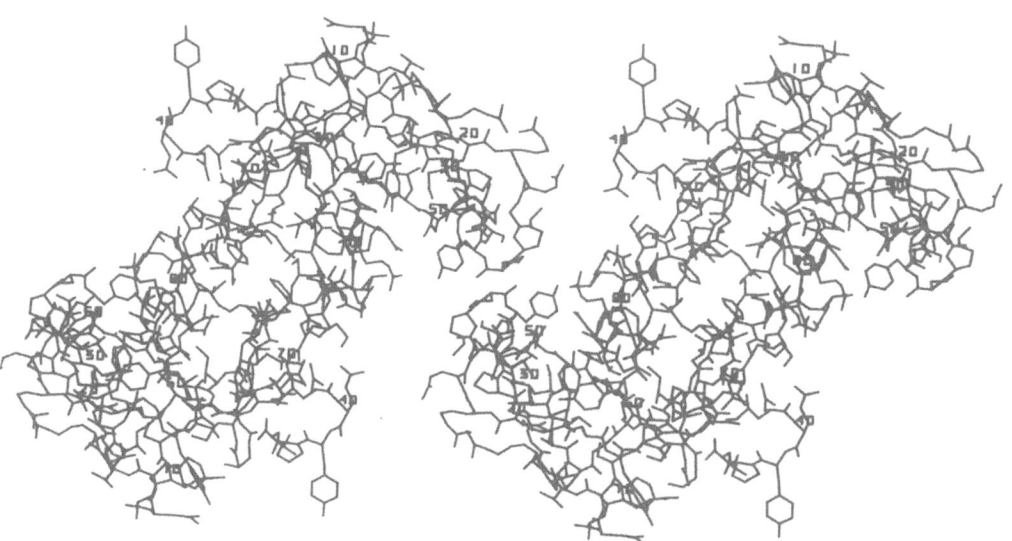

Figure 15.4. Stereo drawing of all the atoms of the G5BP dimer unit. The twofold axis relating G5BP monomers runs directly into the plane of this illustration, and its location is immediately obvious by comparison of common structural elements from each monomer. In addition, every fifth amino acid is numbered to allow comparisons with the amino acid sequence.

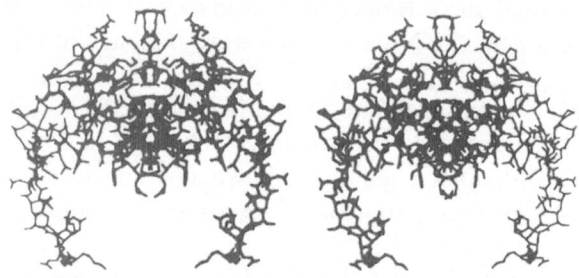

Figure 15.5. A computer-generated stereo image of the gene 5 protein in a second orientation. Here the twofold axis of the dimer is vertical and in the plane of the figure. Note the symmetrical extended DNA-binding loops protruding from the central mass of the molecule.

of the association—the extraordinary degree of intermolecular bonding and complementarity achieved between the dyad-related subunits.

The schematic drawing of the gene 5 protein dimer unit shown in Figure 15.3 is presented in the same orientation as that of the gene 5 protein monomer of Figure 15.1. The twofold axis relating monomers runs directly into the plane of this illustration, and its position is evident upon comparison of common monomer features. The overall dimensions of the dimer, seen in Figure 15.4, are approximately 55 × 45 × 36 Å. This is considerably smaller than one might expect from the dimensions of a gene 5 protein monomer alone. Indeed, by comparing Figures 15.2 and 15.4, the structure can be seen to be a much more compact and globular structure than an isolated gene 5 protein monomer. Examination of the monomer-monomer interface shows that each monomer unit actually encroaches by over 10 Å into the structure of the related monomer by

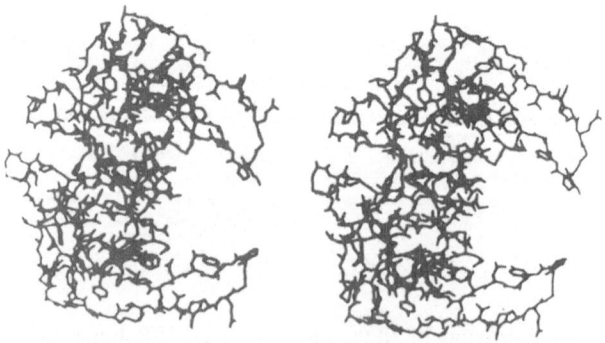

Figure 15.6. A stereo image of the gene 5 protein dimer in a general orientation showing further the disposition of the DNA-binding loops and the DNA-binding clefts.

extension around their common twofold axis. Interactions between monomer units are found to be so exact and cohesive that solvent molecules are completely excluded from the interface region, and no significant interstices are present.

It seems evident from the crystallographic structure of the gene 5 protein and from other physical-chemical techniques [17,18,21,33] that the basic DNA-binding unit is the gene 5 dimer. Indeed, the extensive interbonding and overlap between dyad-related monomers result in a structure as compact as if it had been created from a single polypeptide chain. The formation of a mutual intermolecular hydrophobic β barrel and its continuity with the hydrophobic cores found nucleating each gene 5 molecule suggests that dimer formation is an essential element in the maintenance of the overall stability of each monomer. In addition, the presence of twofold-related monomer polypeptide chains in each of the two opposing DNA binding clefts further implies that the gene 5 protein dimer is the active DNA-binding species.

Given the extensive degree of monomer-monomer overlap and the apparent rigidity of the hydrophobic β barrel so formed, it seems unlikely that subsequent DNA-binding events would substantially alter the observed dimer conformation. Possible exceptions to this, however, are the extended wing or "DNA-binding loop" (residues 20 to 30) and perhaps the tip of the "complex loop" (residues 39 to 43), both of which are

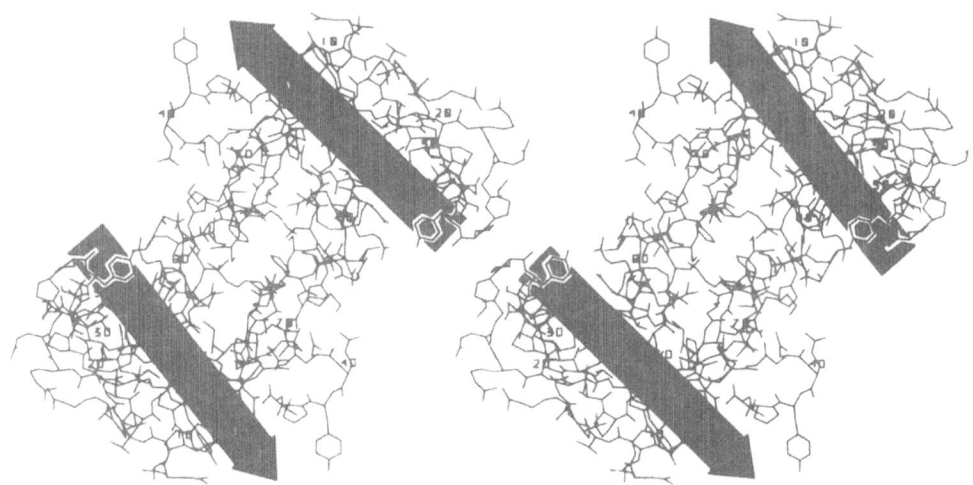

Figure 15.7. Stereo drawing of all the atoms of the G5BP dimer unit in an orientation such that the twofold axis relating monomers is running directly into the plane of the illustration. Also shown schematically are the positions and polarity of the two DNA-binding channels crossing the dimer face. Every 10th amino acid residue has been numbered to allow comparison with the amino acid sequence.

unrestrained by dimer formation. It is important that the two rather well defined binding channels crossing the dimer face include these more mobile loops as structural elements. Figure 15.7 shows schematically the positions of these channels and their relationship to other features of the dimer unit. Essentially, they traverse paths on either side of the molecular twofold axis and are created by the outer extended DNA-binding loops and the ridge of atoms formed at the dyad interface. Note that the environment of each binding cleft is identical as a result of the twofold

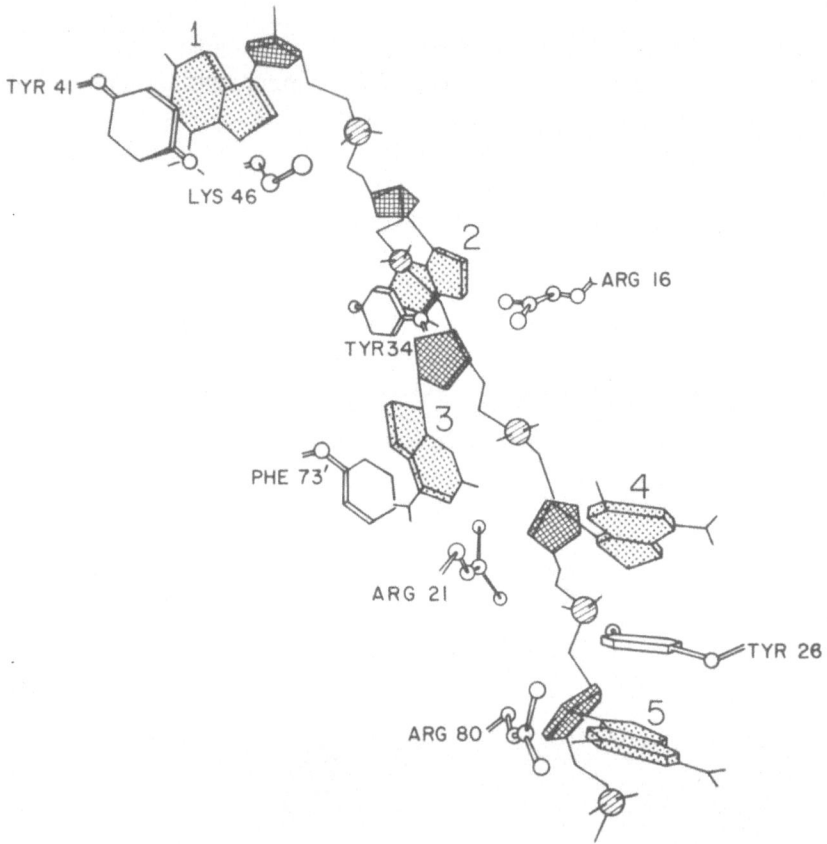

Figure 15.8. Schematic representation of one binding cleft on the G5BP dimer and the major interactions formed with DNA. The twofold-related binding channel exhibits an identical interactive environment. Two types of chemical interactions appear to be responsible for complexation. Aromatic side chains of the protein stack upon base rings of the DNA, and the phosphate backbone is neutralized by a combination of lysyl and arginyl side chains. The distribution of nucleotide-binding sites is such that the DNA must be in a nearly fully extended state. To facilitate discussion in the accompanying text, each nucleotide has been arbitrarily numbered starting at the top of the drawing.

axis, but the polarity or direction of each is reversed from the other by the symmetry element, thus accommodating the antiparallel nature of the two single strands of nucleic acid that are bound.

The Binding of ssDNA to the Gene 5 Protein

By combining results obtained from a number of difference Fourier experiments and interpreting them in a manner consistent with the data from NMR, spectroscopy, chemical modification, and other types of investigations, a mechanism for the binding of the gene 5 protein to single-stranded DNA was proposed [34]. In this model, five nucleotides of each strand of DNA, having opposite polarities, are bound across the three-stranded β-sheets of the protein as shown in Figures 15.5 and 15.6. The DNA strands are in fully extended conformations and make extensive chemical interactions with the surface of the protein. A schematic drawing of the pattern of binding interactions between protein and DNA is shown in Figure 15.8.

The orientation of the single strands of DNA with respect to the surface of the gene 5 protein is shown in Figure 15.9. The two DNA strands are positioned just inside the extended DNA-binding loops and held apart by a smaller protrusion of atoms at the molecular twofold axis. Strand separation is approximately 30 Å, and the two DNA chains run essentially antiparallel to one another.

For the most part, each strand is positioned across the central three-stranded β-sheet which forms the core of the gene 5 protein

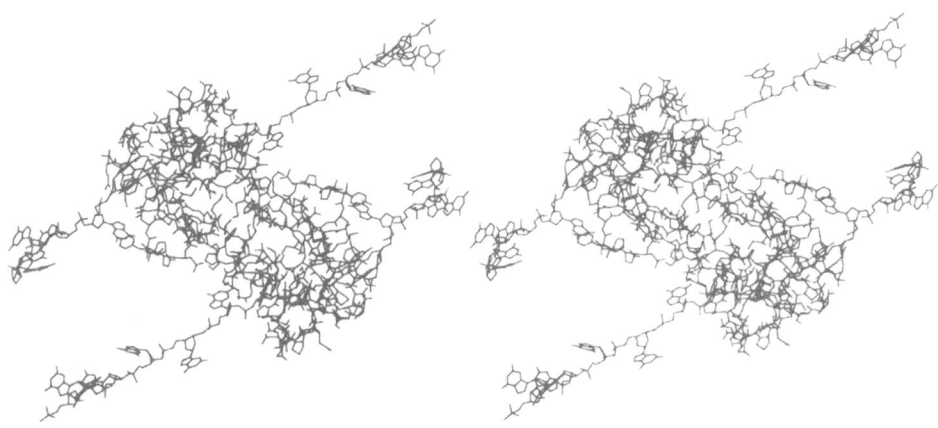

Figure 15.9. Stereo drawings of two consecutive G5BP dimers with strands of DNA bound. These drawings are identical to any two consecutive G5BP dimers seen in the helix of Figure 15.12.

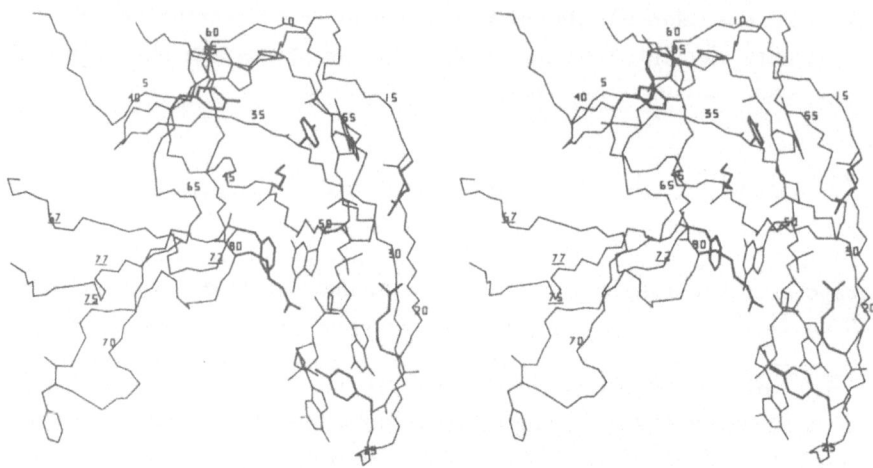

Figure 15.10. Stereo drawing of a DNA binding cleft with a bound oligomer of 5 bases in length. The polypeptide chain of G5BP monomer that forms most of the binding channel surface is drawn. An additional binding element, the dyad loop of a twofold-related G5BP monomer (residue numbers underlined), is illustrated as well. Specific interactive groups are drawn with dark lines along with the bound DNA to accent their positions.

molecule. Major interactions are also made with the "complex" and "DNA-binding" loops. As Figures 15.10 and 15.11 show, the β bend tips of the "dyad" loops are also in the vicinity of the DNA strands. Note, however, that this latter interaction occurs within the twofold-related gene 5 protein monomer-binding channel as opposed to the monomer unit of which the dyad loop is a part. These dyad loop interactions are essential components in DNA association. Cross-linking binding-channel interactions of this sort suggest that the gene 5 protein dimer may be the smallest species capable of binding DNA.

The phosphate backbone of the DNA is bound closely to the surface of the protein throughout its course, while the base rings project somewhat farther from the surface. The DNA conformation is almost fully extended, and bases are unstacked. Even in this extended state it is necessary to utilize five nucleotides to traverse the full length of each binding channel. The average phosphate-to-phosphate distance in the model is very nearly that expected for a fully extended polynucleotide.

The most prominent interactions between bound DNA and the gene 5 protein are detailed in Figure 15.8. Only one binding cleft is illustrated, because its twofold-related counterpart has exactly the same chemical and physical environment. The base of nucleotide 1 stacks coplanar with the aromatic side chain of Tyr 41, and the side chain of Tyr 34 stacks upon

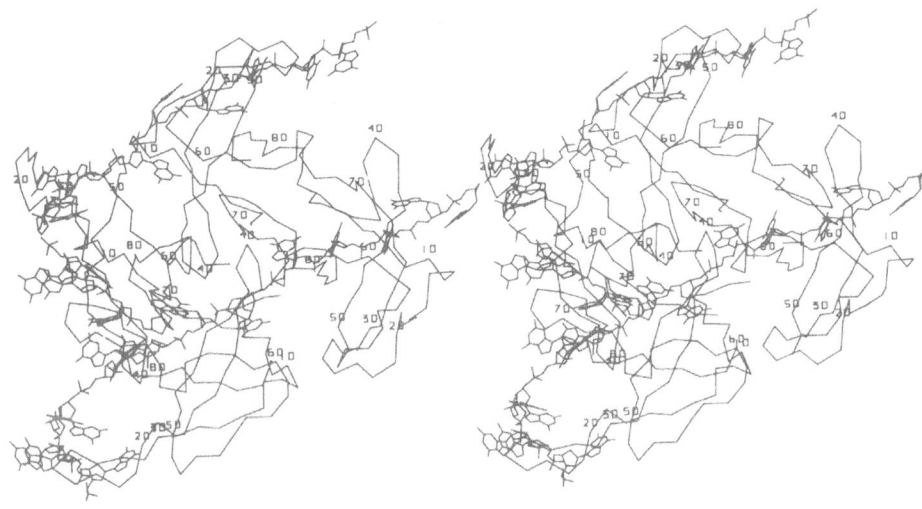

Figure 15.11. Stereo drawing of a gene 5 dimer binding two opposing single strands of DNA. The distance between the two polyphosphate backbones, which have opposite senses, is about 30 Å. The mode of binding and its illustration here are based on model building using the refined gene 5 protein structure and those chemical interactions indicated by spectroscopic and chemical modification experiments.

base 2. Lys 46 is positioned such that its free amino group is approximately equidistant from the phosphates of bases 1 and 2. Nucleotide 3 lies flat against the exposed side of the phenylalanyl side chain of residue 73. Note that this residue is a part of the polypeptide chain of the twofold-related gene 5 protein monomer. The phosphate groups of bases 2 and 3 are located on either side of the side chain of Arg 16. Tyrosine 26, at the tip of the DNA-binding loop, intercalates between base rings 4 and 5. The guanidinium group of Arg 21 lies near phosphate 3, and the same group of Arg 80 is positioned between phosphates 4 and 5. Besides these specific interactions, we observed that there were groupings of hydrophobic side chains clustered on the sides of base rings not stacked against aromatic groups and additional polar residues in the vicinities of bound ribose rings and phosphate groups. Two general mechanisms appear responsible for complexation. Base rings stack on protein aromatic side chains, and the bound phosphate backbone is fixed by appropriately positioned lysyl and arginyl side chains.

Because the binding of gene 5 protein to extended strands of DNA is in a contiguous fashion and is linearly periodic, the model of Figures 15.8 and 15.11 can be extended to linear strands of DNA of indefinite length. Furthermore, correlation of this model with the parameters of the intercellular gene 5–DNA complex derived from electron microscopy, neutron scattering, and optical spectroscopy using a computer-automated search procedure permitted extension to a model of the intercellular complex at the atomic level [35,36]. The protein component of this gene 5–DNA double helical complex is seen in Figure 15.12.

The temptation is always present to generalize from a specific result. From the gene 5 protein–nucleic acid complex structure, it might follow that all or most protein-ssDNA interactions use a similar set of chemical contacts. In particular, they would utilize neutralization of anionic phosphate groups by basic amino acid side chains and the stabilization of purine and pyrimidine groups by hydrophobic stacking and intercalation with aromatic amino acid residues such as tyrosine and phenylalanine. To determine if this is really true, however, it is of course essential to visualize other examples. This led to our analysis of the binding of a second DNA unwinding protein, pancreatic ribonuclease A, and its glycosylated form, ribonuclease B, to nucleic acid.

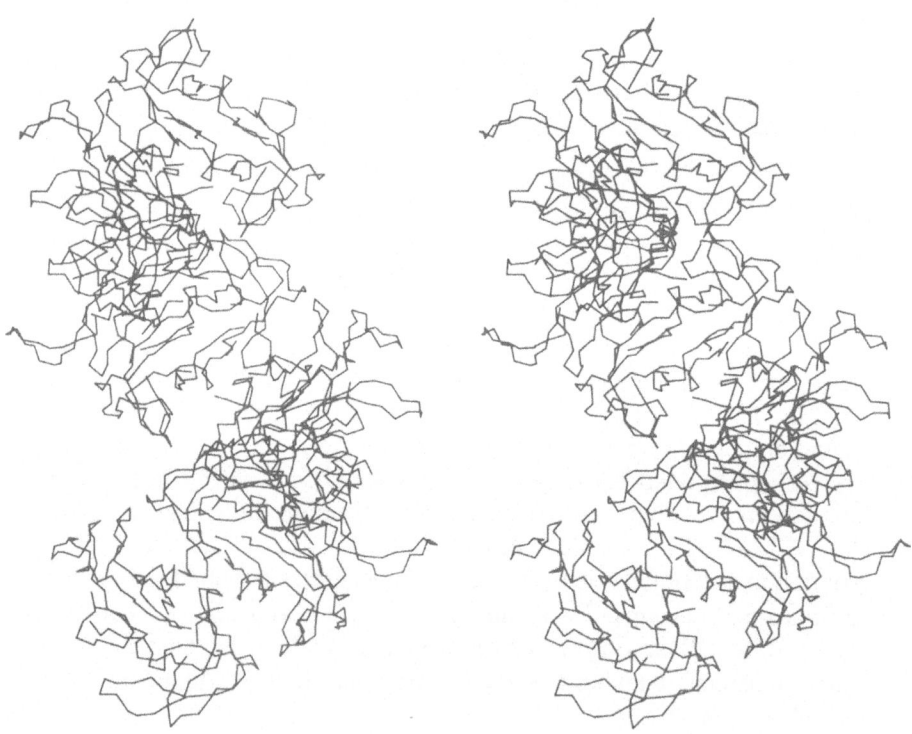

Figure 15.12. The gene 5 protein helices may be unwound, as in the stereo drawing shown here, to produce a continuous ribbon of dimers. Only the α-carbon chain is shown in this drawing. The angle between the ribbon and the helical axis is 113°.

Bovine Ribonuclease as a DNA Unwinding Protein

Bovine ribonuclease A (RNAse) is a protein of 124 amino acids whose function is to cleave RNA into small fragments at phosphodiester bonds following the occurrence of pyrimidine residues [37–39].

It is normally thought of as a digestive enzyme, though its ubiquity would indicate that it might have other physiological roles. RNAse binds single-stranded DNA as well as RNA, presumably as a consequence of the structural similarity of the two nucleic acid polymers, but does not degrade DNA because of the absence of a catalytically essential hydroxyl group on the sugar residues of DNA. As a result, RNAse forms stable protein–nucleic acid complexes with single-stranded DNA [32] and with short deoxyoligomers such as $d(pT)_4$ or $d(pA)_4$.

The structure of RNAse, one of the earliest enzyme crystals to draw the attention of protein crystallographers, was independently determined by X-ray diffraction in three different laboratories, and it has been thoroughly and precisely refined to high resolution. Thus, it is one of the best determined and best understood of all protein molecules in structural terms. It has a bilobal shape, contains both alpha helix and beta structure, and possesses a deep cleft between its two domains. This cleft has been shown by a variety of experiments, both X-ray and otherwise, to contain the active, or catalytic, site of the enzyme [for a review of crystallographic work, see 38]. Curiously, however, it was also shown by a variety of techniques that RNAse, when it complexed with single-stranded DNA, covered or protected up to 12 consecutive nucleotides along the nucleic acid chain [40] and that up to eight "ion pairs" were formed in the course of the association [41]. Neither of these latter findings was explicable in terms of the active site cleft, which is of limited extent. Clearly, then, the enzyme must employ a much more extended set of binding interactions when it binds to long single strands of RNA or DNA, and the nucleic acid must contact the surface of the protein at points outside of the known active site cleft.

Complexes of RNAse with Deoxyoligonucleotides

We found sometime ago that if RNAse was mixed with deoxyoligomers of DNA at ratios of about 1 : 5 or greater and the mixture then combined with polyethylene glycol 4000 to a final concentration of about 17%, crystals readily formed [42]. This could be achieved with $d(pA)_4$, $d(pT)_4$, $d(pA)_6$, or combinations of $d(pA)_4$ and $d(pT)_4$. Chemical tests showed that the crystals contained both protein and nucleic acid and were therefore a complex of the two. Unexpectedly, it further indicated there to be at least two and probably more deoxyoligomers associated with each RNAse molecule in the crystals.

From a Fourier map of the complex crystals at 3.5 Å resolution, the orientation and position of the RNAse protein molecule in the crystalline unit cell was deduced by inspection. The known structure of RNAse—that is, its set of atomic coordinates—was then placed appropriately in the unit cell and refined as a rigid body against the X-ray data to improve the level of precision. Having done this, it was possible to calculate both structure amplitudes and phases for the protein in the unit cell, to form differences between these calculated protein amplitudes and the observed amplitudes from the complex crystals, and to calculate difference Fourier maps. Such maps are, in effect, maps of the difference in electron density between the protein molecule complexed with the deoxyoligonucleotides and the protein molecule alone. One could therefore expect to see the DNA oligomers as they were disposed with respect to the protein. This is, in fact, what was observed. From such difference maps, portions of the nucleic acid present in the crystals appeared and were placed correctly in the unit cell [43,44].

The disposition of the deoxyoligomers in the complex is, of course, not a chance occurrence, but reflects the complementarity of positive charges and nucleotide binding sites on the protein with the negatively charged phosphate groups and bases on an extended nucleic acid chain. The complex of RNAse + $d(pA)_4$ is shown in Figures 15.13 and 15.14. Tetramer I runs from its 3' terminus at the top of Figure 15.13 into the active-site cleft and is positioned with its 5' terminal PO_4 adjacent to

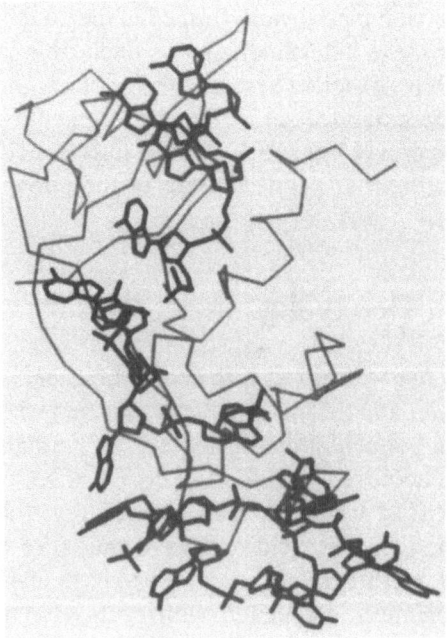

Figure 15.13. A computer-generated image of the four tetramers of $d(pA)_4$ superimposed on the backbone of RNAse A as they are found in the crystalline complex between the two.

histidines 12 and 119, and lysine 41 at the active center. The 5' base is in the purine-binding site identified by previous studies. Tetramer II has its 3' terminal nucleotide abutting but not actually occupying the pyrimidine binding site. It passes over the surface of the protein, extends through a solvent region, and terminates with its 5' PO_4 linked by an electrostatic interaction to another molecule in the crystal lattice. Thus, the first three nucleotides starting from the 3' end are bound to the molecule shown, while the fourth nucleotide at the 5' end is termed "intermolecular" because it links to a different protein molecule in the crystal lattice.

In tetramer III, only the nucleotide at the 5' end is bound to the molecule shown, while the three remaining members of the chain extend through a broad interstitial space in the crystal and terminate with a salt bridge to another protein molecule in the lattice. Thus, fully three nucleotides of tetramer III are "intermolecular." Tetramer IV has its 3' nucleotide adjacent to the 5' PO_4 of tetramer III, curls under the protein, and terminates with its 5' PO_4 in a deep anion trap on the backside of the protein.

The arrangement of tetramers seen in Figures 15.13 and 15.14 suggested to us the course a single polynucleotide strand would assume when bound to RNAse, even though it is in these crystals composed of oligomeric segments. We noted, furthermore, that the polarity of the $d(pA)_4$ strands was consistently 3' → 5' from the top of the figure to the bottom

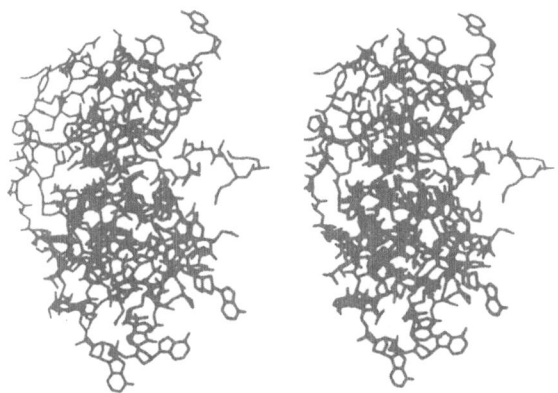

Figure 15.14. Stereo diagram of the ribonuclease A protein molecule associated with the four different $d(pA)_4$ oligonucleotides that comprise the asymmetric unit of the crystal. Running more or less from the top of the figure to the bottom in a consistent 3' to 5' direction, the four fragments of DNA can be seen to trace out an essentially continuous path through the active-site cleft of the enzyme and over the surface of the protein. Electrostatic linkages are made between phosphates on the oligonucleotides and lysines 7, 41, 66, 37, 31, and 98 as well as with arginines 85, 39, and possibly 33. Interactions involving the ribose and base moieties occur in the active-site cleft.

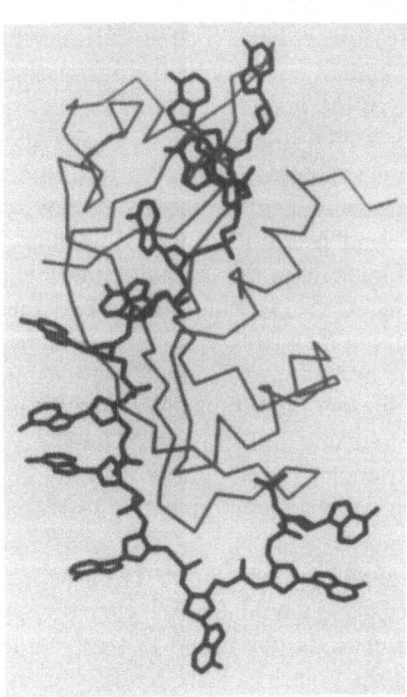

Figure 15.15. The single strand of 12 nucleotides derived from the RNAse + d(pA)₄ complex is shown here superimposed on the backbone of RNAse A. It describes a continuous path from the top of the protein through the active site and around to the opposite side of the enzyme.

(I → II → III → IV) and that a large number of chemical interactions were used that had been previously implicated by other techniques to be important in nucleic acid binding.

The "intermolecular" interactions involving mainly four nucleotides are permissible in the crystal only because the asymmetric unit complex employs discontinuous segments that have additional degrees of freedom. We postulated that the path delineated by the nucleotides bound to this particular RNAse molecule is representative of the course traced by a single strand of RNA or DNA when bound to the protein. That is, the d(pA)₄ oligomers are a "virtual DNA strand."

If the nucleotides that are involved in the crystalline "intermolecular" bonds are ignored, then the remaining oligomers and oligomer fragments can be linked together in a 3' to 5' fashion with no substantial movement of any of the nucleotides, even at the joins, and with no distortion of either protein or deoxyoligomer conformation. Only the single nucleotide of tetramer III requires any significant rearrangement, and this involves only rotations. Figure 15.15 and 15.16 illustrate the result of this connecting process, a continuous single chain of nucleotides, a "virtual strand," extending from the 3' hydroxyl at the top to the 5' phosphate at the bottom. In Figure 15.15, the bases and the conformation of the

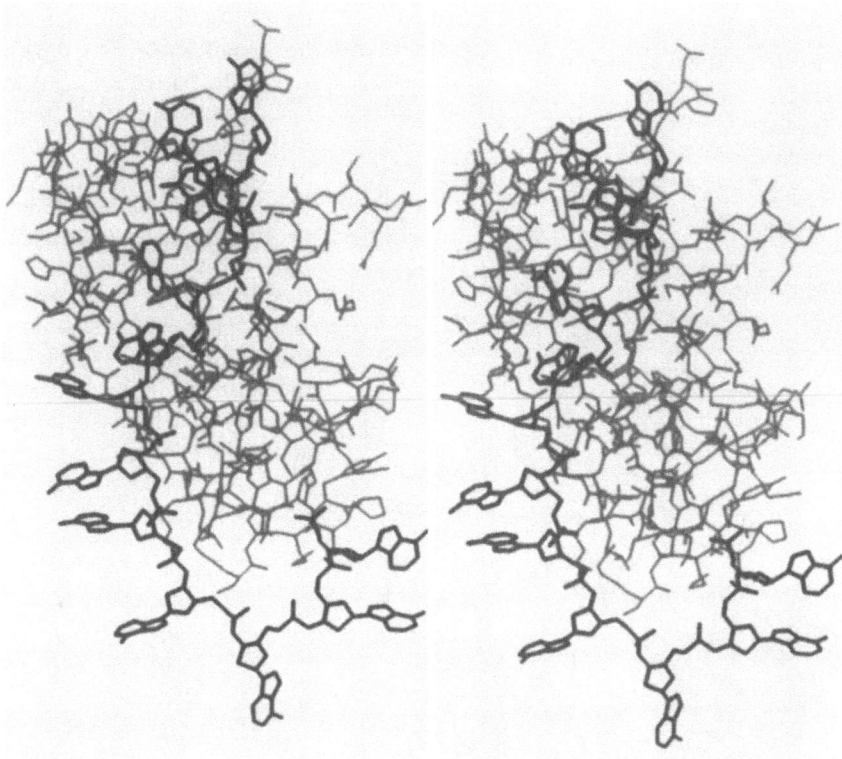

Figure 15.16. The 12 nucleotides that are associated with the same RNAse molecule have been connected together in a consistent 3' to 5' fashion to make an extended polynucleotide chain. This virtual strand maintains all of the chemical interactions with the protein that are observed in the crystal structure of the RNAse plus d(pA)₄ complex (Fig. 15.14). It has been idealized to some extent in terms of base orientations and backbone conformation.

ribose-phosphate backbone have been adjusted to some extent to bring the model into a more ideal conformation, but it still closely resembles the series of deoxyoligomers that were actually observed in the crystal, and it preserves all of those interactions responsible for binding.

If one follows the path taken by the virtual strand of DNA in the model of Figures 15.15 and 15.16, protein–nucleic acid interactions observed in the crystal structure are encountered that would, by inference, be operative in the protein–nucleic acid complex, and these are illustrated schematically in Figure 15.17. Beginning at the 3' end of the DNA chain and proceeding toward the 5' terminus, we encounter salt bridges between phosphate groups and lysines 7, 41, and 66, then arginines 85 and

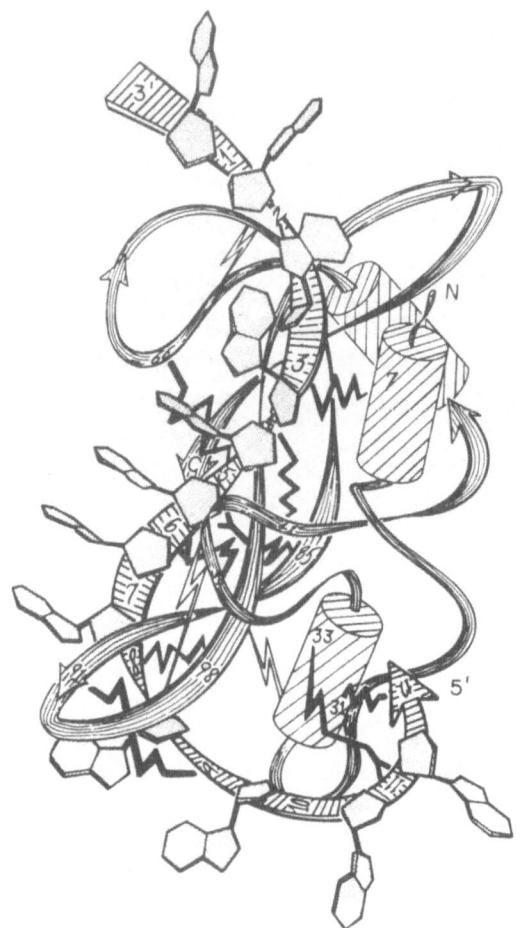

Figure 15.17. The schematic drawing of the model complex between RNAse plus $d(pA)_4$ shows the pattern of eight to nine electrostatic interactions that bind the polynucleotide to the surface of the protein. These include salt bridges between phosphate groups and Lys^7, Lys^{41}, Lys^{66}, Lys^{91}, Lys^{98}, Lys^{31}, and Arg^{85}, Arg^{33}, and Arg^{39}. The only close interactions involving the bases of the nucleic acid with the protein occur in the active-site cleft.

39, followed by lysines 91 and 98, then arginine 33, and finally lysine 31. The only important interactions involving the bases of the DNA occur in the active site where specificity is conferred. The complex between protein and nucleic acid is principally an extended multisite cation-anion interaction. The base and ribose moieties play only a minor role in binding but, clearly, a major role in determining specificity and orientation at the catalytic center.

The complex deduced from the crystal structure utilizes eight to nine electrostatic "ion pairs." This is very close to the number predicted by Record et al. based on cation titrations [41]. Furthermore, many of the lysine and arginine residues involved in binding were predicted by other investigators [for reviews see 37–39]. The protein in this complex engages a total of 12 nucleotides, which corresponds to the protection size determined by Jensen and Von Hippel. The virtual strand passes directly through the active-site cleft in a manner consistent with previous difference Fourier studies using dinucleotides. The conformation of the polynucleotide chain is smooth and extended and exhibits no unacceptable turns or bends.

The nucleic acid components by themselves provide from 20% to 25% of the entire nonsolvent scattering material found in the crystallographic unit cell, which itself contains about 40% solvent by volume. Thus, both the protein and the nucleic acid are, as in most macromolecular crystals, almost fully hydrated.

It is not obvious by inspection of the asymmetric unit of the crystal that there is any particular pattern to the arrangement of the single-stranded DNA. Its disposition appears to be only a consequence of its interaction with the protein. These interactions, except at the active site, are primarily electrostatic and involve phosphates bound to lysine and arginine residues.

If the protein component of the crystals is eliminated, however, and the distribution of deoxyoligomers is displayed as it occurs, not in one, but in several unit cells of the crystal, then the pattern of single stranded DNA, seen in Figure 15.18 for the RNAse + $d(pA)_4$ complex, is observed. It is clear from this figure that the disposition of the deoxyoligomers in the crystal is not arbitrary. The deoxyoligomers are in fact arranged to form long, continuous, helical filaments parallel with the Y axis of the crystal. In addition, some $d(pA)_4$ oligomers and portions of oligomers also cross-link the helical filaments through DNA-DNA interactions. Thus, the DNA components of the RNAse + $d(pA)_4$ crystals visualized alone form a vast, cross-linked network of segmented helical filaments of single-stranded DNA. The protein may, in a sense, be thought of as trapped in this web.

From our analysis of this crystal and others like it using a variety of oligomers of different sequence and length, it appears that this class of crystals may provide a mechanism for the creation of a diverse array of single-stranded DNA networks that display a range of both protein-DNA and DNA-DNA interactions. Because these networks of single-stranded DNA exist in a near fully hydrated environment, they may prove useful in delineating the conformations and arrangements of single-stranded DNA and in defining the kinds of inter- and intramolecular DNA-DNA interactions that govern the formation of secondary and tertiary elements of single-stranded DNA structures.

Because the crystal networks of ssDNA are variable as a function of oligomer length, oligomer sequence, and oligomer-to-protein stoichiometry, and because they change as well when combinations of different oligomers, such as $d(pA)_4 + d(pT)_4$, are crystallized, it appears that a range of such networks may be possible. Thus, when a sufficient number of these networks have been elucidated as a function of the determinants, it may be possible to design protein–nucleic acid crystals composed of

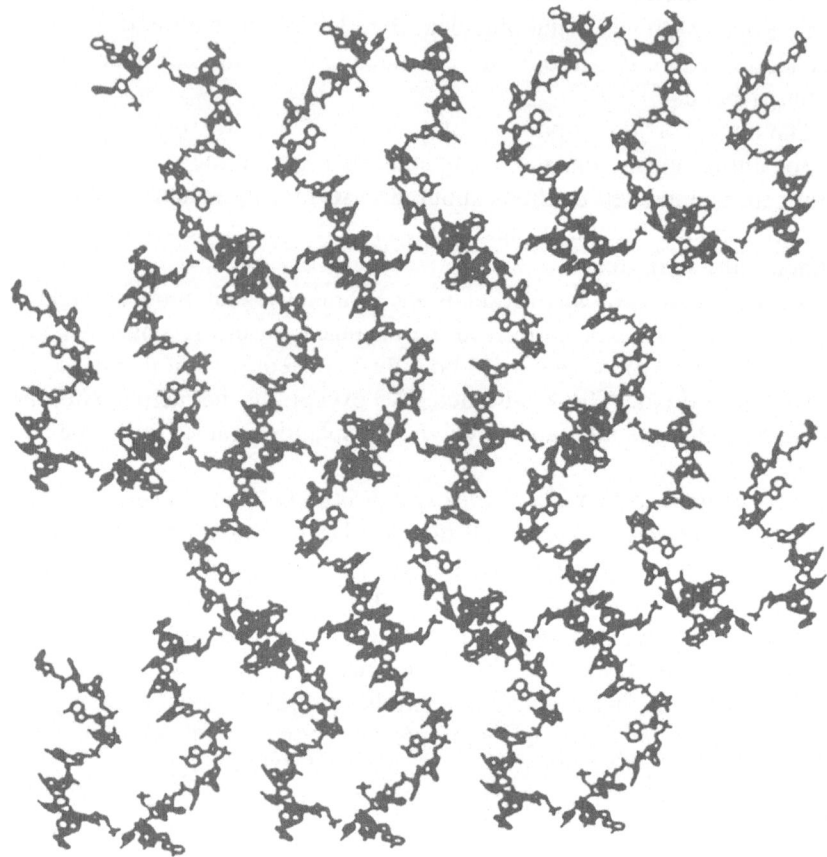

Figure 15.18. The contents of six unit cells of the crystal of RNAse-d(pA)$_4$ with the protein molecules omitted. The individual segments of DNA can be seen to form a network of virtually continuous strands of nucleic acid that run through the crystal and appear to display some degree of helical character. The protein molecules and the DNA strands cross-link one another in the crystal primarily through electrostatic interactions involving the negatively charged phosphate groups and positively charged lysine amino and arginine guanidinium groups on the protein.

specific network structures and, conceivably, networks with unique properties. The possibilities for network architecture, furthermore, are not limited by variations in the DNA components alone, because all of the networks currently being analyzed reflect in different ways the distribution of complementary binding sites on pancreatic ribonuclease A. These are principally electrostatic in nature. By varying the occurrence of lysine and arginine residues, by deletion, addition, or transposition, using a genetically engineered ribonuclease, for example, or using entirely different ssDNA binding proteins, the range of possible ssDNA networks might be expanded.

The DNA networks found in these crystals may, in addition, have practical value. If crystals of RNAse + d(pA)$_4$ are exposed to mutagens, carcinogens, antibiotics, or other ligands known to interact with nucleic acids, it is observed that the crystals absorb and concentrate the compounds. From crystals diffused with different ligands, X-ray diffraction data can be collected and difference Fourier maps computed. These allow us to visualize the bound molecules directly. Thus, aside from other uses they may have, these crystals provide a structural framework for the testing and evaluation at the molecular level of the binding properties of single-stranded DNA interactive drugs.

References

1. McPherson, A. (1982) The Preparation and Analysis of Protein Crystals. John Wiley and Sons, New York.
2. Glusker, J.P., and Trueblood, K.N. (1972) Crystal Structure Analysis: A Primer. Oxford University Press, London.
3. Blundell, T.L., and Johnson, L.N. (1976) Protein Crystallography. Academic Press, New York.
4. McPherson, A. (1987) Cryst. Rev. *1*, 191–250.
5. Schaller, H., Beck, E., and Takanami, M. (1978) In The Single-Stranded DNA Phages, 139–163. Cold Spring Harbor Laboratories, Cold Spring Harbor, New York.
6. Marvin, D.A., and Hohn, B. (1969) Bacteriol. Rev. *33*, 172–204.
7. Ray, D.S. (1977) Comprehens. Virol. *7*, 105–178.
8. Denhardt, D.T. (1975) Crit. Rev. Microbiol. *4*, 161–223.
9. Coleman, J.E., and Oakley, J.L. (1980) CRC Crit. Rev. Biochem. *7*, 247–289.
10. Salstrom, J.S., and Pratt, D. (1971) J. Mol. Biol. *61*, 489–501.
11. Mazur, B.J., and Model, P. (1973) J. Mol. Biol. *78*, 285–300.
12. Mazur, B.J., and Zinder, N.D. (1975) Virology *68*, 490–502.
13. Alberts, B., Frey, L., and Delius, H. (1972) J. Mol. Biol. *68*, 139–152.
14. Pratt, D., Laws, P., and Griffith, J. (1974) J. Mol. Biol. *82*, 425–439.
15. Cuyper, T., Van der Ouderaa, F.J., and De Jong, W.W. (1974) Biochem. Biophys. Res. Commun. *59*, 557–564.
16. Nakashima, Y., Dunker, A.K., Marvin, D.A., and Konigsberg, W. (1974) FEBS Lett. *43*, 125.
17. Pretorius, H.T., Klein, M., and Day, L.A. (1975) J. Biol. Chem. *250*, 9262–9269.
18. Cavalieri, S.J., Neet, K.E., and Goldthwait, D.A. (1976) J. Mol. Biol. *102*, 697–711.
19. Day, L.A. (1973) Biochemistry *12*, 5239–5339.
20. Anderson, R.A., Nakashima, Y., and Coleman, J.E. (1975) Biochemistry *14*, 907–917.
21. Coleman, J.E., Anderson, R.A., Ratliffe, R., and Armitage, I.N. (1976) Biochemistry *15*, 5419–5430.

22. Garssen, G.J., Tesser, G.J., Shoenmakers, J.G.G., and Hilbers, C.W. (1980) Biochim. Biophys. Acta *607*, 361–371.
23. Lica, L., and Ray, D.S. (1977) J. Mol. Biol. *115*, 45–59.
24. Alma-Zeestraten, N.C.M. (1982) Ph.D. Dissertation. University of Nijmegen, Nijmegen, Holland.
25. Gray, D.M., Gray, C.W., and Carlson, R.D. (1982) Biochemistry *221*, 2702–2713.
26. Torbet, J., Gray, D.M., Gray, C.W., Marvin, D.A., and Siegrist, H. (1981) J. Mol. Biol. *146*, 305–320.
27. Oey, J.L., and Knippers, R. (1972) J. Mol. Biol. *67*, 125–138.
28. Gray, C.W., Kneale, G.G., Lenard, K.R., Siegrist, H., and Marvin, D.A. (1982) Virology *116*, 40–52.
29. Lohman, T.M., Green, J.M., and Beyer, R.S. (1986) Biochemistry *25*, 21–25.
30. Herrick, G., and Alberts, B. (1976) J. Biol. Chem. *251*, 2133–2141.
31. Alberts, B.M., Amodio, F.J., Jenkins, M., Gutmann, E.D., and Ferris, F.L. (1968) Cold Spring Harbor Symp. Quant. Biol. *33*, 289–305.
32. Felsenfeld, G., Sandeen, G., and Von Hippel, P.H. (1963) Proc. Natl. Acad. Sci. USA *50*, 644–651.
33. Rasched, I., and Pohl, F.M. (1974) FEBS Lett. *46*, 115–118.
34. Brayer, G.D., and McPherson, A. (1984) Biochemistry *23*, 340–348.
35. McPherson, A., and Brayer, G.D. (1984) In Biological Macromolecules and Assemblies, Vol. II, Nucleic Acids and Their Binding Proteins, 323–392. Jurnak, F.A., and McPherson, A. (eds.). John Wiley and Sons, New York.
36. Brayer, G.D., and McPherson, A. (1985) Eur. J. Biochem. *150*, 287–296.
37. Anfinsen, C.B., and White, F.H. Jr. (1961) In The Enzymes, 2d Ed., Vol. V, Boyer, P.D. (ed.). Academic Press, New York.
38. Richards, F.M., and Wyckoff, H.W. (1971) In The Enzymes, 3d Ed., Vol. IV, Boyer, P.D. (ed). Academic Press, New York.
39. Blackburn, P., and Moore, S. (1982), In The Enzymes, 3d Ed., Vol. XV, Boyer, P.D. (ed.). Academic Press, New York.
40. Jensen, D.E., and Von Hippel, P.H. (1976) J. Biol. Chem. *251*, 7198–7214.
41. Record, M.T., Lohman, T.M., and De Haseth, P. (1976) J. Mol. Biol. *107*, 145–158.
42. Brayer, G.D., and McPherson, A. (1981) J. Biol. Chem. *257*, 3359–3361.
43. McPherson, A., Brayer, G.D., and Morrison, R.D. (1986) J. Mol. Biol. *189*, 305–328.
44. McPherson, A., Brayer, G.D., Cascio, D., and Williams, R. (1986) Science *232*, 765–768.

Chapter 16
Histone and Nucleosome Function in Yeast

Michael Grunstein, Min Han, Ung-Jin Kim,
Tillman Schuster, and Paul Kayne

The nucleosome, which contains two molecules each of histones H2A, H2B, H3, and H4, provides the basis for folding a long, negatively charged DNA polymer into a small nuclear compartment. However, the state of chromosomal condensation is reversible in response to replication, transcription, and chromosomal segregation. How the nucleosome is involved in these functions is a focus of our laboratory. Using reverse genetics we have asked which cellular functions are disturbed when histone proteins are mutated.

The individual histones in a nucleosome may have both unique and redundant functions. For example, histone H1 in higher eukaryotes has been found to be preferentially associated with inactive genes that have been terminally repressed [1,2]. Since H1 increases chromosomal compaction, it may inactivate genes by condensation of specific regions of the chromosome. Different core histones may also accomplish unique tasks. For example, H3 and H4 alone can form a nucleosomelike octamer [3]. H2A and H2B cannot, and they require the other two histones for nucleosome formation. There is also evidence for the disassociation of H2A-H2B dimers from nucleosomes and their reassociation with other H3-H4 "cores" [4,5].

Each of the four core histones contains two distinct domains, an extended hydrophilic N-terminal "tail" containing a number of positively charged amino acids and a globular, hydrophobic C terminus containing a high α-helix content. These latter regions are involved in both histone-histone interactions and in histone-DNA binding and are required for nucleosome assembly in vitro [6]. The function of the hydrophilic N-terminal ends is less well understood. Together, they may be involved in stabilizing nucleosomes since their removal with trypsin alters DNase I resistance as well as thermal and circular dichroism spectra of chromatin [6]. It has also been shown that the flexible N terminus of histone H4 interacts directly with DNA at the sharp bend formed around the nucleosome [7]. Since acetylation of lysine residues, at each of the histone N termini, neutralizes their positive charges, it has long been suggested that this modulates N-terminal interaction with the negatively charged DNA backbone allowing such functions as nucleosome assembly [8] and transcription [9–11]. In fact, chromatin containing active genes is preferentially acetylated [12,13]. However, removal of the histone ends with trypsin does not prevent formation of a nucleosomal structure [14], and extensive acetylation of the histone ends does not lead to nucleosome loss [15]. Nevertheless, since only the N termini undergo reversible interactions, it may be that explaining the dynamic nature of core histone function requires an understanding of the functions of the N termini.

To study histone function directly, our laboratory has concentrated on the yeast *Saccharomyces cerevisiae* with its highly developed genetics and gene replacement capabilities. This has involved a mutational analysis in which individual core histones or histone domains have been deleted in vivo. The consequences of these mutations on the yeast cell cycle and its biochemistry have provided clues regarding histone function.

Yeast Contains the Four Core Histones But Probably Not Histone H1

We approached the characterization of yeast histones by gel electrophoresis of histone proteins and DNA sequence analysis of their genes. In Figure 16.1 is shown an SDS-polyacrylamide gel pattern of histones isolated from *S. cerevisiae*. It contains the four core histones. The H2B band contains two subtypes, H2B1 and H2B2, which differ by four amino acids at the N terminus [16]. The H2A band also contains two subtypes, H2A1 and H2A2, which differ by two amino acids at the C terminus [17]. No subtypes exist for histones H3 and H4 in yeast; however, there are also two gene copies coding for each of these proteins. These genes differ only at silent positions in codons. The yeast histone genes are organized into a total of four, genetically unlinked gene pairs (H2A.1–H2B.1; H2A.2–H2B.2; H3.1–H4.1; and H3.2–H4.2) Each of these pairs is transcribed divergently from a common upstream element of approximately 700 to 800 base pairs [18–20].

The histone H1-like protein (Fig. 16.1) is of the size and charge expected for H1 and copurifies with the other histones. However, when antibodies were raised against this protein, it was discovered by indirect immunofluorescence that it is restricted to the mitochondrion. The yeast H1-like protein is in fact the mitochondrial DNA-binding protein HM [21]. Since this protein was the only candidate for H1, these data leave the presence of H1 in yeast questionable.

Other evidence also supports this negative result. Yeast chromosomes are very small, some no larger than bacteriophage T4 [22], and yeast chromosomal condensation has not been observed in mitosis. Further-

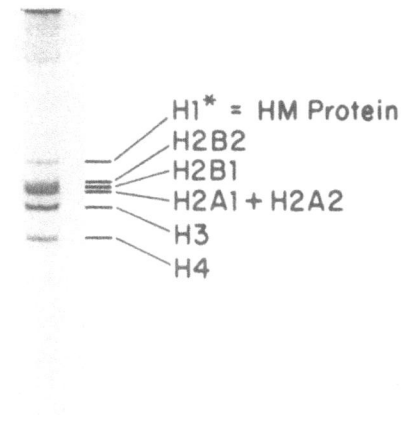

Figure 16.1. Histones of *Saccharomyces cerevisiae.* Proteins were isolated and electrophoresed on SDS-polyacrylamide gels as described by Certa et al. [21]. The H1* or H1-like protein was shown to be the mitochondrial HM protein and not a nuclear protein [21].

more, yeast chromatin is much more sensitive to DNase I than that of higher eukaryotes [23]. This reflects a genome that is more unfolded. Also, a higher percentage of the yeast genome (40%) is transcriptionally active than in higher eukaryotes (5%). Since most yeast genes are not ever terminally repressed, it is hard to escape the conclusion that yeast may have no need for histone H1.

Functions Dependent on Histones

We have previously determined that yeast does not require both H2B subtypes for growth and completion of any part of the life cycle. Nor are both H2A subtypes essential. However, the loss of both H2B or both H2A genes is lethal [24,25]. We then wished to determine why histone loss prevents growth. It has been suggested that histones function to allow chromosomal segregation, DNA replication, and transcription [see reviews in 6,9]. To test these ideas we took the genetic approach described below.

Yeast cell division cycle (*cdc*) mutants conditional for synthesis or function of a particular protein can shed light on its function during the cell cycle [26]. Under nonpermissive conditions, *cdc* mutant cells arrest at a specific point in the cell cycle owing to a stage-specific defect. For example, several of the *cdc* mutants defective in DNA replication arrest after S phase [26]. The topoisomerase 2 *ts* mutant disrupts chromosomal segregation and shows lethality during mitosis [27]. In contrast, mutants affecting general RNA synthesis arrest asynchronously since RNA is both synthesized and required throughout the yeast cell cycle [28]. Histone mRNAs and proteins are synthesized mainly in S phase of the yeast cell cycle [29,30]. However, we do not know whether histones are actually required for processes that occur throughout the cell cycle. A better understanding of histone function would be gained if we knew whether conditional histone mutants cause stage-specific arrest and which cellular processes they disrupt.

Chromosomal Segregation is Blocked by Histone H2B Depletion

We describe here an alternative to obtaining a conditional histone mutation (Fig. 16.2A). A yeast strain was constructed containing a single episomal histone H2B gene whose promoter was replaced by the GAL10 controlling element described by Johnston and Davis [31]. In our yeast strain the chromosomal H2B1 and H2B2 genes were inactivated by replacement with homologous genes containing frameshift mutations. This allows H2B mRNA synthesis to be specifically induced by galactose and repressed by glucose (Fig. 16.2B). Upon inhibiting H2B mRNA synthesis we found that the cells assumed a *cdc* phenotype.

The cells grown on glucose synchronized as large dumbbell shapes, with approximately equal-size mother-bud cells (Fig. 16.2C). After 3 or 4 hours of arrest, over 90% of these budded cells had a single G2 nucleus as determined by DAPI staining and flow cytometry. DAPI staining, used to visualize DNA in the arrested cells, often showed nuclei at the mother-bud neck prevented from or aborted in the process of segregating DNA (Fig. 16.2C). Therefore, histone H2B synthesis is essential for the segregation of chromosomes.

DNA Synthesis Continues in Cells Arrested by Histone H2B Depletion

To determine the rate and extent of DNA synthesis in glucose-arrested MHY102 cells, we first synchronized the cells in S phase with hydroxy-urea, which inhibits DNA synthesis. This was followed by treatment in hydroxyurea + glucose, to deplete cellular H2B mRNA. These cells were then resuspended in various media, all containing ^{3}H uracil to label DNA. As shown in Figures 16.3A and B, the cells resuspended in galactose, released from both blocks, now resumed DNA replication and cell division. In galactose + hydroxyurea, cell division and DNA replication were arrested. In glucose, cell division was arrested, since the histone block prevents further division. However, a full round of DNA replication continued at a relatively normal rate compared to cells shifted into galactose alone. We also examined the chromosomes of MHY102 by pulsed field electrophoresis [22] before and after the histone block was applied to asynchronously growing cells. Twenty hours after arrest there were no changes in the size of any of the chromosomes, no evidence of DNA degradation, nor any obvious decrease in the expected intensity of the chromosomal bands [32]. We conclude that a full round of relatively normal DNA replication occurs in glucose-arrested MHY102 cells in the absence of new histone H2B mRNA synthesis.

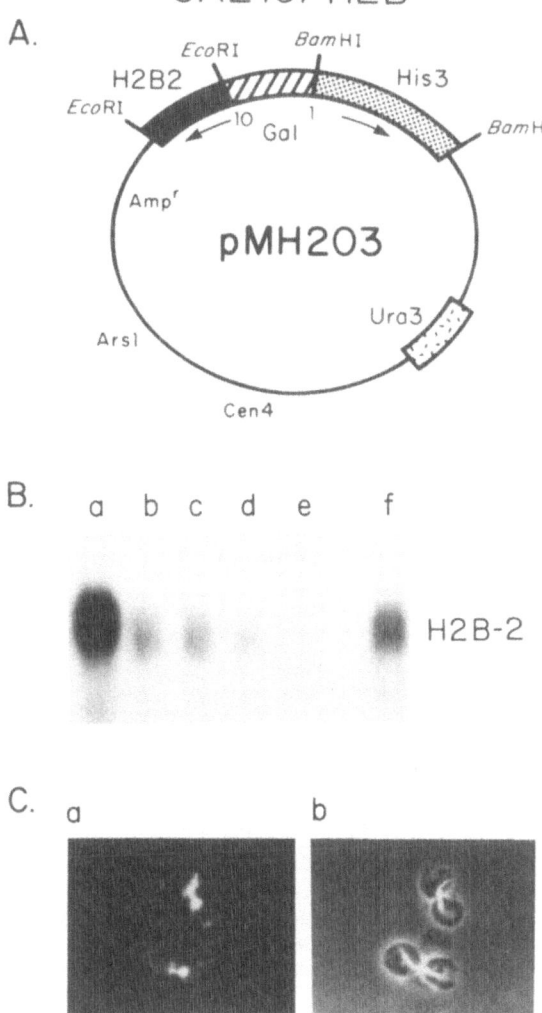

Figure 16.2. Repression of H2B mRNA synthesis results in G2 arrest. (A) Plasmid pMH203 map showing the HTB2 gene, which codes for histone H2B2, cloned next to the *GAL10* transcriptional control element. (B) Northern blot illustrating the repression of the HTB2 gene in the presence of glucose. RNA was isolated from MHY102 cells after various periods of arrest in the presence of glucose. RNA (12 μg) was electrophoresed on each lane of the 1.1% formaldehyde agarose gel for 17 hours at 45 V. A ^{32}P-labeled, 1.1-kb *Hind* 3 HTB2 DNA fragment was used to hybridize to the Northern blot. Lanes a through e show hybridization to cellular RNAs from cells cultured in glucose medium YEPD for 0, 1, 3, 6, and 10 hours, respectively. Lane f shows hybridization to RNA taken from the yeast strain D585-11C, a *GAL+* strain containing wild-type HTB1 and HTB2 genes. (C) (a) DAPI-stained MHY102 cells that have arrested after 6 hours in glucose medium. Note that the block has occurred in the early stages of nuclear segregation. (b) Detection of spindles in these same cells by indirect immunofluorescence using antitubulin antibody. The spindles are blocked in a stage indicative of mitosis.

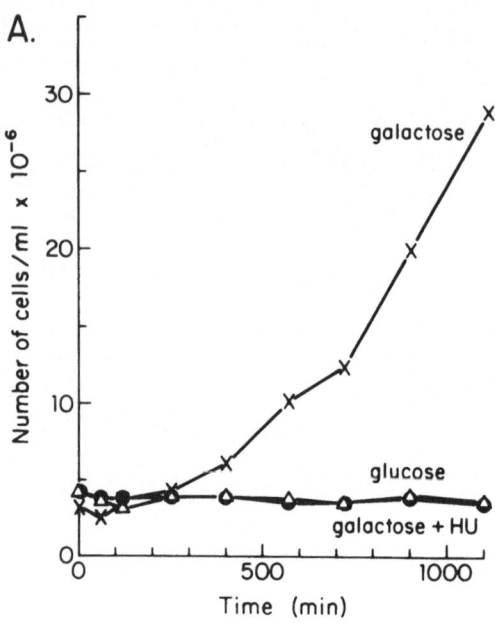

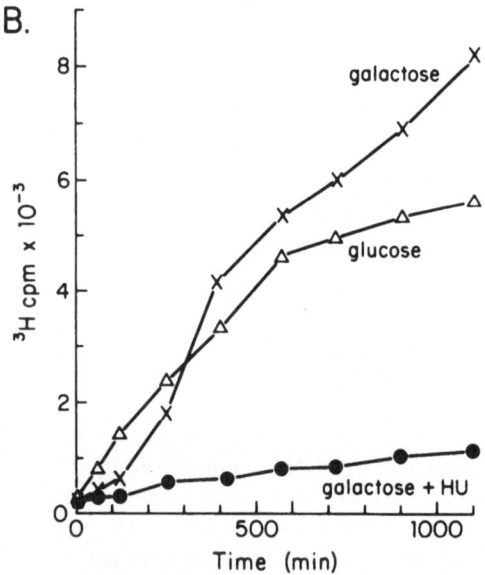

Figure 16.3. DNA synthesis in MHY102 cells arrested by glucose. MHY102 cells were synchronized in hydroxyurea (0.3 M) in galactose medium. This was followed by 2 hours in glucose medium containing hydroxyurea. Cells were then shifted into the three different media shown containing ^{3}H-uracil for the measurement of incorporation into DNA. (A) Growth of MHY102 cells transferred to galactose, glucose, or galactose + hydroxyurea for analysis of DNA synthesis. Note arrest of cell division in glucose or galactose + hydroxyurea. (B) Incorporation of ^{3}H-uracil into DNA of MHY102 cells in the various media. Note that a full round of DNA replication occurs in cells after arrest in glucose medium.

Histone H4 Depletion

Given the more crucial role of histone H4 in assembly of the nucleosome, we extended the analysis to histone H4–depleted cells. A yeast strain (UKY403) was constructed in which the sole functional H4 gene in the cell was fused to the *GAL1* promoter on a centromeric plasmid (Fig. 16.4A). The unlinked histone H4-1 and H4-2 genes were replaced by the marker genes *HIS3* and *LEU2,* respectively [33]. Furthermore, in these experiments we have chosen to concentrate our experiments on UKY403 cells that were presynchronized in G1 with α-mating factor. This allows the analysis of biochemical changes, induced by glucose arrest, in a more homogeneous group of cells. Figure 16.4B shows the results of galactose and glucose addition on H4 mRNA synthesis. As with the GA110/H2B fusions, H4 mRNA synthesis is strongly activated in galactose and repressed in glucose.

UKY403 cells were synchronized with α-factor pheromone in G1 and allowed to incubate in glucose containing α-factor in order to deplete the cell of remaining H4 mRNA. When α-factor was washed away and the cells were released into glucose, we found that virtually all the cells arrested with a single large bud in G2 and that more than 97% could not segregate DNA to the bud even after long periods in glucose (Fig. 16.4C). Therefore, histone H4 is required for the process of chromosomal segregation since, like H2B, it is essential for the formation of the nucleosome.

Lethality Resulting from H4 Depletion Occurs First in S Phase

While histone H4 synthesis is essential for chromosomal segregation, it would seem unlikely that the lethal defect resulting from H4 loss occurs as late as mitosis. For example, new nucleosome assembly occurs concomitant with DNA replication in S phase. Therefore, it seems reasonable that altered chromatin formed in S phase would be the primary lethal defect resulting eventually in aborted chromosomal segregation.

UKY403 cells treated with α-factor and glucose were released into glucose. Their progress during the cell cycle until final arrest was monitored visually and by flow cytometry. Such cells pass synchronously from G1 to G2 in approximately 60 minutes [33]. Cells treated in this manner with glucose were transferred back to galactose plates at various points in the cell cycle, and resultant colonies were counted. As shown in Figure 16.5, lethality is first evident as the cells exit from G1 and pass through S phase and is largely complete by the time the cells reach G2. This period of lethality coincides with the period during which nucleosomes are assembled on newly replicated DNA. We have found that nucleosomes can also assemble on nonreplicating DNA in G2 in vivo [33]. However, the chromatin formed contains fewer nucleosomes and has

GAL 1/H4

A.

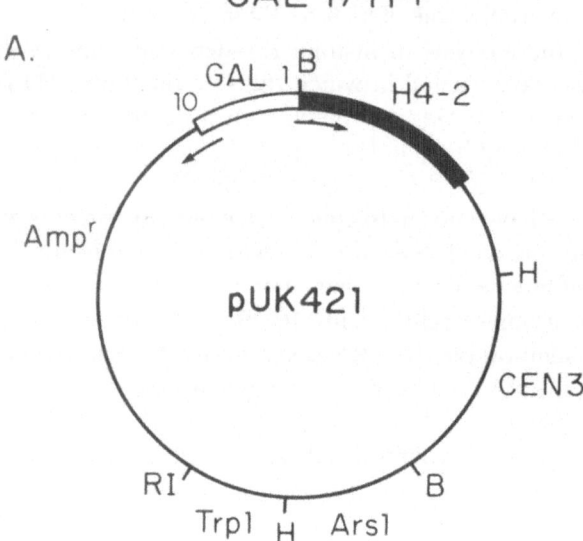

B.

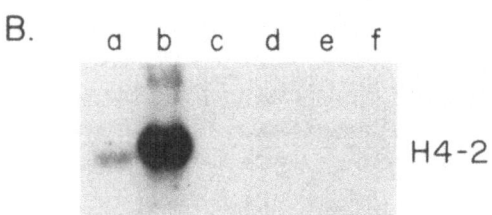

C.

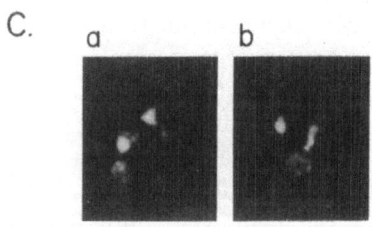

altered nucleosome spacing. This may help explain why nucleosome assembly in G2 gives rise mainly to inviable cells.

Structure of Chromatin Lacking Individual Core Histones

Since the nucleosome protects DNA from digestion by micrococcal nuclease, the ladder of bands observed after MNase digestion is indicative of polynucleosome length. In cells depleted of histone H4 (or H2B) we find that the average number of ethidium bromide–stained bands decreases from approximately 11 bands to three after glucose arrest. Also, there is evidence of greatly increased nuclease sensitivity at all nuclease concentrations, judging from the smearing seen between bands. This suggests that nucleosomes are lost after H4 mRNA (or H2B mRNA) repression, causing decreased protection of DNA [32,33]. These results were further confirmed by measurements of plasmid superhelicity in strains depleted of core histones. Superhelical density measures the presence of nucleosomes on plasmid DNA since, in a covalently closed episomal molecule, a nucleosome induces a single superhelical turn [34,35]. We found that H4 depletion resulted in a decrease of 50% to 60% in the superhelical densities of two different plasmids in UKY403 (2u and pUK421). Similar results were seen after H2B depletion [32,33]. These results strongly suggest that glucose arrest results in a full round of DNA replication in the absence of new nucleosome synthesis. Consequently, the resultant chromosomes contain approximately half the normal number of nucleosomes. As described below, the remaining nucleosomes likely become disordered, presumably owing to the loss of constraints provided by adjacent phased nucleosomes.

◁——

Figure 16.4. (A) Map of plasmid pUK421 containing the H4-2 gene under control of the GAL1 promoter. The plasmid also contains a centromeric DNA fragment CEN3 and the selectable marker gene TRP1. Symbols for restriction enzyme sites: B, *BamHI*; H, *HindIII*; RI, *EcoRI*. (B) Northern blot analysis of H4-2 transcripts in UKY403 cells to illustrate the activation of the *GAL1*-H4-2 fusion gene on galactose and its repression on glucose. The cells were grown in galactose medium YEPG (lane b), and synchronized by adding α-factor for 2 hours. Glucose (2%) was added, and the α-factor arrest was continued for an additional 2 hours (lane c). Then the cells were washed and grown in YEPD glucose medium, and aliquots of the cells were taken after 1, 2, and 4 hours in glucose (lanes d, e, and f, respectively). RNA was isolated from cells as described [33]. RNA (6μg) was loaded per well, electrophoresed, blotted, and hybridized to ^{32}P-labeled H4-2 gene-specific probe. Lane a contains RNA from a wild-type control strain (D585-11C (*MATa,lys1⁻*)) containing both H4-1 and H4-2 genes under control of their own promoters. (C) DAPI-stained nuclei of the terminally arrested UKY403 cells illustrating the defect in chromosomal segregation caused by histone H4 arrest.

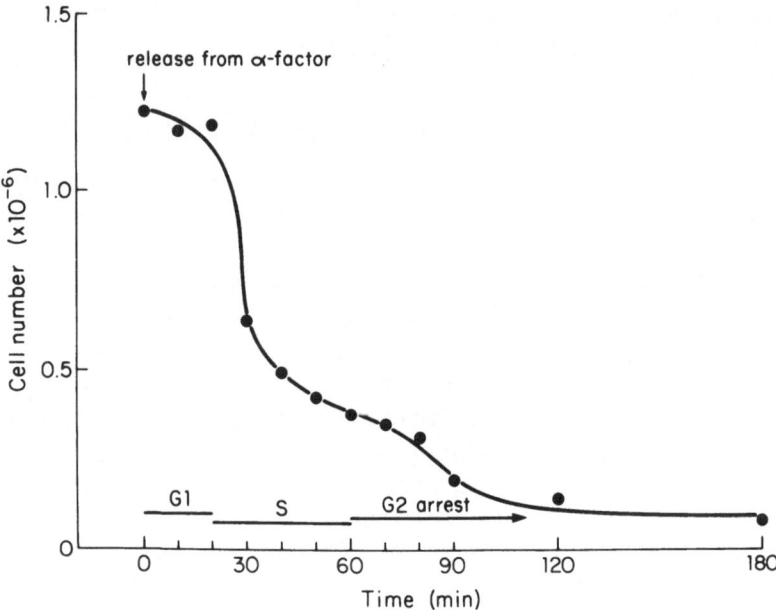

Figure 16.5. Mapping the stage of the cell cycle when histone H4 mRNA repression is found to be lethal. Cells were synchronized with α-factor as described [33] and released into YEPD (glucose medium), and samples were removed at the intervals shown. These were diluted, plated on YEPG (galactose) plates, and assayed for colony formation after 2 days of growth at 30°C. Approximately 500 cells were plated per Petri dish. Lethality is first evident as the cells exit G1 and enter S phase and is largely complete by the time the cells reach G2.

Transcription by RNA Polymerases I, II, and III Continues in Cells Blocked by H4 Depletion

The phasing of nucleosomes has been implicated in the regulation of eukaryotic transcription [36]. However, we don't know to what extent polynucleosome structures are necessary for accurate transcription. Since nucleosome depletion grossly alters nucleosomal density and phasing, we wished to answer this question by repressing the histone H4 gene under *GAL1* control and analyzing for transcription by all three RNA polymerases of yeast. Pulse labeling of 25S and 18S ribosomal RNAs, the products of RNA polymerase I, as well as 5S RNA and tRNAs, the products of RNA polymerase III, showed no difference in transcription rates when nucleosome-depleted chromatin was used as template [33].

We also examined CUP1 transcription [37] in H4-depleted UKY403 cells as one measure of RNA polymerase II activity. Presynchronized cells were analyzed for their ability to be induced by Cu^{++} before and after glucose arrest. We found that the CUP1 gene is activated to similar levels prior to glucose arrest and as long as 8 hours after glucose treatment. Steady-state mRNA levels of certain other yeast genes (*HIS3, HIS4, ARG4*, and *TRP1*) involved in amino acid metabolism were also examined, and no major differences were found in their levels after H4 mRNA repression [33]. Furthermore, H2B repression also allows the continuation of total RNA synthesis and *CUP1* regulation in MHY102 cells [32]. There is clearly no global involvement of nucleosomes in the activation of genes transcribed by polymerases I, II, or III.

The PHO5 Acid Phosphatase Gene Has Ordered Nucleosomes

Chromatin structure for most yeast genes shows distinctive differences from that of higher eukaryotes. Most yeast genes are always in a DNase I–sensitive conformation [23]. There appear to be few cases in which nuclease-hypersensitive sites, indicative of nucleosome-free regions, differ before and after yeast gene activation [38,39]. For example, in the region between the *GAL1* and *GAL10* promoters, the element containing all four UAS regions is equally accessible to nuclease, irrespective of the state of *GAL* gene activation [40,41]. This may explain why nucleosome depletion in yeast has little effect on the activation of *CUP1, GAL1*, and *GAL10* promoters [32,33].

The yeast *PHO5* gene represents an exception, in that there are chromatin structure differences between its active and inactive states. This gene encodes an acid phosphatase which is repressed when the concentration of inorganic phosphate (P_i) is high in the medium and is induced under conditions of low P_i [42]. It has been shown that an array of nucleosomes are positioned precisely over the transcription-controlling region in the repressed condition [43]. Upon induction, four nucleosomes in the promoter region are selectively removed [44] (Fig. 16.6). The *PHO5* gene is controlled through the action of constitutively synthesized negative-controlling factors (*PHO80* and *PHO85*) and positive factors (*PHO4* and *PHO2*). It has been postulated that nucleosomes prevent access of UAS sequences or other promoter sequences to the constitutively synthesized positive factors, *PHO4* and/or *PHO2* [44]. If this is the case, one might predict that removal of nucleosomes from the upstream region of *PHO5* by means of histone H4 repression could allow factor binding and therefore activation of *PHO5*.

Polynucleosomes Are Disrupted in the *PHO5* Upstream Region Upon Glucose Arrest

We found, using micrococcal nuclease digestions, that polynucleosome length and nucleosome spacing in the *PHO5* upstream region are greatly disrupted upon glucose arrest [45] in a similar manner to that of total

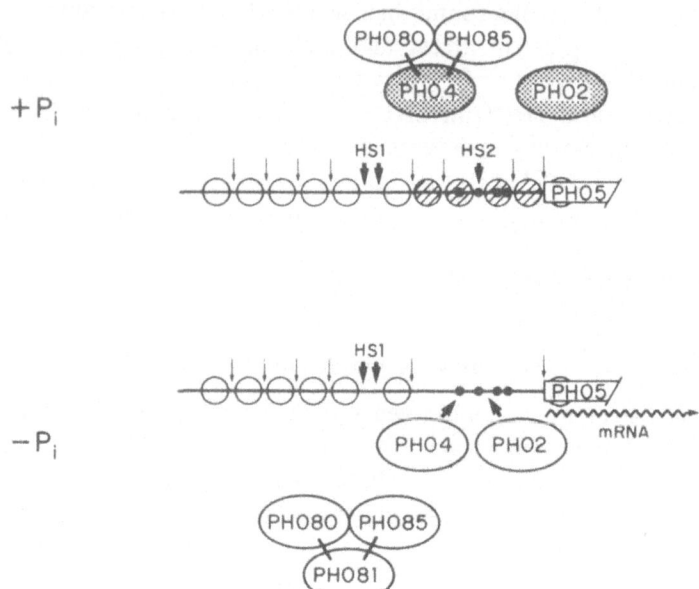

Figure 16.6. Regulation of the *PHO5* gene of *Saccharomyces cerevisiae*. Tran-scription of the acid phosphatase (*PHO5*) *mRNA* is mediated by at least two positive-acting proteins (*PHO4* and *PHO2*) and two that are negative acting (*PHO80* and *PHO85*); all are produced constituitvely at low levels in the repressed, high-P_i condition. Inorganic P_i may act through the mediator *PHO81* gene product. In low P_i, *PHO81* appears to be required for the positively activating function of *PHO4* and/or *PHO2*, which act on a sequence just upstream of the *PHO5* coding region [42]. Therefore, this model suggests a special role for *PHO4* and *PHO2* as positive activators that may act on the UAS sequences shown to be present in the *PHO5* promoter region [43,53]. Almer et al. [43] have found that during *PHO5* repression nucleosomes cover three of the four UAS sequences responsible for gene activation in the region of *PHO5*, although there is some controversy over the number of UAS sequences in this area [53]. All four sites are uncovered during activation. These authors have suggested that basal synthesis of *PHO5* mRNA occurs through the single uncovered UAS sequence under high-P_i conditions. However, in low-P_i medium, all four UAS sequences may be required for maximal activation by the positively acting proteins. Since nucleosome loss in our experiments leads to *PHO5* transcription in the presence of repressive high-P_i conditions, this suggests that the four UAS sequences are now available for interaction with the constitutively produced *PHO4* and/or *PHO2* products. Alter-natively, it is possible that uncovering the TATA promoter sequence upon nucleosome depletion leads to *PHO5* gene activation. In either case, both our data and those of Almer et al. [44] are consistent with a direct role for nucleosomes in repressing transcription in vivo.

chromatin reported previously [32,33]. We also examined by indirect end labeling the position of individual nucleosomes in the *PHO5* promoter region as described by Almer and Horz [43]. We found that H4 depletion in the presence of repressive conditions (high P_i) results in the loss of phased nucleosomes over the upstream region. The resultant micrococcal cleavage pattern resembles that of naked DNA [45], suggesting that nucleosomes remaining after H4 depletion become disordered.

PHO5 Transcription Is Activated When Nucleosomes Are Lost

We next asked whether regulation of *PHO5* was affected by histone H4 and nucleosome depletion. A control strain UKY412 was utilized that is isogenic to UKY403, except that its episomal H4-2 gene (in plasmid pUK499) is under control of its own wild-type promoter instead of the *GAL1* promoter in UKY403. As expected, this yeast strain does not arrest on glucose medium. To measure *PHO5* mRNA levels in the presence and absence of histone H4 synthesis, Northern blot analysis was carried out (Fig. 16.7A). Also, the start points of *PHO5* transcription were examined by S1 nuclease mapping (Fig. 16.7B). It was found that low P_i activates *PHO5* transcription in UKY412 to high levels on either galactose or glucose (lanes a and b). Similarly the *PHO5* gene is repressed in high P_i in either carbon source (lanes c and d). A similar analysis of UKY403 cells shows activation of *PHO5* transcription in low P_i in either galactose or glucose (lanes a and b). Activation of *PHO5* appears unaffected by nucleosome depletion. Under conditions of high P_i, in galactose (lane c), there is repression of *PHO5*. This is expected, since nucleosome structure appears normal under these conditions.

In contrast, in high P_i on glucose (lane d), conditions that deplete nucleosomes and should repress *PHO5*, we see activation of the *PHO5* gene. We conclude that histone H4 and nucleosome depletion leads to *PHO5* activation in the presence of high-P_i repressive conditions. We also found that transcript initiation is relatively normal when induction of *PHO5* occurs through nucleosome depletion (Fig. 16.7B) [45]. Furthermore, *PHO5* activation is not simply a response to arrest in any part of the cell cycle, or even to arrest in G2 or to arrest as a result of changes in chromosome topology. *PHO5* derepression is occurring because of a specific event which appears to be different in nucleosome-depleted cells [45].

Deletion Analysis of Histone Domains

Most of the N Terminus and a Short C-Terminal Sequence in Each Histone Is Dispensable for Growth

Hydropathy analysis establishes that residues are in a hydrophobic or hydrophilic region according to whether they are below or above the

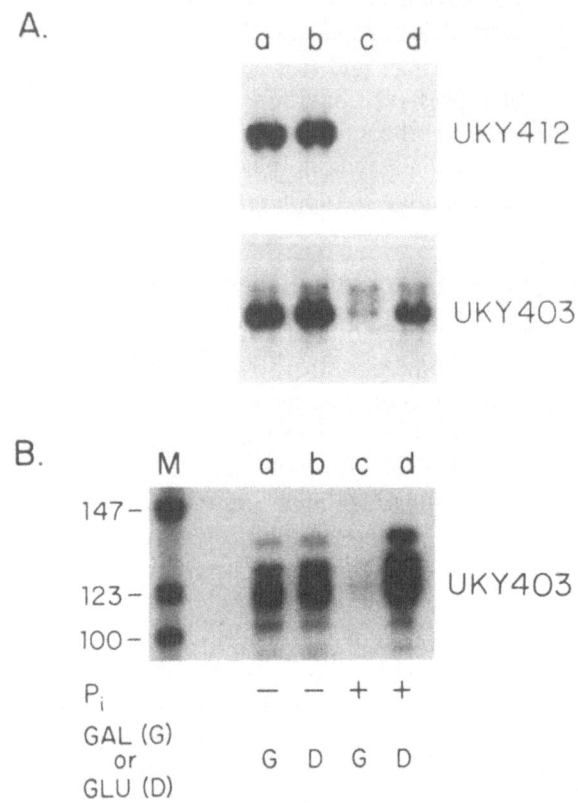

Figure 16.7. Depression of *PHO5* transcription in glucose-arrested UKY403 cells. (A) Northern blot illustrating the synthesis of *PHO5* mRNA in UKY412 and UKY403 yeast strains. The two strains are isogenic containing disrupted H4-1 and H4-2 chromosomal genes. They differ in the UKY412 contains an episomal histone H4-2 gene attached to its own promoter on plasmid pUK499. UKY403 contains an H4-2 gene under the control of the *GAL1* promoter [33]. Cells grown in high-P_i galactose-based medium were then transferred and incubated further for 6 hours in four different media: lane a, low-P_i galactose medium; lane b, low-P_i glucose medium; lane c, high-P_i galactose medium; lane d, high-P_i glucose medium. RNAs were isolated and separated by electrophoresis in 1% agarose, blotted, and probed as described [32]. Under conditions in which the *PHO5* gene is normally repressed (lane d), *PHO5* is seen to be activated as a result of glucose arrest and histone H4 depletion in UKY403 but not in UKY412 whose H4 gene is not under *GAL* control. (B) The transcription start sites of *PHO5* transcripts activated by histone H4 depletion appear normal. RNAs were isolated from UKY403 grown in galactose containing high P_i and transferred to the media described above. Lanes a through d display RNAs isolated from cells in growth conditions as described. S1 nuclease mapping was done with a ^{32}P-labeled fragment as described [45]. Lane M displays the molecular-weight-marker DNA generated by cutting PBP322 plasmid DNA with MspI.

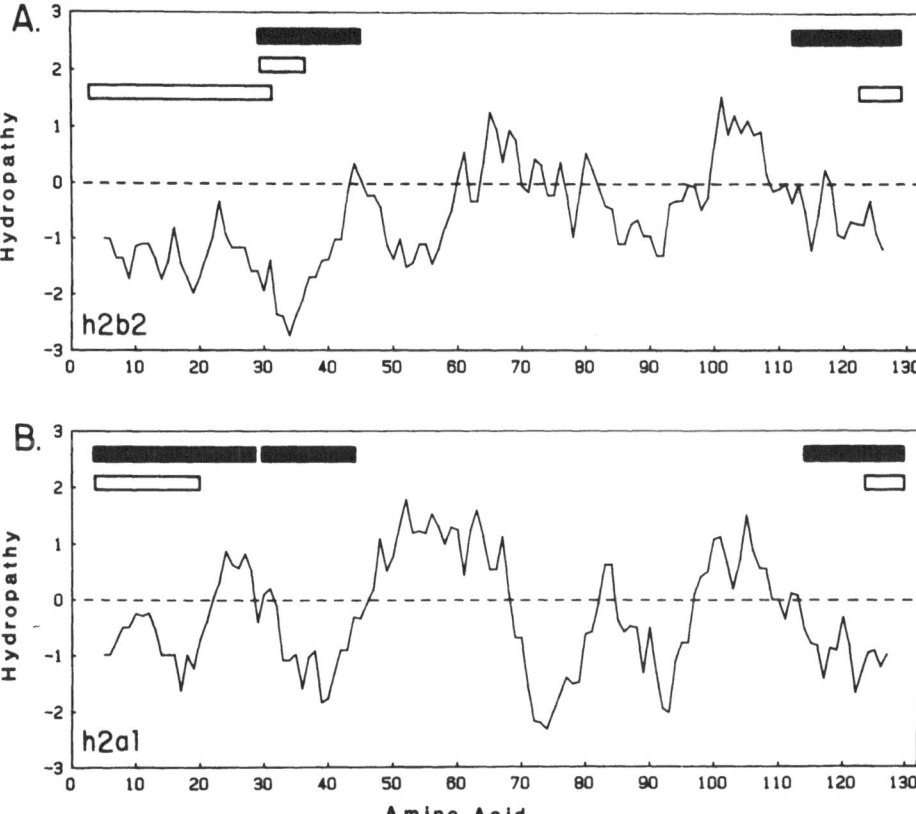

Figure 16.8. Superimposed hydropathy and deletion maps for histones H2B and H2A. White bars represent deletions that were found to function in vivo. Black bars represent deletions that could not rescue the appropriate strain when grown on glucose. The hydropathy plot [46] was generated by scanning the entire length of each protein sequence for relative hydrophobicities. A scanning window of nine residues was used.

hydropathy index of zero. When the deletions made in H2B and H2A [48,49] are mapped onto their respective hydropathy plots (Fig. 16.8), it is evident there is a common pattern relating to the viability of histone deletion mutants. The hydrophilic N terminus of H2B is relatively divergent, while that of H2A is much more conserved [6]. However, for each protein, the hydrophilic region is dispensable for growth. This is also true at the C terminus. Recent evidence utilizing histone H4 shows a similar pattern [50]. Although the H4 protein is extremely conserved in evolution, its hydrophilic sequences are dispensable. Residues 4 to 28 and 100 to 102 have been shown to be nonessential for growth, while its hydrophobic core (residues 29 to 99) is required. Hydrophilic histone domains are thought to extend from the hydrophobic cores [6]. Furthermore, it has been shown that histones whose N-terminal tails have been removed by trypsin cleavage of chromatin are capable of assembling into nucleosomes in vitro [14]. Therefore, viable deletions occur only in the histone domains that extend from the core.

The similarity in the structure and charge characteristics of the histone N termini and the observed pattern of deletions suggest that the histone N termini have similar functions. This is supported further by experiments showing that the extended N termini of H2A and H2B histones can substitute for each other in vivo [49]. Whether this involves fixing nucleosomal position by binding to DNA or allowing higher orders of folding by interactions between nucleosomes is still unclear.

There is now direct evidence that active chromatin is preferentially acetylated [12,13]. Yet we have found that removal of individual histone N termini, which are highly acetylated in yeast [51], has little effect on the repression or activation of yeast genes. This too may argue for redundant functions between the core histones.

Conclusions

We have shown that yeast is not likely to contain histone H1. Therefore, the nucleosomal functions are most probably assumed by the four core histones, H2A, H2B, H3, and H4. To test whether nucleosomes are involved in all chromosomal functions (chromosomal segregation, replication, and transcription), we have repressed histone H2B or H4 synthesis. Alternatively, we have expressed histones that have been mutated to remove N- or C-terminal ends. We believe that nucleosomes are required during the yeast cell cycle in the following manner.

Without new nucleosome synthesis, DNA continues to replicate at a relatively normal rate and produces a full complement of undegraded chromosomal DNAs lacking approximately half the normal number of nucelosomes. This and previous data argue against a function for nucleosomes in replication. If nucleosomes are not provided in S phase to assemble with newly replicated DNA, then growth arrest is irreversible. Nucleosomes can assemble on nonreplicating DNA; however, this assembly is probably inaccurate, producing improperly phased nucleosomes [33]. DNA lacking the normal nucleosomal complement cannot segregate normally in mitosis. Therefore, we conclude that nucleosomes are not required for replication but are essential for chromosomal segregation.

The effect of nucleosomal position on transcription is especially interesting. The rate of transcription by RNA polymerases I, II, and III appears to be relatively normal in the absence of new histone synthesis. There is clearly no global repression of RNA synthesis as a result of nucleosome loss and displacement of remaining nucleosomes. On the contrary, nucleosomes may in some cases be negative regulators of transcription. The evidence is compelling that nucleosomes act to repress *PHO5* gene activity in vivo [45]. This may reflect the normal presence of nucleosomes over the upstream control sequences in the repressed state. Known constitutive *trans*-acting proteins (*PHO4, PHO2*, RNA polymerase II) may then interact with the *PHO5* promoter sequences upon nucleosome loss. More recently, we have been using fusions of the *PHO5* upstream sequences to the *lacZ* gene of *E. coli* in order to identify the promoter sequences that are both necessary and sufficient to generate the response to nucleosome removal [52]. Also, we have found other yeast promoters that are activated upon nucleosome loss. Therefore, either nucleosomes repress transcription by blocking promoter sequences from interacting with activator proteins or the change in topology (superhelicity) caused by nucleosome loss is required for initiating transcription. In either case we have found that nucleosome loss leads to the activation of certain genes in yeast. It seems unlikely that this form of regulation is limited to yeast alone.

One intriguing question that arises from this work is the role played by the flexible N-terminal tails in transcription, given the correlation between histone acetylation and transcription. Our work has shown that irrespective of the sequence conservation of the histone N termini they are *individually* dispensable for vegetative growth and general transcription. This does not necessarily mean that N-terminal acetylation is not important for transcription, especially since the histone N termini may share function. Neutralization of the N-terminal positive charges could be required for allowing nucleosomes to be displaced during RNA synthesis. However, deletion of the N termini could make displacement easier, since deletion reduces the net positive charge of the end (from $+10$ to $+1$ for the H4 N terminus).

Why, then, are the N termini conserved throughout evolution? We would argue that the N termini are required for normal chromosomal folding, a conformation that is loosened by acetylation during transcription. Folding is most likely altered but still functional for growth when a single N-terminal end is deleted, since the other three histone tails have the ability to interact with negatively charged DNA. We do find, for example, that removal of the H4 N-terminal end elongates G2 disproportionately during the cell cycle [50]. This may argue for the involvement of the H4 N terminus in a G2-related function, perhaps chromosomal segregation. We are confident that possibilities for histone function or histone domain function will be testable genetically.

Acknowledgments. This work was supported by Public Health Service grants GM23674 and GM31336 from the National Institutes of Health.

References

1. Schlissel, M.S., and Brown, D.D. (1984) Cell *37*, 903–913.
2. Weintraub, H. (1984) Cell *38*, 17–27.
3. Simon, R.H., Camerini-Otero, R.D., and Felsenfeld, G. (1978) Nucleic Acids Res. *5*, 4850–4858.
4. Jackson, V., and Chalkley, R. (1981) Cell *23*, 121–134.
5. Smith, P.A., Jackson, V., and Chalkley, R. (1984) Biochemistry *23*, 1576–1581.
6. McGhee, J.D., and Felsenfeld, G. (1980) Annu. Rev. Biochem. *49*, 1115–1156.
7. Ebralidse, K.K., Grachev, S.A., and Mirzabekov, A. (1988) Nature *331*, 365–367.
8. Allis, C.D., Chicoine, L.G., Richman, R., and Schulman, I.G. (1985) Proc. Natl. Acad. Sci. USA *82*, 8048–8052.
9. Allfrey, V.G. (1977) In Chromatin and Chromosome Structure, 167–191, Li, H.J., and Eckhardt, R. (eds.). Academic Press, New York.
10. Chahal, S.S., Mathews, H.R., and Bradbury, E.M. (1980) Nature *287*, 76–79.
11. Vavra, K.J., Allis, C.D., and Gorovsky, M.A. (1982) J. Biol. Chem. *257*, 2591–2598.
12. Johnston, E.M., Sterner, R., and Allfrey, V.G. (1987) J. Biol. Chem. *262*, 2943–2946.
13. Hebbes, T.R., Thorne, A.W., and Crane-Robinson, C. (1980) EMBO J. *7*, 1395–1402.
14. Whitlock, J.P., and Stein, A. (1978) J. Biol. Chem. *253*, 3857–3861.
15. Mathis, D.J., Oudet, P., Wasylyk, B., and Chambon, P. (1978) Nucleic Acids Res. *5*, 3523–3547.
16. Wallis, J.W., Hereford, L., and Grunstein, M. (1980) Cell *22*, 799–805.
17. Choe, J., Kolodrubetz, D., and Grunstein, M. (1982) Proc. Natl. Acad. Sci. USA *79*, 1484–1487.
18. Hereford, L.M., Fahrner, K., Wollford, J., Rosbash, M., and Kaback, D.B. (1979) Cell *18*, 1261–1271.
19. Smith, M., and Murray, K. (1983) J. Mol. Biol. *169*, 641–661.
20. Smith, M., and Andresson, O.S. (1983) J. Mol. Biol. *169*, 663–690.
21. Certa, U., Colavito-Shepanski, M., and Grunstein, M. (1984) Nucleic Acids Res. *12*, 7975–7985.
22. Schwartz, D.C., and Cantor, C.R. (1984) Cell *37*, 67–75.
23. Lohr, D., and Hereford, L. (1979) Proc. Natl. Acad. Sci. USA *76*, 4283–4288.
24. Rykowski, M.C., Wallis, J.W., Choe, J., and Grunstein, M. (1981) Cell *25*, 477–487.
25. Kolodrubetz, D., Rykowski, M.C., and Grunstein, M. (1982) Proc. Natl. Acad. Sci. USA *79*, 7814–7818.
26. Pringle, J.R., and Hartwell, L.H. (1981) In The Molecular Biology of the Yeast Saccharomyces: Life Cycle and Inheritance, 97–142, Strathern, L.N., Jones, E.W., and Broach, J.R. (eds.). Cold Spring Harbor Laboratory Press, Cold Spring Harbor, New York.
27. Holm, C., Goto, T., Wang, J.C., and Botstein, S. (1985) Cell *41*, 553–563.
28. Elliott, S.G., and McLaughlin, C.S. (1978) Proc. Natl. Acad. Sci. USA *75*, 4384–4388.

29. Moll, R., and Wintersberger, E. (1976) Proc. Natl. Acad. Sci. USA 73, 1863–1867.
30. Hereford, L., Osley, M.A., Ludwig, J.R., and McLaughlin, C. (1981) Cell 24, 367–375.
31. Johnston, M., and Davis, R.W. (1984) Mol. Cell. Biol. 4, 1440–1448.
32. Han, M., Chang, M., Kim, U.-J., and Grunstein, M. (1987) Cell 48, 589–597.
33. Kim, U., Han, M., Kayne, P., and Grunstein, M. (1988) EMBO J. 7, 2211–2219.
34. Worcel, A., Strogatz, S., and Riley, D. (1981) Proc. Natl. Acad. Sci. USA 78, 1461–1465.
35. Wang, J.C. (1982) Cell 29, 724–726.
36. Gross, D.S., and Garrard, W.T. (1987) Trends Biochem. Sci. 12, 293–297.
37. Fogel, S., and Welch, J.W. (1982) Proc. Natl. Acad. Sci. USA 79, 5342–5346.
38. Struhl, K. (1982) Cold Spring Harbor Symp. Quant. Biol. 47, 901–910.
39. Szent-Gyorgyi, C., Finkelstein, D.B., and Garrard, W.T. (1987) J. Mol. Biol. 193, 71–80.
40. Lohr, D. (1984) Nucleic Acids Res. 12, 8457–8474.
41. Proffitt, J.H. (1985) Mol. Cell. Biol. 5, 1522–1524.
42. Oshima, Y. (1982) In The Molecular Biology of the Yeast Saccharomyces: Metabolism and Gene Expression, 159–180, Strathern, L.N., Jones, E.W., and Broach, J.R. (eds.). Cold Spring Harbor Laboratory Press, Cold Spring Harbor, New York.
43. Almer, A., and Horz, W. (1986) EMBO J. 5, 2681–2687.
44. Almer, A., Rudolph, H., Hinnen, A., and Horz, W. (1986) EMBO J. 5, 2689–2696.
45. Han, M., Kim, U.-J., Kayne, P., and Grunstein, M. (1988) EMBO J. 7, 2221–2228.
46. Kyte, J., and Doolittle, R.F. (1982) J. Mol. Biol. 157, 105–132.
47. Pustell, J., and Kafatos, F.C. (1984) Nucleic Acids Res. 12, 643–655.
48. Wallis, J.W., Rykowski, M., and Grunstein, M. (1983) Cell 35, 711–719.
49. Schuster, T., Han, M., and Grunstein, M. (1986) Cell 45, 445–451.
50. Kayne, P., Han, M., Kim, U.-J., Mullins, J., Yoshizaki, F., and Grunstein, M. (1988) Cell 55, 27–39.
51. Nelson, D.A. (1982) J. Biol. Chem. 257, 1565–1568.
52. Han, M., and Grunstein, M. (1988) Cell 55, 1137–1145.
53. Nakao, J., Miyanohara, A., Toh-e, A., and Matsubara, K. (1986) Mol. Cell Biol. 6, 2613–2623.

Chapter 17
Mammalian Protamines: Structure and Molecular Interactions

Rod Balhorn

During spermiogenesis in most vertebrates, the structure of spermatid chromatin is completely reorganized and the DNA is repackaged by a small, arginine-rich protein called protamine. As a consequence of the interactions that occur between DNA and protamine in late-step spermatids [1], the phosphodiester backbone of DNA is neutralized and the fibers of double-helical DNA and protamine coalesce into a maximally condensed [2], biochemically inert state. The resulting DNA-protamine complex is insoluble, enzymatic processes such as DNA repair are inhibited [3,4], and the genome is rendered genetically inactive [5]. Although we have learned a great deal about the structure, metabolism, and potential function of the protamines in the past two decades, we have learned comparatively little about the actual molecular structure of the DNA-protamine complex. The marked insolubility of sperm chromatin and the high affinity of protamine for DNA have often made it difficult to characterize the structure of sperm chromatin using the biochemical and physical techniques that have been applied to somatic chromatin and other DNA-protein complexes.

Four reviews have recently been written about the protamines [6,7], the protamine genes [8], and the changes that occur in the complement of chromosomal proteins during the process of spermiogenesis [9]. A review on testicular chromosomal proteins and the structure and expression of the mammalian protamine genes, written by Norman Hecht [10], can be found in this same volume. Together, these papers have thoroughly covered much of the historical and contemporary protamine and sperm chromatin literature. Therefore, this chapter is dedicated exclusively to mammalian protamines and describes the most recent data and models that relate specifically to the structure and function of these remarkable proteins.

Primary Structure

Until recently, it appeared that the sperm DNA of diverse mammals could be packaged by a variety of protamine molecules. Electrophoretic and chromatographic studies have indicated that the sperm of the bull [11,12], ram [13,14], boar [15,16], rat [17], rabbit [18], and guinea pig [18] contain only a single type of protamine, protamine 1. Two different protamines have been isolated from mouse [2,19,20], hamster [21], and stallion [22,23] sperm, and three protamines have been found to be present in the sperm of humans [24–26]. After the primary sequences of the mouse [27–29] and human [30–33] protamines were determined and the two mouse protamine genes were isolated and sequenced [34], it became clear that there were really only two types or families of protamine molecules.

Complete amino acid sequences have been reported for protamine 1 molecules isolated from seven different species of mammals, including the bull, ram, boar, stallion, mouse, hamster, and human (Fig. 17.1). A partial sequence for rhesus protamine 1 has also been identified (Fig. 17.1). The primary sequences of bull [11,12], ram [35], boar [16], stallion [22,23], human [30–33], hamster [21], and rhesus protamine 1 were determined by sequencing the isolated proteins. Both the bull and human protamine 1 sequences have been confirmed by sequence analyses of bull and human protamine 1 cDNAs [36–38]. The sequence of mouse protamine 1 was originally derived from a cDNA sequence [27]; the corresponding protein sequence was recently reported by Bellve et al. [29].

One structural feature that appears to be unique to protamine 1 is its amino-terminal sequence; each protamine 1 begins with the hexapeptide sequence ala–arg–tyr–arg–cys–cys (Figs. 17.1, and 17.2). Only three species, the dwarf hamster (*Phodopus sungorus*), beaver (*Castor canadensis*), and rabbit (*Oryctolagus cuniculus*), have been found with mutations in one of these first six residues (Fig. 17.2). The length of each

PROTAMINE 1

BULL	ARYRCCLTHSGSRCRRRRRRRRCRRRRRRRFGRRRRRRVCCRRYTVIRCTRQ
RAM	ARYRCCLTHSRSRCRRRRRRRRCRRRRRRRFGRRRRRRVCCRRYTVVRCTRQ
BOAR	ARYRCCRSHSRSRCRPRRRRRCRRRRRRCCPRRRRRAVCCRRYTVIRCRRC
STALLION	ARYRCCRSQSQSRCRRRRRRRCRRRRRRSVRQRRVCCRRYTVLRCRRRRC*
MOUSE	ARYRCCRSKSRSRCRRRRRRCRRRRRRCCRRRRRRCCRRRRSYTIRCKKY
HAMSTER	ARYRCCRSKSRSRCRRRRRRCRRRRRRCCRRRRRRCCRRRRTYTLRCKRY
HUMAN	ARYRCCRSQSRSRYYRQRQRSRRRRRRSCQTRRRAMRCCRPRYRPRCRRH
RHESUS	ARYRCCRSQSRSRCCRRRRSC......

PROTAMINE 2

HUMAN 2a	RTHGQSHYRRRHCSRRRLHRIHRRQHRSCRRRKRRSCRHRRRHRRGCRTRKRTCRRH
HUMAN 2b	GQSHYRRRHCSRRRLHRIHRRQHRSCRRRKRRSCRHRRRHRRGCRTRKRTCRRH
MOUSE	RGHHHHHRHRRCSRKRLHRIHKRRRSCRRRRRHSCRHRRRHRRGCRR3RRRRRCRCRKCRRHHH
HAMSTER	RGQHHHRRRCSRRKLYRIHRRRRSCRRRRHSCRHRRRHRRGCCRS??RR?C...

Figure 17.1. Amino acid sequences of protamines 1 and 2. The stallion sequence used in this figure contains the proposed carboxy-terminal cysteine residue, cys50.

SPECIES	PROTAMINE 1	PROTAMINE 2
Bull (*Bos tarus*):	ARYRCC......	
Ram (*Ovis aries*):	ARYRCC......	
Boar (*Sus scrofa*):	ARYRCC......	
Moose (*Alces alces*):	ARYRCC......	
Mouse (*Mus musculus*):	ARYRCC......	RGHHHHR....
Human (*Homo sapiens*):	ARYRCC......	RTHGQSH....
Horse (*Equus caballas*):	ARYRCC......	
Rat (*Rattus norvegicus*):	ARYRCC......	
Raccoon (*Procyon lotor*):	ARYRCC......	
Beaver (*Castor canadensis*):	GRYRCC......	
Elephant (*Elephas maximus*):	ARYRCC......	
Domestic Cat (*Felis cattus*):	ARYR??......	
Chinchilla (*Chinchilla laniger*):	ARYRCC......	KGRYH?K....
Rabbit (*Oryctolagus cuniculus*):	VRYRCC......	
Cotton Rat (*Sigmodon hispidus*):	ARYRCC......	RGHHHHR....
Rhesus Monkey (*Macaca mulatta*):	ARYRCC......	RTHSGHS....
Dwarf Hamster (*Phodopus sungorus*):	ARYRCR......	REHHHHR....
Chinese Hamster (*Cricetulus griseus*):	ARYRCC......	
Syrian Hamster (*Mesocricetus auratus*):	ARYRCC......	RGQHHHR....
Armenian Hamster (*Cricetulus migratorius*):	ARYRCC......	
Opossum (*Monodelphis domestica*):	ARYSPH......	
Ring-Tailed Wallaby (*Petrogale xanthopus*):	ARYRHS......	

Figure 17.2. Amino-terminal hexapeptide sequences of protamines 1 and 2.

protamine 1 molecule also appears to be constant at 50 amino acids. Although the primary sequence of stallion protamine 1 was recently reported to contain only 49 amino acids [22,23], this may not be accurate. In our laboratory carboxypeptidase W digests of stallion protamine have indicated that the carboxy-terminal amino acid of this protamine 1 is cysteine. It would appear that the terminal cysteine residue of stallion protamine (or its derivatives in the modified proteins used for sequencing) may be somewhat labile or susceptible to proteolysis, as others have observed for boar protamine. Preparations of boar protamine characterized by Tobita et al. [16,39] were reported to contain two protein species, the intact protamine 1 molecule and a protamine 1 lacking the carboxy-terminal cysteine residue. The instability of this terminal cysteine appears to be a feature common to both boar and stallion protamine and may be related to their carboxy-terminal sequence, arg–arg–cys.

A comparison of the seven complete protamine 1 sequences suggests that this protein may be divided into three structural domains (Fig. 17.3). The central polyarginine-rich region of the molecule is the proposed DNA-binding domain [1]. This sequence is composed of multiple DNA anchoring units (containing three to seven arginine residues per unit) separated by one or more nonarginine amino acids. The central domain is flanked on both sides by smaller, relatively arginine-deficient peptide

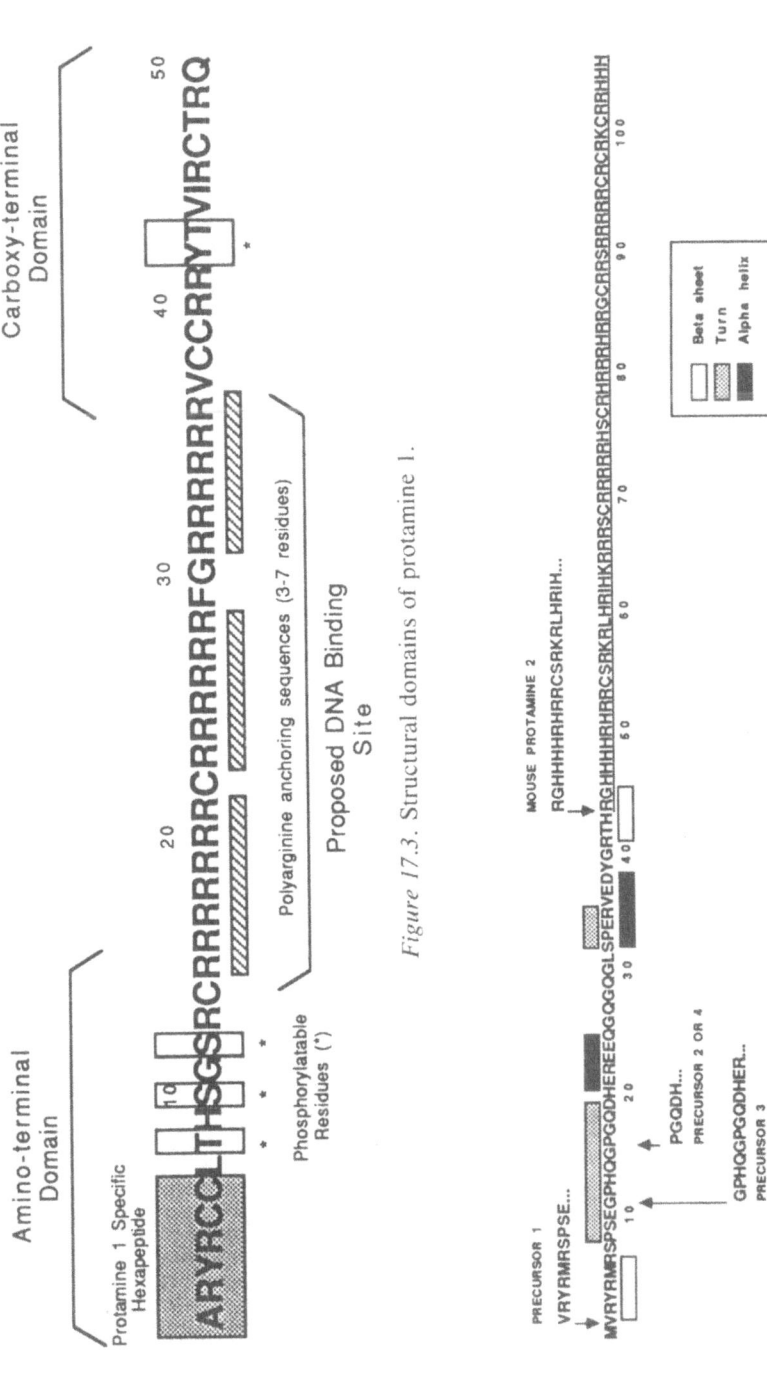

Figure 17.3. Structural domains of protamine 1.

Figure 17.4. Primary sequence of mouse protamine 2 precursor and location of processing sites. Secondary structures were predicted using the method of Chou and Fasman [121].

domains. The amino-terminal domain contains the conserved hexapeptide sequence, while both domains contain serine, threonine, or tyrosine residues located at specific positions that are probable phosphorylation sites.

In contrast to protamine 1, complete amino acid sequences have been reported for only two protamine 2 molecules—human 2 (human 2a in Fig. 17.1) and mouse (Fig. 17.1). A partial sequence (approximately 90% complete) for Syrian hamster protamine 2 has also been reported [21]. The human protamine 2 sequence was determined by sequencing the isolated protein [30–33] as well as its cDNA [40]. The sequence of mouse protamine 2 was originally determined by combination of cDNA and protein sequencing [28] and has been confirmed by Bellve et al. [29]. One difference was observed between the two mouse sequences, however. The recent protein sequence was found to be one histidine residue shorter than the sequence determined from the mouse cDNA and genomic sequences. The significance of this difference is currently unclear.

Protamine 2 sequences differ from protamine 1 in several ways. Protamine 2 is substantially larger than protamine 1 and is variable in length. In addition, the sequence of protamine 2 is not as rigorously conserved as the sequence of protamine 1. While certain amino acid sequences do appear to be conserved between species, these minidomains are irregularly spaced. The carboxy-terminal region of each protein is also variable, both in sequence and length. Another notable feature that distinguishes protamine 2 is its high content of histidine (Table 17.1). Protamine 1 molecules rarely contain histidine. A further difference is that the arginine clusters of protamine 2 appear to be distributed more evenly throughout the length of the molecule.

The sperm of humans [24–26,33] and certain primates contain a third protamine that has been designated protamine 2b (or 2″). The amino acid sequence of human protamine 2b was found to be nearly identical to that of human protamine 2 (hereinafter protamine 2a); protamine 2b only lacked the first three amino-terminal amino acids, arg–thr–his, of protamine 2a (Fig. 17.1) [26,31–33]. At present, it is not clear whether human protamines 2a and 2b are derived from the same or different genes. Mouse protamine 2 has been shown to be synthesized as a precursor protein that is shortened in late-step spermatids to the final 63 amino acid sequence found in mature sperm. The final processing site occurs on the carboxy-terminal side of the sequence arg–thr–his between histidine 43 and arginine 44 (Fig. 17.4). The observed three–amino acid difference between human protamines 2a and 2b, by comparison, might be explained by an incomplete or differential processing of the protamine 2 precursor. The results of preliminary analyses of rhesus protamine 2a and 2b, on the other hand, suggest an alternative explanation. As observed for the human proteins, the amino-terminal sequence arg–thr–his is present in rhesus protamine 2a and absent in protamine 2b (Fig. 17.5). Rhesus

Table 17.1. Amino acid compositions (in mole %) of protamine 2 molecules.

Amino acid	Mouse	Human	Rhesus	Cotton rat	Syrian hamster	Dwarf hamster	Turkish hamster
Arginine	50.8	47.8	61.5	55.7	58.4	55.8	60.7
Histidine	20.6	15.0	16.3	18.6	14.7	20.8	13.4
Lysine	0	3.5	0	5.0	1.8	3.4	2.2
Alanine	0.	0	4.1	0	0	0	0.4
Cysteine	11.1	6.9	5.1	9.0	9.4	8.2	8.2
Glu/Gln	4.8	4.6	0	0	1.5	0	0
Glycine	3.2	5.0	0	3.0	3.1	2.3	2.8
Isoleucine	1.6	1.2	1.5	1.2	1.3	1.8	1.0
Leucine	1.6	1.9	1.9	1.7	1.8	1.9	1.8
Methionine	0	0	0	0	0	0	0
Phenylalanine	0	0	0	0	0	0	0
Proline	0	0	0	0	0	0	0
Serine	6.3	7.5	6.7	5.7	5.8	5.8	6.2
Threonine	0	5.0	2.9	0	0.4	0	0.9
Tyrosine	0	1.6	0	0	1.8	0	2.4
Valine	0	0	0	0	0	0	0

protamine 2a and 2b, however, differ in amino acid sequence at three other positions within the amino-terminal peptide domain. Because the rhesus genome appears to contain two different protamine 2 genes, the possibility exists that human protamines 2a and 2b may also represent the expression products of two very similar, but different human genes. The recent publication of a sequence for a human protamine 2 cDNA [40] and amino-terminal protein sequences for two human protamine 2 precursors [41] supports this latter possibility. Neither the HPS1 nor HPS2 precursor sequences reported by Sautiere et al. [41] correspond to the human protamine 2 cDNA sequence.

While it has always been assumed that both protamines 1 and 2 are present in every sperm produced by mammals expressing the protamine 2

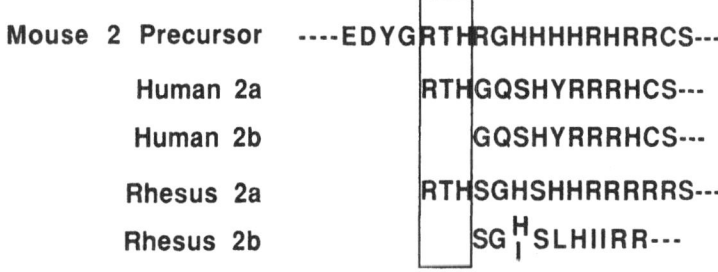

Figure 17.5. Comparison of primary sequence around final mouse protamine 2 processing site with amino-terminal sequences of human and rhesus protamines 2a and 2b. Residue 3 in rhesus 2b is either histidine or isoleucine.

gene, the recent costaining of human sperm with two fluorescently labeled monoclonals, one specific for human protamine 1 and the other specific for human protamine 2, and their analysis by flow cytometry have provided the only direct evidence that both protamines are actually present in all normal human sperm [42]. Previous biochemical analyses of protamine 1/protamine 2 distributions in the sperm of numerous fertile human males displaying widely variable sperm morphology profiles [43; unpublished results] indicated that both protamines should be present in equal ratios in each major morphological class of human sperm. Determinations of the protamine 1/protamine 2 contents of flow sorted chinchilla X and Y chromosome bearing sperm (Table 17.2) have provided evidence that these two subpopulations of sperm also contain equivalent ratios of protamine 1 and 2.

Synthesis and Metabolism

Several studies have demonstrated that protamine synthesis is initiated late in spermiogenesis [44–48], coincident with the onset of nuclear sonication resistance in the mouse [46] and rat [49]. The binding of protamine to DNA displaces the bulk of the histones and transition proteins and eliminates the nucleosomal organization characteristic of somatic chromatin [2,50–57]. Analyses of the human [33], mouse [58], and rat [59] proteins have shown that protamine 1 is postsynthetically modified by phosphorylation at multiple sites within the same molecule [59]. The phosphorylated residues of rat protamine 1 appear to be maintained during spermatid maturation in the testis, but the level of phosphorylation progressively declines as the sperm pass through the epididymis [58]. By the time the sperm reach the vas, the majority of the phosphate groups have been removed.

The synthesis of protamine 2 appears to coincide temporally with that of protamine 1 [60], but studies in the mouse have revealed that protamine 2 is synthesized and incorporated into spermatid chromatin in the form of a precursor protein. A recent sequence published for a human protamine 2 cDNA indicates that the human protein is also synthesized as a precursor [40]. The complete amino acid sequence of the mouse precursor

Table 17.2. Distribution of protamines 1 and 2 in chinchilla X and Y chromosome-bearing sperm.

	Protamine 2 (percent of total protamine)
Total sperm	25.2 ± 2.2
X sperm	25.3 ± 1.1
Y sperm	26.2 ± 1.2

is shown in Figure 17.4. After binding to DNA, the mouse precursor is slowly processed by a series of proteolytic cleavages that remove specific segments of the precursor's amino terminus. Two intermediate-processed forms of the precursor (Fig. 17.6) have been isolated from mouse late-step spermatids and characterized by amino acid analysis and sequencing [61]. The cleavage sites responsible for their formation were identified by locating the amino-terminal sequences of the two proteins within the sequence of the intact precursor (Fig. 17.4). The larger of the two proteins was shown to be produced as a result of a cleavage between residues glu10 and gly11. The smaller intermediate was produced by a cleavage between gly15 and pro16. Both sites are located within a predicted turn in the peptide chain (Fig. 17.4). A third intermediate was observed, but it has not yet been identified.

In vivo radiolabeling studies in the mouse [61] and in vitro experiments with mouse testes [58] have demonstrated that the precursor is deposited onto DNA intact and that processing does not commence until several hours after its deposition. Both the newly synthesized mouse precursor and the fully processed protamine 2 molecule (mouse [58] and human [62]) appear to be phosphorylated. The relative abundance of the precursor and each intermediate processed form and the sequential movement of radiolabel from the intact precursor to each intermediate [61] suggest that precursor processing occurs in an ordered fashion and that each intermediate protein persists for several hours before being processed further. The final cleavage, yielding "mature" protamine 2, occurs between his43 and arg44 approximately 24 to 30 hours after precursor synthesis and deposition onto DNA.

The function of the 43 amino-terminal residues of the mouse protamine 2 precursor is unknown. Because protamine 1 is not similarly synthesized with an expendable amino-terminal sequence, it seems unreasonable to assume that this sequence might be required for transport of the newly synthesized protamine 2 molecule into the spermatid nucleus. That the precursor is not processed until some time after it is deposited onto DNA

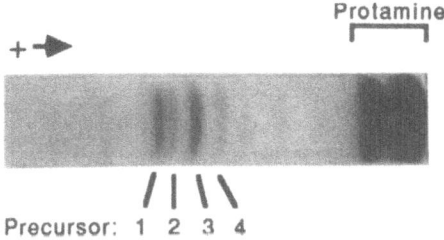

Figure 17.6. Mouse protamine 2 precursors isolated from late-step spermatids. The intact precursor (1) and two intermediate processed forms (2 and 3) were isolated by HPLC, and their amino-terminal sequences were determined.

suggests that the additional amino-terminal sequence may perform some function related to the integration of the protamine 2 molecule into sperm chromatin, such as directing the proper binding of protamine 2 to DNA, facilitating precursor or protamine 2 folding, or establishing intermolecular contacts between protamine 2 and protamine 1 (or some other sperm chromatin protein). Once the proper conformation or interaction is achieved, specific proteases may remove the superfluous amino acid residues.

Disulfide Formation

In addition to detailed studies of the primary structure, synthesis, and metabolism of protamines, other experiments have been conducted to identify the cross-links that interconnect neighboring protamines. Numerous studies have provided evidence that the protamines of eutherian mammals are cross-linked around DNA through intermolecular disulfide bridges [63–68]. Analyses of the disulfide contents of human [65], rat [66,67], boar [68], and mouse protamine isolated from late-step spermatids and sperm of the caput, corpus, and cauda epididymis have shown that disulfide formation is initiated in testicular sperm and continues as the sperm traverse the epididymis (Fig. 17.7). By the time the sperm finally reach the cauda epididymis in bull and mouse, all the cysteine residues of protamine have formed disulfide cross-links.

Progressive disulfide reduction and cross-linking experiments have

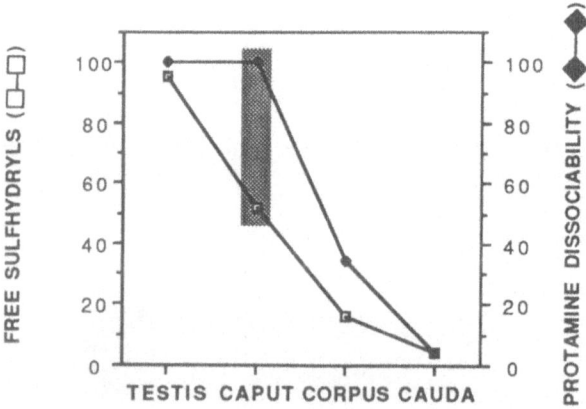

Figure 17.7. Disulfide formation in protamines of mouse late-step spermatids and epididymal sperm, and solubility of protamine and the sperm nucleus in guanidine hydrochloride. Free sulfhydryls: the percent of cysteine residues that can be alkylated with ^{3}H-iodoacetate. Protamine dissociability: the percent of protamine that dissociates from the sperm chromatin in 5 M guanidine hydrochloride.

revealed that each protamine 1 molecule in bull sperm participates in the formation of two intramolecular and three intermolecular disulfide crosslinks (Fig. 17.8) [69]. All five cross-links have been identified by titrating the disulfides of intact bull sperm with dithiothreitol (DTT) and following the progressive carboxymethylation of paired cysteine residues with radiolabeled sodium iodoacetate. The derivatization of each cysteine was monitored as a function of DTT concentration by a combination of HPLC peptide mapping and protein sequencing. The least stable bull protamine disulfide is reduced at a DTT/cysteine ratio of 0.3, while the most resistant disulfide requires a DTT/cys ratio in excess of 2.0 (this range is substantially lower than that reported for the disulfides of boar sperm [70]). Intra- and intermolecular disulfides have been identified by correlating the reduction of specific disulfides with the ability of guanidine hydrochloride (GuCl) to dissociate protamine from the DNA in partially reduced sperm and sperm treated with bismaleimidoethane, a bifunctional sulfhydryl cross-linking agent. Two equivalent interprotamine disulfide cross-links that join cys5 to cys22 and cys22 to cys5 (in neighboring molecules) were found to be the first disulfides to be reduced. The intermolecular disulfide linking the cys38 residues of two different molecules was reduced second. The two final disulfides to be reduced were the intramolecular disulfides (one between cys6 and cys14 and one between cys39 and cys47) that lock the folded amino- and carboxy-terminal peptide domains in place.

The normal formation of these disulfides in vivo appears to occur in an ordered fashion. Intraprotamine disulfide formation begins in late-step spermatids and nears completion by the time the sperm leave the testis and enter the caput epididymis. Nearly half of the cysteines of mouse

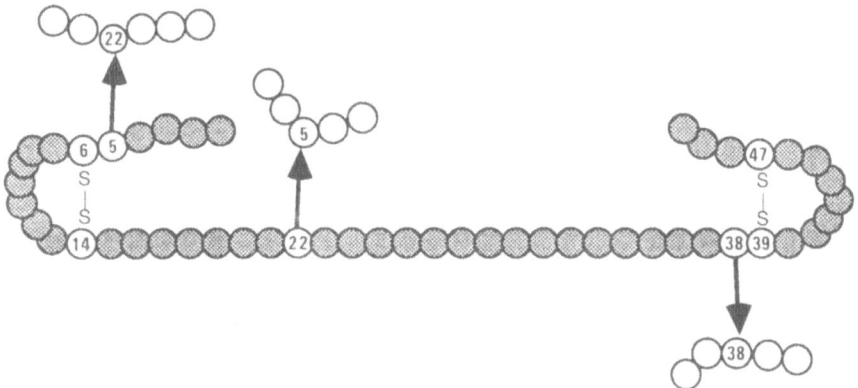

Figure 17.8. Disulfide cross-linking pattern in bull protamine. Schematic shows cysteine residues (numbered) that interact to form intra- and intermolecular disulfides.

protamine are cross-linked as disulfides in caput epididymal sperm (Fig. 17.7). These first disulfides do not, however, prevent the dissociation of protamine from DNA with GuCl, nor do they cross-link neighboring molecules together to form protamine multimers; only protamine mono-mers are observed on electrophoresis of the isolated proteins under nonreducing conditions. As the sperm enter and traverse the corpus epididymis, however, the intermolecular disulfides begin forming, and the protamines become resistant to dissociation from DNA as adjacent molecules are cross-linked around the DNA helix.

Although it is not yet clear how protamine disulfide formation is organized or regulated during this final stage of sperm chromatin "matu-ration," the sequential formation of intra- and intermolecular disulfides observed in vivo certainly suggests that some mechanism must exist for dictating which cysteine residues participate in the formation of disulfides at any particular time and for precluding those cysteines that form disulfides later. The results of other experiments appear to support this hypothesis. The intermolecular disulfides of protamine in reduced bull sperm, for example, will spontaneously reform in vitro at rates that are determined by the type and number of disulfides that are initially reduced. Sperm treated under conditions (DTT/cys = 0.2 to 0.5) that reduce only the intermolecular disulfides will completely reform these disulfides within 120 minutes at pH 8 [69]. On treating the sperm with a DTT concentration sufficient to reduce the first intramolecular disulfide (DTT/cys = 0.7 to 1.0), the disulfides continue to reform, but they do so at a substantially reduced rate. Once the sperm are treated with a DTT concentration sufficiently high to begin reducing the final intraprotamine disulfide (DTT/cys~2), the reformation of intermolecular disulfides that reestablish GuCl insolubility requires hours to complete. Many of the cross-links that are reformed involve cysteines that originally participated in the formation of intramolecular disulfides. In contrast, analyses of disulfide formation in vitro in isolated bull late-step spermatids suggest that some feature of protamine or sperm chromatin structure must prevent the formation of interprotamine disulfides between newly synthe-sized or newly deposited proteins. Under conditions that permit the complete reformation of interprotamine disulfides in reduced caudal sperm, only one third of the late-step spermatid protamines form interpro-tamine disulfides following incubation of isolated spermatids in vitro (unpublished results). It is expected that these disulfides are cross-links that form only between the protamines of the most mature testicular sperm.

While the participation of particular cysteine residues in disulfide formation may be modulated by some feature of protamine or sperm chromatin structure, the actual formation of the disulfide bond in vivo could be driven by air oxidation. Experiments with purified bull pro-tamine have shown that protamine 1 molecules "fold" rapidly in solution [71] and that stable disulfide cross-links are formed within 10 minutes

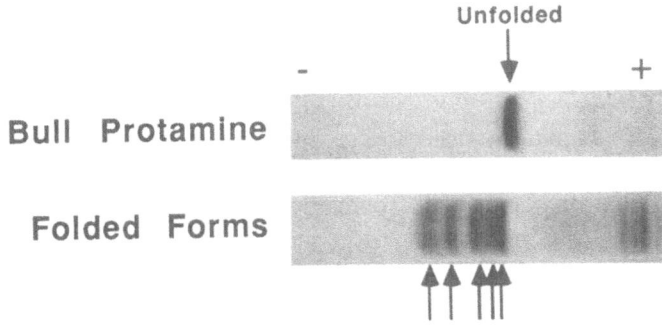

Figure 17.9. Folding of reduced bull protamine in vitro. The five folded forms have been separated by electrophoresis in acid-urea gels are marked by arrows.

after solubilization of the protein (Fig. 17.9). The in vitro formation of these disulfides is catalyzed by air oxidation. Multiple folded forms of protamine are produced, each representing the formation of a different disulfide crosslink (or combination of cross-links). Both the rate of formation of the disulfides and the structure of the protamine molecule also indicate that the folding of the amino- and carboxy-terminal ends of the protamine is not driven by specific amino acid interactions. Folding in vitro would appear to be controlled simply by the kinetics of motion of the two terminal peptide domains. Infrared studies have shown that the protamines of fish adopt an extended conformation in solution [72]. It seems reasonable to assume, based on amino acid sequence similarities between the polyarginine core of protamine 1 and fish protamine and the results of protamine folding experiments using synthetic DNA-protamine complexes, that the central polyarginine DNA binding domain of protamine 1 also exists in an extended conformation both in solution and when bound to DNA. Under such conditions, the only intramolecular disulfides that could be formed are those that link cysteine residues within the amino- or carboxy-terminal half of the molecule. Disulfides should not form between one cysteine residue located in the amino-terminal domain and another located in the carboxy-terminal domain of the same protein; the distance between these two domains is far too great. In practice, this is exactly what we have observed. Folded forms of the size expected from cross-links between the amino- and carboxy-terminal domains are not observed either with partially reduced protamine preparations isolated from sperm chromatin or with in vitro studies of protamine folding in solution and in synthetic DNA-protamine complexes [71].

DNA-Protamine Interactions

The analyses of protamine cross-linking and folding also yield information about DNA-protamine interactions. The in vitro studies of protamine folding and disulfide formation in synthetic DNA-protamine complexes have provided the only direct experimental evidence that the DNA-binding domain of protamine 1 is limited to the central polyarginine region of the protamine molecule [71]. In these experiments, protamine was allowed to fold and form disulfide cross-links in solution, and the resulting products were analyzed by gel electrophoresis. Five different folded forms of bull protamine were resolved by electrophoresis (Fig. 17.9), each representing a different size class of folded, disulfide-cross-linked protamine. The relative size of the various folded forms matched the expected molecular lengths that should be obtained for folded molecules containing different intramolecular disulfide cross-links in either (or both) the amino- or carboxy-terminal domains of the protamine. When the folding experiments were repeated after binding the protamine to DNA in synthetic DNA-protamine complexes, identical folded forms were observed. The results of these experiments, combined with the observation that similar folded forms are extracted from bull sperm following minimal reduction with DTT, demonstrate that the amino- and carboxy-terminal domains of protamine 1 do not participate in binding to DNA. The central polyarginine region of protamine 1 must remain in an extended conformation both in solution and when bound to DNA, and the only intramolecular disulfides that form are those located outside this region of the molecule. Such observations are also consistent with the known structure of other vertebrate protamines. The entire fish protamine molecule, for example, is comparable in size and sequence [73–77] to the central polyarginine "core" of mammalian protamine 1. Fish protamines lack the amino- and carboxy-terminal peptide domains found in mammalian protamines and appear to be little more than polyarginine interspersed with a few carefully positioned nonbasic or phosphorylatable amino acids. Based on purely electrostatic considerations, we would expect that both the polyarginine-rich "core" of mammalian protamines and the entire fish protamine molecule would interact intimately with DNA.

Although it has not been demonstrated unequivocally that the protamines bind in either groove of DNA, X-ray diffraction studies provide the most convincing evidence that the binding site is located in or above the minor groove [72,78–81]. Computer graphics models of polyarginine binding to DNA also support this hypothesis [82]. The phosphate groups of the two phosphodiester chains positioned on either side of the major groove are too far apart to accommodate the binding of polyarginine in an extended conformation. The central polyarginine domain of mammalian protamine binds snugly inside the minor groove, however, and adjacent arginine residues can be appropriately positioned to interact with the phosphates of alternate DNA strands on either side of the groove. Results obtained from DNA-binding experiments with actinomycin D [83] and its fluorescent analog, 7-amino actinomycin D [84], are consistent with the

X-ray data and computer-generated models. On binding a free DNA, the pseudopeptide of actinomycin D is positioned in the minor groove of DNA, and the aminoacridine ring intercalates between base pairs [85,86]. Actinomycin D binds readily to DNA in somatic nuclei, but its binding is markedly diminished in late-step spermatid and epididymal sperm nuclei. This difference would appear to be related to the position of the protamine molecule in the minor groove (blocking the binding of the actinomycin peptide) and the immobilization of the two phosphodiester chains on binding to alternating arginine residues in each protamine [1]. Other studies also suggest that the minor groove of DNA is physically blocked by protamine in sperm. The alkylation of sperm DNA by B[a]P is not observed either in vivo [87] or in vitro [88]. These results indicate that the primary DNA alkylation site for B[a]P, the N-2 position of guanine, either interacts with protamine or is blocked by the protein in the minor groove.

Interprotamine Interactions

Protamine molecules interact not only with DNA, but also with other protamine molecules. Two different types of interprotamine interactions have been observed: interactions among protamine 1 molecules (in the sperm of mammals that use only protamine 1 to package their DNA) and intermolecular interactions between protamine 1 and protamine 2 molecules (in the sperm of mammals that contain both protamines).

The specific intermolecular disulfides that have been identified for the protamines of the bull (a species that uses only protamine 1 to package DNA) are compatible with two very different intermolecular disulfide cross-linking models (Fig. 17.10). One model involves the formation of interstrand disulfide cross-links between adjacent protamine molecules positioned along the same strand of DNA, as shown in Figure 17.10A. In this model, the presence of the cys39–cys39 cross-link between the carboxy-terminal domains of neighboring molecules and the two cys5–cys22 cross-links between the amino-terminal domains of neighboring molecules requires that the protamines be positioned in alternating (amino-carboxyl, carboxyl-amino) orientations along the DNA strand. The other model, which yields identical interprotamine disulfides, is one that involves the formation of disulfide cross-links between protamine molecules on different DNA strands. Interestingly, in this second model, either orientation of protamine molecules is consistent with the identified disulfides. If the protamines bind along the DNA strand in the same orientation (Fig. 17.10B), the observed intermolecular disulfides can only be formed between molecules on neighboring DNA strands oriented in the same direction. If the protamines alternate orientation along the strand (Fig. 17.10C), the correct disulfide cross-links can be achieved with protamines on neighboring DNA strands running in either direction.

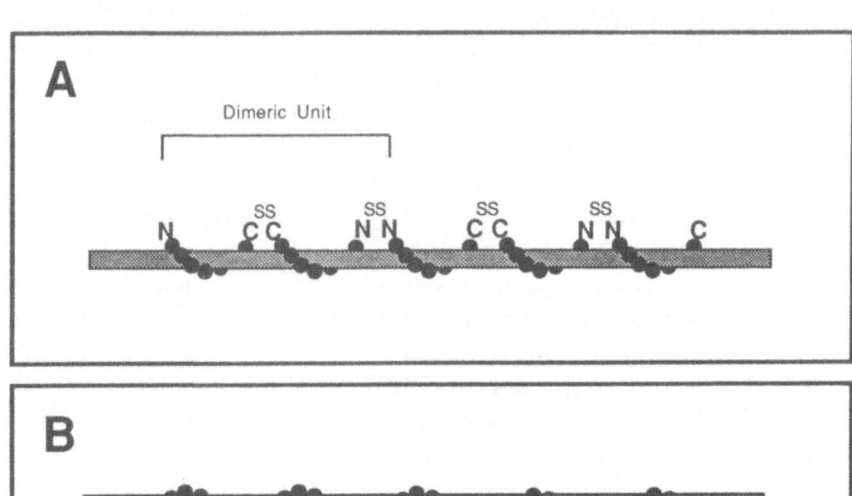

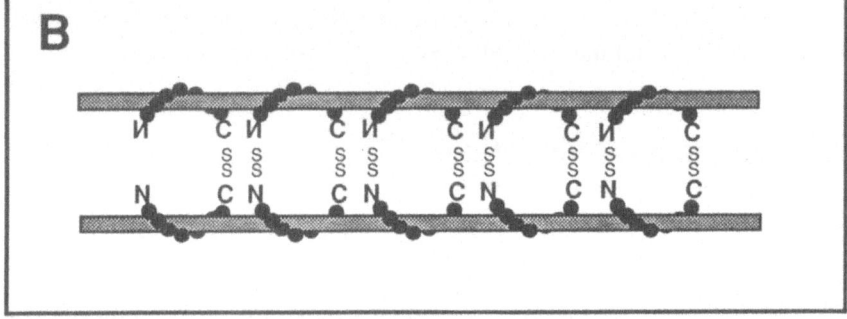

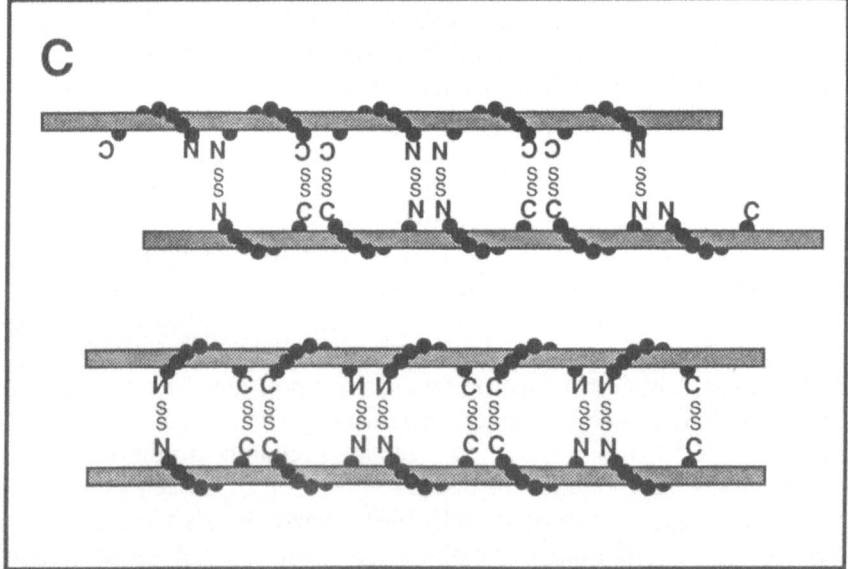

Figure 17.10. Models describing the formation of interprotamine disulfide cross-linking patterns between neighboring bull protamine molecules in sperm chromatin. (A) The orientation of each protamine molecule alternates along the same DNA strand. Cross-links are formed between neighbors along the entire length of ▷

There are no experimental data available, at present, that exclude either model. Both appear feasible from examinations of two-dimensional models. However, unpublished observations of changes in sperm chromatin viscosity as a function of interprotamine disulfide reduction suggest that interstrand cross-links may exist. As the DTT/cys ratio is increased from 0 to 1.0 in experiments monitoring the solubility of bull sperm chromatin in GuCl, the viscosity appears to increase to a maximum and then decrease to a somewhat lower level and plateau. Such changes in viscosity would be consistent with the progressive reduction of interprotamine disulfide cross-links linking neighboring DNA strands. Incomplete reduction of these interprotamine disulfide cross-links would result in the formation of partially cross-linked DNA strands with a larger hydrodynamic volume than fully reduced samples containing only individual, unlinked DNA molecules.

The recent isolation of two electrophoretically distinct protamine 1 dimers from bull sperm [89] provides evidence that pairs of protamine 1 molecules may actually be bound to DNA in alternating orientations as a dimeric unit. The extracted dimer resolves into two distinct protein bands upon electrophoresis in acid-urea gels, and both proteins have been isolated and purified by HPLC. The amino acid compositions and amino-terminal sequences of the two dimers are indistinguishable from protamine 1.

The protamine components of these dimers cannot be linked through disulfide bridges because both forms of the dimer are stable in boiling mercaptoethanol. Treatment of the dimers with 6 N HCl at 22°C for 3 hours, on the other hand, cleaves the linkage and yields only monomeric protamine 1. The susceptibility of the linkage to HCl suggests that the two proteins may be linked through some type of ester linkage. Because methylamine and hydroxylamine do not cleave the link, the two proteins are not linked through a thiol ester bond. Digestions of HPLC-enriched dimer with carboxypeptidase have recently revealed that the two dimers must contain linkages at different positions within the protamine components. One dimer is partially digested by carboxypeptidase, while the other is not. These results also indicate that the linkage in the undigested dimer must be located near the carboxy-terminal end of both protamine components.

Our knowledge of interactions between protamine 1 and protamine 2 in

---

the DNA. (B) Each protamine is oriented in the same direction along the DNA strand, and the intermolecular disulfides are formed between protamines on two different strands running in the same direction. (C). The orientation of each protamine molecule alternates along the same DNA strand. Correct intermolecular cross-links can form between protamines on neighboring DNA strands running in either direction. Individual protamine molecules are schematically represented with only the carboxy- or amino-terminal intermolecular disulfides.

the sperm of species that express the protamine 2 gene (primates and certain rodents) is extremely limited; nonetheless, some information is available. For example, dimers such as those detected in mammals using only protamine 1 to package their DNA have not been detected in the sperm of mammals that use both protamine 1 and protamine 2. It also appears that protamines 1 and 2 are cross-linked to each other in mature sperm through intermolecular disulfide bridges; neither protamine 1 nor protamine 2 can be extracted from sperm chromatin without first reducing these disulfides [63,64]. Whether these cross-links represent direct links between protamine 1 and protamine 2 molecules or links mediated through some other peptide or protein is unknown. Preliminary protein disulfide reduction cross-linking experiments performed with hamster sperm suggest that the interprotamine interaction between hamster protamine 1 and 2 may involve only a single disulfide cross-link. Both hamster and human protamine 1 and 2 also appear to interact readily in solution, forming protamine complexes containing both proteins. These complexes have not been examined in any detail, but their formation has made it particularly difficult to separate and purify the two underivatized human protamines by HPLC [26].

The relative proportions of protamine 1 and protamine 2 molecules present in mature sperm differ widely among species, unlike the histones that package the DNA of somatic cells. The sperm of many species contain only protamine 1, but no species has been found that uses only protamine 2 to package its DNA. The sperm of those mammals that express both protamine genes also contain highly variable amounts of protamine 2. At the two extremes, macaque sperm contain three times as much protamine 2 as protamine 1, while chinchilla sperm contain three times as much protamine 1 as protamine 2. All species examined fall within this range (Table 17.3). These results clearly demonstrate that protamine 1 and protamine 2 do not interact with a particular or universal stoichiometry.

While the expression of the protamine 2 gene varies widely between different mammalian genera, the level of expression appears to be more tightly controlled within a genus. Only minor differences are observed in the protamine 2 contents of closely related species (Table 17.3). The proportion of protamine 1 and protamine 2 in the hamsters *Mesocricetus auratus* and *M. brandti*, the macaques *Macaca mullata* and *M. fascicularis*, and several species of closely related mice (*Mus musculus, M. spretus, M. hortulanus*) were found to differ only slightly. Significant variability was observed, however, in more distantly related species, such as *Mus musculus, M. saxicola,* and *M. minutoides.*

These results suggest that some mechanism must exist, at least within a species, to control the production of the two different protamines and coordinate their deposition onto DNA. Recent studies by Peschon et al. [90] have suggested that this control may operate at the level of the gene

Table 17.3. Protamine 2 contents of sperm isolated from various eutherian mammals (percent of total protamine).

Chinchilla (*Chinchilla laniger*)	25.0 ± 2.2
Nutria(*Myocastor coypus*)	30.0 ± 1.1
Cotton-top tamarin (*Saguinus oedipus*)	30.4 ± 1.6
Vole (*Microtus pennsylvanicus*)	31.7 ± 3.6
Syrian hamster (*Mesocricetus auratus*)	34.0 ± 0.7
Turkish hamster (*Mesocricetus brandti*)	36.5 ± 0.9
Marmoset (*Callithrix jacchus*)	38.8 ± 1.4
Human (*Homo sapiens*)	43.1 ± 3.3
Dwarf hamster (*Phodopus sungorus*)	45.8 ± 2.1
Egyptian spiny mouse (*Acomys cahirinus*)	46.3 ± 0.4
Spiny pocket mouse (*Perognathus* sp.)	50.5
Mouse (*Mus minutoides*)	57.9 ± 1.9
White-footed deermouse (*Peromyscus leucopus*)	58.1 ± 1.2
Mouse (*Mus saxicola*)	59.8 ± 2.4
Cotton rat (*Sigmodon hispidus*)	60.9 ± 1.9
African green monkey (*Cercopithecus aethiops*)	61.0 ± 9.9
Deermouse (*Peromyscus maniculatus*)	63.8 ± 1.6
Mouse (*Mus caroli*)	64.5 ± 1.2
Mouse (*Mus musculus*)	67.0 ± 0.9
Mouse (*Mus spretus*)	68.1 ± 1.6
Mouse (*Mus hortulanus*)	68.3 ± 1.4
Mouse (*Mus pahari*)	69.9 ± 1.8
Mouse (*Mus cookii*)	71.0 ± 1.4
Rhesus monkey (*Macaca mulatta*)	72.3 ± 2.0
Cynomolgous monkey (*Macaca fascicularis*)	77.4 ± 5.6

and not at the level of protamine deposition. Two transgenic mouse lines containing different numbers of protamine 1 gene copies [91] were analyzed for protamine 1 and protamine 2 in late-step spermatids and caudal sperm. The analyses showed the protamine 1 content of late-step spermatids isolated from the transgenic lines 1688 and 2011 to be 1.5 to twentyfold higher, respectively, than in the wild-type control mice. As the sperm of these animals passed through the epididymis and matured, the proportion of protamine 1 in the sperm decreased, but it never returned to normal. While excess protamine 1 remained tightly attached to the chromatin in the sperm and spermatids of these two transgenic lines, the nature of the interaction between the protein synthesized by the inserted genes and the proteins expressed by the endogenous genes could not be ascertained.

Function and Significance of Protamine 2

Perhaps the most surprising result of the studies of protamine 2 gene expression in various genera and species of mammals has been the extreme variability in protamine 2 contents of their sperm chromatin. We had naively imagined, in the beginning, that the two protamines might interact with a particular stoichiometry that could be exploited to help characterize the interactions that occur between these two proteins in sperm chromatin. This has not been observed. In addition, the difference in protamine 2 contents of mouse and Syrian hamster sperm have been particularly difficult to reconcile. The sperm of the mouse contains twice as much protamine 2 as protamine 1, while the sperm of this hamster contains twice as much protamine 1 as protamine 2. And yet the protamine 1 sequences of these two species differ by only three amino acids, all conservative changes that would not be expected to make a difference in how the protein might interact with protamine 2 or with DNA. Similarly, only subtle differences have been detected in the primary sequences of mouse and hamster protamine 2.

These findings would appear to suggest that protamines 1 and 2 are used interchangeably in sperm chromatin. While certain features of protamine 1 and 2 structure are similar, such as the relative location of the cysteine residues within the molecules (Fig. 17.11), others, such as the unusually high histidine content of protamine 2 and the histidine deficiency of protamine 1, suggest that the two protamines may interact differently with other proteins or ligands. Histidine and cysteine are routinely used by enzymes and other proteins to coordinate metals such as zinc [92–97]. Reports by Kvist et al. [98–101] have shown that zinc is essential for the maintenance of normal human sperm structure and function. While the peptide sequences containing histidine in protamine 2 do not resemble the zinc fingers [94–97] identified in other proteins, the proximity of multiple histidine and cysteine residues could easily result in metal chelation.

Other studies suggest that the presence of protamine 2 alters the way the protamines interact with each other and with DNA in sperm chromatin. Comparisons of sperm decondensation rates and pronucleus formation in hamster eggs microinjected with heterologous sperm have demonstrated that the presence of protamine 2 in sperm chromatin is essential for pronucleus formation in the hamster egg [102]. Sperm containing both protamine 1 and 2 (hamster, human, mouse, vole, and chinchilla) decondensed rapidly after microinjection into hamster oocytes, well within the 60-minute window required for pronucleus formation. Bull and rat sperm, in contrast, showed little evidence of decondensation during this period and rarely formed pronuclei. Additional experiments with late-step spermatids and reduced bull and rat sperm revealed that the disulfides of bull and rat sperm could not be reduced properly by the oocyte. The presence of protamine 2 in sperm chromatin appeared to alter the interprotamine interactions, rendering the intermolecular disulfides more susceptible to reduction by the oocyte.

Biochemical studies of sperm chromatin in normal and infertile human males have provided additional evidence that protamine 2 may play a special role in sperm chromatin organization and fertility in humans [103].

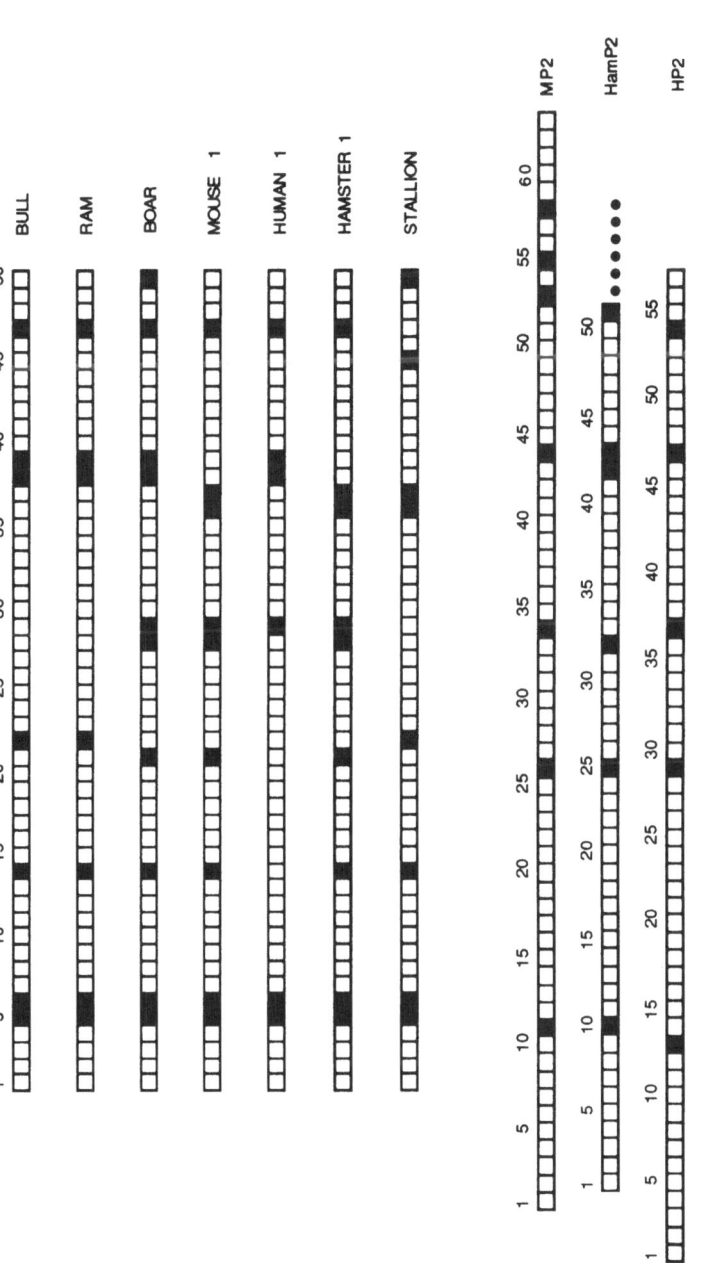

Figure 17.11. Location of cysteine residues in mammalian protamines. Cysteine resides are denoted as filled blocks.

Quantitative determinations of the protamine 1, 2a, and 2b contents of sperm obtained from fertile human males have shown that the proportion of the three protamines does not vary significantly between different individuals. The distribution of protamines in sperm obtained from infertile males producing an elevated level of large sperm heads, in contrast, was found to be dramatically different from the distribution observed for the fertile males. In every case but one, the proportion of protamine 2 was markedly reduced. While the link between this observed protamine 2 deficiency and a particular type of infertility may be tenuous, the similarities of the human and hamster results are intriguing. The biochemical studies suggest that the expression of the protamine 1 and protamine 2 genes is uncoupled in the developing spermatids of certain human males and that this uncoupling correlates with infertility. The hamster microinjection studies show that sperm lacking protamine 2 do not decondense properly once inside the hamster oocyte.

The most convincing evidence that protamine 2 must perform some unique function in sperm chromatin, however, involves its synthesis and metabolism. Protamine 2 is synthesized as a precursor that is deposited onto DNA and subsequently processed by a series of selective proteolytic cleavages. Protamine 1 is not. The protamine 2 precursor binds to DNA while still intact, and the subsequent processing steps occur sequentially over a relatively long period of time (24 to 30 hours). All intermediate products are sufficiently long-lived that they appear as prominent components of the basic chromatin proteins of late-step spermatids.

None of these observations, unfortunately, provide conclusive evidence that protamine 1 and protamine 2 are either interchangeable or unique. A definitive statement on this matter must await the results of additional studies.

Sperm Chromatin Organization and Evolution in Mammals

The observation that expression of the protamine 2 gene is conserved within a genus presents the possibility of using protamine expression as a method to evaluate mammalian evolution. Comparisons of protamine 1 and 2 mutation rates by Bellve et al. [29] have also indicated that the primary sequences of these two proteins appear to be evolving much more rapidly than any other protein previously studied. While analyses of protamine 2 expression may facilitate the phylogenetic placement of certain species, the rapid rate of mutation observed in the protamine 1 and 2 genes should also prove useful in identifying essential structural domains within the two protamine molecules.

Marsupials, for example, are reported to have diverged from eutherian mammals nearly 100 million years ago. Recent electrophoretic and HPLC analyses of the basic nuclear proteins of the ring-tailed wallaby (*Petrogale xanthopus*) and a South American opossum (*Monodelphis domestica*) have revealed that marsupials also use a single type of protamine to package DNA. The amino termini of these two protamines (Table 17.4) begin with the sequence ala–arg–tyr, suggesting that both proteins are protamine 1 molecules. The most unusual feature of wallaby and opossum protamine is that they lack cysteine residues. As in fish, the absence of disulfide cross-links between marsupial protamines is apparent; marsupial sperm readily dissolve in GuCl, and the protamines can be dissociated from DNA with GuCl or acids without prior treatment with reducing agents.

A comparison of partial amino acid sequences of wallaby and opossum protamine (Table 17.4) suggests that the sequence and structure of the protamines of Australian marsupials within the family Macropodoidea are more closely related to those of eutherian mammals than are the protamines of *Monodelphis*. Opossum protamine contains a slightly larger proportion of arginine residues and a ser–ser–ser tripeptide sequence found in several fish protamines [75,77]. This observation is consistent with our current knowledge of marsupial evolution [104]; *Petrogale xanthopus* represents one of the more recently derived groups of marsupials, and *Monodelphis domestica* belongs to the oldest family of marsupials to survive to the present day, the family Didelphoidea.

Comparisons of protamine sequences may also provide information that change the interpretation of the time that the mammalian order

Table 17.4. Amino acid composition and amino-terminal sequences of opossum, wallaby, and elephant protamines.

Amino acid	Opossum (*monodelphis domestica*)	Ring-tailed wallaby (*petrogale xanthopus*)	Elephant (*elephas maximus*)
Arginine	70.2	67.1	69.6
Lysine	0	0	0
Histidine	3.6	2.9	2.2
Phenylalanine	0	0	0
Tyrosine	3.6	8.3	5.7
Leucine	0	0.4	2.1
Isoleucine	0	0	0
Valine	0	0.4	0
Alanine	1.8	1.8	5.7
Glycine	7.7	6.0	3.9
Proline	1.5	0	0
Glutamine/glutamic acid	0	0.6	0
Asparagine/aspartic acid	0	0	0
Serine	11.7	12.6	8.7
Threonine	0	0	1.6
Cysteine	0	0	0.5

Opossum: ARYRRRSRSRSRSRYGRRRRRSSS . . .
Wallaby: ARYRHSRSRSRSRYRRRRRRRSRYRSRRRRY . . .

Proboscoidea diverged from other mammals. Recent primary sequence analyses of lens crystallins suggest that the two remaining members of this order, the Asian elephant *Elephas maximus* and the African elephant *Loxodonta africana,* represent a group of mammals that diverged from other mammals as early as 75 to 80 million years ago [105]. Based on morphological criteria and studies of various blood proteins, elephants were previously classified as being closely related to the artiodactyls, a group that diverged from other mammals only 30 to 50 million years ago. The protamines of one of these species, the Asian elephant, have been isolated and partially characterized. The amino acid composition of one of these proteins, shown in Table 17.4, is similar to the protamines of other eutherian mammals. The arginine content is high (66%), and all the amino acids normally found in the protamines of other mammals are present in the elephant protein, including a small amount of cysteine. While the amino-terminal sequence indicates that this protein is definitely protamine 1, preliminary analyses of the two other major sperm chromatin proteins suggest that the elephants may not be as closely related to artiodactyls as others previously thought.

An enigma that has continued to plague comparative studies of sperm chromatin structure among eutherian mammals is the variability in protamine composition of chromatins isolated from the sperm of different species. While the sperm of every mammal examined to date has been shown to contain protamine 1, only selected species contain protamine 2. Using cDNA probes for the two mouse protamine genes, Johnson et al. [106] have determined that DNA sequences homologous to both the protamine 1 and protamine 2 genes appear to be present in all mammals. Our own studies, however, have indicated that the protamine 2 gene is expressed at a detectable level in only a very specific subset of these species. To identify the relationship between species evolution and expression of the protamine 2 gene, the protamine content of a large number of mammalian species has been examined for protamine 1 and protamine 2, and the proteins have been identified by gel electrophoresis, HPLC, amino acid analysis, and amino-terminal sequencing. The latest results of this survey, which is still incomplete, are shown in Figure 17.12.

Perhaps the most obvious finding of this study is that the expression of the protamine 2 gene appears to relate directly to the evolution of mammalian species. Using protein [107–113] and DNA sequence mutations [114,115], other investigators have determined that primates and other eutherian mammals radiated and began evolving along separate lines approximately 70 million years ago. Our analyses of protamine 2 expression show that the presence and absence of the second protamine in sperm chromatin correlate well with the phylogenetic maps generated from studies of myoglobin [107], hemoglobin [108], fibrinogen peptides [109], cytochrome c, and various other protein [110–113] and DNA [114,115] sequences. The sperm of lagomorphs, carnivores, artiodactyles,

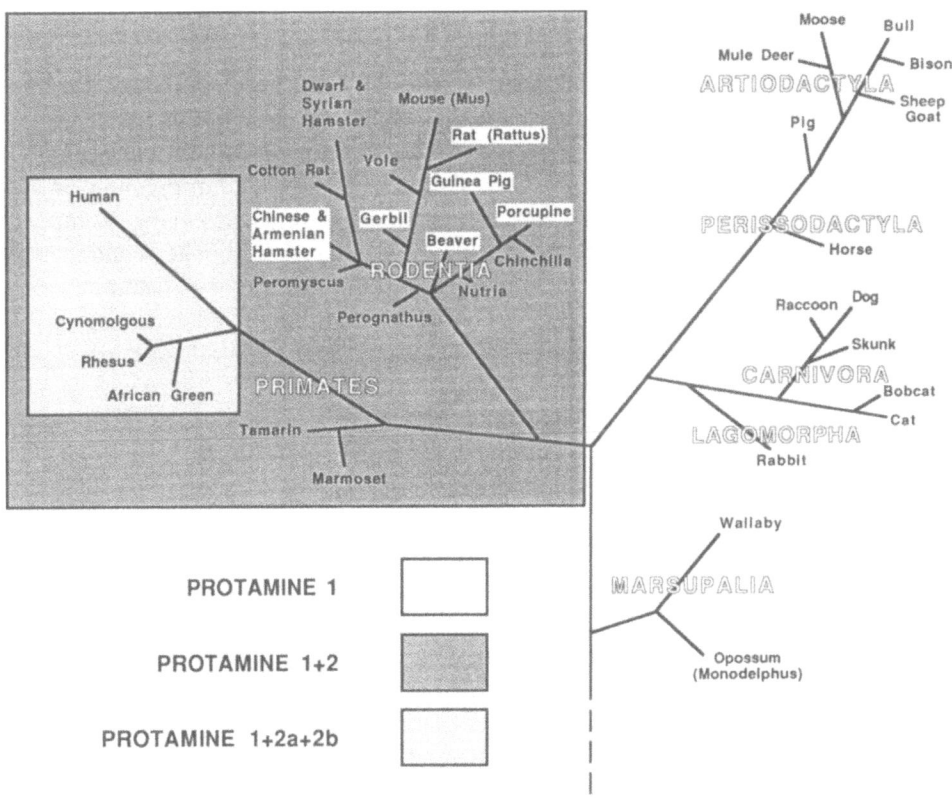

Figure 17.12. Phylogenetic map showing the position of those mammals that express both the protamine 1 and 2 genes and those that express only the protamine 1 gene. The species that contain protamine 2b in their sperm are also shown. The sperm of the Chinese and Armenian hamsters, gerbil, beaver, porcupine, guinea pig, and rat (*Rattus norvegicus*) do not contain sufficient protamine 2 to be detected by HPLC or gel electrophoresis.

and perissodactyles (Fig. 17.12) contain only protamine 1. The sperm of primates and rodents contain both protamines 1 and 2. The observation that the sperm of many rodents contain protamine 2 while others do not may indicate that the current phylogenetic map that has been constructed for the order Rodentia is incorrect. An alternative and more attractive possibility is that the protamine 2 gene is expressed in all rodents but the protamine 2 precursor is not processed to protamine 2 in certain species. The recent observation that a protamine-2-specific monoclonal antibody cross-reacts with a protein found in rat (*Rattus norvegicus*) and gerbil protamine preparations would appear to support the latter possibility. ELISA analyses performed on HPLC-fractionated rat sperm basic proteins using this antibody (unpublished results) have identified a protein that displays an HPLC retention time and electrophoretic mobility similar to the mouse protamine 2 precursor.

Although only a limited number of species have been examined, the current pattern also suggests an explanation for the observed presence of protamine 2b in human sperm and its absence in other mammals expressing the protamine 2 gene. As shown in Figure 17.12, protamine 2b has been detected only in the sperm of primates. Its presence in the sperm of man and the rhesus, cynomolgous, and African green monkeys and its absence in the marmoset and tamarin suggest that the use of protamine 2b to package DNA in sperm evolved very late within the order Primates. The specific branch point appears to occur near the radiation point between Old and New World monkeys [116,117].

These collective results suggest that what we once perceived as random differences in the protamine contents of mammalian sperm may really be indicators of sperm chromatin evolution. The earliest mammals and their contemporary descendants, the marsupials, may have begun packaging their sperm DNA using a primitive protamine 1 molecule that could not, like the protamines of its predecessors, form disulfide cross-links. By the time the Proboscoidea diverged from the main line, the arginine content of the protamine molecule had increased, and cysteine residues appeared. About 70 million years ago, the packaging of sperm DNA in the collective species we now look on as eutherian mammals appears to have changed near the point of radiation of Rodentia and Primates, and mammals began evolving along two separate lines. The protamine molecules of eutherian mammals contained numerous cysteine residues, and the proteins were cross-linked around DNA by disulfide bridges. One line continued to organize its sperm chromatin using a single protamine, protamine 1, while the other evoked a mechanism that required the presence of two different proteins, protamine 1 and protamine 2. At a some later point in time, perhaps 30 to 40 million years ago, an additional twist in chromatin packaging appeared within the order Primates. As the Old World monkeys diverged from other primates, they began using a third protamine, the variant protamine 2b, to organize sperm DNA while the New World primates continued to use only protamines 1 and 2.

While these observations appear to help clarify the differences we have observed among mammals, their implications with respect to the evolution of the protamine genes are still puzzling. The DNA hybridization studies of Johnson et al. [106] have suggested that the protamine 2 gene may be present in all mammalian species, including marsupials. Our analyses of the sperm chromatin proteins, on the other hand, indicate that neither marsupials nor those species evolving along the right branch (Fig. 17.12) of contemporary mammals produce protamine 2 in sufficient quantity to be detected by gel electrophoresis or HPLC. This apparent discrepancy could be explained in several ways. One possibility is that the protamine 2 gene may be present in all species but that either it is not expressed or its expression occurs at such a low level that the protamine 2 gene product cannot be detected by the techniques we have applied. Our HPLC analyses of bull and rat protamine indicate that protamine 2 would be missed if it were present in sperm chromatin at a level of 0.5% or less of the total protamine. The rat (*Rattus norvegicus*) genome clearly contains the protamine 2 gene, since it has been isolated and sequenced [118]. Analyses of the messenger RNAs also indicate that the rat gene may be expressed at 10% to 20% of the level of the mouse gene and that at least some of the RNA binds to polyribosomes [119,120]. Whether these RNAs are translated or not is unknown. If they are translated, the protamine 2 precursor that is produced may not be processed to protamine 2. An alternative possibility, of course, is that the protamine 2 gene is not present in all species but that other sufficiently homologous DNA sequences are responsible for the observed hybridization signal. This explanation would seem less likely because it would require the appearance of a completely new gene set near the radiation point of rodents and primates, the protamine 2 gene and a complement of accessory genes required to produce the various protamine 2 precursor processing enzymes. While the lateral transmission of a new gene set is conceivable, it would appear to be the less probable of the two possibilities.

Acknowledgments. Most of the new data presented in this review could not have been obtained without the expertise and efforts provided by Shelley Corzett, Joe Mazrimas, Christine Campos, Mishelle Ashmore, Terri Lee, and Lynda Vicars. I thank them for their input and enumerable discussions. I would also like to thank Norman Hecht, Paula Johnson, Larry Stanker, Susan Tanhauser, Andy Wyrobek, and Pam Yelick for the many insightful hours of discussion and speculation that helped inspire and direct this research. This work was performed under the auspices of the U.S. Department of Energy by Lawrence Livermore National Laboratory under contract W-7405-Eng-48 and an IR&D grant provided by the Lawrence Livermore National Laboratory.

References

1. Balhorn, R. (1982) J. Cell Biol. *93*, 298–305.
2. Pogany, G.C., Corzett, M., Weston, S., and Balhorn, R. (1981) Exp. Cell Res. *136*, 127–136.
3. Sega, G.A. (1974) Proc. Natl. Acad. Sci. USA *71*, 4955–4959.
4. Chandley, A.C., and Kofman-Alfaro, S. (1971) Exp. Cell Res. *69*, 45–48.
5. Kierszenbaum, A.L., and Tres, L. L. (1975) J. Cell Biol. *65*, 258–270.
6. Dixon, G.H. (1989) In Histones and Other Basic Nuclear Proteins, Stein, G., Hnilica, L.S., and Stein, J. (eds.). CRC Press, Boca Raton, FL (in press).

7. Risley, M.S. (1989) In Chromosomes: Eukaryotic, Prokaryotic, and Viral, Adolph, K. (ed.). CRC Press, Boca Raton, FL (in press).

8. Hecht, N.B. (1989) In Histones and Other Basic Nuclear Proteins, Stein, G., Hnilica, L.S., and Stein, J. (eds.). CRC Press, Boca Raton, FL (in press).

9. Meistrich, M.L. (1989) In Histones and Other Basic Nuclear Proteins, Stein, G., Hnilica, L.S., and Stein, J. (eds.). CRC Press, Boca Raton, FL (in press).

10. Hecht, N.B. (1989) In Molecular Biology of Chromosome Function, Adloph, K. (ed). Springer-Verlag, New York (in press).

11. Coelingh, J.P., Monfoort, C.H., Rozijn, T.H., Geversleuven, A., Schiphof, R., Steyne-Parve, E.P., Braunitzer, G., Schrank, B., and Ruhfus, A. (1972) Biochim. Biophys. Acta *285*, 1–14.

12. Mazrimas, J.A., Corzett, M., Campos, C., and Balhorn, R. (1986) Biochim. Biophys. Acta *872*, 11–15.

13. Loir, M., and Lanneau, M. (1980) Biochem. Biophys. Res. Commun. *80*, 975–982.

14. Sautiere, P., Belaiche, D., Martinage, A., and Loir, M. (1984) Eur. J. Biochem. *44*, 121–125.

15. Tobita, T., Nomoto, M., Nakano, M., and Ando, T. (1982) Biochim. Biophys. Acta *707*, 252–258.

16. Tobita, T., Tsutsumi, H., Kato, A., Suzuki, H., Nomoto, M., Nakano, M., and Ando, T. (1983) Biochim. Biophys. Acta *744*, 141–146.

17. Kistler, W.S., Keim, P.S., and Heinrikson, R.L. (1976) Biochim. Biophys. Acta *427*, 752–757.

18. Calvin, H. (1976) Biochim. Biophys. Acta *434*, 377–389.

19. Balhorn, R., Gledhill, B.L., and Wyrobek, A.J. (1977) Biochemistry *16*, 4074–4080.

20. Bellve, A.R., and Carraway, R. (1978) J. Cell Biol. *79*, 177a.

21. Corzett, M., Blacher, R., Mazrimas, J.A., and Balhorn, R. (1987) J. Cell Biol. *105*, 152a.

22. Belaiche, D., Loir, M., Kruggle, W., and Sautiere, P. (1987) Biochim. Biophys. Acta *913*, 145–149.

23. Ammer, H., and Henschen, A. (1987) Biol. Chem. Hoppe Seyler *368*, 1619–1626.

24. Pongsawasdi, P., and Svasti, J. (1976) Biochim. Biophys. Acta *434*, 462–473.

25. Tanphaichitr, N., Sobhon, P., Taluppeth, N., and Chalermisarachai, P. (1978) Exp. Cell Res. *117*, 347–356.

26. Balhorn, R., Corzett, M., Mazrimas, J.A., Stanker, L.H., and Wyrobek, A. (1987) Biotechnol. Appl. Biochem. *9*, 82–88.

27. Kleene, L.C., Distel, R.J., and Hecht, N.B. (1985) Biochemistry *24*, 719–722.

28. Yelick, P.C., Balhorn, R., Johnson, P.A., Corzett, M., Mazrimas, J.A., Kleene, K.C., and Hecht, N.B. (1987) Mol. Cell. Biol. *7*, 2173–2179.

29. Bellve, A.R., McKay, D.J., Renaux, B.S., and Dixon, G.H. (1988) Biochemistry *27*, 2890–2897.

30. McKay, D.J., Renaux, B.S., and Dixon, G.H. (1985) Biosci. Rep. *5*, 383–391.

31. McKay, D.J., Renaux, B.S., and Dixon, G.H. (1986) Eur. J. Biochem. *156*, 5–8.

32. Ammer, H., Henschen, A., and Lee, C.-H. (1986) Biol. Chem. Hoppe-Seyler *367*, 515–522.

33. Gusse, M., Sautiere, P., Belaiche, D., Martinage, A., Roux, C., Dadoune, J.-P., and Chevaillier, P. (1986) Biochim. Biophys. Acta 884, 124–134.
34. Johnson, P.A., Peschon, J.J., Yelick, P.C., Palmiter, R.D., and Hecht, N.B. (1988) Biochim. Biophys. Acta 950, 45–53.
35. Sautiere, P., Belaiche, D., Martinage, A., and Loir, M. (1984) Eur. J. Biochem. 144, 121–125.
36. Krawetz, S.A., Connor, W., and Dixon, G.H. (1987) DNA 6, 47–57.
37. Lee, C.-H., Mansouri, A., Hecht, W., Hecht, N.B., and Engel. W. (1987) Biol. Chem. Hoppe-Seyler 368, 131–135.
38. Lee, C.-H., Hoyer-Fender, S., and Engel, W. (1987) Nucleic Acids Res. 15, 7639.
39. Tobita, T., Nomoto, M., Nakano, M., and Ando, T. (1982) Biochim. Biophys. Acta 707, 252–258.
40. Luerssen, H., Hoyer-Fender, S., and Engel, W. (1988) Nucleic Acids Res. 16, 7723.
41. Sautiere, P., Martinage, A., Beliache, D., Arkhis, A., and Chevaillier, P. (1988) J. Biol. Chem. 263, 11059–11062.
42. Wyrobek, A.J., Stanker, L.H., Bigbee, W.L., and Balhorn, R. (1989) (submitted).
43. Wyrobek, A.J., Gordon, L., Watchmaker, G., and Moore, D., II (1983) Hum. Genet. 35, 123a.
44. Soderstrom, K.-O. (1985) Hereditas 102, 63–69.
45. Bellve, A.R., Anderson, E., and Hanley-Bowdoin, L. (1975) Dev. Biol. 47, 349–365.
46. Mayer, J.F., Chang, T.S.K., and Zirkin, B.R. (1981) Biol. Reprod. 25, 1041–1051.
47. Lam, D.M.K., and Bruce, W.R. (1971) J. Cell. Physiol. 78, 13–24.
48. Platz, R.D., Grimes, S.R., Meistrich, M.L., and Hnilica, L.S. (1975) J. Biol. Chem. 250, 5791–5800.
49. Grimes, S.R. Jr., Meistrich, M.L., Platz, R.D., and Hnilica, L.S. (1977) Exp. Cell Res. 110, 31–39.
50. Evenson, D.P., Darzynkiewicz, Z., and Melamed, M.R. (1980) Chromosoma 78, 225–238.
51. Lung, B. (1972) J. Cell Biol. 52, 179–186.
52. Evenson, D.P., Witkin, S.S., De Harven, E., and Bendich, A. (1978) J. Ultrastruct. Res. 63, 178–187.
53. Tanphaichitr, N., Sobhon, P., Chalermisarachai, P., and Patilantakarnkool, M. (1981) Gamete Res. 4, 297–315.
54. Loir, M., and Courtens, J.L. (1979) J. Ultrastruct. Res. 67, 309–324.
55. Loir, M., Bouvier, D., Fornells, M., Lanneau, M., and Subirana, J.A. (1985) Chromosoma 92, 304–312.
56. Young, R.J., and Sweeney, K. (1979) Gamete Res. 2, 265–283.
57. Kierszenbaum, A.L., and Tres, L.L. (1975) J. Cell Biol. 65, 258–270.
58. Green G.R., Balhorn, R., Yelick, P.C., and Hecht, N.B. (1987) J. Cell Biol. 105, 167a.
59. Marushige, Y., and Marushige, K. (1975) J. Biol. Chem. 250, 39–45.
60. Balhorn, R., Weston, S., Thomas, C., and Wyrobek, A. (1984) Exp. Cell Res. 150, 298–308.
61. Balhorn, R., Corzett, M., and Mazrimas, J.A. (1989) (submitted).

62. Pruslin, F.H., Imesch, E., Winston, R., and Rodman, T.C. (1987) Gamete Res. *18*, 179–190.

63. Marushige, Y., and Marushige, K. (1974) Biochim. Biophys. Acta *340*, 498–508.

64. Bedford, J.M., and Calvin, H.I. (1974) J. Exp. Zool. *188*, 137–156.

65. Saowaros, W., and Panyim, S. (1979) Experientia *35*, 191–192.

66. Calvin, H.I., Yu, C.C., and Bedford, J.M. (1973) Exp. Cell Res. *81*, 333–341.

67. Calvin, H.I., and Bedford, J.M. (1971) J. Reprod. Fertil. (Suppl.) *13*, 65–75.

68. Tobita, T., Tanaka, H., Tanaka, K., Tanaka, T., Kojima, S., and Nakano, M. (1984) Biochem. Int. *9*, 161–168.

69. Balhorn, R., Corzett, M., and Mazrimas, J.A. (1987) J. Cell Biol. *105*, 151a.

70. Tobita, T., Suzuki, H., Soma, K., and Nakano, M. (1983) Biochim. Biophys. Acta *748*, 461–464.

71. Balhorn, R., Corzett, M., and Mazrimas, J.A. (1989) (submitted).

72. Herskovits, T.T., and Brahms, J. (1976) Biopolymers *15*, 687–706.

73. Ando, T., and Suzuki, K. (1966) Biochim. Biophys. Acta *121*, 427–429.

74. Ando, T., and Suzuki, K. (1967) Biochim. Biophys. Acta *140*, 375–376.

75. Ando, T., and Watanabe, S. (1969) J. Protein Res. *1*, 221–224.

76. Iwai, K., Nakahara, C., and Ando, T. (1971) J. Biochem. (Tokyo) *69*, 493–509.

77. McKay, D.J., Renaux, B.S., and Dixon, G.H. (1986) Eur. J. Biochem. *158*, 361–366.

78. Feughelman, M., Langridge, R., Seeds, W.E., Stokes, A.R., Wilson, H.R., Hooper, C.W., Wilkins, M.H.F., Barclay, R.K., and Hamilton, L.D. (1955) Nature *175*, 834–838.

79. Wilkins, M.F.H. (1956) Cold Spring Harbor Symp. Quant. Biol. *21*, 75–90.

80. Suwalsky, M., and Traub, W. (1972) Biopolymers *11*, 2223–2231.

81. Suau, P., and Subirana, J.A. (1977) J. Mol. Biol. *117*, 909–926.

82. Max, N., Gwilliam, W., and Balhorn, R. (Color Film, 1988) Lawrence Livermore National Laboratory.

83. Darynkiewicz, Z., Gledhill, B.L. and Ringertz, N.R. (1969) Exp. Cell Res. *58*, 435–438.

84. Balhorn, R., Kellaris, K., Corzett, M., and Clancy, C. (1985) Gamete Res. *12*, 411–422.

85. Sobell, H.M. (1973) Prog. Nucl. Acid Res. *13*, 153–190.

86. Sobell, H.M., and Jain, S.C. (1972) J. Mol. Biol. *68*, 21–34.

87. Balhorn, R., Mazrimas, J.A., and Corzett, M. (1985) Environ. Health Perspect. *62*, 73–79.

88. Vicars, L., and Balhorn, R. (unpublished data).

89. Balhorn, R., Ashmore, M., Corzett, M., and Mazrimas J.A. (1989) (submitted).

90. Peschon, J.J. (1988) Thesis, Department of Biochemistry, University of Washington, Seattle.

91. Peschon, J.J., Behringer, R.R., Brinster, R.L., and Palmiter, R. D. (1987) Proc. Natl. Acad. Sci. USA *84*, 5316–5319.

92. Eklund, H., Nordstrom, B., Zeppezauer, E., Soderlund, G., Ohlsson, I., Boiwe, T., Soderberg, B.-O., Tapia, O., Branden, C.-I., and Akeson, A. (1976) J. Mol. Biol. *102*, 27–59.

93. Garbett, K., Partridge, G.W., and Williams, R.J.P. (1972) Bioinorg. Chem. *1*, 309–329.

94. Klug, A., and Rhodes, D. (1987) Trends Biochem. Sci. *12*, 464–469.
95. Miller, J., McLachlan, A.D., and Klug, A. (1985) EMBO J. *4*, 1609–1614.
96. Berg, J.M. (1986) Science *232*, 485–487.
97. Vincent, A. (1986) Nucleic Acids Res. *14*, 4385–4391.
98. Bjorndahl, L., and Kvist, U. (1985) Acta Physiol. Scand. *124*, 189–194.
99. Kvist, U., and Bjorndahl, L. (1985) Acta Physiol. Scand. *124*, 195–200.
100. Kvist, U., Bjorndahl, L., Roomans, G.M., and Lindholmer, C. (1985) Acta Physiol. Scand. *125*, 297–303.
101. Kvist, U., Bjorndahl, L., Kjellberg, S., Lindholmer, C., and Roomans G.M. (1985) Acta Physiol. Scand. *124* (Suppl. 542), 389.
102. Perreault, S.D., Barbee, R.R., Elstein, K.H., Zucker, R.M., and Keefer C.L. (1988) Biol Reprod. *39*, 157–167.
103. Balhorn, R., Reed, S., and Tanphaichitr, N. (1988) Experientia *44*, 52–55.
104. Austad, S.N. (1988) Sci. Am. *258*, 98–104.
105. Goodman, M., Czelusniak, J., and Beeber, J.E. (1985) Cladistics *1*, 171–185.
106. Johnson, P.A., Yelick, P.C., Liem, H., and Hecht, N.B. (1988) Gamete Res. *19*, 169–175.
107. Dene, H., Goodman, M., Walz, D.A., and Romero-Herrera, A.F. (1983) Hoppe-Seyler's Z. Physiol. Chem. *364*, 1585–1595.
108. Liu, H.-X., Kleinschmidt, T., Braunitzer, G., and Goltenboth, R. (1988) Biol. Chem. Hoppe-Seyler *369*, 209–216.
109. Nakamura, S., Takenaka, O., and Takahashi, K. (1983) J. Biochem. *94*, 1975–1978.
110. Dayhoff, M.O. (1972) Atlas of Protein Sequence and Structure, Vol. 5., National Biomedical Research Foundation, Georgetown University Medical Center, Washington, D.C.
111. DeJong, W.W., and Goodman, M. (1982) Zeit. Saugetierkunda *47*, 257–276.
112. Florkin, M., and Bricteux-Gregoire, S. (9172) In Molecular Evolution, Prebiological and Biological, 319–329, Rohlfing, D.L., and Oparin, A.I. (eds.). Plenum Press, New York.
113. Bonhomme, F., Iskandar, D., Thaler, L., and Petter, F. (1985) In Evolutionary Relationships Among Rodents. A Multidisciplinary Analysis, 671–684, Luckett W.P., and Jartenberger, J.L. (eds.). Plenum Press, New York.
114. Brown, W.M., Prager, E.M., Wang, A., and Wilson, A.C. (1982) J. Mol. Evol. *18*, 225–239.
115. Hardies, S.C. (1986) In Current Topics in Microbiology and Immunology, Vol. 127, The Wild Mouse in Immunology, 145–52, Potter, M., Nadeau, J.H., and Cancro, M.P.P. (eds.). Springer-Verlag, New York.
116. Goodman, M. (1975) In The Phylogeny of Primates. A Multidisciplinary Approach, 219–248, Luckett, W.P., and Szalay, F.S. (eds.). Plenum Press, New York.
117. Wooding, G.L., and Doolittle, R.F. (1972) J. Hum. Evol. *1*, 553–563.
118. Tanhauser, S., Johnson, P.A., Yelick, P.C., and Hecht, N.B. (1987) J. Cell Biol. *105*, 166a.
119. Bower, P.A., Yelick, P.C., and Hecht, N.B. (1987) Biol. Reprod. *37*, 479–488.
120. Bunick, D., Balhorn, R., Stanker, L.H., and Hecht, N.B. (1989) (submitted).
121. Chou, P.Y., and Fasman, G.D. (1974) Biochemistry *13*, 222–245.

Chapter 18
Molecular Biology of Structural Chromosomal Proteins of the Mammalian Testis

Norman B. Hecht

The term spermatogenesis describes the complex series of events in which a terminally differentiated cell, the spermatozoon, is produced from a stem cell [1–3]. The events leading to the formation of the species-specific-shaped spermatozoon are dependent on the products from a large number of temporally expressed genes, many unique to the testis [4]. Prominent among the organ-specific proteins expressed during spermatogenesis are a group of structural DNA-binding proteins.

In mammals the nuclei of testicular cells change markedly during spermatogenesis as the nucleosome structure of the male germ cell is replaced by the DNA-protamine complex of the spermatozoon. A group of novel DNA-binding proteins are involved in this reorganization of the chromatin. The chromatin of premeiotic and meiotic germ cells contains a group of histone variants in addition to the usual complement of somatic histones. As spermatogenesis proceeds, a class of testis-specific basic nuclear proteins called transition proteins replaces the histones. These temporally regulated chromatin changes occur during spermiogenesis, the interval of spermatogenesis when the transition proteins are in turn displaced by protamines as the haploid spermatid differentiates into the spermatozoon.

In this chapter I shall present our current knowledge of the testicular DNA-binding proteins known to be essential in the formation of a mammalian spermatozoon. Following a brief summary of the histones and high-mobility-group proteins of the testis, I shall focus primarily on the molecular biology of the testicular transition proteins and protamines. Special emphasis will be placed on the structure, comparative sequences, regulation, and evolution of the protamines, the predominant DNA-binding structural proteins of the mammalian spermatozoan nucleus. Nonmammalian protamines and other testicular proteins will only be mentioned in passing since excellent recent reviews discussing these proteins are available [5–10].

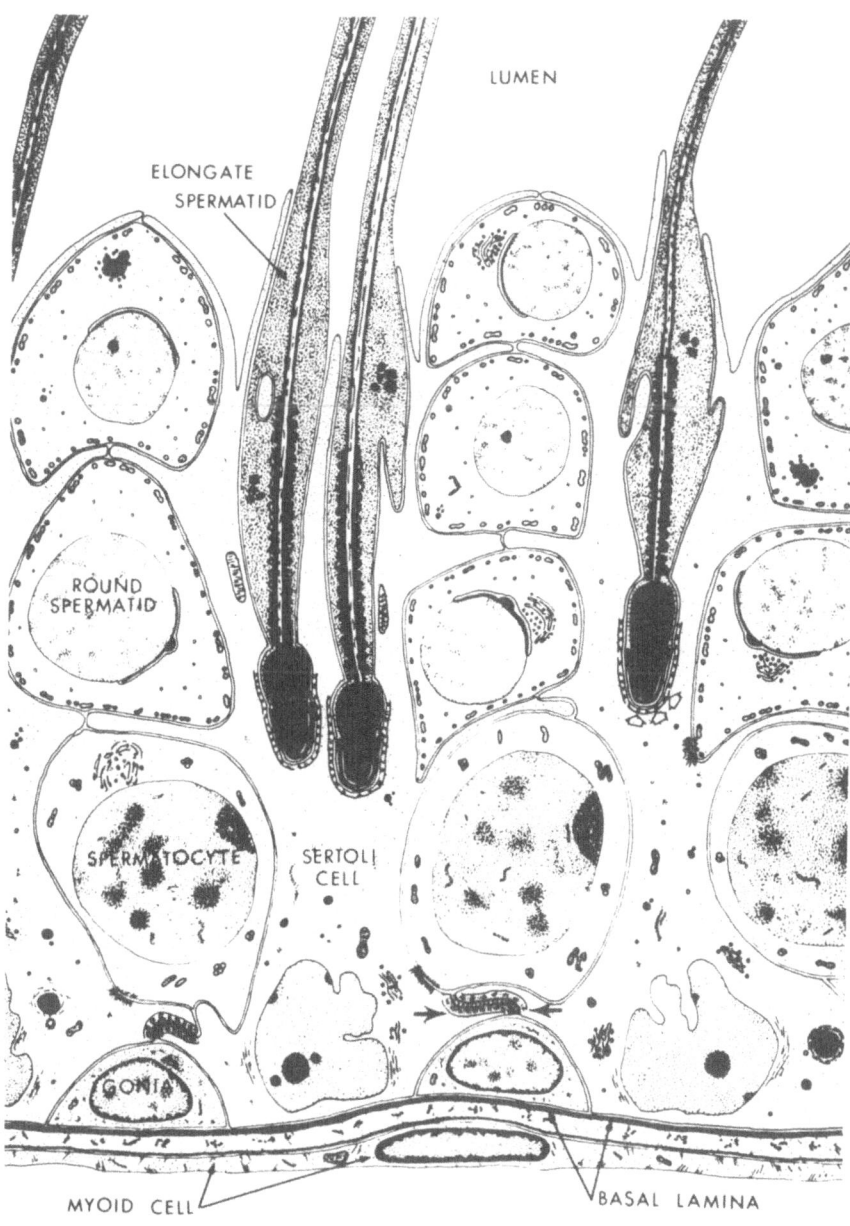

Figure 18.1. Diagram of spermatogenesis. The cell types and sequence of events of spermatogenesis initiate at the basement membrane (gonia) and proceed upward to the lumen. (Kindly provided by Dr. L. Russell, Southern Illinois University. Reprinted with permission from *Reproductive and Developmental Toxicity of Metals*, edited by Clarkson et al., © Plenum Press.)

```
                                    1
                             Met Ser Thr Ser Arg Lys Leu Lys Thr
TP1  Mouse          GAAAGTACC ATG TCG ACC AGC CGC AAG CTA AAG ACT
     Rat    ATTTTGGC A----T---A --- --- --- --- --- --A --- --- ---
                  -9            1
```

```
          10                              20
   His Gly Met Arg Arg Gly Lys Asn Arg Ala Pro His Lys Gly Val
   CAT GGC ATG AGG AGA GGC AAG AAC CGA GCT CCT CAC AAG GGC GTC
   --- --- --- --- --- --- --- --- --- --- --- --- --- --- ---
    30                                  60
```

```
                         30
   Lys Arg Gly Gly Ser Lys Arg Lys Tyr Arg Lys Ser Val Leu Lys
   AAG AGA GGT GGA AGC AAG AGA AAA TAC CGG AAG AGC GTC CTG AAA
   --- --- --A --- --- --- --- --- --- --- --- --- AG- --- --G
                         90                      Ser
```

```
      40                              50
   Ser Arg Lys Arg Gly Asp Asp Ala Ser Arg Asn Tyr Arg Ser His
   AGT AGG AAA CGG GGC GAT GAT GCA AGT CGC AAT TAC CGA TCC CAC
   --- --- --- --- --- --- --- --- --- --- --- --- --- --- ---
   120                              150
```

```
   Leu End
   TTG TGA TGCGGCAATG AGCTCTGCCC TGGTGGTCTT CAAACAACAC GGGGCAGGAG
   --- -A- ---------- ------AG-- ---------- -----T---- CA-----C-A
                     180                            210
```

```
   CATGAGGACA TCAGAGGGGG ACTGCCAAAG AGATCTGAAG TTAGACCAAA
   --**-*---- --------  240
```

```
    AGCCAAAGAT CCTATCAGAG TGGGTAAATG CCAGTCGTGA CGAAATTCGG
   270                           300
```

```
   AATGTATATG TTGGCTGTTT CTCCCCAACA TCTCAATAAC ATTTTGAAAA
               330                      360
```

```
   CAAATAAAAT TGTGAAAAAC AAAAAAAA
               390
```

```
                                       1
                              Met Asp Thr Lys Met Gln
TP2  Mouse  GGAAGTCTC TGCCCCGAGT GTGGCCTCCC ATG GAC ACC AAG ATG CAG
                 -20                   1
```

```
               10                                  20
   Ser Leu Pro Thr Thr His Pro His Pro His Ser Ser Ser Arg Pro Gln
   AGC CTT CCC ACC ACT CAT CCC CAC CCC CAC AGC TCC TCG CGG CCT CAA
               30                                  60
```

```
                      30
   Ser His Thr Ser Asn Gln Cys Asn Gln Cys Thr Cys Ser His His Cys
   AGT CAC ACC AGT AAC CAG TGC AAT CAG TGC ACC TGC AGC CAC CAC TGC
                      90
```

```
          40                              50
   Arg Ser Cys Ser Gln Ala Gly His Ala Gly Ser Ser Ser Ser Pro Ser
   CGG AGC TGC AGC CAG GCA GGC CAC GCG GGC TCT AGC TCC AGC CCC AGC
          120                             150
```

```
                      60                              70
   Pro Gly Pro Pro Met Lys His Pro Lys Pro Ser Val His Ser Arg His
   CCT GGC CCG CCC ATG AAG CAC CCC AAG CCA TCC GTG CAC TCT CGA CAC
                      180                             210
```

Continued

```
                                      80
Ser Pro Ala Arg Pro Ser His Arg Gly Ser Cys Pro Lys Asn Arg Lys
TCA CCT GCA AGA CCC AGC CAC CGC GGG AGC TGC CCC AAG AAC AGG AAG
                                     240


       90                                    100
Thr Phe Glu Gly Lys Val Ser Lys Arg Lys Ala Val Arg Arg Arg Lys
ACC TTT GAA GGG AAA GTG AGC AAG AGA AAG GCC GTC AGG AGG CGG AAA
       270                                   300


                  110                             117
Arg Thr His Arg Ala Lys Arg Arg Ser Ser Gly Arg Arg Tyr Lys End
CGG ACT CAC AGA GCT AAG AGG CGT AGC TCA GGG CGA AGA TAC AAG TGA
                  330


CGCACTCCAG GATGTTCCTG TGTCCATTTG ATCCCAAAAT GAGATAGCCA TCACTAGGGG
   360                        390


ACTGTTGGGA TGATGTCACA GGAACATGTC ACTGCAGCAA TTTCTATGCA ACATGGATTA
   420                        450


AAGCTTGTAC CCTGGAAGAC TAAAAAAAAA AAAAAA
   480
```

Figure 18.2. Nucleotide sequence and predicted amino acids of mouse and rat TP1 and mouse TP2. The predicted amino acid sequences are listed above the nucleotide sequence [51,52]. The nucleotides are numbered below the nucleotide sequence, and the amino acids are numbered above the amino acid sequence. For the TP1 proteins, nucleotide No. 1 has been assigned to the A of the ATG codon and amino acid No. 1 is assumed to be serine. For mouse TP2, nucleotide No. 1 has been assigned to the A of the first ATG codon, and amino acid No. 1 is the first methionine. The initiation codons and the putative polyadenylation signal in the nucleotide sequences have been underlined. For TP2 a sequence of 27 amino acids (amino acids 91 through 117) perfectly match rat TP2 [53,59]. With mouse TP2, both initiation codons are predicted to be used, generating two polypeptides differing by four amino acids at the amino terminus.

Spermatogenesis

In the testes of sexually mature males, spermatozoa are continually produced from a population of stem cells. Spermatogenesis can be divided into three intervals: (1) stem cell differentiation and renewal; (2) meiosis; and (3) spermiogenesis (Fig. 18.1). In the initial phase of spermatogenesis, primitive spermatogonia after several mitotic divisions serve as progenitor cells for type A, intermediate, and type B spermatogonia. Responding to currently unknown regulatory signals, a subpopulation of the diploid stem cells initiate differentiation. The resultant preleptotene primary spermatocytes enter meiosis, the second major interval of spermatogenesis. During the lengthy interval of meiosis, a period of 11 to 12 days in the mouse, chromosomes condense and homologous chromosomes pair. By morphological criteria, the events leading to chromosome pairing appear to commence during the leptotene stage of meiotic prophase when axial elements of the chromosomes can be first detected. Days later, during the zygotene stage of meiotic prophase, the synaptonemal complex, a structure believed necessary for effective eukaryotic chromosome synapsis, begins to form.

Chromosome pairing is completed at pachynema, the longest stage of meiotic prophase. During the pachytene stage of meiosis, genetic recombination occurs, producing chromosomes with the genetic diversity critical to the survival of a species. Pachynema is followed by diplonema, the stage when the paired chromosomes begin to separate and diakinesis, the interval when chiasma, structures believed to be the morphological evidence for recombination, are seen between chromosome homologues. Following meiotic prophase, two cellular divisions without any DNA synthesis occur. During the first meiotic division homologous chromosomes separate, producing two diploid secondary spermatocytes. A second, "mitoticlike" division reduces the nuclear DNA content of each male germ cell to a haploid level in the round spermatid. Spermiogenesis, the final interval of spermatogenesis, lasts around 14 days in the mouse, culminating in the production of the spermatozoon. In the mouse, spermiogenesis has been divided by morphological criteria into 16 steps. As spermiogenesis proceeds the nucleus changes from a round transcriptionally active form to an elongated structure with highly condensed chromatin. Despite the fact that transcription terminates during midspermiogenesis, many changes, including the assembly of a sperm tail, formation of an acrosome in the nucleus, and extensive remodeling and reshaping of the nucleus, occur during spermiogenesis. For additional information on the molecular biology and genetics of spermatogenesis, information can be obtained in several recent reviews [2,4,11–15]. In this chapter, discussion will be restricted to the structural DNA-binding proteins that produce the uniquely compacted nuclei of spermatozoa.

Histones and High-Mobility-Group Proteins of the Testis

Histone metabolism in the mammalian testis is complex and seemingly quite variable among species [reviewed in 16]. In addition to the usual complement of the ubiquitous H1, H2A, H2B, H3, and H4 histones, the mammalian testis also contains numerous histone variants that are temporally expressed during spermatogenesis. Some of these variants appear to be germ cell enriched, i.e., are present in the testis at a far higher level than in other organs, while some may be germ cell specific, having been found to date only in the testis. However, care should be exercised before declaring any protein variant to be organ specific since usually only a limited number of organs have been examined. Testicular variants have been reported for the H1t [17–27], H2A [24–26, 28–31],

H2B [24–26,32–34], and H3 histones [24–26,35]. Some of these germ cell–enriched histone variants are especially abundant in the testes of rodents and primates. In addition to the quantitative differences, the testicular histone variants show temporal differences of accumulation in spermatogonia and spermatocytes.

A major limitation to our understanding of the functional roles of testicular histone variants is the fact that most of our current knowledge of these proteins is based on their identification by electrophoretic mobilities. Although much useful information has been obtained with this methodology, it is difficult to establish whether gel patterns that differ among species reflect the absence of a particular variant or simply a modification that results in a shift in its electrophoretic mobility in the acid/urea/Triton X-100 polyacrylamide gels. Moreover, the presence of a group of transition proteins in the mammalian testis (see below) and the existence of processing intermediates of precursor proteins like the P2 protamine [36,37] add to the difficulty of identifying protein bands on polyacrylamide gels. The recent publication of amino acid sequences for the H1t histone of boar and rat [22,27,121] and a rat H2B histone variant [34] provides an important start for a database to assess qualitative and quantitative differences of testicular histones among species. Immunocytochemical localization of histones, as has been done for an H2A variant, can also help define the precise temporal appearance of these proteins [29–31]. In the near future, antisense, insertional mutagenesis, and transgenic techniques are likely to help to define the role of the germ cell–specific histones in chromatin structure during spermatogenesis. A better understanding of the localization of the histone variants in the testicular chromatin with respect to specific genes needs to also be undertaken [37a]. For a more complete analysis of the testicular histones, the reader is referred to a recent review by Meistrich [16].

At least four high-mobility-group proteins (HMG)—1, 2, 14, and 17—have been described in mammalian somatic and germ cells [38–42]. Since much of our knowledge of testicular HMG proteins is again based on electrophoretic mobilities, the precise number of distinct HMG proteins and their distribution among species await further analysis. In light of the many isozymic and structural variants of proteins in the mammalian testis [43], it would not be surprising to find testicular HMG variants. The mammalian testis offers an excellent system to test the hypothesis that HMG proteins may induce or maintain transcriptionally active regions of a genome.

Transition Proteins

The transition proteins are a group of small basic proteins that are present in the testicular cells of sexually mature mammals but not in epididymal or ejaculated spermatozoa [44–50]. Transition proteins (TPs) are generally less than 20,000 daltons in molecular weight and are present in elongating spermatids. They are relatively widespread in eutherian mammals, sharing similar properties of electrophoretic mobility, acid solubility, and general size and amino acid composition. They do, however, differ significantly in sequence, and the number of variants found among species appears to also differ. This uncertainty is again based on the electrophoretic mobility shift differences among species. Taking into consideration that some TP variants detected by gel electrophoresis could be posttranslational modifications of a single gene product while additional variants could be masked by comigration with other acid-soluble proteins (as occurs with rat TP4 and high-mobility-group proteins 14 and 17) [45], there appears to be three or four distinct transition proteins in the rat. At present, sequence information has confirmed that at least TP1 and TP2 are unique [51–53].

Transition Protein 1 (TP1)

TP1 was the first transition protein isolated and sequenced [44]. It is one of the first nuclear proteins to appear during spermiogenesis in condensing spermatids. Amino acid sequence analysis of the purified rat TP1 protein and sequence analysis of a cDNA coding for rat TP1 reveal a 54 amino acid protein that contains 50% arginine and lysine [51]. By the criterion of electrophoretic mobility, TP1 appears to be the smallest of the transition proteins [45]. Isolation and sequence analysis of TP1 clones from a mouse testis cDNA library reveal that the amino acid sequence of TP1 is highly conserved with the mouse protein, differing from rat TP1 by one amino acid, a substitution of a valine for a serine (Fig. 18.2) [52]. Greater differences are likely to be found when the sequences of TP1 of Bovidae species such as the bull, ram, and goat are known, since they have been reported to contain cysteine [50], an amino acid absent in TP1 of mouse and rat. Aside from the species mentioned above, the extent to which TP1 is utilized is not known. Hybridization analysis does, however, detect DNA sequences with homology to the mouse TP1 cDNA in rat, hamster, boar, bull, human, turtle, and American opossum [Kwon and Hecht, unpublished observations].

The mRNA for TP1 first appears postmeiotically in rat and mouse in late round spermatids [54–56]. The mRNA is stored in the cytoplasm, remaining translationally inactive for 3 or 4 days before the protein begins to be synthesized in elongating spermatids. In both the rat and mouse, the TP1 mRNA has a discrete size of about 600 nucleotides, including a poly(A) tail, before it is translated. Later in spermatogenesis, the polysome-associated mouse and rat TP1 mRNAs are shorter and more heterogeneous in size. This decrease in size of the TP1 mRNA is the result of partial deadenylation of the poly(A) terminus [55], a common feature of testicular DNA-binding proteins (see below). Recently, the

```
                    10              20              30            40            50
Mouse I   A R Y R C C R S K S R S R C R R R R R  R C R R R R R  C R R R R R S  Y T  I R C K K Y
Rat       - - - - - - - - - - - - - - - - - - -  - - - - - - -  - - - - - - -  - -  - - - -
Boar      - - - - - H - - - - - - P - - - - - -  - - - - - - -  - - P - - - -  A V  - - - - V - - R R C
Bull*     - - - - L T H - G - - - - - - - - - -  - R C - - - -  - R F G - - -  - R V  - - V - - T R Q
Bull**    - - - - L T H - G - - - - - - - - - -  - R C - - - -  - R F G - - -  - R V  - - V - - T R Q
Stallion  - - - - - Q - Q - - - - - - - - - - -  - R C - - - -  R S V - Q - -  V - -  - V L - - R R R R
Ram       - - - - L T H - R - - - - - - - - - -  - R C - - - -  - R F G - - -  - R V  - - V V - - T R Q
Human I   - - - - - Q - - Y Y - Q - Q - - - - -  - S - - - - -  S - Q T - - -  A M R  - - P - - R P - - R R H
Trout     M P - - - - - - - - - - - - S***P V - - - - - * R V S - - - - - - -  G G - - -
```

Figure 18.3. Amino acid sequence homologies of P1 protamines. — = Identical amino acid; * = deleted amino acid; A = Ala; C = Cys; D = Asp; E = Glu; F = Phe; G = Gly; H = His; I - Ile; K = Lys; L = Leu; M = Met; N = Asn; P = Pro; Q = Gln; R = Arg; S = Ser; T = Thr; V = Val; W = Trp; Y = Tyr. Sources of sequences: mouse [85,89]; rat [100a]; boar [82]; bull* [79]; bull** [81,97,98]; stallion [87,88]; ram [83]; human [84,85]; trout [6].

mouse TP1 gene has been mapped by Southern blotting of hamster-mouse cell hybrid DNAs to mouse chromosome 1 [57,58].

Transition Protein 2 (TP2)

TP2, a protein that migrates slightly in front of the H4 histone in acid-urea polyacrylamide gels, is a small protein that in the mouse contains about 33% basic amino acids [53]. The mouse TP2 is hydrophilic with a highly basic domain comprising about one third of the polypeptide chain at the carboxyl terminus and a much less basic domain comprising the remaining two thirds of the molecule. Although a mouse cDNA clone encoding TP2 has been identified and sequenced [53], the exact size of the proteins(s) of mouse TP2 has not yet been established, since the 5' end of the mouse TP2 mRNA contains two in-phase initiation codons, both of which could be used to generate polypeptides differing in length at the amino terminus of the molecule. Partial amino acid sequence analysis of a rat TP2 cDNA reveals a perfect match for the 27 amino acid sequence at the carboxyl terminus of mouse TP2 [59].

Additional TPs

Although no sequence information exists at present for other transition proteins, a partial characterization of TP4 from rat reveals a protein of molecular weight of about 20,000 daltons with an arginine-lysine composition of about 30% [48]. Rat TP4 contains cysteine and comigrates with TP2, migrating a little faster than the H4 histone. The presence of cysteine in rat TP4 distinguishes it from TP1 but not TP2. Whether it represents a third unique mammalian transition protein remains to be established. The existence of an additional transition protein, TP3, which is found in the sonication-resistant condensed spermatids of the rat, awaits confirming sequence analysis [45].

Biological Functions of the Transition Proteins

Although the temporal expression of transition proteins in the condensing chromatin of differentiating male germ cells has been well established, the reasons for the existence and presumed evolutionary advantage of transition proteins comes from the physicochemical studies of Singh and important start to our understanding of possible mechanisms of action of transition proteins comes from the physiocochemical studies of Singh and Rao [61,62]. Applying fluorescence quenching, UV difference absorption spectroscopy, and thermal melting techniques to the nucleic acid–binding properties of TP1, they conclude that TP1 is a DNA-melting protein probably acting through the interaction of its tyrosine residues with DNA. Interestingly, at the low ionic strength of 1 mM NaCl, TP1 brings about a slight stabilization of DNA against thermal melting, whereas at the higher ionic strength of 50 mM NaCl, a destabilization of the DNA occurs. The presence of the amino acid tyrosine in TP2 and in the protein called TP4 suggests they could serve a similar role [45,53].

Additional efforts to explain the function(s) of the transition proteins have been based on their temporal appearances during spermiogenesis. Soon after the first transition proteins enter the chromatin of the spermatid, the nucleus becomes resistant to breakage by sonication [63]. In the rat, the major classes of histone appear to be completely replaced by the TPs by the elongated spermatid stage, a time when transcription has terminated. The inability of the elongated spermatid to transcribe RNA is likely due to the loss of nucleosomal structure with its associated nonhistone proteins essential for RNA synthesis. Although the replacement of the histones by the TPs clearly locks in the nuclear DNA-protein changes of spermiogenesis, any direct effects of the transition proteins on transcription in the spermatid nucleus need to be proved.

Concomitant with termination of transcription, chromosome compaction initiates in elongating spermatids. Morphological examination of maturing germ cells during spermiogenesis reveals that the knobby (nucleosomal) chromatin fibers of the round spermatid are replaced by smooth fibers coincident with TP synthesis. However, care must be taken in ascribing function to the few characterized testicular DNA-binding proteins since it is apparent that there are many additional tightly bound nuclear proteins in male germ cells [64–66]. It is likely that the transition proteins, interacting with other less abundant chromatin proteins, facilitate chromatin condensation during spermiogenesis. It is not known whether transition proteins initiate or simply facilitate this process. In rodents, the nuclei of the spermatozoa are shaped before most of the TPs are synthesized, suggesting that the TPs do not play a significant role in this species-specific event. Alteration of the programmed expression of the structural chromosomal proteins by transgenic constructs could help define the functional roles of some of these proteins.

Protamines

Toward the end of spermiogenesis, the protamines, a class of small, highly basic proteins, are synthesized [24,67–70]. The high content of the amino acids arginine and cysteine in protamines of eutherian mammals allows an effective neutralization of DNA charge and the formation of intra- and intermolecular cross-links of the proteins [71–73]. The resultant DNA-protamine complex produces a highly compacted haploid spermatozoan genome in a state where the DNA is better protected against cleavage by endonucleases and exonucleases. The rigid nature of the sperm nucleus may also facilitate movement of the spermatozoon in the female reproductive tract and aid fertilization. Although attractive in concept, there is no evidence that the protamines determine the species-specific shape of the nucleus of spermatozoa [74,75], and their synthesis at the end of spermatogenesis suggests no significant role.

Bloch has proposed five major classes of sperm nuclear proteins ranging from nonbasic proteins in the crab to basic keratin-type protamines in the mouse [113]. Protamines have been isolated from the spermatozoa of a large number of species of both plants and animals [reviewed in 5-10,76]. Vertebrate protamine sequences are known for a number of species ranging from fish and birds to man. Trout and the dogfish have been shown to express a family of protamine genes [6,124-126], whereas the mature spermatozoa of most mammals are believed to contain only one detectable form of protamine [77-88]. A second protamine variant has been found in the spermatozoa of the mouse [37,89], human [84,86,87,90-96], horse [87,88], and certain species of hamsters [Balhorn, personal communication]. The protamines represent the predominant, if not the sole, highly basic protein of the mammalian spermatozoon. Their structure, regulation, and evolution will be discussed in the remainder of this chapter. Models of nucleoprotamine organization will not be considered here since they are discussed elsewhere [76].

P1 Protamine

Spermatozoa of the bull [81,97-99], boar [82], ram [83], mouse [85], and human [84,86,100] contain a 50-amino-acid-long protamine molecule with a central highly basic core, an N-terminal sequence of Ala–Arg–Tyr–Arg–Cys–Cys, a serine or threonine at positions 8, 10, and 12, and a highly variable C terminus (Fig. 18.3). A similar P1 protamine of 49 amino acids has been isolated from stallion spermatozoa [87,88], although a stallion P1 protamine containing cysteine as its 50th amino acid has also been found [Balhorn, personal communication].

Although the mammalian P1 protamines differ markedly in sequence, they all contain an identical N terminus of Ala–Arg–Tyr–Arg–Cys–Cys, and four arginine tracts in the core domain, also present in trout, are highly conserved [6]. In light of the detection of two boar P1 protamines of 50 and 49 amino acids, the latter being a minor component deficient in the C-terminal cysteine, analysis of the P1 protamine genes for stallion and boar is needed. Genomic DNA sequence analysis would define the primary sequence and would determine whether the 49 amino acid form of the P1 protamine is coded in the genome or is due to specific proteolysis during preparation or a posttranslational modification. To date, a shortened form of the P1 protamine has only been found in spermatozoa of stallion and boar.

In addition to the amino acid sequence data for mammalian protamines,

```
               1                                  10                                      20
              Ala Arg Tyr Arg Cys Cys Arg Ser Lys Ser Arg Ser Arg Cys Arg Arg Arg Arg Arg Arg Cys Arg Arg Arg Arg Arg
Mouse     ATG GCC AGA TAC CGA TGC TGC CGC AGC AAA AGC AGG AGC AGA TGC CGC CGT CGC AGA CGA AGA TGT CGC AGA CGG AGG AGG

                                              Gln                 Tyr Tyr     Gln     Gln     Ser
Human     --- --- --G --- A-- --- --T --- --- C-G --- C-- --- --- -AT TA- --C -AG --- -A- --- A-- --- --- --A --- ---

                                      Leu Thr His     Gly                                     Arg Cys
Bull      --- --- --- --- --- --- --- -T- -C- C-T --- G-- --- --- --- --- --C --- C-C --- --- A-A T-T C-C A-A C-A ---

                      30                                      40                                      50
          Arg Cys Cys Arg Arg Arg Arg Arg Arg Cys Cys Arg Arg Arg Arg Ser Tyr Thr Ile Arg Cys Lys Lys Tyr
Mouse     CGA TGC TGC CGG CGG AGG AGG CGA AGA TGC TGC CGT CGC CGC CGC TCA TAC ACC ATA AGG TGT AAA AAA TAC TAG

              Ser     Gln Thr             Ala Met Arg Cys Cys     Pro Arg     Arg Pro         Arg Arg His
Human     --G A-- --- -A- ACA C-- --- A-- GCC ATG A-G T-C T-- --- -C- AGG --- -GA CCG --A --- -G- -G- C-- --A

              Arg Phe Gly                     Arg Val Cys Cys         Tyr Thr Val             Thr Arg Gln
Bull      A-G C-- -TT G-T --- --- C-C A-G --G A-A GTG T-C T-- --T --- -AC AC- GT- --- --- --- -C- -G- C-G --A
```

Figure 18.4. Nucleotide and predicted amino acid sequences of mouse, human, and bull P1 protamines. Both the human and bull sequences are compared to the mouse. Sequence sources are mouse [85], human [100], and bull [97,98].

P2 PROTAMINES

```
                            10              20            30              40
Mouse:   V R Y R M R S P S E G P H Q G P G G Q D H E R E E Q G Q G G L S P E R V E D Y G
Rat:     - - - - - - - - - - - Q - - - - - - - - - - - - - E - - - - - - - - - - - - -

                 Cleavage
                     ▾
                            50              60            70              80
Mouse:   R T H R G H H H H R H R R C S R K R L H R I H K R R * R S C R R R R R H S C R H R
Rat:     - - E - - - - * - - - - - * * - - - - - - * - - - - - - - - - - - - - - - - -
Human (2a): - - - * - Q S - Y - R - H - - - R - - - - - R - Q H - - - K - R - - - - - - -
                 ▴
                 H2b

                            90             100
Mouse:   R R H R R G C R R S R R R R R C R C R K C R R H H H
Rat:     - - - - - - - - - - - - - - S - - - - - - W - Y Y
Human (2a): - - - - - - T * * * * * * * * - K - T - - - -
```

Figure 18.5. Amino acid sequence homologies of P2 protamines. — = identical amino acid; * = deleted amino acid; A = Ala; C = Cys; D = Asp; E = Glu; F = Phe; G = Gly; H = His; I = Ile; K = Lys; L = Leu; M = Met; N = Asn; P = Pro; Q = Gln; R = Arg; S = Ser; T = Thr; V = Val; W = Trp; Y = Tyr. Sources of sequences: mouse [37], rat [105], human [86,94].

cDNA and genomic DNA sequences are now available for the P1 protamines of the mouse [85,101], human [100], and bull [98,99] (Fig. 18.4). Comparison of these DNA sequences reveals a strongly conserved region at the N terminus of the protein but substantial variation throughout most of the coding region of the P1 protamine. Compared to histones, the mammalian protamines are evolving very rapidly.

P2 Protamine

In the spermatozoa of the mouse [36,37,89], human [86,90–95], stallion [87,88], and certain species of hamsters, an additional protamine, the P2 protamine, has been detected. It differs considerably from the P1 protamines in size, composition, and sequence (Fig. 18.5). The P2 protamine has been isolated in two related forms from human spermatozoa and may also be heterogeneous in the horse. In the human the two P2 protamine variants are 57 and 54 amino acids in length [84,86]. The shorter P2 protamine, P2b, lacks the three N-terminal amino acids of the P2a variant. In the mouse cDNAs and the gene for protamine 2 have been

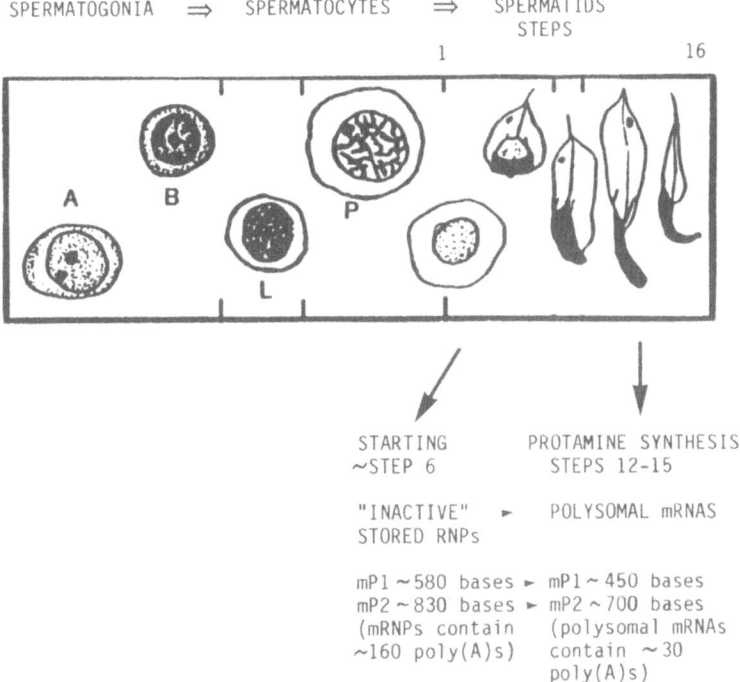

Figure 18.6. Summary of the translational regulation of protamine synthesis in the mouse. A, type A spermatogonia; B, type B spermatogonia; L, leptotene stage spermatocyte; P, pachytene stage spermatocyte.

isolated [55,101]. They code for a mature protein 63 amino acids in length, which is synthesized as a precursor protein of 106 amino acids (excluding the N-terminal methionine) [37]. The precursor is sequentially processed in the nucleus of late-step spermatids [36]. Sequence analysis of a P2 protamine isolated from mouse spermatozoa confirms the mouse cDNA and gene sequences with the exception that the isolated protein lacks a C-terminal histidine [89]. A rat gene for the precursor for protamine 2 has also been isolated (Fig. 18.5) [105].

The physiological reason is not known why the P2 protamine is synthesized as a precursor and requires proteolytic cleavage while the P1 protamine is not cleaved. Protein stability differences may be partially responsible since arginine, the N terminus of the mature mouse P2 protamine and the human P2a protamine, is believed to give a polypeptide a very short half-life, whereas the valine of the precursor of the mouse P2 protamine allows for a longer half-life [102]. The precursor form of protamine 2 may also be essential for proper interaction with the P1 protamine, allowing correct deposition onto DNA. It is likely that the P2 protamine will also be posttranslationally processed in other mammals since rat and hamster protamine 2 mRNAs are the same length as the mouse precursor protamine 2 mRNA [105].

Significance of Multiple Protamine Variants

The relative levels of the P1 and P2 protamines in mature spermatozoa vary considerably among mammals, ranging from the mouse, where the P2 protamine is the predominant (about 70% of total) protamine, to undetectable levels in species such as the rat and bull. Although both P1 and P2 protamines are arginine rich, recent microinjection studies of sperm nuclei into hamster oocytes indicate that the sperm nuclei containing the P2 protamine decondense more quickly than those that do not [103]. Perreault and colleagues have demonstrated that the rate of decondensation of spermatozoan nuclei microinjected into hamster oocytes decreases as follows: human > chinchilla > mouse > hamster > rat and bull [103]. With the exception of human spermatozoa, which contain about 50% P2 protamine but also contain a sizable amount of histone [90–92], these rates of decondensation correlate with the amount of the P2 protamine present in the spermatozoa of each species. These results suggest that the relative structural stabilities of sperm nuclei will vary with the ratio of the P1 and P2 protamines.

Although the distribution of the P2 protamine in spermatozoa of mammals appears limited, its apparent absence in spermatozoa does *not* reflect the absence of a P2 protamine gene or even the lack of P2 protamine synthesis. For instance, in the rat, where no P2 protamine has been detected in spermatozoa [77], genomic DNA sequences with an apparently high level of homology to the mouse P2 protamine cDNA have been detected by molecular hybridization [104]. This homology has been confirmed by the isolation of a rat protamine 2 gene with over 90% homology to the mouse protamine 2 gene (Fig. 18.5) [105a]. Moreover, hybridization of rat testicular RNA with the mouse protamine 2 precursor cDNA reveals that mRNAs of protamine 2 precursor size are present in the rat testis at a level 1% to 2% of rat protamine 1 or mouse protamine 2 mRNAs [104]. The presence of rat protamine 2 mRNAs on polysomes further argues that the P2 protamine protein is synthesized in the rat but at a level too low to produce enough protein to be detected in spermatozoa or the P2 protamine is not deposited in the nucleus. In the golden Syrian hamster, protamine 2 mRNA is present at a level intermediate between mouse and rat [104].

Hybridization analysis of cDNAs of mouse protamine 1 and 2 with genomic DNA from a large number of both eutherian and marsupial mammals reveals that many other species also contain DNA sequences with homology to the mouse P2 protamine gene [106]. The lack of the P2 protamine in the spermatozoa of species such as the rat suggests that an important regulatory role in DNA protamine organization could be exerted at the chromatin level during spermiogenesis, since low levels of this P2 protamine appear not to be incorporated.

Regulation of Protamine Expression

The protamine genes have proved to be an excellent model system to study the regulation of gene expression in the mammalian testis because of their abundant mRNAs and their precise temporal expression (Fig. 18.6). In the mouse the protamines are synthesized toward the end of spermatogenesis between steps 12 and 15 of spermiogenesis. In addition to being representative of a large number of genes that are uniquely transcribed from the haploid genome, the timing of protein synthesis, long after transcription has terminated, allows translational regulation to be studied in a system where transcription does not occur. The numerous morphological changes occurring in the differentiating spermatid indicate that many additional proteins will be similarly regulated.

POSTMEIOTIC APPEARANCE

The synthesis of the protamines toward the end of spermatogenesis suggests that the protamine mRNAs should be abundant in postmeiotic cell types. This has been substantiated by the isolation of cDNAs coding for the two mouse protamines following a screening of a mouse testis cDNA library with radiolabeled cDNAs prepared from cytoplasmic poly (A)$^+$ RNA from postmeiotic round spermatids [107]. No protamine cDNAs were detected with radiolabeled cDNAs prepared from meiotic pachytene spermatocytes. Sequence analysis of a group of round-spermatid-enriched cDNAs identified several cDNA clones that coded for mouse protamine 1 and for the precursor of protamine 2 [37,52]. The protamine 1 and 2 cDNAs hybridized to cytoplasmic RNA isolated from round spermatids, elongating spermatids, and residual bodies, but not to cytoplasmic poly(A)$^+$ RNA from liver, brain, prepuberal testes of 16- to-17-day-old mice (spermatogenesis has reached meiosis at this age), total RNA from pachytene spermatocytes or cultured Sertoli cells, or poly(A)$^-$ RNA from adult testis. In addition to the evidence from Northern blot hybridizations with RNA from isolated populations of testicular cells [107], analysis of testicular extracts of prepuberal mice [108] and in situ hybridization of testicular sections [109,110] confirm that the mouse protamine mRNAs are only found in postmeiotic (haploid) male germ cells. The rooster protamine, galline [129], and a bull protamine [123] are similarly expressed during spermiogenesis. In situ hybridization of rat testis first detects protamine 1 mRNA at stage VII of the spermatogenic cycle [110]. The growing number of DNA probes and availability of antibody to protamine will allow a more precise determination of the temporal regulatory relationship of RNA and protamine protein synthesis to be made [127–130].

TRANSLATIONAL REGULATION

The presence of protamine mRNA in round spermatids approximately 1 week before the protamines are synthesized in elongating spermatids indicates that protamines are translationally regulated (Fig. 18.6). Analysis of protamine mRNAs in isolated populations of cells reveals that round spermatids contain P1 and P2 protamine mRNAs of 580 and 730 nucleotides in length, respectively [111]. In elongating spermatids where the protamine protein is synthesized, the P1 and P2 protamine mRNAs are about 450 and 600 nucleotides in length, respectively. Moreover, the shortened mRNAs are found in polysomes, whereas the longer protamine mRNAs are primarily in the nonpolysomal (ribonucleoprotein) fraction. Experiments using RNase H digestion and thermal chromatography with poly(U) Sepharose have demonstrated that the shortening of protamine mRNA is due to a partial deadenylation [111]. A similar shortening of the poly(A) tail has been noted for trout [112], rat, and hamster polysomal protamine mRNAs [104] and also for TP1 mRNAs [54,55]. In the trout, protamine mRNAs are transcribed during meiosis and stored as ribonucleoprotein particles until translation during spermiogenesis [6]. From in vitro translation studies, Sinclair and Dixon have demonstrated that conformational changes probably caused by a release of protein or other RNP component allow "in vitro activation" of the protamine mRNA [114]. Whether similar factors regulate mammalian protamine synthesis is not known.

CHROMOSOME LOCATION

Bloch has speculated that genes that are only active during spermatogenesis are localized on the Y chromosome [113]. To test this hypothesis, the chromosome location of mouse protamines 1 and 2 has been determined using the DNA from a series of hamster-mouse somatic cell hybrids that contain a complete set of hamster chromosomes and subsets of the mouse chromosomes. Both mouse protamine 1 and 2 are located on chromosome 16 [115]. Further mapping using DNA restriction fragment length polymorphisms has revealed the two protamines are tightly linked in mouse and hamster [116]. In the mouse the P1 and P2 protamines are approximately seven map units from the centromere on the proximal part of chromosome 16. Human protamine 1 is encoded by a gene on human chromosome 16 [116].

GENOMIC REGULATORY SEQUENCES

To identify *cis*-acting DNA sequences that could regulate the organ- and cell-specific expression of the two coordinately expressed protamine genes, genomic clones for protamine 1 and 2 have been isolated and the sequences of the 5′ flanking regions and coding regions compared [101]. Both mouse protamine genes contain one intron. In protamine 1 the intron is 94 nucleotides in length whereas it is 105 nucleotides in protamine 2. The P1 protamine gene from bull also contains one intron of 101 nucleotides [99]. Several flanking promoter elements are conserved in the 5′ flanking sequences of mP1 and mP2 including the TATAA box and several upstream elements. In the 3′ untranslated regions of both mouse protamine genes, the sequence GCCACCTG is found in a larger conserved sequence. This is a consensus sequence that is found in four binding sites within a mouse immunoglobulin enhancer. In the mouse protamine genes it is present immediately adjacent to the poly(A) addition signal in a larger conserved region [37,85]. This 3′ sequence is not unique to the mouse protamines as it is also present in the protamine DNA sequences of bull [97,98], human [100], and rat [105]. Considering the substantial evolutionary distance between rodents and human, the conservation of such a sequence in the 3′ untranslated region of every P1 and P2 protamine examined to date argues it plays an essential function in the transcription or, more likely, the translation of the protamines.

To functionally determine *cis*-acting DNA elements involved in protamine gene regulation, constructs using flanking regions of the P1 and P2 protamine genes were attached to reporter genes, and transgenic mice were produced [117,118]. For the protamine 1 gene, it has been shown that the *cis*-acting sequences residing on a 2.4-kb restriction fragment, of which 880 base pairs is 5′ to the start of transcription, conferred tissue-specific protamine expression [117]. A 5′ flanking region of 859 nucleotides of the P2 protamine gene was sufficient to obtain haploid-specific transcription of two fusion genes in transgenic mice [118]. The mRNAs for the c-myc and T-antigen reporter genes were not translated in constructs lacking the conserved 3' UT sequence of the protamines, suggesting that all or part of this sequence may be essential for translation [118]. The delayed expression during spermatogenesis of a human growth hormone construct containing the P1 protamine 3' UT in a transgenic mouse line further argues for a translational regulatory role for this sequence [Braun, personal communication]. Recently, additional transgenic mice were generated with a chimeric gene containing a SV40 T antigen gene fused to 5' and 3' flanking sequences of mouse protamine 1 [119]. Although this construct failed to produce testicular pathology, atrial and bone tumors were produced. The authors suggest that novel regulatory elements may have been generated by the juxtaposition of the protamine 1 and SV40 sequences [119]. Future efforts using histone, TP, and protamine promoters in transgenic mice promise to provide important insights into the functional roles of DNA-binding proteins of testicular chromatin.

Evolutionary Considerations

In the animal and plant kingdoms, a great diversity of DNA-binding proteins ranging from ubiquitously expressed histones to the organ-specific protamines are utilized to compact the DNA of the spermatozoon [113]. This variability in structural testicular DNA-binding proteins extends to mammals. Among the P1 protamines, sequences differ in as much as one third of the amino acids (compare mouse and human in Fig. 18.3). Based on the highly conserved sequences of somatic histones and the critical need to produce functional spermatozoa for species survival, it is surprising that the protamines are evolving at such a rapid rate. The inconsistent presence of the P2 protamine in mammalian spermatozoa makes it even more difficult to consider one unified structure for the chromatin of mammalian spermatozoa.

In light of the many differences, it is important to consider which protamine determinants remain constant. Both P1 and P2 protamines contain tracts of polyarginine and a scattering of cysteines. Although arginine can be encoded by six different codons, at both the protein and nucleotide levels, the core region of the P1 protamines is conserved (Fig. 18.4). The P1 protamines also contain an invariant N terminus, and, with proper alignment, 27 of the amino acid positions appear essential [87]. Other regions of the P1 protamine appear less critical for DNA binding, a conclusion supported by the trout protamines that are coded for by a multigene family of proteins that contain the central arginine cores but lack the amino and carboxyl termini of the mammalian P1 protamine (Fig. 18.3) [6].

Although the database for the P2 protamines is restricted to the human, rat, and mouse at present, it is evident that P2 protamines also contain polyarginine clusters. The P2 protamine sequences of man, mouse, and rat are much more closely related to each other than to the P1 protamines. The evolutionary relationship and differential utilization of the P1 and P2 protamines are puzzling. The overall structure and functional similarities (clusters of arginine and dispersed cysteine) suggest a common evolutionary origin for the two protamines. The chromosomal linkage and relatively similar sizes and sites of the intron in the P1 and P2 protamine genes support this. Assuming both protamines evolved from one ancestral gene, the marked differences between the P1 and P2 protamine sequences indicate that this divergence occurred long ago and/or at a rapid rate. This time scale is supported by the sequence differences among the P2 protamines of mouse, human, and rat. An understanding of the regulation of the P2 protamines at both the molecular and chromatin attachment levels may help clarify the apparently selective use of this variant in the mammalian sperm nucleus.

Although the shape and the nuclear stability of spermatozoa differ substantially among eutherian mammals, all species analyzed to date have sclerotic nuclei that require a strong reducing agent for decondensation in vitro. The tightly compacted nucleus of eutherian mammal spermatozoa results from the large number of cysteine residues in the P1 and P2 protamines allowing intra- and interdisulfide bond formation [72]. In contrast, the nuclei of marsupial spermatozoa appear to be much less compacted, decondensing in 0.3 M salts of calcium or magnesium [120]. Preliminary sequence analysis studies indicate that a protamine from the wallaby, a marsupial, lacks cysteine, suggesting considerable structural differences in the protamines of eutherian and marsupial mammals [Balhorn, personal communication]. However, hybridization analysis of genomic DNA of marsupials with the mouse protamine cDNAs detects sequence homologies [106]. These DNA cross-hybridizations are not simply due to the polyarginine tracts of eutherian and marsupial protamines, because no cross-hybridization was detected between the protamine 1 and 2 cDNAs under the same hybridization stringencies. Sequence analysis comparisons of marsupial and eutherian mammal protamine genes are needed. Minimally, such studies will provide insights into critical shared structural and regulatory elements of this class of testicular chromosomal protein. In addition, a detailed comparative study of protamine gene expression in eutherian and marsupial mammals offers a unique experimental system with which to understand the regulation of an essential, rapidly evolving mammalian gene in highly differentiated cell types.

Acknowledgments. The research described in this chapter from the author's laboratory was supported by NIH grant GM 29224. Thanks are extended to Kimberly Calderwood and Valerie Ricciardone for the excellent preparation of this chapter and to Drs. Jacquetta Trasler and Pamela Yelick for a critical reading of this chapter.

References

1. Fawcett, D.W. (1975) Dev. Biol. *44*, 394–436.
2. Bellvé, A.R. (1979) In Oxford Reviews of Reproductive Biology, 159–261, Finn, C.A. (ed.). Oxford University Press (Clarendon), London.
3. Clermont, Y. (1972) Physiol. Rev. *52*, 198–236.
4. Hecht, N.B. (1986) In Experimental Approaches to Mammalian Embryonic Development, 151–193, Rossant, J., and Pedersen, R.A. (eds.). Cambridge University Press, New York.
5. Subirana, J.A. (1983) In The Sperm Cell, 197–213, J. Andre (ed.). Martinus Nijhoff Publishers, The Hague.
6. Dixon, G.H., Aiken, J.M., Jankowsky, J.M., McKenzie, D.I., Moir, R., and States, J.C. (1985) In Chromosomal Proteins and Gene Expression, 287–314, Reeck, G., Goodwin, G., and Puigdomenech, P. (eds.). Plenum Press, New York.
7. Kasinsky, H.E., Mann, M., Lemke, M., and Huang, W.Y. (1985) In Chromosomal Proteins and Gene Expression, 333–359, Reeck, G., Goodwin, G., and Puigdomenech, P. (eds.). Plenum Press, New York.
8. Mezquita, C. (1985) In Chromosomal Proteins and Gene Expression, 315–332, Reeck, G., Goodwin, G., and Puigdomenech, P. (eds.). Plenum Press, New York.
9. Poccia, D. (1986) Int. Rev. Cytol. *105*, 1–65.
10. Kasinksy, H.E. (1989) In Histones and Other Basic Nuclear Proteins, Hnilica, L., Stein, G., and Stein, J. (eds.). CRC Press, Boca Raton, FL (in press).
11. Bellvé, A.R., and O'Brien, D.A. (1983) In Mechanism and Control of Animal Fertilization, 55–137, Hartman, J.F. (ed.). Academic Press, London.
12. Handel, M. (1987) In Results and Problems in Cell Differentiation in Spermatogenesis: Genetic Aspects, 1–62, Hennig, W. (ed.). Springer-Verlag, Berlin.
13. Willison, K., and Ashworth, A. (1987) Trends Genet. *3*, 351–355.
14. Eddy, E.M. (1988) In The Physiology of Reproduction, 27–68, Knobil, E., and Neill, J. (eds.). Raven Press, New York.
15. Wolgemuth, D. (1988) In Molecular Approaches to Fertilization, Schatten, G., and Schatten, H. (eds.). Academic Press, New York.
16. Meistrich, M.L. (1989) In Histones and Other Basic Nuclear Proteins, Hnilica, L., Stein, G., and Stein, J. (eds.). CRC Press, Boca Raton, FL (in press).
17. Shires, A., Carpenter, M.P., and Chalkley, R. (1975) Proc. Natl. Acad. Sci. USA. *71*, 2714–2718.
18. Levinger, L.F., Carter, C.W., Kumaroo, K.K., and Irvin, J.L. (1978) J. Biol. Chem. *253*, 5232–5234.
19. Seyedin, S.M., and Kistler, W.S. (1979) Biochemistry *18*, 1371–1375.
20. Seyedin, S.M., Cole, R.D., and Kistler, W.S. (1981) Exp. Cell Res. *136*, 399–405.

21. Seyedin, S.M., and Kistler, W.S. (1983) Exp. Cell Res. *143*, 451–454.
22. Cole, K.D., York, R.G., and Kistler, W.S. (1984) J. Biol. Chem. *259*, 13695–13702.
23. Lennox, R.W., and Cohen, L.H. (1984) Dev. Biol. *103*, 80–84.
24. Bhatnagar, Y.M., Romrell, L.J., and Bellve, A.R. (1985) Biol. Reprod. *32*, 599–609.
25. Meistrich, M.L., Brock, W.A., Grimes, S.R., Platz, R.D., and Hnilica, L.S. (1985) Fed. Proc. *37*, 2522–2525.
26. Meistrich, M.L., Bucci, L.R., Trostle-Weige, P.K., and Brock, W.A. (1985) Dev. Biol. *112*, 230–240.
27. Cole, K.D., Kandala, J.C., and Kistler, W.S. (1986) J. Biol. Chem. *261*, 7178–7183.
28. Trostle-Weige, P.K., Meistrich, M.L., Brock, W.A., Nishioka, K., and Bremer, J.W. (1982) J. Biol. Chem. *257*, 5560–5567.
29. Bhatnagar, Y.M., McCullar, M.K., Faulkner, R.D., and Ghai, R.D. (1983) Biochim. Biophys. Acta *760*, 25–33.
30. Bhatnagar, Y.M., Faulkner, R.D., and McCullar, M.K. (1985) Biochim. Biophys. Acta *827*, 14–22.
31. Bhatnagar, Y. (1985) Biol. Reprod. *32*, 957–968.
32. Shires, A., Carpenter, M.P., and Chalkley, R.A. (1976) J. Biol. Chem. *251*, 4155–4158.
33. Wattanaseree, J., and Svasti, J. (1983) Arch. Biochem. Biophys. *225*, 892–895.
34. Kim, Y.-J., Hwang, I., Tres, L.L., Kierszenbaum, A.L., and Chae, C.-B. (1987) Dev. Biol. *124*, 23–34.
35. Trostle-Weige, P.K., Meistrich, M.L., Brock, W.A., and Nishioka, K. (1984) J. Biol. Chem. *259*, 8769–8776.
36. Balhorn, R., Weston, S., Thomas, C., and Wyrobek, A. (1984) Exp. Cell Res. *150*, 298–308.
37. Yelick, P.C., Balhorn, R., Johnson, P.A., Corzett, M., Mazrimas, J.A., Kleene, K.C., and Hecht, N.B. (1987) Mol. Cell Biol. *7*, 2173–2179.
37a. Rao, M.R.S., Rao, B.J., and Ganguly, J. (1982) Biochem. J. *205*, 15–21.
38. Seyedin, S.M., and Kistler, W.S. (1979) J. Biol. Chem. *254*, 11264–11271.
39. Goodwin, G.H., Brown, E., Walker, J.M., and Johns, E.W. (1980) Biochim. Biophys. Acta *623*, 329.
40. Bucci, L.R., Brock, W.A., Goldknopf, I.L., and Meistrich, M.L. (1984) J. Biol. Chem. *259*, 8840–8846.
41. Bucci, L.R., Brock, W.A., and Meistrich, M.L. (1985) Biochem. J. *229*, 233–240.
42. Loir, M., Dupressoir, T., Lanneau, M., Le Gac, F., and Sautiere, P. (1986) Exp. Cell Res. *165*, 441–449.
43. Goldberg, E. (1977) In Isozymes: Current Topics in Biological Research, Vol. 1, 79–124, Rattazzi, M.C., Scandalios, J.C., and Whitt, G.S. (eds.). Alan R. Liss, New York.
44. Kistler, W.S., Geroch, M.E., and Williams-Ashman, H.G. (1973) J. Biol. Chem. *248*, 4532–4543.
45. Meistrich, M.L., Trostle, P.K., and Brock, W.A. (1981) In Bioregulators of Reproduction, 151–166, Jagiello, G., and Vogel, H.J. (eds.). Academic Press, New York.

46. Platz, R.D., Grimes, S.R., Meistrich, M.L., and Hnilica, L.S. (1975) J. Biol. Chem. *250*, 5791–5800.
47. Grimes, S.R., Meistrich, M.L., Platz, R.D., and Hnilica, L.S. (1977) Exp. Cell Res. *110*, 31–39.
48. Meistrich, M.L., Bucci, L.R., Brock, W.A., Trostle, P.K., Platz, R.D., Grimes, S.R., and Burleigh, B.D. (1980) Fed. Proc. *39*, 1884.
49. Lanneau, M., and Loir, M. (1982) J. Reprod. Fertil. *65*, 163–170.
50. Dupressoir, T., Sautiere, P., Lanneau, M., and Loir, M. (1985) Exp. Cell Res. *161*, 63–74.
51. Heidaran, M.A., and Kistler, W.S. (1987) Gene *54*, 281–284.
52. Kleene, K.C., Bozorgzadeh, A., Flynn, J.F., Yelick, P.C., and Hecht, N.B. (1988) Bichim. Biophys. Acta *950*, 215–220.
53. Kleene, K.C., and Flynn, J.F. (1987) J. Biol. Chem. *262*, 17272–17277.
54. Heidaran, M.A., and Kistler, W.S. (1987) J. Biol. Chem. *262*, 13309–13315.
55. Yelick, P.C., Kwon, Y., Flynn, J.G., Borzorgzodeh, A., Kleene, K.C., and Hecht, N.B. (1989) Molec. Reprod. and Devel., 13 in press.
56. Heidaran, M.A., Showman, R.M., and Kistler, W.S. (1988) J. Cell Biol. *106*, 1427–1433.
57. Yelick, P.C., Kwon, Y., Kozak, C., and Hecht, N.B. (1989) (submitted).
58. Heidaran, M.A., Kozak, C.A., and Kistler, W.S. (1989) Gene *75*, 39–46.
59. Cole, K.D., and Kistler, W.S. (1987) Biochem. Biophys. Res. Commun. *147*, 437–442.
60. Ashikawa, I., Kinosita, K. Jr., Ikegami, A., and Tobita, T. (1987) Biochim. Biophys. Acta *908*, 263–267.
61. Singh, J., and Rao, M.R.S. (1987) J. Biol. Chem. *262*, 734–740.
62. Singh, J., and Rao, M.R.S. (1987) Indian J. Biochem. Biophys. *24*, 181–188.
63. Meistrich, M.L. (1977) Methods Cell Biol. *15*, 15–54.
64. O'Brien, D.A., and Bellvé, A.R. (1980) Dev. Biol. *75*, 386–404.
65. O'Brien, D.A., and Bellvé, A.R. (1980) Dev. Biol. *75*, 405–418.
66. Avramova, Z., and Tasheva, B. (1987) Mol. Cell. Biochem. *74*, 67–75.
67. Geremia, R., Goldberg, R.B., and Bruce, W.R. (1975) Andrology *8*, 147–156.
68. Bellvé, A.R., Anderson, E., and Hanley-Bowdoin, L. (1975) Dev. Biol. *47*, 349–365.
69. Mayer, J.F., and Zirkin, B.R. (1979) J. Cell Biol. *81*, 403–410.
70. Mayer, J.F., Chang, T.S.K., and Zirkin, B.R. (1981) Biol. Reprod. *25*, 1041–1051.
71. Balhorn, R., Gledhill, B.L., and Wyrobek, A.J. (1977) Biochemistry *16*, 4074–4080.
72. Balhorn, R. (1982) J. Cell Biol. *93*, 298–305.
73. Bedford, J.M., and Calvin, H.I. (1974) J. Exp. Zool. *188*, 137–156.
74. Fawcett, D.W., Anderson, W.A., and Phillips, D.M. (1971) Dev. Biol. *26*, 220–251.
75. Risley, M.S., Eckhardt, R.A., Mann, M., and Kasinsky, H.E. (1982) Chromosoma *84*, 557–569.
76. Hecht, N.B. (1989) In Histones and Other Basic Nuclear Proteins, Hnilica, L., Stein, G., and Stein, J. (eds.). CRC Press, Boca Raton, FL (in press).
77. Calvin, H.I. (1976) Biochim. Biophys. Acta *434*, 377–389.
78. Coelingh, J.P., Rozijn, T.H., and Monfoort, C.H. (1969) Biochim. Biophys. Acta *188*, 353–356.

79. Coelingh, J.P., Monfoort, C.H., Rozijn, T.H., Gevers Leuven, J.A., Schiphof, R., Steyn-Parve, E.P., Braunitzer, G., Schrank, B., and Ruhfus, A. (1972) Biochim. Biophys. Acta *285*, 1–14.

80. Monfoort, C.H., Schiphof, R., Rozijn, T.H., and Steyn-Parve, E.P. (1973) Biochim. Biophys. Acta *322*, 173–177.

81. Mazrimas, J.A., Corzett, M., Campos, C., and Balhorn, R. (1986) Biochim. Biophys. Acta *872*, 11–15.

82. Tobita, T., Tstutsumi, H., Kato, A., Suzuki, H., Nomoto, M., Nakano, M., and Ando, T. (1983) Biochim. Biophys. Acta *744*, 141–146.

83. Sautiere, P., Belaiche, D., Martinage, A., and Loir, M. (1984) Eur. J. Biochem. *144*, 121–125.

84. McKay, D.J., Renaux, B.S., and Dixon, G.H. (1985) Biol. Reprod. *5*, 383–391.

85. Kleene, K.C., Distel, R.J., and Hecht, N.B. (1985) Biochemistry *24*, 719–722.

86. Ammer, H., Henschen, A., and Lee, C. (1986) Biol. Chem. Hoppe-Seyler *367*, 515–522.

87. Ammer, H., and Henschen, A. (1987) Biol. Chem. Hoppe-Seyler *368*, 1619–1626.

88. Belaiche, D., Loir, M., Kruggle, W., and Sautiere, P. (1987) Biochim. Biophys. Acta *913*, 145–149.

89. Bellvé, A.R., McKay, D.J., Renaux, B.S., and Dixon, G.H. (1988) Biochemistry *27*, 2890–2897.

90. Puwaravutipanich, T., and Panyim, S. (1975) Exp. Cell Res. *90*, 152–158.

91. Pongsawasdi, P., and Svasti, J. (1976) Biochim. Biophys. Acta *434*, 462–468.

92. Tanphaichitr, N., Sobhon, P., Taluppeth, N., and Chalermisarachai, P. (1978) Exp. Cell Res. *117*, 347–356.

93. Rodman, T.C., Litwin, S.D., Romani, M., and Vidali, G. (1979) J. Cell Biol. *80*, 605–620.

94. McKay, D.J., Renaux, B.S., and Dixon, G.H. (1986) Eur. J. Biochem. *156*, 5–8.

95. Gusse, M., Sautiere, P., Belaiche, D., Martinage, A., Roux, C., Dadoune, J.-P., and Chevaillier, P. (1986) Biochim. Biophys. Acta *884*, 124–134.

96. Rodman, T.C., Pruslin, F.H., Chauhan, Y., To, S.E., and Winston, R. (1988) J. Exp. Med. *167*, 1228–1246.

97. Lee, C.-H., Mansouri, A., Hecht, W., Hecht, N.B., and Engel, W. (1987) Biol. Chem. Hoppe-Seyler *368*, 131–135.

98. Krawetz, S.A., Connor, W., and Dixon, G.H. (1987) DNA *6*, 47–57.

99. Krawetz, S.A., Connor, W., and Dixon, G.H. (1988) J. Biol. Chem. *263*, 321–326.

100. Lee, C.-H., Hoyer-Fender, S., and Engel, W. (1987) Nucleic Acids Res. *15*, 7639.

100a. Kistler, W.S., Keim, P.S., and Heinrikson, R.L. (1976) Biochim. Biophys. Acta *427*, 752–757.

101. Johnson, P.A., Peschon, J.J., Yelick, P.C., Palmiter, R.D., and Hecht, N.B. (1988) Biochim. Biophys. Acta *950*, 45–53.

102. Bachmair, A., Finley, D., and Varshavsky, A. (1986) Science *234*, 179–186.

103. Perreault, S.D., Barbee, R.R., Elstein, K.H., Zucker, R.M., and Keefer, C.L. (1988) Biol. Reprod. *39*, 157–167.

104. Bower, P.A., Yelick, P.C., and Hecht, N.B. (1987) Biol. Reprod. *37*, 479–488.
105. Tanhauser, S., and Hecht, N.B., (1989) Nucleic Acids Res. (in press).
105a. Bunick, D., Balhorn, R., Starker, L.H., and Hecht, N.B. (1989) (submitted).
106. Johnson, P.A., Yelick, P.C., Liem, H., and Hecht, N.B. (1988) Gamete Res. *19*, 169–176.
107. Kleene, K.C., Distel, R.J., and Hecht, N.B. (1983) Dev. Biol. *98*, 455–464.
108. Hecht, N.B., Bower, P.A., Waters, S.H., Yelick, P.C., and Distel, R.J. (1986) Exp. Cell Res. *164*, 183–190.
109. Hecht, N.B., and Penschow, J.D. (1987) Exp. Cell Res. *173*, 274–281.
110. Mali, P., Sandberg, M., Vuorio, E., Yelick, P.C., Hecht, N.B., and Parvinen, M. (1988) J. Cell Biol. *107*, 407–412.
111. Kleene, K.C., Distel, R.J., and Hecht, N.B. (1984) Dev. Biol. *105*, 71–79.
112. Iatrou, K., and Dixon, G.H. (1978) Fed. Proc. *37*, 2526–2533.
113. Bloch, D.P. (1969) Genetics (Suppl.) *61*, 93–111.
114. Sinclair, G.D., and Dixon, G.H. (1982) Biochemistry *21*, 1869–1877.
115. Hecht, N.B., Kleene, K.C., Yelick, P.C., Johnson, P.A., Pravtcheva, D.D., and Ruddle, F.H. (1986) Somat. Cell Mol. Genet. *12*, 191–196.
116. Reeves, R.H., Gearhart, J.D., Hecht, N.B., Yelick, P., Johnson, P., and O'brien, S.J. (1989) (submitted).
117. Peschon, J.J., Behringer, R.R., Brinster, R.L., and Palmiter, R.D. (1987) Proc. Natl. Acad. Sci. USA *84*, 5316–5319.
118. Stewart, T.A., Hecht, N.B., Hollingshead, P.G., Johnson, P.A., Leong, J.A.C., and Pitts, S.L. (1988) Mol. Cell. Biol. *8*, 1748–1755.
119. Behringer, R.R., Peschon, J.J., Messing, A., Gartside, C.L., Hauschka, S.D., Palmiter, R.D., and Brinster, R.L. (1988). Proc. Natl. Acad. Sci. USA *85*, 2648–2652.
120. Cummins, J.M. (1980) Gamete Res. *3*, 351–367.
121. Grimes, S., Weisz-Carrington, P., Daum, H. III, Smith, J., Green, L., Wright, K., Stein, G., and Stein, J. (1987) Exp. Cell Res. *173*, 534–545.
122. Oliva, R., Mezquita, J., Mezquita, C., and Dixon, G.H. (1988) Dev. Biol. *125*, 332–340.
123. Lee, C.-H., Bartels, I., and Engels, W. (1987) Biol. Chem. Hoppe-Seyler *368*, 807–811.
124. Gusse, M., Sautiere, P., Chauviere, M., and Chevaillier, P. (1983) Biochim. Biophys. Acta *748*, 93–98.
125. Berlot-Picard, F., Vodjdani, G., and Doly, J. (1986) Eur. J. Biochem. *160*, 305–310.
126. Berlot-Picard, F., Vodjdani, G., and Doly, J. (1987) Eur. J. Biochem. *165*, 553–557.
127. Courtens, J.L., Delaleu, B., Dubois, M.P., Lanneau, M., Loir, M., and Rozinek, J. (1983) Gamete Res. *8*, 21–28.
128. Courtens, J.L., Ploen, L., and Loir, M. (1988) J. Reprod. Fertil. *82*, 635–643.
129. Gusse, C.R.M., Chevaillier, P., and Dadoune, J.P. (1988) J. Reprod. Fertil. *82*, 35–42.
130. Stanker, L.H., Wyrobek, A., and Balhorn, R. (1987) Hybridoma *6*, 293–303.

Index